The Theory of Queuing Systems with Correlated Flows

Alexander N. Dudin • Valentina I. Klimenok •
Vladimir M. Vishnevsky

The Theory of Queuing Systems with Correlated Flows

Springer

Alexander N. Dudin
Dept. of Applied Mathematics
and Informatics
Belarusian State University
Minsk, Belarus

Valentina I. Klimenok
Dept. of Applied Mathematics
and Informatics
Belarusian State University
Minsk, Belarus

Vladimir M. Vishnevsky
V.A. Trapeznikov Institute of Control
Sciences
Russian Academy of Sciences
Moscow, Russia

ISBN 978-3-030-32074-4 ISBN 978-3-030-32072-0 (eBook)
https://doi.org/10.1007/978-3-030-32072-0

This Springer imprint is published by the registered company Springer Nature Switzerland AG.
The registered company address is: Gewerbestrasse 11, 6330 Cham, Switzerland

Foreword

It is my pleasure and honor to write this Foreword. As is known, every product or process we face in day-to-day activities involves either a manufacturing aspect or a service aspect (and possibly both). Queuing theory plays a major role in both these aspects. There are a number of excellent books written on queuing theory in the last many decades starting with the classical book by Takacs in 1960s to Kleinrock's two volumes in the 1970s to Neuts' ground-breaking ones on matrix-analytic methods in the 1980s. Along the way a number of notable books (undergraduate and graduate level) were published by Alfa, Bhat, Cohen, Gross and Harris, He, Latouche and Ramaswami, and Trivedi, et al., among others. This book, written by well-known researchers on queueing and related areas, is unique in the sense that the entire book is devoted to queues with correlated arrivals. Since the introduction of Markovian arrival processes by Neuts, possible correlation present in the inter-arrival times as well as correlation in the service times has been getting attention. A number of published research articles have pointed out how correlation cannot be ignored when analyzing queuing systems. Further, the input to any system comes from different sources and unless one assumes every single source to be Poisson, it is incorrect to assume that the inter-arrival times of customers to the system are independent and identically distributed. Thus, it is time that there is a book devoted to presenting a thorough analysis of correlated input in queuing systems. After setting up the basic tools needed in the study of queuing theory, the authors set out to cover a wide range of topics dealing with batch Markovian arrival processes, quasi-birth-and-death processes, GI/M/1-type and M/G/1-type queues, asymptotically quasi-Toeplitz Markov chains, multidimensional asymptotically quasi-Toeplitz Markov chains, and tandem queues. There is a chapter discussing the characteristics of hybrid communication systems based on laser and radio technologies. This book will be an excellent addition to the collection of books on queuing theory.

Professor of Industrial Engineering & Statistics
Departments of Industrial and Manufacturing
Engineering & Mathematics
Kettering University, Flint, MI, USA

Srinivas R. Chakravarthy

Introduction

Mathematical models of networks and queuing systems (QS) are widely used for study and optimization of various technical, physical, economic, industrial, administrative, medical, military, and other systems. The objects of study in the queuing theory are situations when there is a certain resource and customers needing to obtain that resource. Since the resource is restricted, and the customers arrive at random moments, some customers are rejected, or their processing is delayed. The desire to reduce the rejection rate and delay duration was the driving force behind the development of QS theory. The QS theory is especially important in the telecommunications industry for solving the problems of optimal distribution of telecommunication resources between numerous users which generate requests at random time.

This development began in the works of a Danish mathematician and engineer A.K. Erlang published in 1909–1922. These works resulted from his efforts to solve design problems for telephone networks. In these works, the flow of customers arriving at a network node was modeled as a Poisson flow. This flow is defined as a stationary ordinary flow without aftereffect, or as a recurrent flow with exponential distribution of interval durations between arrival moments. In addition, Erlang presumed that the servicing time for a claim also has an exponential distribution. Under these assumptions, since the exponential distribution has no memory the number of customers in systems considered by Erlang is defined by a one-dimensional Markov process, which makes it easy to find its stationary distribution. The assumption of an exponential form of the distribution of interval durations between arrival moments appears rather strong and artificial. Nevertheless, the theoretical results of Erlang agreed very well with the results of practical measurements in real telephone networks. Later this phenomenon was explained in the works of A.Ya. Khinchine, G.G. Ososkov, and B.I. Grigelionis who proved that under the assumption of uniform smallness of flow intensities a superposition of a large number of arbitrary recurrent flows converges to a stationary Poisson flow. This result is a counterpart of the central limit theorem in probability theory, which states that under uniform smallness of random values their normalized sum in the limit (as the number of terms tends to infinity) converges in distribution to a random value with standard normal distribution. Flows of customers arriving to a telephone station node represent a superposition of a large number of flows with small intensity arriving from individual network users. Therefore, the stationary Poisson flow model describes well enough the flows in real world telephone networks.

The introduction of computer data transmission networks, whose packet-switching approach was more efficient than the channel switching used in telephone networks, motivated the need of a new mathematical formalism for their optimal design. Like Erlang's work, initial studies in the field of computer networks (L. Kleinrock [67], M. Schwartz [101], G.P. Basharin et al. [5], V.M. Vishnevsky et al. [107, 111], and others) were based on simplified assumptions regarding the character of information flows (Poisson flows) and exponential distributions of the packet translation time. This let them use the queuing theory formalism, which was well developed by the time computer networks appeared, to evaluate performance, design topological structures, control routing and the network as a whole, choose optimal parameters for network protocols, and so on.

Further development of theoretical studies was related to the introduction of integrated service digital networks (ISDN), which represent an improvement on earlier packet switching networks. A characteristic feature of these networks is that the same hardware and software is used for joint transmission of various multimedia information: speech, interactive information, large volumes of data, faxes, video, and so on. Since the flows are highly non-uniform, these networks are significantly non-stationary (they have the so-called "bursty traffic") and correlated. Numbers of customers arriving over non-intersecting time intervals can be dependent, and the dependence can be preserved even for intervals spread far apart. It is especially important to take these factors into account in the modeling of modern wideband 4G networks [81, 112] and the new generation 5G networks that are under active development now and expected to be introduced by 2020 (see, e.g., [83, 95, 115, 118]).

The complex character of flows in modern telecommunication networks can be captured with so-called self-similar flows (see, e.g., [82, 106]). The main drawback of models with self-similar flows from the point of view of their use for analytic modeling of information transmission processes in telecommunication networks is the following. The model of an input flow is just one of a QS's structural elements. Therefore, keeping in mind the prospect of an analytic study of the entire QS, we need to aim to have as simple a flow model as possible. A self-similar flow is a complex object by itself, and even the most successful methods of defining it, e.g., via a superposition of a large number of on/off sources with heavy-tailed distributions of on and off periods presume the use of asymptotics. Therefore, an analytic study of not only the flow itself but also a QS where it comes to appears to be virtually impossible.

The simplest flow model is a stationary Poisson flow. The simplicity of a stationary Poisson flow is in that it is defined with a single parameter, which has the meaning of flow intensity (expectation of the number of customers arriving per unit of time) or the parameter of the exponential distribution of intervals between arrival moments of the customers. However, the presence of only one parameter also causes obvious drawbacks of the stationary Poisson flow model. In particular, the variation coefficient of the distribution of interval durations between arrival moments equals 1, and the correlation coefficient between the lengths of adjacent intervals equals zero. Therefore, if in the modeling of some real world object the results of statistical data processing about the input flow show that the value of variation coefficient is different from 1, and the correlation coefficient is significantly different from 0, it is clear that a stationary Poisson flow cannot represent an acceptable model for a real flow. Thus, it was obviously important to introduce models of the input flow that would let one describe flows with variation coefficient significantly different from 1 and correlation coefficient significantly different from 0. At the same time, this model should allow the result of QS studies with correlated flows to be relatively easily obtained as transparent counterparts of the results of studying the corresponding QS with stationary Poisson flows. Studies on the development of such models were conducted independently in the research group of G.P. Basharin in the USSR (the corresponding flows were called MC-flows, or flows controlled by a Markov chain) and the research group of M. Neuts in the USA. The name of these flows in the USA changed with time from versatile flows [92] to N-flows (Neuts flows) [98] and later Markovian arrival processes (MAPs) and their generalization—batch Markovian arrival processes (BMAPs). The shift from stationary Poisson flows to BMAP-flows was gradual, first via the so-called Interrupted Poisson Process (IPP) flows, where intervals between arrivals of a stationary Poisson flow alternate with intervals when there are no claims, and the distribution of these intervals is exponential, and then Switched Poisson Process (SPP) flows, where intervals between arrivals from a stationary Poisson flow of one intensity alternate with intervals between arrivals of a stationary Poisson flow with a different intensity. Finally, assuming that there are not two but rather a finite number of intensities, researchers arrived at the MMPP (Markov Modulated Poisson Process) flows, and then generalized it to the BMAP-flow model.

It is important to account for correlations in the input flow because both measurement results in real networks and computation results based on QS models considered in this book show that the presence of a positive correlation significantly deteriorates the characteristics of a QS (increases average waiting time, rejection probability, and so on). For example, for the same average rate of customer arrival, same distribution of servicing time and same buffer capacity the probability of losing a claim may differ by several orders of magnitude [31]. Thus, if the results of studying a real flow of customers indicate that it is a correlated flow then it is a bad idea to apply the wellstudied models with recurrent flows and, of course, stationary Poisson flow. For this reason, the theory of QS with correlated arrival processes developed in the works of M. Neuts, D. Lucantoni, V. Ramaswami, S. Chakravarthy, A.N. Dudin, V.I. Klimenok and others has had numerous applications in the studies of telecommunication networks, in particular wideband wireless networks operating under the control of IEEE802.11 and 802.16 protocols (see, e.g., [85, 86, 94]), hybrid highspeed communication systems based on laser and radiotechnologies (see, e.g., [17, 87, 100]), cellular networks of the European standard UMTS and its latest version LTE (Long Term Evolution) (see, e.g., [18–20, 27, 66, 108]), wireless networks with linear topology (see, e.g., [13–15, 21, 36, 37]).

The works of Lucantoni [71, 79] describe in detail the history of the problem and main results of studying single-server QS with BMAP-flows obtained before 1991. Further progress in this area up to the end of the twentieth century is shown in a detailed survey by Chakravarthy [78]. In the present survey, whose purpose is to describe and systematize new results of studying complex QS with incoming BMAP-flows and their applications to telecommunication networks, we mostly mention the papers published after the survey [78] appeared in 2001.

Active research in the field of queueing systems with correlated input, conducted for over a quarter of a century, is described in numerous papers by A.S. Alfa, S. Asmussen, A.D. Banik, D. Baum, L. Breuer, H. Bruneel, S.R. Chakravarthy, M.L. Chaudhry, B.D. Choi, A.N. Dudin, D.V.Efrosinin, U.S. Gupta, Q.M. He, C.S. Kim, A. Krishnamoorty, V.I. Klimenok, G.Latouche, H.W. Lee, Q-L. Li, L. Lakatos, F. Machihara, M. Neuts, S. Nishimura, V. Ramaswami, Y. Takahashi, T. Takine, M. Telek, Y.Q. Zhao, V.M. Vishnevsky. A brief description of some issues on queuing theory with correlated input is given in separate sections of the monographs on queueing theory and its applications [6, 9, 12, 43, 45, 110].

However, there is a lack of a systematic description of the theory of QS with correlated arrivals in the world literature. The sufficient completeness of the mathematical results obtained in the recent years, and the practical needs of developers of modern telecommunications networks have made it expedient to write the present monograph to close this gap.

The content of the book is based on original results obtained by the authors; links to their recent publications are listed in the "Bibliography" section. Here we also use the materials of lecture courses delivered to students and post-graduates of the Belarusian State University and the Moscow Institute of Physics and Technology.

The book includes an introduction, six chapters and an appendix.

Chapter 1 sets out the fundamentals of the queuing theory. The basic concepts of this theory are given. The following notions which play an important role in study of queuing systems are described: Markov and semi-Markovian (SM) stochastic processes, Laplace-Stieltjes transforms, and generating functions. We describe the classical methods for evaluation of characteristics of one-server, multi-server and multi-stage QS with limited or unlimited buffers, different service disciplines and different service time distributions.

Chapter 2 (Sect. 2.1) describes the *BMAP* (Batch Markov Arrival Process), discusses its properties in detail, and communicates with some of the simpler types of input flows known in the recent literature. The so-called Marked Markovian flow, which is an extension of the *BMAP* concept to the case of heterogeneous applications, and the SM input flow is briefly defined, where the time intervals between customer arrivals can be dependent and distributed according to an arbitrary law. Further we describe the phase type distributions (PH).

Section 2.2 describes multidimensional (or vector) Quasi-Birth-and-Death Processes which are the simplest and most well-studied class of multidimensional Markov chains. An ergodicity criterion and a formula to calculate vectors of stationary probabilities of the system states are obtained in the so-called matrix-geometric form. As an example of application of vector Birth-and-Death process, a queue of the type $MAP/PH/1$, is analyzed, i.e., a single-server queue with an infinite volume buffer, the Markov arrival process (which is a special case of *BMAP*) and a phase type distribution of the service time. Hereafter it is assumed that the reader is familiar with J. Kendall's notation for describing queuing models. For the sake of completeness of this system description, we also obtained (in terms of the Laplace-Stieltjes transform) the distribution of the waiting time in the system for a query. A spectral approach to the analysis of vector Birth-and-Death Processes is described.

In Sect. 2.3 we give brief results for multidimensional Markov chains of type $G/M/1$. As an example of applications of multidimensional Markov chains of type $G/M/1$, we analyze the embedded chain with moments of arrivals as embedded moments for the queueing system of type $G/PH/1$. To make the description more complete we also obtained the system state probability distribution for this system at an arbitrary moment of time and distribution of the waiting time of an arbitrary query in the system.

In Sect. 2.4 we give results for multidimensional Markov chains of type $M/G/1$ (or quasi-Toeplitz upper-Hessenberg Markov chains). A necessary and sufficient condition for ergodicity of the chain is proved. Two methods are described for obtaining the stationary probability distribution of the chain states: a method that utilizes vector generating functions and considerations of their analyticity on a unit circle of the complex plane, and the method of M.Neuts and its modification, obtained on the basis of the censored Markov chains. The pros and cons of both methods are discussed.

Section 2.5 briefly describes the procedure for obtaining the stationary probability distribution of states of multidimensional Markov chains of type $M/G/1$ with a finite state space.

In Sect. 2.6 we describe the obtained results for discrete-time asymptotically quasi-Toeplitz Markov chains (AQTMC), and in Sect. 2.7—the results for continuous-time asymptotically quasi-Toeplitz Markov chains. We prove the sufficient conditions for ergodicity and nonergodicity of an AQTMC. An algorithm for calculating the stationary probabilities of an AQTMC is derived on the basis of the apparatus of the theory of sensor Markov chains.

As an example of applying the results presented for multidimensional Markov chains of type $M/G/1$, in Chap. 3 we study $BMAP/G/1$ and $BMAP/SM/1$ queues. For the $BMAP/G/1$ system in Sect. 3.1 an example is given of calculating the stationary state probability distribution of the embedded Markov chain with service completion moments as embedding points. The derivation of equations for the stationary state probability distribution of the system at an arbitrary instant of time is briefly described on the basis of the results in the theory of the Markovian renewal processes. The distribution of the virtual and real waiting time in the system and queueing time of a request is obtained in terms of the Laplace-Stieltjes transform. For a more general system $BMAP/SM/1$ with semi-Markovian service, in Sect. 3.2 we have derived in a simple form the ergodicity condition for the embedded Markov chain with service completion moments as embedding points, and discussed the problem of finding block matrices of the transition probabilities for this chain. A connection is established between the stationary distribution of the system state probabilities at an arbitrary instant of time with the stationary distribution of the state probabilities of the embedded Markov chain with moments of termination of service as embedding points. Section 3.3 is devoted to the analysis of the $BMAP/SM/1/N$ system with different disciplines for accepting a batch of requests in a situation where the batch size exceeds the number of vacant spaces in the buffer at the moment of the batch's arrival.

In Sect. 3.4, a multi-server system of type $BMAP/PH/N$ is investigated, and in Sect. 3.5 we consider a multi-server loss system of type $BMAP/PH/N/0$, which is a generalization of Erlang's model $M/M/N/0$, with the study of which queuing theory has started its history. Different disciplines for accepting a batch of requests for service are examined in a situation when the batch size exceeds the number of vacant servers at the moment of the batch's arrival. It is numerically illustrated that the known property of the invariance of the steady state probability distribution with respect to the service time distribution (under a fixed mean service time) inherent in the system $M/G/N/0$ is not satisfied when the arrival flow is not a stationary Poisson one, but a more general *BMAP* flow.

In Chap. 4 (Sect. 4.1), using the results obtained for asymptotically quasi-Toeplitz Markov chains, we consider a $BMAP/SM/1$ system with retrial requests. The classical retrial strategy (and its generalization—a linear strategy) and a strategy with a constant retrial rate are considered. Conditions for the existence of a stationary probability distribution of system states at moments of termination of service are given. A differential-functional equation is obtained for the vector generating function of this distribution. A connection is established between the stationary distribution of the system state probabilities at an arbitrary instant of time and the stationary distribution of the state probabilities of the Markov chain, embedded at the moments of service termination. And in Sect. 4.2, the $BMAP/PH/N$ system with retrial calls is considered. We considered the retrial strategy with an infinitely increasing rate of retrials from the orbit with an increase in the number of requests in the orbit and a strategy with a constant retrial rate. The behavior of the system is described by a multidimensional Markov chain, the components of which are the number of requests in the orbit, the number of busy servers, the state of the *BMAP* flow control process, and the state of the PH service process in each of the busy servers. The conditions for the existence of a stationary probability distribution of the system states are given. Formulas for computing the most important characteristics of system performance are obtained. The case of non-absolutely insistent requests is also considered, in which, after each unsuccessful attempt to get the service, the request from the orbit is permanently dismissed from the system. We provide the results of numerical calculations illustrating the operability of the proposed algorithms and the dependence of the basic system performance characteristics on its parameters. The possibility of another description of the dynamics of the system functioning is illustrated in terms of the multidimensional Markov chain, the blocks of the infinitesimal generator of which, in the case of a large number of servers, have a much smaller dimension than that of the original chain.

Chapter 5 provides an analysis of the characteristics of hybrid communication systems based on laser and radio technologies in terms of two-server or three-server queueing systems with unreliable servers. The basic architectures of wireless systems of this class are being investigated that provide high-speed and reliable access to information resources in existing and prospective next-generation networks. Sections 5.1–5.3 contain description of models and methods for analyzing the characteristics of two-server queuing systems with unreliable servers and *BMAP* arrival flows, adequately describing the functioning of hybrid communication systems where the laser channel is backed up by a broadband radio channel (cold and hot standby). In Sect. 5.4, methods and algorithms for assessing performance characteristics of the generalized architecture of a hybrid system are given: an atmospheric laser channel and a millimeter-wave (71–76 GHz, 81–86 GHz) radio channel operating in parallel are backed up by a broadband IEEE 802.11 centimeter-wave radio channel. For all the queuing models considered in this chapter, we give a description of the multidimensional Markov chain, the conditions for the existence of a stationary regime, and the algorithms for calculation of stationary distributions, and basic performance characteristics.

Chapter 6 discusses multiphase queuing systems with correlated arrivals and their application assessing the performance of wireless networks with a linear topology. In Sect. 6.1 a brief overview of studies on multi-stage QSs with correlated arrivals is given. Section 6.2 generalizes the results of the previous section to the case when service times at the first phase are dependent random variables and are described by an SM service process.

Section 6.3 is devoted to the study of the tandem queue, first station of which is a $BMAP/G/1$ system and the second one is a multi-server system without a buffer with exponentially distributed service tine.

In Sects. 6.4 and 6.5, we consider the methods of analysis of the stationary characteristics of tandem queues $BMAP/G/1 \to \bar{M}/\bar{N}/\bar{R}$ and $BMAP/G/1 \to M/N/O$ with retrials and group occupation of servers of the second-station.

In Sect. 6.6 a tandem queue with MAP arrivals is considered which consists of an arbitrary finite number of stations, defined as multi-server QSs without buffers. The service times at the tandem's stations are PH-distributed, which allows us to adequately describe real-life service processes. The theorem is proved stating that the output flow of customers at each station of the tandem queue belongs to the MAP-type. A method is described for exact calculation of marginal stationary probabilities for the tandem's parts, as well as for the tandem itself, and the corresponding loss probabilities. With the use of the Ramaswami-Lucantoni approach an algorithm is proposed for computation of the main characteristics of a multi-stage QS under a large values of the number of servers at the stages and a large state space of the PH service control process.

Section 6.7 considers a more general case of QS, than described in the previous section, which has a correlated cross-traffic incoming at each station in addition to the traffic from the previous station of the tandem queue.

Section 6.8 is devoted to the study of a tandem queue with an arbitrary finite number of stations, represented by single-server QS with limited buffers, which adequately describes the behavior of a broadband wireless network with linear topology.

In the appendix we give some information from the theory of matrices and the theory of functions of matrices, which are essentially used in presenting the mathematical material of the book.

The authors are grateful to the colleagues from the Belarusian State University and the V.A. Trapeznikov Institute of Control Sciences of the Russian Academy of Sciences, who have helped in the preparation of the book. Special thanks should be given to Prof. S. Chakravarthy who have written the preface to this book's foreword, as well as to Russian Foundation for Basic Research and to Belarusian Republic Foundation for Fundamental Research for financial support in the framework of the grant № 18-57-00002–№ F18R-136.

Contents

Notation

AQTMC	Asymptotically quasi-Toeplitz Markov chain
QTMC	Quasi-Toeplitz Markov chain
MC	Markov chain
LST	Laplace-Stieltjes Transform
PGF	Probability generating function
MAP	Markovian Arrival Process
$MMAP$	Marked Markovian Arrival Process
$MMPP$	Markov Modulated Poisson Process
$BMAP$	Batch Markov Arrival Process
PH	Phase Type
QBD	Quasi-Birth-and-Death Process
SM	Semi-Markovian Process
A_n	Square matrix A of size $n \times n$
$\det A$	Determinant of matrix A
$(A)_{i,k} = a_{ik}$	The entry in row i and column k of matrix A
A_{ik}	(i, k) Cofactor of matrix $A = (a_{ik})$
$\mathrm{Adj}A = (A_{ki})$	Adjugate matrix of the matrix A
$\mathrm{diag}\{a_1, \ldots, a_M\}$	Order M diagonal matrix with diagonal entries $\{a_1, \ldots, a_M\}$
$\mathrm{diag}^+\{a_1, \ldots, a_M\}$	Order $M + 1$ square matrix with superdiagonal $\{a_1, \ldots, a_M\}$, the rest of the entries are zero
$\mathrm{diag}^-\{a_1, \ldots, a_M\}$	Order $M + 1$ square matrix with subdiagonal $\{a_1, \ldots, a_M\}$, the rest of the entries are zero
$\delta_{i,j}$	Kronecker delta, which is 1 if $i = j$, and 0 otherwise
$\mathbf{e}$ $(\mathbf{0})$	Column vector (row vector) consisting of 1's (0's). If necessary, the vector size is determined by the subscript
$\hat{\mathbf{e}}$	Row vector $(1, 0, \ldots, 0)$
$\mathbf{E}$	Symbol of expected value of a random variable
I	Identity matrix. If necessary, the matrix size is determined by the subscript
O	Zero matrix. If necessary, the matrix size is determined by the subscript

$\tilde{I}$	Matrix diag$\{0, 1, \ldots, 1\}$
^{T}	Matrix transposition symbol
$\otimes$	Kronecker product of matrices
$\oplus$	Kronecker sum of matrices
$\bar{W} = W + 1$	
$f^{(j)}(x)$	The order j derivative of function $f(x)$, $j \geq 0$
$f'(x)$	Derivative of function $f(x)$
$\square$	End of proof

Bold lower-case letters denote row vectors, unless otherwise defined (e.g., column vector $\mathbf{e}$).

Chapter 1
Mathematical Methods to Study Classical Queuing Systems

This book is devoted to analysis of various kinds of queues with correlated arrival process. Investigation of such queues is essentially more complicated than the analysis of the corresponding classical queues with renewal or stationary Poisson arrival process. However, to make the text more comprehensive and reader-friendly, in the beginning we present in brief the known results for some classical systems.

1.1 Introduction

Queuing theory originated in the early twentieth century. The Danish scientist A.K. Erlang, who is considered to be its founder, worked in the Copenhagen Telephone Company and dealt with the design of telephone networks. Later, the theory was intensively developed and applied in various fields of science, engineering, economics, and production. This is due to the fact that this theory studies the situations prevalent in human life when there is some limited resource and multiple customers to use it, which results in service delays or rejection for some customers. The desire to understand the objective reasons for these delays or rejections and the possibility of reducing their impact is an incentive to develop queuing theory.

Usually, customers (or groups of them) arrive at random moments of time and their service requires a random part of the limited resource (or random time of its use). Therefore, the process of satisfying customer needs (service process) is usually conducted within the framework of the theory of random processes as a special field of probability theory. Sometimes the service process analysis requires the use of more sophisticated theory and serious mathematical tools. This makes the results almost impracticable for an engineer potentially interested in their application to study real objects. This, in turn, deprives a mathematician of feedback, which is very important for informing the choice of the direction of study for further generalization of the results and objects to study. This serious problem was noticed

A. N. Dudin et al., *The Theory of Queuing Systems with Correlated Flows*,
https://doi.org/10.1007/978-3-030-32072-0_1

in the review [105] by the well-known specialist R. Syski, who notes the danger of a possible split of unified queuing theory into theoretical and engineering parts. The direct consequence of this problem when writing an article or book is usually the question of choosing the language and the corresponding level of rigor in the presentation of the results. This book is aimed both at specialists in the field of queuing theory and specialists in the field of its application to the study of real objects (first of all, telecommunications networks). Therefore, in this chapter, we give a brief overview of methods for analysing queuing systems on the average (compromise) level of rigor. It is assumed that the reader is familiar with probability theory within the framework of a technical college course. Where necessary, some information is provided directly in the text.

In queuing theory applications, an important stage in studying a real object is a formal description of the object's operation in terms of a particular queuing system. The queuing system is considered preset if the following components are fully described:

- input of customers (requests, messages, calls, etc.);
- number and types of servers;
- buffer size, or number of waiting places (waiting space) where the customers arriving at the system when all servers are busy wait before their service starts;
- service time of customers;
- queue discipline, which determines the order in which customers are served in the queue.

A commonly used shorthand notation for a queuing system called Kendall notation and presented in 1953 is a set of four symbols separated by the slash: $A/B/n/m$. The symbol n, $n \geq 1$ specifies the number of identical parallel servers. The symbol m, $m \geq 0$, means the number of places to wait in the buffer in literature in Russian and the maximum number of customers in the system in literature in English. If $m = \infty$, then the fourth position in the queuing system description is usually omitted. Symbol A describes the incoming flow of customers, and symbol B characterizes the service time distribution. Some possible values for these symbols will be given and explained in the next section.

1.2 Input Flow, Service Time

The incoming flow of customers significantly determines the performance characteristics of queuing systems. Therefore, a correct description of the input flow of customers arriving at random times into the real system and the identification of its parameters are very important tasks. A rigorous solution of this problem lies in the mainstream of the theory of point random processes and is out of the scope of our book. Here we give only brief information from the theory of homogeneous random flows that is necessary to understand the subsequent results.

Customers arrive at the queue at random time points $t_1, t_2, \ldots, t_n, \ldots$. Let $\tau_k = t_k - t_{k-1}$ be the length of the time interval between the arrivals of the $(k-1)$th and kth customers, $k \geq 1$ (t_0 is supposed to be 0) and x_t be the number of moments t_k lying on the time axis to the left of point t, $t \geq 0$.

The stochastic arrival process (input flow) is assumed to be determined if we know the joint distribution of random values τ_k, $k = 1, \ldots, n$ for any n, $n \geq 1$ or the joint distribution of the random values x_t for all t, $t \geq 0$.

Definition 1.1 A stochastic arrival process is called *stationary* if for any integer m and any non-negative numbers $u_1, \ldots, u_m$, the joint distribution of random variables $(x_{t+u_k} - x_t)$, $k = 1, \ldots, m$ is independent of t.

On the intuitive level, this means that the distribution of the number of customers arriving during a certain time interval depends on the length of this interval and does not depend on the location of this interval on the time axis.

Definition 1.2 A stochastic arrival process is called *ordinary* if for any t it holds

$$\lim_{\Delta \to 0} \frac{P\{x_{t+\Delta} - x_t > 1\}}{\Delta} = 0.$$

On the intuitive level, this means that the probability of more than one arrival during a short time interval is of higher order of vanishing compared to the interval length. Roughly speaking, it means that simultaneous arrival of two or more customers is almost impossible.

Definition 1.3 A stochastic arrival process is said to be a flow *without aftereffect* if the numbers of customers arriving in non-overlapping time intervals are mutually independent random variables.

Definition 1.4 A stochastic arrival process is said to be a flow *with limited aftereffect* if the random variables τ_k, $k \geq 1$ are mutually independent.

Definition 1.5 A stochastic arrival process is said to be *recurrent* (or renewal) if it is a flow with limited aftereffect and the random values τ_k, $k \geq 1$ are identically distributed.

Let $A(t) = P\{\tau_k < t\}$ be the distribution function, which completely determines the renewal arrival process.

If $A(t)$ is the exponential distribution, $A(t) = 1 - e^{-\lambda t}$ then the first symbol is M in Kendall's notation.

If distribution $A(t)$ is degenerate, i.e., the customers arrive at regular intervals, then the first symbol is D in Kendall's notation.

If $A(t)$ is hyperexponential,

$$A(t) = \sum_{k=1}^{n} q_k (1 - e^{-\lambda_k t}),$$

where $\lambda_k \geq 0$, $q_k \geq 0$, $k = 1, \ldots, n$, $\sum_{l=1}^{n} q_l = 1$ then in Kendall's notation the first symbol is HM_n.

If $A(t)$ is Erlang with parameters (λ, k),

$$A(t) = \int_0^\infty \lambda \frac{(\lambda t)^{k-1}}{(k-1)!} e^{-\lambda t} dt,$$

then the first symbol is E_k, k is called the order of the Erlang distribution.

A more general class of distributions including hyperexponential and Erlang as special cases is the so-called phase-type distribution. In Kendall's notation, it is denoted by PH. See [9] and the next chapter for detailed information about the PH distribution, its properties and the probabilistic interpretation.

If no assumptions are made about the form of the distribution function $A(t)$ then the symbols G (General) or *GI* (General Independent) are used as the first symbol in Kendall's notation. Strictly speaking, the symbol G usage does not require an input flow to be recurrent but symbol *GI* means exactly a recurrent flow. But in the literature, sometimes there is no difference between these symbols.

Definition 1.6 The *intensity* (the rate) λ of the stationary stochastic arrival process is the expectation (the mean number) of customers arriving per time unit

$$\lambda = M\{x_{t+1} - x_t\} = \frac{M\{x_{t+T} - x_t\}}{T}, \; T > 0.$$

Definition 1.7 *Parameter* α of a stationary stochastic arrival process is called the positive value, determined as

$$\alpha = \lim_{\Delta \to 0} \frac{P\{x_{t+\Delta} - x_t \geq 1\}}{\Delta}.$$

Definition 1.8 A stationary ordinary input flow without aftereffect is called *simple*.

The following propositions are valid.

Proposition 1.1 *The input flow is simple if and only if it is stationary Poisson, i.e.,*

$$P\{x_{t+u} - x_t = k\} = \frac{(\lambda u)^k}{k!} e^{-\lambda u}, \; k \geq 0.$$

Proposition 1.2 *The input flow is simple if and only if it is recurrent with exponential distribution of time intervals between customer arrivals,* $A(t) = 1 - e^{-\lambda t}$.

Proposition 1.3 *If n customers of a simple input flow arrived during a time interval of length T then the probability that a tagged customer arrived during time interval*

τ inside the interval of length T does not depend on when the other customers arrived and how the time interval τ is located inside T. This probability is τ/T.

Proposition 1.4 *The intensity and the parameter of a simple flow are equal. The mean number of customers arriving during the time interval T is equal to λT.*

Proposition 1.5 *Superposition of two independent simple flows having parameters λ_1 and λ_2 is again a simple flow with parameter $\lambda_1 + \lambda_2$.*

Proposition 1.6 *A flow obtained from a simple flow of rate λ as a result of applying the simplest procedure of recurrent sifting (an arbitrary customer is accepted to the sifted flow with probability p and ignored with probability $1 - p$) is a simple flow of rate $p\lambda$.*

Proposition 1.7 *The flow obtained as a result of the superposition of n independent recurrent flows having a uniformly small rate converges to a simple flow as n increases.*

In the literature, a simple flow is often referred to as Poisson arrivals. The most well-studied queuing systems are the ones with the Poisson input flow of customers. This is largely explained by Proposition 1.2 and the well-known memoryless property of the exponential distribution. For the exponentially distributed random variable ν, this property is expressed in terms of conditional probabilities as follows:

$$P\{\nu \geq t + \tau | \nu \geq t\} = P\{\nu \geq \tau\}.$$

Let us explain this equation. The definition of conditional probability and the fact that $P\{\nu < x\} = 1 - e^{-\lambda x}$ yield

$$P\{\nu \geq t + \tau | \nu \geq t\} = \frac{P\{\nu \geq t + \tau, \nu \geq t\}}{P\{\nu \geq t\}}$$

$$= \frac{P\{\nu \geq t + \tau\}}{P\{\nu \geq t\}} = \frac{e^{-\lambda(t+\tau)}}{e^{-\lambda t}} = e^{-\lambda \tau} = P\{\nu \geq \tau\}.$$

From this property, it follows that the distribution of time intervals from an arbitrary moment to the moment when the next customer of the Poisson input arrives is independent of when the previous customer arrived. The property considerably simplifies the analysis of queues with Poisson arrivals.

Proposition 1.7 explains the fact that the Poisson input often may be realistic in practical systems (for example, the input of customers arriving at the automatic telephone station is the sum of a large number of independent small flows coming from individual subscribers of the telephone network, and therefore close to a simple one), and therefore the use of a Poisson input for the simulation of real flow not only facilitates the investigation of queuing systems but is also justified.

Below we present useful information about recurrent flows. Let an arbitrary moment of time be fixed. We are interested in the distribution function $F_1(t)$ of the time passed to an arbitrary moment from the previous customer arrival of the recurrent flow, and the distribution function $F_2(t)$ of the time interval between the given moment and the next customer arrival. It can be shown that

$$F_1(t) = F_2(t) = F(t) = a_1^{-1} \int_0^t (1 - A(u))du$$

where a_1 is the mean inter-arrival time: $a_1 = \int_0^\infty (1 - A(t))dt$. The value a_1 and the input rate λ are related as $a_1\lambda = 1$.

Function $F(t)$ is identical to $A(t)$ if and only if the input flow is Poisson.

Definition 1.9 The instantaneous intensity $\mu(t)$, $\mu(t) > 0$ of a recurrent input flow is the value

$$P\{x_{t+\Delta} - x_t \geq 1|E_t\} = \mu(t)\Delta + o(\Delta)$$

where E_t is a random event that a customer arrived at time 0 and no one arrived during $(0, t]$. Here $o(\Delta)$ means a value with order of vanishing higher than Δ (as $\Delta \to 0$).

Instantaneous intensity $\mu(t)$ is related to the distribution function $A(t)$ as

$$A(t) = 1 - e^{-\int_0^t \mu(u)du}, \qquad \mu(t) = \frac{\frac{dA(t)}{dt}}{1 - A(t)}.$$

In the case of Poisson arrivals, the instantaneous intensity coincides with the intensity.

In modern integrated digital communication networks (unlike traditional telephone networks), data traffic no longer represents a superposition of a large number of uniformly small independent recurrent flows. As a result, these flows are often not only not simple but not recurrent. To describe such flows, D. Lucantoni proposed the formalism of batch Markov flows, see [85]. To denote them in Kendall's symbols, the abbreviation BMAP (Batch Markovian Arrival Process) is used. For more information on BMAP flows and related queuing systems see [26, 85].

Note that the so-called input from a finite source (primitive flow, second-order Poisson flow) is also popular in telephony. It is defined as follows.

Let there be a finite number of n objects generating requests independently of others, called request sources. A source can be in a busy state for some random time during which it is not able to generate requests. After the busy state finishes, the source goes into a free state. In this state, the source can generate a new request during time exponentially distributed with parameter λ. After generating the request, the source immediately goes into a busy state.

A primitive input flow of n request sources is a flow with a limited aftereffect whose intensity λ_i is directly proportional to the number of currently available sources at a given time: $\lambda_i = \lambda(n-i)$ where i is the number of busy sources.

Concerning the customer service process in the system, it is usually assumed that the service is recurrent, i.e., the consecutive service times of customers are independent and identically distributed random variables. Let us denote their distribution function by $B(t)$. It is usually assumed that $B(+0) = 0$, i.e., customer service cannot be instantaneous. It is also usually assumed that a random variable having the distribution function $B(t)$ has the required number of finite initial moments.

To determine the type of service time distribution in Kendall's classification, the same symbols are used as for the type of arrival process. So, the symbol G means either the absence of any assumptions about the service process, or it is identified with the symbol GI meaning a recurrent service process, the symbol M means the assumption that the service time is exponentially distributed, that is, $B(t) = 1 - e^{-\mu t}, \mu > 0, t > 0$, the symbol D denotes deterministic service times, etc.

The symbol *SM* (Semi-Markovian), see e.g., [94], means the service process that is more general than a recurrent one, i.e., consecutive service times are the consecutive times a semi-Markovian process stays in its states. An *SM* process is supposed to have a finite state space and a fixed kernel.

1.3 Markov Random Processes

Markov random processes play an important role in analysis of queuing systems. Therefore, we give some information from the theory of such processes that will be used below.

Definition 1.10 A random process $y_t,\ t \geq 0$ determined on a probability space and taking values in a certain numerical set Y is called *Markov* if for any positive integer n, any $y, u, u_n, \ldots, u_1 \in Y$ and any $\tau, t, t_n, \ldots, t_1$, ordered such that $\tau > t > t_n > \cdots > t_1$, the following relation holds:

$$P\{y_\tau < y | y_t = u, y_{t_n} = u_n, \ldots, y_{t_1} = u_1\} = P\{y_\tau < y | y_t = u\}. \tag{1.1}$$

Process parameter t is considered to be the time. If τ is a future time point, t is a present time point, and $t_k, k = 1, \ldots, n$ are past points, the condition (1.1) can be interpreted as follows: the future behavior of a Markov process is completely determined by its state at the present time. For a non-Markov process, its future behavior depends also on states of the process in the past.

In case a state space Y of a Markov process $y_t, t \geq 0$ is finite or countable, the process is called a *Markov chain* (MC). If parameter t takes values in a discrete set, the MC is called a discrete-time MC. If the parameter t takes values in some continuous set then the MC is called a continuous-time MC.

An important special case of the continuous-time MC is a so-called birth-and-death process.

1.3.1 Birth-and-Death Processes

Definition 1.11 A random process $i_t,\ t \geq 0$, is called a *birth-and-death process* if it satisfies the following conditions:

- the process state space is a set (or a subset) of non-negative integers;
- the sojourn time of the process in state i is exponentially distributed with parameter $\gamma_i, \gamma_i > 0$ and is independent of the past states of the process;
- after the process finishes its stay in state i, it transits to state $i - 1$ with probability $q_i, 0 < q_i < 1$ or to state $i + 1$ with probability $p_i = 1 - q_i$. The probability p_0 is supposed to be 1.

A state of the process $i_t, t \geq 0$ at time t is considered to be the size of some population at this point in time. Transition from state i to state $i + 1$ is assumed to be the birth of a new population member, and transition to state $i - 1$ is considered to be population member death. Thus, this interpretation of the process explains its name.

Denote by $P_i(t)$ the probability that the process i_t is in state i at time t.

Proposition 1.8 *The probabilities $P_i(t) = P\{i_t = i\}, i \geq 0$ satisfy the following system of linear differential equations:*

$$P_0'(t) = -\lambda_0 P_0(t) + \mu_1 P_1(t), \tag{1.2}$$

$$P_i'(t) = \lambda_{i-1} P_{i-1}(t) - (\lambda_i + \mu_i) P_i(t) + \mu_{i+1} P_{i+1}(t), i \geq 1 \tag{1.3}$$

where $\lambda_i = \gamma_i p_i, \mu_i = \gamma_i q_i, i \geq 0$.

Proof We apply the so-called Δt-method. This method is widely used in the analysis of continuous-time MCs and Markov processes. □

The main point of this method is as follows. We fix the time t and a small time increment Δt. The probability state distribution of the Markov process at time point $t + \Delta t$ is expressed by its probability state distribution at an arbitrary time point t and the probability of the possible process transitions during time Δt. The result is a system of difference equations for probabilities $P_i(t)$. By dividing both sides of these equations by Δt and taking a limit as Δt tends to 0, we obtain a system of differential equations for the process state probabilities.

Below we apply the Δt-method to derive (1.3). Denote by $R(i, j, t, \Delta t)$ the probability that process i_t transits from state i to state j during the time interval $(t, t + \Delta t)$. Since the duration of time when the process i_t is in state i has an exponential distribution possessing the property of no aftereffect, the time passing from time point t to the completion of the process sojourn time in state i is also exponentially distributed with parameter γ_i. The probability that the process i_t will exit the state i during the time interval $(t, t + \Delta t)$ is the probability that time exponentially distributed with parameter γ_i expires during time Δt. Taking into account the definition of a distribution function, this probability is

$$1 - e^{-\gamma_i \Delta t} = \gamma_i \Delta t + o(\Delta t). \tag{1.4}$$

That is, the probability that the process changes its state during time Δt is a quantity of order Δt. Hence, the probability that the process i_t makes two or more transitions during time Δt is of order $o(\Delta t)$. Taking into account that the birth-and-death process makes one-step transitions just to the neighboring states, we obtain $R(i, j, t, \Delta t) = o(\Delta t)$ for $|i - j| > 1$.

Using the above arguments and the formula of total probability, we obtain the relations

$$P_i(t + \Delta t) = P_{i-1}(t)R(i - 1, i, t, \Delta t) + P_i(t)R(i, i, t, \Delta t) \tag{1.5}$$

$$+P_{i+1}(t)R(i + 1, i, t, \Delta t) + o(\Delta t).$$

From the description of the process and formula (1.4) it follows that

$$R(i - 1, i, t, \Delta t) = \gamma_{i-1}\Delta t p_{i-1} + o(\Delta t),$$

$$R(i, i, t, \Delta t) = (1 - \gamma_i \Delta t) + o(\Delta t), \tag{1.6}$$

$$R(i + 1, i, t, \Delta t) = \gamma_{i+1}\Delta t q_{i+1} + o(\Delta t).$$

Substituting relations (1.6) into (1.5) and using the notation λ_i and μ_i, we rewrite (1.5) in the form

$$P_i(t + \Delta t) - P_i(t) = P_{i-1}(t)\lambda_{i-1}\Delta t - P_i(t)(\lambda_i + \mu_i)\Delta t$$

$$+ P_{i+1}(t)\mu_{i+1}\Delta t + o(\Delta t). \tag{1.7}$$

Dividing both sides of this equation by Δt and tending Δt to 0, we obtain relation (1.3). Formula (1.2) is derived similarly.

Proposition 1.8 is proved. □

To solve an infinite system of differential equations (1.2) and (1.3) by means of Laplace transforms $p_i(s)$ of probabilities $P_i(t)$ (see below), the system is reduced to an infinite system of linear algebraic equations. However, this system can also be solved explicitly only for special cases, e.g., when the tridiagonal matrix of this system has additional specific properties (for example, $\lambda_i = \lambda,\ i \geq 0,\ \mu_i = \mu,\ i \geq 1$). Situations when the resulting system of equations for the probabilities $P_i(t)$ cannot be explicitly solved are rather typical in queuing theory. Therefore, in spite of the fact that sometimes these probabilities depending on t and characterizing the process dynamics (for the known initial state i_0 of the process or the known probability distribution of i_0) are of significant practical interest, usually we have to deal with the so-called stationary probabilities of the process

$$\pi_i = \lim_{t\to\infty} P_i(t), i \geq 0. \tag{1.8}$$

Positive limit (stationary) probabilities π_i may not always exist, and the conditions for their existence are usually established by so-called ergodic theorems.

For the birth-and-death process we are considering, the following result can be proved.

Proposition 1.9 *The stationary probability distribution (1.8) of the birth-and-death process exists if the following series converges:*

$$\sum_{i=1}^{\infty} \rho_i < \infty, \tag{1.9}$$

where

$$\rho_i = \prod_{l=1}^{i} \frac{\lambda_{l-1}}{\mu_l}, i \geq 1, \rho_0 = 1,$$

and the following series diverges:

$$\sum_{i=1}^{\infty} \prod_{l=1}^{i} \frac{\mu_l}{\lambda_l} = \infty. \tag{1.10}$$

Note that the stationary probabilities $\pi_i, i \geq 0$ are calculated as

$$\pi_i = \pi_0 \rho_i, i \geq 1, \pi_0 = \left(\sum_{i=0}^{\infty} \rho_i\right)^{-1}. \tag{1.11}$$

Proof The last part of the Proposition is elementary to prove. We assume that conditions (1.9) and (1.10) are valid, and the limits (1.8) exist. Let t tend to infinity in (1.2), (1.3). The derivatives $P_i^{'}(t)$ tend to zero. The existence of these derivative limits follows from the existence of limits on the right-hand side of system (1.2), (1.3). The zero limits of the derivatives follow from the fact that the assumption that the limits are non-zero contradicts the boundedness of probabilities: $0 \leq P_i(t) \leq 1$.

As a result, from (1.2), (1.3) we obtain the system of linear algebraic equations for the distribution $\pi_i, i \geq 0$:

$$-\lambda_0 \pi_0 = \mu_1 \pi_1, \tag{1.12}$$

$$\lambda_{i-1}\pi_{i-1} - (\lambda_i + \mu_i)\pi_i + \mu_{i+1}\pi_{i+1} = 0, i \geq 1. \tag{1.13}$$

Using the notation $x_i = \lambda_{i-1}\pi_{i-1} - \mu_i\pi_i, i \geq 1$, the system (1.12), (1.13) can be rewritten in the form

$$x_1 = 0, x_i - x_{i+1} = 0$$

resulting in $x_i = 0, i \geq 1$, which in turn implies the validity of relations

$$\lambda_{i-1}\pi_{i-1} = \mu_i\pi_i, i \geq 1. \tag{1.14}$$

It follows that $\pi_i = \rho_i\pi_0, i \geq 1$. The formula for the probability π_0 follows from the normalization condition. □

1.3.2 Method of Transition Intensity Diagrams

Note that there is an effective method to obtain equations of type (1.12), (1.13) (so-called equilibrium equations) for continuous-time MCs (and the birth-and-death processes, in particular) without using equations for non-stationary probabilities.

This alternative method is called the method of transition intensity diagrams. The main idea of the method is as follows. The behavior of a continuous-time MC is described by an oriented graph. The nodes of the graph correspond to the possible states of the chain. Arcs of the graph correspond to possible one-step transitions between chain states. Each arc is supplied with a number equal to the intensity of the corresponding transition. To obtain a system of linear algebraic equations for stationary probabilities, the so-called flow conservation principle is used. This principle is as follows. If we make a cross-section of the graph, that is we remove some arcs in such a way that a disjoint graph is obtained, then the flow out of one part of the cut graph to the other is equal to the flow in the opposite direction. The flow is assumed to be the sum (over all removed arcs) of the stationary probabilities of the nodes which are the origins of the removed arcs multiplied by the corresponding transition intensities. Equating the flows over all cross-sections selected in an appropriate way, we obtain the required system of equations.

Note that we get exactly the equilibrium equations obtained by applying the Δt method (called the equations of global equilibrium) if the cross-section in the graph is done by cutting all arcs around the chosen graph node. Sometimes, due to a lucky choice of cross-section in the graph, it is possible to obtain a much simpler system of equations (called local equilibrium equations). In particular, if we depict the behavior of a birth-and-death process considered in the form of a string graph and make a cut not around the node corresponding to the state i (in this case we obtain Eqs. (1.12) and (1.13)) but between nodes corresponding to the states i and $i+1$ then we immediately obtain simpler equations (1.14) from which formulas (1.11) automatically follow.

1.3.3 Discrete-Time Markov Chains

Below we give brief information from the theory of discrete-time MCs. More detailed information can be found, for example, in [40].

Without loss of generality, we assume that the MC state space is the set of non-negative integers (or some subset of it).

Definition 1.12 Homogeneous discrete-time MC $i_k,\ k \geq 1$ is determined if

- the initial probability distribution of the chain states is determined:

$$r_i = P\{i_0 = i\}, i \geq 0;$$

- the matrix P of the one-step transition probabilities $p_{i,j}$ defined as follows:

$$p_{i,j} = P\{i_{k+1} = j | i_k = i\}, i, j \geq 0,$$

  is determined.

The matrix P of the one-step transition probabilities $p_{i,j}$ is stochastic, that is, its elements are non-negative numbers, and the sum of the elements of any row is equal to one. The transition probability matrix for m steps is P^m.

Denote by $P_i(k) = P\{i_k = i\}, k \geq 1, i \geq 0$ the probability that the MC is in state i after the kth step. Potentially, the probabilities $P_i(k)$ characterizing the non-stationary behavior of the MC can be very interesting in solving the problem of finding the characteristics of an object described by a given MC. However, the problem of finding such probabilities is complicated. Therefore, the so-called stationary probabilities of states of the MC are usually analyzed:

$$\pi_i = \lim_{k\to\infty} P_i(k), i \geq 0. \tag{1.15}$$

These probabilities are also called limiting, final, or ergodic. We shall only deal with irreducible non-periodic chains for which the positive limits (1.15) exist, and the listed alternative names of stationary probabilities express practically the same properties of chains: the existence of the limits (1.15), the independence of the limits of the initial probability distribution of the states of the MC, and the existence of the unique positive solution of the following system of linear algebraic equations (equilibrium equations) for the stationary state probabilities:

$$\pi_j = \sum_{i=0}^{\infty} \pi_i p_{i,j}, \tag{1.16}$$

$$\sum_{i=0}^{\infty} \pi_i = 1. \tag{1.17}$$

There are a number of results (theorems of Feller, Foster, Moustafa, Tweedy, etc.) that allow us to determine whether the stationary probabilities π_i (limits in (1.15)) exist or not for a specific type of transition probabilities. If, as a result of analysis of the transition probabilities $p_{i,j}$, it is established that the limits (1.15) exist under the specific conditions on the parameters of the MC, then it is assumed that these conditions are fulfilled and the system of equilibrium equations (1.16), (1.17) is solved. The investigation of the MC is considered finished.

In case the chain state space is infinite, the solution of the system (1.16), (1.17) becomes a tough task, and its effective solution is possible only if the matrix of one-step transition probabilities has certain specific properties.

1.4 Probability Generating Function, Laplace and Laplace-Stieltjes Transforms

In queuing theory, the apparatus of Laplace and Laplace-Stieltjes transforms and probability generating functions (PGFs) is intensively used. In particular, we have already mentioned the possibility of using Laplace transforms to reduce the problem of solving a system of linear differential equations to the solution of a system of linear algebraic equations. We give the basic information about these functions and transforms.

Definition 1.13 The *Laplace-Stieltjes transform* (LST) of the distribution $B(t)$ is a function $\beta(s)$ determined as follows:

$$\beta(s) = \int_0^\infty e^{-st} dB(t),$$

and the *Laplace transform* $\phi(s)$ (LT) is a function

$$\phi(s) = \int_0^\infty e^{-st} B(t)dt.$$

If s is a pure imaginary variable, the LST coincides with the characteristic function that corresponds to the distribution function $B(t)$. The right half-plane of the complex plane is usually considered to be the analyticity domain for functions $\beta(s), \phi(s)$. However, without substantial loss of generality, within the framework of this chapter we can consider s to be a real positive number.

Let us note some of the properties of the Laplace-Stieltjes transform.

Property 1.1 If both LSTs $\beta(s)$ and $\phi(s)$ exist (that is, the corresponding improper integrals converge), then they are related as $\beta(s) = s\phi(s)$.

Property 1.2 If two independent random variables have LSTs $\beta_1(s)$ and $\beta_2(s)$ of their distribution functions, then the LST of the distribution function of the sum of these random variables is the product $\beta_1(s)\beta_2(s)$.

Property 1.3 The LST of the derivative $B^{'}(t)$ is $s\beta(s) - sB(+0)$.

Property 1.4

$$\lim_{s\to 0}\beta(s) = \lim_{t\to\infty} B(t).$$

Property 1.5 Let b_k be the kth initial moment of the distribution $b_k = \int\limits_0^\infty t^k dB(t)$, $k \geq 1$. It is expressed using the LST as

$$b_k = (-1)^k \frac{d^k\beta(s)}{ds^k}\bigg|_{s=0}. \tag{1.18}$$

Property 1.6 The LST $\beta(s)$ can be given a probabilistic meaning as follows. We assume that $B(t)$ is a distribution function of a length of some time interval, and a simple input of catastrophes having parameter $s > 0$ arrives within this time interval. Then it is easy to see that $\beta(s)$ is the probability that no catastrophe arrives during the time interval.

Property 1.7 LST $\beta(s)$ considered as a function of the real variable $s > 0$ is a completely monotone function, i.e., it has derivatives $\beta^{(n)}(s)$ of any order and $(-1)^n\beta^{(n)}(s) \geq 0$, $s > 0$.

Definition 1.14 The *probability generating function* (PGF) of the probability distribution q_k, $k \geq 0$, of a discrete random variable ξ is a function

$$Q(z) = Mz^\xi = \sum_{k=0}^{\infty} q_k z^k, \; |z| < 1.$$

Let us mention the main properties of PGFs.

Property 1.8

$$|Q(z)| \leq 1, \; Q(0) = q_0, \; Q(1) = 1.$$

Property 1.9 Random variable ξ has the mth initial moment $\mathbf{E}\xi^m$ if and only if there exists a finite left-side derivative $Q^{(m)}(1)$ of the PGF $Q(z)$ at the point $z = 1$. The initial moments are easily calculated through the factorial moments

$$\mathbf{E}\xi(\xi-1)\ldots(\xi-m+1) = Q^{(m)}(1).$$

In particular, $\mathbf{E}\xi = Q^{'}(1)$.

Property 1.10 In principle, PGF $Q(z)$ allows calculation (generation) of probabilities q_i using the following formula:

$$q_i = \frac{1}{i!} \frac{d^i Q(z)}{dz^i}\bigg|_{z=0}, \quad i \geq 0. \tag{1.19}$$

Property 1.11 The PGF $Q(z)$ can be given a probabilistic sense as follows. We interpret the random variable ξ as the number of customers arriving during a certain period of time. Each customer is colored red with probability z, $0 \leq z \leq 1$, and is colored blue with the complementary probability. Then it follows from the formula of the total probability that $Q(z)$ is the probability that only red customers arrive during the considered time interval.

Thus, if the PGF $Q(z)$ of the probabilities $q_k, k \geq 0$ is known, we can easily calculate the moments of this distribution and, in principle, can calculate the probabilities $q_k, k \geq 0$. If the direct calculation by (1.19) is difficult, we can use the method of PGF inversion by expanding it into simple fractions or by numerical methods (see, for example, [23, 97]). When solving practical problems, it is possible to try to approximate this distribution by smoothing over a given number of coinciding moments of the distribution.

1.5 Single-Server Markovian Queuing Systems

The most well-studied queuing systems are the single-server systems with a Poisson input or (and) an exponential distribution of the service time. This is explained by the fact that the processes that researchers are primarily interested in are one-dimensional and, in addition, are either Markovian or easily treatable by a Markovization procedure by considering them only at embedded time moments or by expansion of the process phase space. These processes are the number i_t of customers in the system at time t, the waiting time w_t of a customer that arrives (or may arrive) at time t, etc.

1.5.1 $M/M/1$ Queue

Let's consider the system $M/M/1$, that is a single-server queuing system with unlimited buffer (or unlimited waiting space), a simple flow of customers (a stationary Poisson input) with parameter λ and exponentially distributed service time with parameter μ.

Analyzing the behavior of the system, we easily see that the process i_t of the queue length at the time t is a birth-and-death process with parameters

$$\gamma_0 = \lambda, \gamma_i = \lambda + \mu, i \geq 1,$$

$$p_i = \int_0^\infty e^{-\mu t} \lambda e^{-\lambda t} dt = \frac{\lambda}{\lambda + \mu}, i \geq 1.$$

Therefore, Propositions 1.8 and 1.9 are valid with parameters $\lambda_i = \lambda, i \geq 0$, $\mu_i = \mu, i \geq 1$. So the value ρ_i in the statement of Proposition 1.9 is defined as $\rho_i = \rho^i$ where $\rho = \lambda/\mu$.

Parameter ρ characterizing the ratio of the input flow intensity and the service intensity is called the *system load* and plays an important role in queuing theory.

Checking the condition of the process $i_t, t \geq 0$ stationary state distribution existence given in the statement of Proposition 1.9, we easily see that the stationary distribution of the number of customers in the system exists if the following condition holds:

$$\rho < 1, \tag{1.20}$$

which is often called the *stability condition*. Below we assume that this condition holds.

Note that for most single-server queuing systems, the stability condition also has the form (1.20), which agrees well with the following intuitive considerations: in order that the system does not accumulate an infinite queue, it is necessary that the customers are served, on average, faster than they arrive.

Thus, we can formulate the following corollary of Proposition 1.9.

Proposition 1.10 *The stationary distribution $\pi_i, i \geq 0$ of the number of customers in $M/M/1$ is obtained by*

$$\pi_i = \rho^i (1 - \rho), \; i \geq 0. \tag{1.21}$$

It follows that the probability π_0 that the system is idle at any time is $1 - \rho$, and the average number L of customers in the system is given by the formula

$$L = \sum_{i=0}^{\infty} i \pi_i = \frac{\rho}{1 - \rho}. \tag{1.22}$$

The average queue length L_o is defined by the formula

$$L_o = \sum_{i=1}^{\infty} (i - 1) \pi_i = L - \rho = \frac{\rho^2}{1 - \rho}. \tag{1.23}$$

In situations where the distribution of inter-arrival times and the service time distribution are unknown and we know only their averages, formulas (1.22) and (1.23) are sometimes used to estimate (roughly) the average queue length in the system and the average queue length at an arbitrary time.

As stated above, another interesting characteristic of the queuing system is the waiting time distribution w_t of a customer arriving at time t (i.e., the time from the moment the customer arrives at the queue to the time its service starts).

Denote by $W(x)$ the stationary distribution of the process w_t,

$$W(x) = \lim_{t\to\infty} P\{w_t < x\}, x \geq 0.$$

We assume that customers are served in the order they arrive at the system. Sometimes such a discipline to choose customers for service is encoded as FIFO (First In First Out) or the same FCFS (First Come First Served).

Proposition 1.11 *The stationary distribution $W(x)$ of the waiting time in an $M/M/1$ queue is defined as*

$$W(x) = 1 - \rho e^{(\lambda-\mu)x}. \tag{1.24}$$

Proof The waiting time of an arbitrary customer depends on the number of customers present in the system upon its arrival. For an $M/M/1$ queue, the queue length distributions both at an arbitrary time and at an arbitrary time of a customer arrival coincide and are given by formula (1.21). A customer arriving at an empty system (with probability π_0) has zero waiting time. An arriving customer that sees i customers in the system (with probability π_i) waits for a time having an Erlang distribution with parameters (μ, i). The last conclusion follows from the facts that, firstly, due to the lack of aftereffect in the exponential distribution, the service time remaining after the arrival moment of an arbitrary customer has the same exponential distribution with parameter μ as the total service time, and secondly, the sum of i independent exponentially distributed random variables with parameter μ is an Erlang random variable with parameters (μ, i).

From these arguments and (1.21), it follows that

$$W(x) = 1 - \rho + \sum_{i=1}^{\infty} \rho^i (1-\rho) \int_0^x \mu \frac{(\mu t)^{i-1}}{(i-1)!} e^{-\mu t} dt$$

$$= 1 - \rho + (1-\rho)\lambda \int_0^x e^{(\lambda-\mu)t} dt.$$

This immediately implies (1.24). □

The average waiting time W in the system is calculated as

$$W = \int_0^\infty (1 - W(x))dx = \lambda^{-1} \frac{\rho^2}{1-\rho}. \tag{1.25}$$

The average sojourn time V in the system (i.e., the time from the moment a customer arrives at the system until the end of its service) is given by the formula

$$V = W + \mu^{-1} = \lambda^{-1} \frac{\rho}{1-\rho}. \tag{1.26}$$

Comparing expression (1.25) for the mean waiting time W and formula (1.23) for the mean queue length L_o, as well as formula (1.26) for the mean sojourn time V with formula (1.22) for the mean number L of customers in the system, we see that

$$L_o = \lambda W, L = \lambda V. \tag{1.27}$$

Note that these formulas hold for many queuing systems that are more general than an $M/M/1$ queues they are called the Little formulas. The practical significance of these formulas is that they eliminate the necessity for a direct W, V calculation given that L_o, L are known, and vice versa.

1.5.2 $M/M/1/n$ Queue

Now we consider an $M/M/1/n$ queuing system, i.e., a single-server queue with a buffer of capacity $n - 1$. An incoming customer, which sees a server busy, waits for service in the buffer if there is a free space. If all $n - 1$ waiting places for waiting are busy the customer leaves the system unserved (is lost).

Denote by $i_t, t \geq 0$ the number of customers in the system at time t. This process can take values from $\{0, 1, \ldots, n\}$. It is easy to see that the process $i_t, t \geq 0$ is a birth-and-death process, and non-zero parameters λ_i, μ_i are defined as $\lambda_i = \lambda, 0 \leq i \leq n - 1$, $\mu_i = \mu$, $1 \leq i \leq n$. Then it follows from the formula for the stationary state probabilities of a birth-and-death process that the stationary probabilities of the number of customers in the system under consideration have the form

$$\pi_i = \rho^i \frac{1-\rho}{1-\rho^{n+1}}, 0 \leq i \leq n. \tag{1.28}$$

One of the most important characteristics of a system with possible loss of customers is the probability P_{loss} that an arbitrary customer is lost. For the considered queuing system, it can be shown that this probability coincides with the probability that all waiting places are occupied at an arbitrary time, that is the following formula holds:

$$P_{loss} = \rho^n \frac{1-\rho}{1-\rho^{n+1}}. \tag{1.29}$$

Formula (1.29) can be used to choose the required buffer size depending on the system load and the value of the permissible probability of customer loss.

Note that, unlike the $M/M/1$ queue, the stationary distribution of the number of customers in $M/M/1/n$ exists for any finite values of the load ρ. For $\rho = 1$ calculations by formulas (1.28), (1.29) can be performed using L'Hospital's rule.

1.5.3 System with a Finite Number of Sources

In Sect. 1.2, we introduced the definition of an input flow from a finite number of sources. Let us briefly consider the queuing model describing service of this flow. This model was first investigated by T. Engset.

There is a single-server queuing system with a size m buffer and an input flow of customers from m identical sources. Any source is in a busy state (and therefore cannot generate customers) until its previous customer is served. The service time of any customer from any source has an exponential distribution with parameter μ. In a free state, the source can generate the next customer after a time exponentially distributed with parameter λ and then it goes into a busy state.

Denote by $i_t, t \geq 0$ the number of customers in the system (in the buffer and on the server) at time point t. This process has state space $\{0, 1, \ldots, m\}$. It is easy to see that the process $i_t, t \geq 0$ is a birth-and-death process, and its parameters (the birth rate λ_i and the death rate μ_i), are defined as

$$\lambda_i = \lambda(m-i), 0 \leq i \leq m-1, \mu_i = \mu, 1 \leq i \leq m.$$

From formula (1.11) for the stationary probabilities of a birth-and-death process, we obtain the following expressions for stationary state probabilities $\pi_i, i = 0, \ldots, m$ of the number of customers in the system in an obvious manner:

$$\pi_i = \pi_0 \rho^i \frac{m!}{(m-i)!}, 1 \leq i \leq m, \tag{1.30}$$

where probability π_0 is defined by the normalization condition

$$\pi_0 = \left(\sum_{j=0}^{m} \rho^j \frac{m!}{(m-j)!} \right)^{-1}. \tag{1.31}$$

Using formulas (1.30), (1.31), it is easy to calculate the average number of customers in the system and in the queue. It is also possible to calculate the so-called stationary coefficient k_R of a source's readiness (the probability that a source is ready to generate a customer at an arbitrary time):

$$k_R = \sum_{i=0}^{m-1} \frac{m-i}{m}\pi_i = \frac{\mu(1-\pi_0)}{\lambda m}.$$

1.6 Semi-Markovian Queuing Systems and Their Investigation Methods

As noted in the previous section, the process $i_t, t \geq 0$, which is the number of customers in an $M/M/1$ queue at time t, is a birth-and-death process that is a special case of a continuous-time MC. A similar process for $M/G/1$ -type queuing systems with a non-exponential service time distribution and $GI/M/1$-type queues with non-Poisson input is no longer a Markov process.

The obvious reason for this fact is that the process behavior after some fixed time point t is not completely determined in general by the state of the process at that point but also depends on how long ago the current customer service began or the previous arrival happened.

Nevertheless, the study of the process $i_t, t \geq 0$ can be reduced to the investigation of a Markov process. The first method of Markovization (the so-called embedded MC method) is illustrated by the example of an $M/G/1$ queuing system in Sect. 1.6.1 and a $GI/M/1$ queue in Sect. 1.6.2. One of the variety of other methods of Markovization is the method of introducing an additional variable and it is illustrated in Sect. 1.6.3. In Sect. 1.6.4, one more powerful method to investigate queues is briefly described and illustrated with examples, namely the method of introducing an additional event.

1.6.1 Embedded Markov Chain Method to Study an $M/G/1$

We consider a single-server queue with an unlimited buffer and a Poisson input with parameter λ. The customer service time has an arbitrary distribution with distribution function $B(t)$, Laplace-Stieltjes transform $\beta(s)$, and finite initial moments $b_k,\ k = 1, 2$.

As noted above, the process $i_t,\ t \geq 0$, which is the number of customers in the system at time t, is not Markovian because we cannot describe the behavior of the process after an arbitrary time without looking back. At the same time, it is obvious that if we know the state $i, i > 0$ of the process i_t at the time t_k of the service completion of the kth customer then we can predict the state of the process i_t at the time of the $(k+1)$th customer service completion, which occurs after a random time u with distribution $B(t)$. During this time, a random number of customers (distributed according to a Poisson law with parameter λu) can arrive at the system and one customer (which is served) leaves the system.

We can now introduce the idea of the method of an embedded MC. In general, the method is as follows. For the non-Markovian process $i_t, \ t \geq 0$, we find a sequence of time moments $t_k, k \geq 1$ such that the process $i_{t_k}, k \geq 1$ forms an MC. Using the methods of MC theory, we investigate the stationary distribution of the embedded MC and then, by means of this distribution, we reconstruct the stationary state distribution of the initial process. To this end, usually the theory of renewal processes or Markovian renewal processes is applied.

Let t_k be the kth customer service completion moment in an $M/G/1$ queue. The process $i_{t_k}, k \geq 1$, is a discrete-time homogeneous MC.

It was noted above that an effective investigation of a discrete-time MC with a countable state space is possible only if the matrix of its one-step transition probabilities has a special structure. The matrix P of one-step transition probabilities $p_{i,j}$ of the embedded MC $i_{t_k}, k \geq 1$ has such a structure. Below we find the entries of the matrix P.

Let the number of customers i_{t_k} in the system be i when $i > 0$ at the moment t_k of the kth customer service completion. Since the number of customers in the system makes a jump at time t_k, we will assume that $i_{t_k} = i_{t_k+0}$, i.e., the served customer exits the system and is no longer counted. Since $i > 0$, the next customer is immediately accepted for service, which leaves the system at the next moment t_{k+1} of service completion. Therefore, in order that j customers are in the system at time $t_{k+1}+0$ it is necessary that $j-i+1$ customers arrive at the queue during the time interval (t_k, t_{k+1}). The probability of this event is f_{j-i+1} where the quantities f_l are given by the formula

$$f_l = \int_0^\infty \frac{(\lambda t)^l}{l!} e^{-\lambda t} dB(t), l \geq 0. \tag{1.32}$$

Thus, the transition probability $p_{i,j}, i > 0, j \geq i-1$ is determined by

$$p_{i,j} = f_{j-i+1}, \ i > 0, \ j \geq i-1. \tag{1.33}$$

Suppose now that the system becomes empty after the moment t_k, i.e., $i_{t_k} = 0$. Obviously, the system remains empty until the next customer arrival moment. From now on, the system behaves exactly the same as after a service completion moment with one customer left in the queue. Therefore, $p_{0,j} = p_{1,j}$, which implies that

$$p_{0,j} = f_j, \ j \geq 0.$$

Thus, we have completely described the non-zero elements of the matrix of one-step transition probabilities of the embedded MC. This matrix P has a special structure

$$P = \begin{pmatrix} f_0 & f_1 & f_2 & f_3 & \dots \\ f_0 & f_1 & f_2 & f_3 & \dots \\ 0 & f_0 & f_1 & f_2 & \dots \\ 0 & 0 & f_0 & f_1 & \dots \\ 0 & 0 & 0 & f_0 & \dots \\ \vdots & \vdots & \vdots & \vdots & \ddots \end{pmatrix}. \tag{1.34}$$

Such a structure significantly facilitates analysis of this chain. Using the well-known ergodicity criteria, it is easy to verify that the embedded MC under consideration has a stationary distribution if and only if

$$\rho < 1, \tag{1.35}$$

where the system load ρ is equal to λb_1.

The equilibrium equations (1.16) can be rewritten by taking into account the form (1.34) of the one-step transition probability matrix as follows:

$$\pi_j = \pi_0 f_j + \sum_{i=1}^{j+1} \pi_i f_{j-i+1},\ j \geq 0. \tag{1.36}$$

To solve the infinite system of linear algebraic equations (1.36) we use the PGF. Introduce the generating functions

$$\Pi(z) = \sum_{j=0}^{\infty} \pi_j z^j,\ F(z) = \sum_{j=0}^{\infty} f_j z^j,\ |z| < 1.$$

Taking into account the explicit form (1.32) of the probabilities $f_l, l \geq 0$, we can obtain an explicit expression for the PGF $F(z)$

$$F(z) = \int_0^{\infty} e^{-\lambda(1-z)t} dB(t) = \beta(\lambda(1-z)). \tag{1.37}$$

Multiplying the equations of system (1.36) by the corresponding powers of z and summing them, we obtain

$$\Pi(z) = \pi_0 F(z) + \sum_{j=0}^{\infty} z^j \sum_{i=1}^{j+1} \pi_i f_{j-i+1}.$$

Changing the order of summation, we rewrite this relation in the form

$$\Pi(z) = \pi_0 F(z) + \sum_{i=0}^{\infty} \pi_i z^{i-1} \sum_{j=i-1}^{\infty} f_{j-i+1} z^{j-i+1}$$

$$= \pi_0 F(z) + (\Pi(z) - \pi_0) F(z) z^{-1}. \quad (1.38)$$

Note that we succeeded in reducing the double sum in (1.38) due to the specific properties of the transition probability matrix (1.34), namely, due to the fact that the transition probabilities $p_{i,j}$ of the embedded MC for $i > 0$ depend only on $j - i$ and do not depend on i and j separately. This property of the matrix is called quasi-Toeplitz. It is also essential that all elements of the matrix below its subdiagonal are zero.

Taking (1.37) into account, we can rewrite formula (1.38) in the following form:

$$\Pi(z) = \pi_0 \frac{(1-z)\beta(\lambda(1-z))}{\beta(\lambda(1-z)) - z}. \quad (1.39)$$

Formula (1.39) defines the required PGF of the embedded MC up to the value of the yet unknown probability π_0 that the system is empty at an arbitrary moment of service completion. To find this probability, we recall that the system of equilibrium equations also contains Eq. (1.17) (the normalization condition). It follows from the normalization condition that $\Pi(1) = 1$. Therefore, to find the probability π_0, we must substitute $z = 1$ in (1.39). However, a simple substitution does not give a result since both the numerator and the denominator in (1.39) vanish.

To evaluate an indeterminate form, we can use L'Hospital's rule. However, when calculating the average number of customers $\Pi'(1)$ in the system at service completion moments, we need to apply L'Hospital's rule twice or even three times to get the variance of the number of customers in the system and so on. In order to avoid repeated application of L'Hospital's rule, it is recommended to expand the numerator and denominator of the fraction on the right-hand side of (1.39) in the Taylor series as powers of $(z - 1)$ (if we are interested in calculating the kth initial moment of the queue length distribution then the values should be expanded in a series up to $(z - 1)^{k+1}$). Then we divide the numerator and denominator by $(z - 1)$ and then perform the operations of taking the derivatives and substituting $z = 1$. Note that for calculation of the higher-order moments, one can use the *Mathematica* package.

Using the considerations above, we obtain the following expressions for the probability π_0, the average number of customers L in the system, and the average queue length L_o at service completion moments:

$$\pi_0 = 1 - \rho, \quad (1.40)$$

$$L = \Pi'(1) = \rho + \frac{\lambda^2 b_2}{2(1-\rho)}, \qquad L_o = L - \rho. \tag{1.41}$$

Substituting expression (1.40) into formula (1.39), we obtain

$$\Pi(z) = (1-\rho)\frac{(1-z)\beta(\lambda(1-z))}{\beta(\lambda(1-z)) - z}. \tag{1.42}$$

Formula (1.42) is called the Pollaczek-Khinchine formula for the PGF of the number of customers in an $M/G/1$ queue.

Note that the expressions for L, L_o include the second initial moment b_2 of the service time distribution. Therefore, with the same average service times, the mean queue lengths can differ substantially. So, the average queue length L_o is $\frac{\rho^2}{1-\rho}$ for an exponential service time distribution (in this case $b_2 = \frac{2}{\mu^2}$), and half that for a deterministic service time ($b_2 = \frac{1}{\mu^2}$). With the Erlang service time distribution, the average queue length L_o takes an intermediate value. And with hyperexponential service, it can take significantly larger values. Therefore, estimating the type of the service time distribution for a real model is very important. Taking into account only the averages can result in a significant error when evaluating the system performance characteristics.

Thus, the problem of finding the stationary state distribution of the embedded MC is solved. It should be mentioned, however, that we are not interested in this MC but in the non-Markov process $i_t,\ t \geq 0$, which is the number of customers in the system at an arbitrary time. Let us introduce the stationary state distribution of this process:

$$p_i = \lim_{t\to\infty} P\{i_t = i\}, i \geq 0.$$

From the theory of renewal processes (see, for example, [22]), this distribution exists under the same conditions as the embedded distribution (i.e., when condition (1.35) is fulfilled) and is calculated through the embedded distribution as follows:

$$p_0 = \tau^{-1}\pi_0 \int_0^\infty e^{-\lambda t} dt, \tag{1.43}$$

$$p_i = \tau^{-1}\Big[\pi_0 \int_0^\infty \int_0^t e^{-\lambda v}\lambda dv \frac{(\lambda(t-v))^{i-1}}{(i-1)!} e^{-\lambda(t-v)}(1 - B(t-v))dt$$

$$+ \sum_{l=1}^{i} \pi_l \int_0^\infty \frac{(\lambda t)^{i-l}}{(i-l)!} e^{-\lambda t}(1 - B(t))dt\Big], i \geq 1. \tag{1.44}$$

Here τ is the average interval between the customer departure time points. For our system (without loss of requests) $\tau = \lambda^{-1}$.

Introduce into consideration the PGF $P(z) = \sum\limits_{i=0}^{\infty} p_i z^i$.

Multiplication of Eqs. (1.43) and (1.44) by the corresponding powers of z and further summation yield

$$P(z)=\tau^{-1}\Bigg\{\pi_0\Bigg[\lambda^{-1}+\int\limits_0^{\infty}\int\limits_0^{t} e^{-\lambda v}\lambda dv \sum_{i=1}^{\infty}\frac{(\lambda(t-v))^{i-1}z^i}{(i-1)!}e^{-\lambda(t-v)}(1-B(t-v))dt\Bigg]$$

$$+\sum_{i=1}^{\infty} z^i \sum_{l=1}^{i} \pi_l \int\limits_0^{\infty} \frac{(\lambda t)^{i-l}}{(i-l)!} e^{-\lambda t}(1-B(t))dt\Bigg\}.$$

Changing the order of integration in the double integral and the order of summation in the double sum and counting the known sums, we obtain

$$P(z) = \tau^{-1}\Bigg\{\pi_0\Bigg[\lambda^{-1} + z\lambda \int\limits_0^{\infty} e^{-\lambda v} dv \int\limits_v^{\infty} e^{-\lambda(1-z)(t-v)}(1 - B(t-v))dt\Bigg]$$

$$+\sum_{l=1}^{\infty} \pi_l z^l \int\limits_0^{\infty} e^{-\lambda(1-z)t}(1 - B(t))dt\Bigg\}.$$

Making the change of the integration variable $u = t - v$ and taking into account the equality $\tau^{-l} = \lambda$ and the relation between the Laplace and Laplace-Stieltjes transforms, we have

$$P(z) = \pi_0[1 + \frac{z}{1-z}(1 - \beta(\lambda(1-z)))] + (\Pi(z) - \pi_0)\frac{1}{1-z}(1 - \beta(\lambda(1-z))).$$

By applying expression (1.39) for the PGF $\Pi(z)$, elementary transformations yield

$$P(z) = \pi_0 \frac{(1-z)\beta(\lambda(1-z))}{\beta(\lambda(1-z)) - z}.$$

This results in the validity the equation

$$P(z) = \Pi(z). \tag{1.45}$$

Thus, for the considered $M/G/1$ system, the PGF of the number of customers in the system at the end of service and at arbitrary times coincide. Khinchine [57] called this statement the basic law of the stationary queue.

We consider now the problem of finding the stationary distribution of the waiting time and the sojourn time in the system. We assume that customers are served in order of their arrival at the queue (the FIFO discipline). Let w_t be the waiting time and v_t be the sojourn time for a customer arriving at the queue at the time t. Denote by

$$W(x) = \lim_{t\to\infty} P\{w_t < x\}, V(x) = \lim_{t\to\infty} P\{v_t < x\} \tag{1.46}$$

and

$$w(s) = \int_0^\infty e^{-sx} dW(x), v(s) = \int_0^\infty e^{-sx} dV(x).$$

The limits (1.46) exist if the inequality (1.35) holds true. Since the sojourn time in the system is the sum of the waiting time and the service time, and customer service time in the classical queuing models is assumed to be independent of the state of the system (and of the waiting time), then the LST Property 1.2 yields

$$v(s) = w(s)\beta(s). \tag{1.47}$$

A popular method to obtain an expression for LSTs $w(s)$ and $v(s)$ is to derive the integro-differential Takach equation for the distribution of the virtual waiting time (i.e., the time a virtual customer would wait if it entered the system at a given moment), see, for example, [45].

We get these expressions in a different, simpler way. It is easy to see that for the FIFO discipline, the number of customers left in the system at a service completion time coincides with the number of customers that arrived at the system during the time that the departing customer spent in the system. Hence the following equations hold:

$$\pi_i = \int_0^\infty \frac{(\lambda x)^i}{i!} e^{-\lambda x} dV(x), i \geq 0. \tag{1.48}$$

Multiplying the equations (1.48) by the corresponding powers of z and summing them up, we get

$$\Pi(z) = \int_0^\infty e^{-\lambda(1-z)x} dV(x) = v(\lambda(1-z)).$$

Substituting the explicit form (1.42) of the PGF $\Pi(z)$, taking into account (1.47) and making the change of variable $s = \lambda(1 - z)$ in the relation above, we obtain the following formula:

$$w(s) = \frac{1-\rho}{1-\lambda\frac{1-\beta(s)}{s}}. \tag{1.49}$$

Formula (1.49) is called the Pollaczek-Khinchine formula for the LST of the waiting time distribution in an $M/G/1$ queue.

Using (1.18) and formula (1.49) it is easy to derive the following expression for the mean waiting time W:

$$W = \frac{\lambda b_2}{2(1-\rho)}.$$

The average sojourn time V in the system is as follows:

$$V = b_1 + \frac{\lambda b_2}{2(1-\rho)}. \tag{1.50}$$

The comparison of Eqs. (1.41) and (1.50) yields Little's formula again:

$$L = \lambda V.$$

If it is necessary to find the form of the waiting time distribution function $W(x)$ and not just its LST, the inversion of the transform defined by (1.49) can be performed by expanding the right-hand side of (1.49) into simple fractions (if possible) or by numerical methods (see, e.g., [23] and [97]). The so-called Benes's formula can also be useful

$$W(x) = (1-\rho)\sum_{i=0}^{\infty}\rho^i \tilde{B}_i(x) \tag{1.51}$$

where $\tilde{B}_i(x)$ is the order i convolution of the distribution function

$$\tilde{B}(x) = b_1^{-1}\int_0^x (1 - B(u))du$$

and the convolution operation is defined recurrently:

$$\tilde{B}_0(x) = 1,\ \tilde{B}_1(x) = \tilde{B}(x),$$

$$\tilde{B}_i(x) = \int_0^x \tilde{B}_{i-1}(x-u)d\tilde{B}(u),\ i \geq 2.$$

1.6.2 The Method of Embedded Markov Chains for a $G/M/1$ Queue

We briefly describe the application of the embedded-MC method to the analysis of $GI/M/1$ systems with recurrent input flow and exponentially distributed service time. Let $A(t)$ be the distribution function of inter-arrival times, $\alpha(s)$ be its LST, and $\lambda = a_1^{-1}$ be the intensity of arrivals. The intensity of the exponentially distributed service time is denoted by μ.

The random process $i_t,\ t \geq 0$, which is the number of customers in the system at an arbitrary time, is non-Markovian here since the process behavior after an arbitrary time moment $t, t \geq 0$ is not completely determined by its state at that moment but also depends on the time that has passed since the last customer arrived. Let the time points $t_k,\ k \geq 1$ of the customer arrivals at the system be the embedded points. Since the process $i_t,\ t \geq 0$ makes a jump at these moments, we assume to be definite that $i_{t_k} = i_{t_k - 0}, k \geq 1$, i.e., a customer entering the queue at this moment is excluded from the queue length.

It is easy to see that the process $i_{t_k},\ k \geq 1$ is a discrete MC with a state space that coincides with the set of non-negative integers. The one-step transition probabilities are calculated as follows:

$$p_{i,j} = \int_0^\infty \frac{(\mu t)^{i-j+1}}{(i-j+1)!} e^{-\mu t} dA(t),\ 1 \leq j \leq i+1,\ i \geq 0, \tag{1.52}$$

$$p_{i,0} = 1 - \sum_{j=1}^{i+1} p_{i,j}\ , i \geq 0, \tag{1.53}$$

$$p_{i,j} = 0,\ j > i+1,\ i \geq 0. \tag{1.54}$$

Formula (1.52) is obtained from the following considerations. Since $j \geq 1$, the customers were served constantly between embedded moments, and the number of customers served during this time interval is $i - j + 1$. Since the service time has an exponential distribution with parameter μ, the time points of service completions can be considered a Poisson input (see Proposition 1.2), therefore (see Proposition 1.1) the probability that $i - j + 1$ customers will be served during time t is $\frac{(\mu t)^{i-j+1}}{(i-j+1)!} e^{-\mu t}$. By averaging over all possible values of the inter-arrival times, we obtain formula (1.52). Formula (1.53) follows from the normalization condition.

Using the well-known ergodicity criteria, we can verify that a necessary and sufficient condition for the existence of the stationary state distributions

$$p_i = \lim_{t \to \infty} P\{i_t = i\}, r_i = \lim_{k \to \infty} P\{i_{t_k} = i\},\ i \geq 0,$$

is the fulfillment of the already known inequality

$$\rho < 1, \tag{1.55}$$

where the system load ρ is defined as $\rho = \frac{\lambda}{\mu}$. We assume below that this condition is fulfilled.

Equilibrium equations for the stationary state probabilities $r_i, i \geq 0$ of the embedded MC are written by means of (1.52)–(1.54) as

$$r_j = \sum_{i=j-1}^{\infty} r_i \int_0^{\infty} \frac{(\mu t)^{i-j+1}}{(i-j+1)!} e^{-\mu t} dA(t), \ j \geq 1. \tag{1.56}$$

Let us find the solution of the system of linear algebraic equations (1.56) in the form

$$r_j = C\sigma^j, \ j \geq 0. \tag{1.57}$$

Substituting (1.57) into (1.56) results in the following equation for the unknown parameter σ:

$$\sigma = \int_0^{\infty} e^{-\mu(1-\sigma)t} dA(t) = \alpha(\mu(1-\sigma)). \tag{1.58}$$

We can show that if condition (1.55) holds, Eq. (1.58) has a unique real root $\sigma, 0 < \sigma < 1$. Denote by $y(\sigma) = \sigma - \alpha(\mu(1-\sigma))$. To make sure that equation $y(\sigma) = 0$ has a unique real root $\sigma, 0 < \sigma < 1$, we study the properties of function $y(\sigma)$:

$$y(0) = -\alpha(\mu) < 0;$$

$$y(1) = 0;$$

$$y'(\sigma) = 1 + \mu\alpha'(\mu(1-\sigma));$$

$$y'(1) = 1 - \rho^{-1} < 0 \text{ by virtue of (1.55)};$$

$$y''(\sigma) = -(\mu)^2\alpha''(\mu(1-\sigma)) \leq 0.$$

The last inequality is valid by virtue of the LST Property 1.7.

Thus, the function $y(\sigma)$ is concave, negative at the point $\sigma = 0$, and decreasing at the point $\sigma = 1$. This implies the uniqueness of the root.

The unknown constant C in Eq. (1.57) is easily found from the normalization condition $\sum_{i=0}^{\infty} r_i = 1$ and has the form $C = 1 - \sigma$.

Thus, we have obtained the following expression for the stationary probabilities $r_i, i \geq 0$ of the embedded MC distribution:

$$r_i = \sigma^i (1 - \sigma), \ i \geq 0, \tag{1.59}$$

where the constant σ is a root of Eq. (1.58).

Now we find the stationary probabilities $p_i, \ i \geq 0$ that there are i customers in the system at an arbitrary moment. We fix an arbitrary moment of time. At this moment, the system has exactly i customers, $i \geq 1$, if there were $j, \ j \geq i-1$ customers at the previous customer arrival moment and $i - j + 1$ customers got served and left the system during the time, say u, passed since the previous arrival. Taking into account that the time u has the distribution function $F(t) = \lambda \int\limits_0^t (1 - A(y))dy$ (see Sect. 1.2), we conclude that

$$p_i = \sum_{j=i-1}^{\infty} r_j \lambda \int\limits_0^{\infty} \frac{(\mu t)^{j-i+1}}{(j-i+1)!} e^{-\mu t} (1 - A(t))dt, \ i \geq 1. \tag{1.60}$$

Substituting the probability r_i in the form (1.59) into Eq. (1.60) and summing up, we obtain

$$p_i = (1 - \sigma)\sigma^{i-1} \lambda \int\limits_0^{\infty} e^{-\mu(1-\sigma)t} (1 - A(t))dt,$$

which results in

$$p_i = \rho(1 - \sigma)\sigma^{i-1}, \ i \geq 1 \tag{1.61}$$

taking into account the relation between the LST and the LT and Eq. (1.58).

The probability p_0 is calculated by the normalization condition as

$$p_0 = 1 - \rho. \tag{1.62}$$

The mean number L of customers in the system and the mean queue length L_o at an arbitrary time are defined as

$$L = \sum_{i=1}^{\infty} i p_i = \frac{\rho}{1 - \sigma}, \qquad L_o = \sigma \frac{\rho}{1 - \sigma}. \tag{1.63}$$

Now we find the stationary distribution $W(x)$ of an arbitrary customer waiting time. A customer arriving at an empty system (with probability r_0) has a zero waiting time. Since the sum of the i independent random variables exponentially distributed with parameter μ is an Erlang random variable of order i, then a customer that sees $i, i \geq 1$ customers in the system upon its arrival has to wait for time having the Erlang distribution with parameters (i, μ). Thus we obtain

$$W(x) = 1 - \sigma + (1 - \sigma) \sum_{i=1}^{\infty} \sigma^i \int_0^x \frac{\mu(\mu t)^{i-1}}{(i-1)!} e^{-\mu t} dt$$

$$= 1 - \sigma e^{-\mu(1-\sigma)x}. \tag{1.64}$$

The mean waiting time W is $\frac{\sigma}{\mu(1-\sigma)}$. Comparing this expression with (1.63), we see that Little's formula is also valid for the system considered.

1.6.3 Method of Supplementary Variables

The idea of this method of investigating the non-Markovian processes consists of expanding the process state space by introducing some supplementary components so that the resulting multidimensional process is Markovian. If it is possible to study this Markov process (for example, using the Δt method) then the distribution of the original non-Markov process is usually obtained in a simple way.

For illustration, as in Sect. 1.6.1, we consider an $M/G/1$ queue, i.e., a single-server queuing system with Poisson input with parameter λ and arbitrary service time distribution function $B(t)$, its LST $\beta(s)$, and finite initial moments $b_k,\ k = 1, 2$.

We have already noted that in the case of an $M/G/1$ queue the process $i_t,\ t \geq 0$, which is the number of customers in the system at time t, is not Markovian. Analyzing the reason for this fact, we see that if we include an additional variable $v_t,\ t \geq 0$ (which is either the service time passed to the given point of time t or the time until the service finishes) into the description of the process $i_t,\ t \geq 0$ the resulting two-dimensional random process $\{i_t, v_t\},\ t \geq 0$, is Markovian. These two variants of Markovization are approximately equally popular in the literature. We will now describe the second variant.

Thus, we consider the two-dimensional Markov random process $\{i_t, v_t\},\ t \geq 0$, where $v_t, v_t \geq 0$, is the time from the point t to the time point when the current service will complete. Note that if $i_t = 0$ there is no need to use a supplementary variable since no customer is being served. Consider the following functions:

$$\varphi_t(0) = P\{i_t = 0\},$$

$$\varphi_t(i, x) = P\{i_t = i, v_t < x\}, i \geq 1, x > 0.$$

Proposition 1.12 *Functions* $\varphi_t(0), \varphi_t(i,x), i \geq 1, x > 0$ *satisfy the following system of equations:*

$$\frac{\partial \varphi_t(0)}{\partial t} = -\lambda\varphi_t(0)\frac{\partial \varphi_t(1,x)}{\partial x}|_{x=0}, \tag{1.65}$$

$$\frac{\partial \varphi_t(1,x)}{\partial t} - \frac{\partial \varphi_t(1,x)}{\partial x} = -\lambda\varphi_t(1,x) - \frac{\partial \varphi_t(1,x)}{\partial x}|_{x=0}$$

$$+\frac{\partial \varphi_t(2,x)}{\partial x}|_{x=0}B(x) + \lambda\varphi_t(0)B(x), \tag{1.66}$$

$$\frac{\partial \varphi_t(i,x)}{\partial t} - \frac{\partial \varphi_t(i,x)}{\partial x} = -\lambda\varphi_t(i,x) - \frac{\partial \varphi_t(i,x)}{\partial x}|_{x=0}$$

$$+\frac{\partial \varphi_t(i+1,x)}{\partial x}|_{x=0}B(x) + \lambda\varphi_t(i-1,x), i \geq 2. \tag{1.67}$$

The proof consists of the use of the formula of total probability and analysis of all possible transitions of the process during time Δt and the probabilities of the corresponding transitions. As a result, we get the following system of difference equations:

$$\varphi_{t+\Delta t}(0) = \varphi_t(0)(1-\lambda\Delta t) + \varphi_t(1,\Delta t) + o(\Delta t),$$

$$\varphi_{t+\Delta t}(1,x) = (\varphi_t(1,x+\Delta t) - \varphi_t(1,\Delta t))(1-\lambda\Delta t) + \varphi_t(2,\Delta t)B(x) \tag{1.68}$$

$$+\varphi_t(0)\lambda\Delta t B(x) + o(\Delta t),$$

$$\varphi_{t+\Delta t}(i,x) = (\varphi_t(i,x+\Delta t) - \varphi_t(i,\Delta t))(1-\lambda\Delta t) + \varphi_t(i+1,\Delta t)B(x)$$

$$+\varphi_t(i-1,x+\Delta t)\lambda\Delta t + o(\Delta t), i \geq 2.$$

By dividing both sides of Eqs. (1.68) by Δt and making Δt vanish, we obtain the system (1.65)–(1.67).

As we noted above, the problem of finding the non-stationary (time t dependent) state probability distribution can be solved analytically only in rare cases. Therefore, let us obtain the stationary distribution of the process $\{i_t, \nu_t\}$:

$$\varphi(0) = \lim_{t\to\infty} \varphi_t(0), \varphi(i,x) = \lim_{t\to\infty} \varphi_t(i,x), i \geq 1, x > 0. \tag{1.69}$$

The limits (1.69) exist if the inequality

$$\rho = \lambda b_1 < 1$$

holds. Below this condition is assumed to hold.

Taking a limit in (1.65)–(1.67) as $t \to \infty$, we obtain the following system of equations for the stationary state distribution of the process $\{i_t, \nu_t\},\ t \geq 0$:

$$\lambda\varphi(0) = \frac{\partial\varphi(1,x)}{\partial x}|_{x=0}, \tag{1.70}$$

$$\frac{\partial\varphi(1,x)}{\partial x} - \frac{\partial\varphi(1,x)}{\partial x}|_{x=0} - \lambda\varphi(1,x) + \frac{\partial\varphi(2,x)}{\partial x}|_{x=0}B(x)$$

$$+ \lambda\varphi(0)B(x) = 0, \tag{1.71}$$

$$\frac{\partial\varphi(i,x)}{\partial x} - \frac{\partial\varphi(i,x)}{\partial x}|_{x=0} - \lambda\varphi(i,x)\frac{\partial\varphi(i+1,x)}{\partial x}|_{x=0}B(x) \tag{1.72}$$

$$+\lambda\varphi(i-1,x) = 0, i \geq 2.$$

To solve this infinite system of equations, we use the PGF method and introduce the generating function

$$\Phi(z,x) = \varphi(0) + \sum_{i=1}^{\infty}\varphi(i,x)z^i,\ |z| < 1.$$

Multiplying Eqs. (1.70)–(1.72) by the corresponding powers of z and summing them up, we get the following equation for the generating function $\Phi(z, x)$:

$$\frac{\partial\Phi(z,x)}{\partial x} - \lambda(1-z)\Phi(z,x) = \frac{\partial\Phi(z,x)}{\partial x}|_{x=0}\left(1 - \frac{B(x)}{z}\right) - \lambda\varphi(0)(1-z)(1-B(x)). \tag{1.73}$$

To solve the differential-functional equation (1.73) we introduce the LT

$$\phi(z,s) = \int_0^{\infty} e^{-sx}\Phi(z,x)dx,\ Re\ s > 0.$$

Applying the LT to both sides of Eq. (1.73) and using the relation between the LT and the LST, and also the LST Property 1.3, we obtain an equation of the form

$$(s - \lambda(1-z))\phi(z,s) \tag{1.74}$$

$$= \frac{1}{s}\left[\frac{\partial\Phi(z,x)}{\partial x}|_{x=0}\left(1 - \frac{\beta(s)}{z}\right) - \lambda\varphi(0)(1-z)(1-\beta(s))\right] + \varphi(0).$$

It is known that the generating function is analytic (that is, representable in the form of a convergent power series) for $|z| < 1$ and the LT is analytic in the domain

$Re\ s > 0$. Therefore, for any $z, |z| < 1$, the left-hand side of (1.74) vanishes at $s = \lambda(1 - z)$. Consequently, for such s, the right-hand side of (1.74) also vanishes. From this condition, after simple transformations we obtain

$$\frac{\partial \Phi(z, x)}{\partial x}|_{x=0} = \varphi(0)\lambda z \frac{\beta(\lambda(1 - z))(1 - z)}{\beta(\lambda(1 - z)) - z}. \tag{1.75}$$

Substituting (1.75) into (1.74) yields

$$(s - \lambda(1 - z))\phi(z, s) \tag{1.76}$$

$$= \frac{\lambda\varphi(0)(1 - z)}{s}\left[\frac{z\beta(\lambda(1 - z))\lambda(1 - z)}{\beta(\lambda(1 - z)) - z}\left(1 - \frac{\beta(s)}{z}\right) - 1 + \beta(s)\right] + \varphi(0).$$

Formula (1.76) gives the form of the stationary state distribution up to the value of probability $\varphi(0)$. Now it is worth mentioning that we are forced to consider the two-dimensional Markov process $\{i_t, \nu_t\},\ t \geq 0$ because the process $i_t,\ t \geq 0$ we are interested in, which is the number of customers in the system at an arbitrary time t, is non-Markovian. It is easy to see that the stationary distributions of the processes $\{i_t, \nu_t\},\ t \geq 0$ and $i_t,\ t \geq 0$ are related by

$$p_0 = \lim_{t\to\infty} P\{i_t = 0\} = \varphi(0),$$

$$p_i = \lim_{t\to\infty} P\{i_t = i\} = \lim_{x\to\infty} \varphi(i, x), i \geq 1.$$

Therefore, the PGF $P(z) = \sum_{i=0}^{\infty} p_i z^i$ is determined in a similar way:

$$P(z) = \lim_{x\to\infty} \Phi(z, x).$$

Recalling the relation between the LT and the LST and also the LST Property 1.4, we obtain

$$P(z) = \lim_{x\to\infty} \Phi(z, x) = \lim_{s\to 0} s\phi(z, s),$$

which (by taking (1.76) into account) results in

$$P(z) = p_0 \frac{(1 - z)\beta(\lambda(1 - z))}{\beta(\lambda(1 - z)) - z}. \tag{1.77}$$

Calculating the constant $p_0 = 1 - \rho$ from the normalization condition $P(1) = 1$, we finally obtain the Pollaczek-Khinchine formula

$$P(z) = (1 - \rho)\frac{(1 - z)\beta(\lambda(1 - z))}{\beta(\lambda(1 - z)) - z}. \tag{1.78}$$

Note that (1.78) easily results in the well-known formula for the stationary distribution of the number of customers in an $M/M/1$ queue

$$p_i = (1-\rho)\rho^i, \; i \geq 0.$$

For an $M/D/1$ system with a constant service time, explicit expressions for the stationary state probabilities are as follows:

$$p_0 = 1-\rho, \, p_1 = (1-\rho)(e^{\rho}-1),$$

$$p_i = (1-\rho)\sum_{k=1}^{i}(-1)^{i-k}e^{k\rho}\left[\frac{(k\rho)^{i-k}}{(i-k)!} + \frac{(k\rho)^{i-k-1}}{(i-k-1)!}\right], i \geq 2.$$

1.6.4 Method of Supplementary Events

The method proposed by van Dantzig and by Kesten and Runnenburg (method of collective marks) and subsequently developed by G.P. Klimov (method of catastrophes) makes it easy to obtain analytical results when other known methods lead to cumbersome calculations. It is especially effective when analyzing unreliable and priority queuing systems.

The essence of this method is as follows. Suppose we are required to find some distribution that characterizes the operation of a queuing system. The PGF of the distribution (if it is discrete) or its LST is given a probabilistic sense by *coloring* the requests or introducing a flow of catastrophes. Then some (additional) random event is introduced and its probability is calculated in terms of the PGF or LST of the distribution to be found in two different ways, which results in an equation whose solution is the function the researcher is interested in.

We illustrate this method by applying it to analysis of the $M/G/1$ queue probabilistic characteristics. An important performance characteristic of many real systems is the busy period distribution. The busy period is the time interval from the moment a customer arrives at the empty system until the system becomes empty again. If we know the distribution of the period then we are able to solve problems associated, for example, with planning preventive works in the system, investigating the possibility of the server handling additional load by performing some minor background work, etc.

Denote by $\Pi(t), \; t \geq 0$ a busy period stationary distribution and by $\pi(s), \; s > 0$ its LST.

We assume that the inequality

$$\rho < 1$$

holds and this guarantees that a busy period stationary distribution exists for the queuing system considered.

Proposition 1.13 *The LST $\pi(s)$ of the busy period stationary distribution for an $M/G/1$ queue satisfies the following functional equation:*

$$\pi(s) = \beta(s + \lambda(1 - \pi(s))). \tag{1.79}$$

Proof It is easy to see that the busy period stationary distribution does not depend on the order in which the customers are served. To facilitate the analysis of the busy period structure, suppose that the customers are served in the inverse order, i.e., the server always takes the customer that came last. This discipline of selection from the queue is denoted LIFO (Last In First Out) or LCFS (Last Come First Served). With this service discipline, a customer arrival generates a busy period for the customers that arrive after it. And the structure and consequently the busy period distribution generated by a certain customer are the same as the structure and the distribution of the busy period for the whole system. Using the same reasoning, we come to the understanding that the busy period of the system consists of the service time of the first customer, which initiates the busy period, and the random number of further busy periods generated by customers that came into the system during the first customer service time.

Now suppose that there is an additional Poisson input of catastrophes of intensity s. Introduce into consideration the (additional) event A that no catastrophe arrives during the given busy period.

Recall that according to the probabilistic interpretation of the LST, the quantity $h(s) = \int\limits_0^\infty e^{-st} dH(t), s > 0$ is the probability that no catastrophe arrives during a random time having distribution function $H(t)$. So it is easy to understand that the probability of the event A is

$$P(A) = \pi(s). \tag{1.80}$$

Let us now find the probability of the event A in another way. Let an arbitrary customer be called *bad* if a catastrophe occurs during the busy period generated by this customer. Using the understanding of the busy period structure, it is easy to see that the customer generating the busy period is not bad (with probability $P(A)$) if and only if no catastrophe and no bad customer arrived during its service time.

The catastrophe input flow is Poisson with parameter s. The flow of bad customers is obtained from the original input flow with parameter λ as a result of applying the simplest sifting procedure (an arbitrary customer is accepted to the sifted flow with probability $1 - P(A) = 1 - \pi(s)$ independently of other customers). Therefore, according to Proposition 1.6, the sifted flow is Poisson of intensity $\lambda(1 - \pi(s))$. According to Proposition 1.5, the joint input of catastrophes and bad customers is Poisson of intensity $s + \lambda(1 - \pi(s))$.

Thus, using again the probabilistic interpretation of the LST, we obtain the following formula for the probability of the event A:

$$P(A) = \beta(s + \lambda(1 - \pi(s))). \tag{1.81}$$

Comparing expressions (1.80) and (1.81), we see that (1.79) is valid. Proposition 1.13 is proved. □

Equation (1.79), obtained by D. Kendall in 1951, has a unique solution in the domain $Re s > 0$ such that $|\pi(s)| \leq 1$.

If the service time distribution is exponential, then $B(t) = 1 - e^{-\mu t}$, the considered queue is $M/M/1$, and the LST of the service time distribution $\beta(s)$ has the form $\beta(s) = \frac{\mu}{\mu+s}$. In this case, the functional equation (1.79) goes into the quadratic one for the unknown LST $\pi(s)$:

$$\rho\pi^2(s) - (s\mu^{-1} + \rho + 1)\pi(s) + 1 = 0. \tag{1.82}$$

Solving Eq. (1.82) yields

$$\pi(s) = \frac{1 + s\mu^{-1} + \rho \pm \sqrt{(1 + s\mu^{-1})^2 - 4\rho}}{2\rho}. \tag{1.83}$$

In the formula above, we choose the root with minus because the derived solution satisfies the condition $|\pi(s)| \leq 1$. By inverting the LST $\pi(s)$ we get the following expression for the derivative of the function $\Pi(t)$ of the busy period distribution for an $M/M/1$ queue:

$$\Pi'(t) = \frac{1}{t\sqrt{\rho}} e^{-(\lambda+\mu)t} I_1(2t\sqrt{\lambda\mu}) \tag{1.84}$$

where $I_1(x)$ is a modified Bessel function of the first kind.

In general, Eq. (1.79) can be solved in iterative way by indexing the function $\pi(s)$ by $n+1$ on the left-hand side of the equation and by n on the right-hand side. This procedure has a geometric convergence rate of the sequence $\pi_n(s), n \geq 1$ to the value $\pi(s)$ under the fixed value of the argument s.

In addition, by successively differentiating equation (1.79) with subsequent substitution of the argument $s = 0$ and taking into account the LST Property 1.5, we can obtain the recursive formulas to calculate the initial moments of the busy period distribution. Thus, the mean value π_1 of the busy period and the second initial moment π_2 of its distribution are defined by formulas

$$\pi_1 = \frac{b_1}{1-\rho}, \quad \pi_2 = \frac{b_2}{(1-\rho)^3}. \tag{1.85}$$

As expected, the average length of the busy period tends to infinity as the system load ρ tends to 1.

Now consider the other performance characteristic for an $M/G/1$ queue, namely, the number ξ of customers served during a busy period.

Denote by $\gamma_i = P\{\xi = i\},\ i \geq 1$ and $\Gamma(z) = \sum\limits_{i=1}^{\infty} \gamma_i z^i$.

Proposition 1.14 *The PGF* $\Gamma(z), |z| < 1$ *satisfies the following functional equation:*

$$\Gamma(z) = z\beta(\lambda(1 - \Gamma(z))). \tag{1.86}$$

Proof The PGF $\Gamma(z)$ is given a probabilistic sense as follows. Each customer independently of the others is called *red* with probability z and *blue* with complementary probability. An arbitrary customer is called *dark red* if it is red and only red customers were served in the system during the busy period initiated by it. Introduce the event A that the customer initiating a busy period is dark red. Let us find the probability of this event. On the one hand, it is obvious that

$$P(A) = \Gamma(z). \tag{1.87}$$

On the other hand, from the above analysis of the busy period, it is clear that a customer is dark red if and only if it is red (with probability z) and only dark red customers arrived during its service. Since the customers input is Poisson with parameter λ, and an arbitrary customer is dark red with probability $\Gamma(z)$, then the input of non-dark-red customers (as it follows from Proposition 1.6) is Poisson with parameter $\lambda(1 - \Gamma(z))$. Recalling the probabilistic interpretation of the LST, we derive the following alternative formula for the probability of the event A from the above arguments:

$$P(A) = z\beta(\lambda(1 - \Gamma(z))). \tag{1.88}$$

Comparing formulas (1.87) and (1.88), we see that (1.86) holds. □

Equation (1.86) defines a unique analytic function in the domain $|z| < 1$ such that $|\Gamma(z)| < 1$.

Corollary 1.1 *The mean number* $\mathbf{E}\xi$ *of customers served during a busy period in an* $M/G/1$ *queue is given by*

$$\mathbf{E}\xi = \frac{1}{1 - \rho}.$$

Now we give one more proof of the Pollaczek-Khinchine formula for the PGF of the number of customers in an $M/G/1$ queue at service completion moments. Each customer arriving at the system is called red with probability z and blue with complementary probability independently of the others. We introduce the event A that a customer leaving the server at the given moment is red and all the customers waiting in the queue at that moment are red as well.

It follows from the probabilistic interpretation of the generating function that

$$P(A) = z\Pi(z), \tag{1.89}$$

where $\Pi(z)$ is the required PGF of the number of customers in an $M/G/1$ queue at service completion moments.

On the other hand, an event A holds if and only if all customers that waited in the queue at the previous service completion moment (if the queue is non-empty) were red and no blue customers arrived during the service time. If the system is empty after the service completion the first arriving customer should be red and no blue customer should arrive during its service time. From these considerations it follows that

$$P(A) = (\Pi(z) - \pi_0)\beta(\lambda(1-z)) + \pi_0 z\beta(\lambda(1-z)).$$

From this relation and (1.89), the Pollaczek-Khinchine formula follows in an obvious way:

$$\Pi(z) = \pi_0 \frac{(1-z)\beta(\lambda(1-z))}{\beta(\lambda(1-z)) - z}$$

obtained earlier by means of the embedded-MC method.

Concluding the subsection, we find the characteristics of an $M/G/1$ system with LIFO discipline.

As was noted above, the busy time distribution for an $M/G/1$ queue does not depend on the service discipline. Therefore, Eq. (1.79) defines the LST of the busy period distribution for all disciplines. In addition, it is easy to see that the queue length distributions for an $M/G/1$ queue with FIFO and LIFO disciplines coincide and are given by formula (1.78).

The waiting time distributions for FIFO and LIFO disciplines are different. For the FIFO discipline, the LST $w(s)$ of the stationary waiting time distribution is given by (1.49).

Proposition 1.15 *For the FIFO discipline, the LST $w(s)$ has the following form:*

$$w(s) = 1 - \rho + \frac{\frac{\lambda}{s}(1-\pi(s))}{1 + \frac{\lambda}{s}(1-\pi(s))} \tag{1.90}$$

where function $\pi(s)$ is the solution of (1.79).

Proof Introduce the flow of catastrophes and the concept of a bad customer as we did in the proof of Proposition 1.13. In this case, the function $w(s)$ is the probability that during the customer waiting time no catastrophe occurs, and the function $\pi(s)$ is the probability that an arbitrary customer is not bad, i.e., a catastrophe does not occur during the busy period initiated by this customer.

Given the nature of the LIFO discipline and the reasoning used in the proof of Proposition 1.13, we obtain the formula

$$w(s) = p_0 + (1 - p_0)\tilde{\beta}(s + \lambda - \lambda\pi(s)), \tag{1.91}$$

where $\tilde{\beta}(s)$ is the LST of the residual service time distribution for the customer being served. The residual time is a time period starting from the considered customer arrival to the end of the current service. By analogy with the distribution function $F(t)$ of the time remaining to the next customer arrival moment for the recurrent input flow given in Sect. 1.2, we have the following formula for the distribution function $\tilde{B}(t)$ of the residual service time:

$$\tilde{B}(t) = b_1^{-1} \int_0^t (1 - B(u))du. \tag{1.92}$$

Note that this function has already been used in Benes's formula (1.51). It is easy to see that the LST of this function has the form

$$\tilde{\beta}(s) = \frac{1 - \beta(s)}{sb_1}.$$

Taking this into account, Eq. (1.79) and the formula $p_0 = 1 - \rho$, from (1.91) we obtain the relation (1.90). □

Remark 1.1 Comparing the formulas (1.49) and (1.90), we conclude that waiting time distributions in systems with FIFO and LIFO service disciplines are different but the average waiting times are the same.

1.7 Multi-Server Queues

In the previous section, we did not consider the general single-server $G/G/1$ queue since it does not yield accurate analytical results even for average values of the queue length and waiting time in the system. For these characteristics, only a series of lower and upper bounds are obtained, which make it possible to calculate their values approximately. A rather accurate approximation of the characteristics of this system is possible by means of the characteristics of a $PH/PH/1$ system, which can be investigated analytically.

For similar reasons, we do not consider multi-server $G/G/n$ and $M/G/n$ queues. Note that the average characteristics of the latter of kind system are usually estimated by summing the corresponding known average characteristics with some weights for $M/M/n$ and $M/D/n$ queues. A $GI/M/n$-type queue can be easily analyzed using the embedded-MC method, and the results are close to those obtained for a $GI/M/1$ queue in the previous section. Therefore, we also do not consider it.

In Sect. 1.7.1, we investigate $M/M/n$ and $M/M/n/n+m$ queues. In Sect. 1.7.2, we present the results for $M/M/n/n$ queues (Erlang systems) and their generalizations the $M/G/n/n$ queues. In the last Sect. 1.7.3, an $M/M/\infty$ queue is considered.

1.7.1 $M/M/n$ and $M/M/n/n+m$ Queues

Consider n parallel identical servers (channels) and an infinite buffer for waiting. The input of customers is Poisson with parameter λ, and the customer service time at each server has an exponential distribution with parameter μ. A customer arriving at the system and finding at least one server free immediately occupies a free server and starts its service. If all servers are busy at a customer arrival moment, it joins the queue. Customers are selected for service from the queue according to the FIFO discipline.

Let a random process i_t be the number of customers in the system at time t, $i_t \geq 0,\ t \geq 0$. It is easily verified that i_t is a birth-and-death process with parameters

$$\gamma_0 = \lambda, \gamma_i = \lambda + i\mu, 1 \leq i \leq n, \gamma_i = \lambda + n\mu, i > n,$$

$$p_i = \frac{\lambda}{\lambda + i\mu}, 1 \leq i \leq n, p_i = \frac{\lambda}{\lambda + n\mu}, i > n.$$

Thus, the stationary state probability distribution of the process $i_t,\ t \geq 0$ satisfies the system of differential equations (1.2) and (1.3).

In this case, the birth rates λ_i and death rates μ_i are defined as

$$\lambda_i = \lambda,\ i \geq 0,\ \ \mu_i = i\mu,\ 1 \leq i \leq n,\ \mu_i = n\mu,\ i > n.$$

Thus, the value ρ_i has the form

$$\rho_i = \frac{(\rho n)^i}{i!}, 0 \leq i \leq n, \rho_i = \frac{n^n}{n!}\rho^i, i > n,$$

where $\rho = \frac{\lambda}{n\mu}$.

Parameter ρ, which characterizes the ratio of the input intensity to the total intensity of service by all servers, is called the system load factor.

Checking the condition for existence of the stationary state distribution of the process $i_t,\ t \geq 0$, which is given in Proposition 1.9, it is easy to see that the stationary distribution of the number of customers in the system under consideration

$$\pi_i = \lim_{t\to\infty} P\{i_t = i\}, i \geq 0,$$

exists if the following inequality holds:

$$\rho < 1.$$

Below this condition is assumed to hold.

Proposition 1.16 *Stationary state probabilities $\pi_i, i \geq 0$, are defined as*

$$\pi_i = \begin{cases} \pi_0 \frac{(n\rho)^i}{i!}, & 1 \leq i \leq n, \\ \pi_0 \frac{n^n}{n!} \rho^i, & i > n, \end{cases} \tag{1.93}$$

where

$$\pi_0 = \left[\sum_{j=0}^{n-1} \frac{(n\rho)^j}{j!} + \frac{(n\rho)^n}{n!(1-\rho)} \right]^{-1}. \tag{1.94}$$

The validity of formulas (1.93), (1.94) follows immediately from Proposition 1.9.

Corollary 1.2 *The mean number L of customers in the system at an arbitrary time is defined as follows:*

$$L = \pi_0 \left[\sum_{j=1}^{n-1} \frac{(n\rho)^j}{(j-1)!} + \frac{(n\rho)^n}{n!} \frac{n(1-\rho)+\rho}{(1-\rho)^2} \right]. \tag{1.95}$$

Corollary 1.3 *The mean number L_o of customers waiting in the queue at an arbitrary time is defined as*

$$L_o = \pi_0 \frac{(n\rho)^n \rho}{n!(1-\rho)^2}. \tag{1.96}$$

Corollary 1.4 *The probability P_o that an arbitrary customer arrives at the system at a moment when all servers are busy is defined as*

$$P_o = \frac{(n\rho)^n}{n!(1-\rho)} \left[\sum_{j=0}^{n-1} \frac{(n\rho)^j}{j!} + \frac{(n\rho)^n}{n!(1-\rho)} \right]^{-1}. \tag{1.97}$$

Formula (1.97) is sometimes called the C-formula of Erlang.

The function $W(x)$ of the stationary waiting time distribution is defined as follows:

$$W(x) = 1 - P_o e^{-n\mu(1-\rho)x}, \ x \geq 0. \tag{1.98}$$

The derivation of the last formula is similar to the derivation of (1.24). The average waiting time W has the form

$$W = \frac{P_o}{n\mu(1-\rho)}. \tag{1.99}$$

Comparing formulas (1.96) and (1.99), we note that for the systems under consideration the following version of Little's formula holds:

$$L_o = \lambda W.$$

We now briefly consider $M/M/n/n+m$ queuing systems, where the queue size is limited by a number m, $m \geq 1$. A customer arriving at the system when all servers and all places for waiting in the buffer are busy leaves the system without any effect on the further system operation.

Stationary probabilities π_i, $1 \leq i \leq n+m$ that there are i customers in the system are defined by formulas (1.93), and probability π_0 is calculated as

$$\pi_0 = \left[\sum_{j=0}^{n-1} \frac{(n\rho)^j}{j!} + \frac{(n\rho)^n}{n!} \frac{1-\rho^{m+1}}{1-\rho}\right]^{-1}. \tag{1.100}$$

The values L, L_o, and P_o are defined by formulas

$$L = \sum_{i=1}^{n+m} i\pi_i, \; L_o = \sum_{i=n+1}^{n+m} (i-n)\pi_i, \; P_o = \sum_{i=n}^{n+m} \pi_i.$$

The function $W(x)$ of the stationary waiting time distribution in the system under consideration is calculated according to the scheme given in Sect. 1.3, and is a mixture of Erlang distributions with a jump at zero.

1.7.2 $M/M/n/n$ and $M/G/n/n$ Queues

Let the system have n identical parallel servers and no waiting space. The input of customers is Poisson with parameter λ, and the customer service time has the distribution function $B(t)$ with a finite mean. Below we first consider the case when $B(t)$ is an exponential distribution with parameter μ.

A customer that has arrived at the system with at least one server free immediately occupies any free server and starts service. If all servers are busy at the customer arrival moment, the customer is lost, i.e., leaves the system immediately without any effect on the system's future operation.

Note that the history of queuing theory starts with the study of this model by A.K. Erlang. The practical importance of the model is due to the fact that it adequately describes the functioning of a bundle of telephone channels that receives a flow of connection requests.

Let a random process i_t be the number of customers in the system at an arbitrary time t, $0 \leq i_t \leq n$, $t \geq 0$. It is easy to see that the process i_t is a birth-and-death process with parameters

$$\gamma_0 = \lambda, \gamma_i = \lambda + i\mu, 1 \leq i \leq n-1, \gamma_i = n\mu, i = n,$$

$$p_i = \frac{\lambda}{\lambda + i\mu}, 1 \leq i \leq n-1, p_n = 0.$$

In contrast to the previous subsection, parameter ρ is defined here as the ratio of the input intensity to the service intensity of one server: $\rho = \frac{\lambda}{\mu}$.

It can be shown that due to the finiteness of the process $i_t,\ t \geq 0$ state space, the stationary state distribution of the number of customers in the system

$$\pi_i = \lim_{t \to \infty} P\{i_t = i\}, 0 \leq i \leq n$$

exists for any finite values of the intensities of the incoming flow and service.

Proposition 1.17 *The stationary probabilities $\pi_i, 0 \leq i \leq n$, are defined as*

$$\pi_i = \frac{\frac{\rho^i}{i!}}{\sum_{j=0}^{n} \frac{\rho^j}{j!}}, 0 \leq i \leq n. \tag{1.101}$$

The validity of this proposition follows from Proposition 1.9.

Corollary 1.5 *The probability P_{loss} of an arbitrary customer loss has the form*

$$P_{loss} = \pi_n = \frac{\frac{\rho^n}{n!}}{\sum_{j=0}^{n} \frac{\rho^j}{j!}}. \tag{1.102}$$

This formula is called the Erlang B-formula and plays a very important role in telephony. With its help, we can calculate the probability of loss of customer (blocking channels) under the fixed number of servers, the intensity of the incoming traffic and the intensity of customer service. We can also solve dual tasks, for example, the calculation of the required number of channels or the allowable flow of customers based on the specified maximum permissible probability of a customer loss.

In applications of Erlang's B-formula, it was noted that the loss probability calculated by formula (1.102) coincides in practice with the value of this probability calculated as the average fraction of lost customers in a real system. This seemed somewhat strange since the probability (1.102) was calculated on the assumption that the input flow was Poisson, and the distribution of service time was exponential. And if the good quality of the information flow approximation in telephone networks by the Poisson input can be explained by taking into account Proposition 1.7, the insensitivity of the loss probability to the form of the service time distribution elicited a question. The most probable value of an exponentially distributed random variable is 0, which is poorly consistent with the real statistics of the duration of phone calls.

Therefore, the efforts of many specialists in queuing theory were aimed at proving the invariance of the loss probability equation (1.102) with respect to the form of the distribution function of the service time under the fixed average service time. The strict proof of this fact belongs to B.A. Sevastyanov, who established that the stationary state distribution for an $M/G/n/n$ queue is indeed invariant with respect to the service time distribution.

1.7.3 $M/M/\infty$ Queue

Let the system have an infinite number of parallel identical servers (channels), that is, any customer arriving at the system immediately starts its service. The input flow is supposed to be Poisson with intensity λ, and the service time of a customer has an exponential distribution with parameter μ.

The study of such a system is of interest from several points of view. First, with a small intensity of the input flow and a large (but finite) number of servers, simultaneous employment of all servers is practically impossible, and therefore the number of servers can be considered infinite. Therefore, this model can serve to approximate a model with a large number of servers. Secondly, this model is one of the few for which the non-stationary probability distribution of the number of customers in the system can be obtained in a relatively simple form, and due to this fact the system is sometimes called the simplest queuing system.

Let the random process i_t be the number of customers in the system at time t, $i_t \geq 0,\ t \geq 0$, which is a birth-and-death process with parameters

$$\gamma_0 = \lambda, \gamma_i = \lambda + i\mu, i \geq 1,$$

$$p_i = \frac{\lambda}{\lambda + i\mu}, i \geq 1.$$

Denote $P_i(t) = P\{i_t = i\}, i \geq 0$. Taking into account that the birth intensities λ_i and the death intensities μ_i have the form $\lambda_i = \lambda, i \geq 0, \mu_i = i\mu, i \geq 1$ for the process i_t, we can rewrite the system of differential equations (1.2) and (1.3) for the probabilities $P_i(t), i \geq 0$, as follows:

$$P_0'(t) = -\lambda P_0(t) + \mu P_1(t), \tag{1.103}$$

$$P_i'(t) = \lambda P_{i-1}(t) - (\lambda + i\mu)P_i(t) + (i+1)\mu P_{i+1}(t), i \geq 1. \tag{1.104}$$

Since we are interested in the non-stationary probability distribution, the initial state of the queuing system has to be fixed. Suppose that at time 0, k requests were being served in the system. Then the initial condition for system (1.103), (1.104) has the form

$$P_k(0) = 1,\ P_i(0) = 0,\ i \neq k. \tag{1.105}$$

In order to solve (1.103)–(1.104), consider the PGF

$$\Pi(z,t) = \sum_{i=0}^{\infty} P_i(t) z^i,\ \ |z| < 1.$$

Multiplying the equations of the system (1.103), (1.104) by the corresponding powers of z and summing them up, we obtain the following partial differential equation of the first order:

$$\frac{\partial \Pi(z,t)}{\partial t} = \lambda(z-1)\Pi(z,t) - \mu(z-1)\frac{\partial \Pi(z,t)}{\partial z}. \tag{1.106}$$

The initial condition (1.105) takes the form

$$\Pi(z,0) = z^k. \tag{1.107}$$

In addition, we obtain the following boundary condition from the normalization condition:

$$\Pi(1,t) = 1,\ \ t \geq 0. \tag{1.108}$$

A direct substitution shows that the solution of the system (1.103), (1.104) with the initial condition (1.107) and the boundary condition (1.108) is the function

$$\Pi(z,t) = \left[1 + (z-1)e^{-\mu t}\right]^k e^{\rho(z-1)(1-e^{-\mu t})}, \tag{1.109}$$

where $\rho = \frac{\lambda}{\mu}$.

Thus we have the following proposition.

Proposition 1.18 *The PGF of the non-stationary (depending on time t) state probability distribution for the considered queuing system given the initial state $i_0 = k$ is defined by formula (1.109).*

Corollary 1.6 *The average number L of customers in the system at time t with the initial system state $i_0 = k$ is given by the formula*

$$L = \rho + (k - \rho)e^{-\mu t}.$$

Corollary 1.7 *The stationary state distribution*

$$\pi_i = \lim_{t\to\infty} P_i(t), i \ge 0,$$

of the considered system has the form

$$\pi_i = \frac{\rho^i}{i!} e^{-\rho}, i \ge 0, \tag{1.110}$$

i.e., it obeys the Poisson law with the parameter ρ.

Proof Equation (1.109) yields $\Pi(z) = e^{\rho(z-1)}$ as $t \to \infty$. Expanding this function in the Maclaurin series, we obtain (1.110). □

Corollary 1.8 *If the initial state of the queuing system is not fixed but is chosen randomly in accordance with a certain probability distribution and this distribution is taken to be the stationary distribution (1.110), then for any t it holds*

$$\Pi(z, t) = e^{\rho(z-1)}.$$

We now give two generalizations of Proposition 1.18.

Let $Q_k(i, j, t)$ be the probability that i customers are in the system at time t and exactly j customers are served during time interval $(0, t)$ given that k customers were at the system at time 0. Denote by

$$Q_k(z, y, t) = \sum_{j=0}^{\infty} \sum_{i=\max(0,k-j)}^{\infty} Q_k(i, j, t) z^i y^j, |z| < 1, |y| < 1.$$

Proposition 1.19 *The PGF* $Q_k(z, y, t)$ *satisfies the formula*

$$Q_k(z, y, t) = \left[y + (z - y)e^{-\mu t} \right]^k e^{\lambda t(y-1)+\rho(z-y)(1-e^{-\mu t})}.$$

Suppose now that the customer service time has an arbitrary distribution function $B(t)$ with finite mean b_1.

Proposition 1.20 *The PGF of the non-stationary (depending on time t) state probability distribution for an* $M/G/\infty$ *queuing system given the initial state* $i_0 = k$ *is defined by the formula*

$$\Pi(z, t) = \left[1 + (z - 1)(1 - \tilde{B}(t)) \right]^k e^{\rho(z-1)\tilde{B}(t)},$$

where $\rho = \lambda b_1$, *and the function* $\tilde{B}(t)$ *has the form (1.92).*

The derivation of this formula is due to Riordan and Benes.

To conclude this subsection, we investigate a more general queuing system with an infinite number of servers for which the intensity λ_n of the input flow depends on the number n of customers in the system, $n \geq 0$. We assume that $\lambda_0 = \lambda, \lambda_n = n\lambda, n \geq 1$, and the service time distribution function has density $b(x)$ defined as $B(x) = \int\limits_0^x b(u)du$. When a new customer arrives and the system already provides service to n customers, the busy servers are re-enumerated, and the new customer will be served in the channel with the number $i, i = 1, \ldots, n+1$ with probability $\frac{1}{n+1}$. When a customer at server j completes its service, all servers with numbers greater than j are re-enumerated by decreasing their numbers by one.

Let i_t be the number of customers in the system at time t, and $\xi_t^{(k)}$ be the residual service time at server $k, k = 1, \ldots, i_t$.

The process $\{i_t, \xi_t^{(1)}, \ldots, \xi_t^{(i_t)}, t \geq 0\}$ is Markovian.

Denote

$$q_0 = \lim_{t \to \infty} P\{i_t = 0\},$$

$$Q_i(x_1, \ldots, x_i) = \lim_{t \to \infty} P\{i_t = i, \xi_t^{(1)} < x_1, \ldots, \xi_t^{(i)} < x_i\}, x_l > 0, l = \overline{1, i}, i \geq 1.$$

Let $q_i(x_1, \ldots, x_i)$ be the density of the probability distribution $Q_i(x_1, \ldots, x_i)$,

$$Q_i(x_1, \ldots, x_i) = \int\limits_0^{x_1} \ldots \int\limits_0^{x_i} q_i(u_1, \ldots, u_i) du_1 \ldots du_i.$$

By using the Δt method, we can obtain the following system of equations for the probability q_0 and densities $q_i(x_1, \ldots, x_i)$:

$$\lambda_0 q_0 = q_1(0),$$

$$-\left(\frac{\partial q_i(x_1, \ldots, x_i)}{\partial x_1} + \cdots + \frac{\partial q_i(x_1, \ldots, x_i)}{\partial x_i}\right) = -\lambda_i q_i(x_1, \ldots, x_i)$$

$$+\sum_{j=1}^{i+1} q_{i+1}(x_1, \ldots, x_{j-1}, 0, x_{j+1}, \ldots, x_{i+1})$$

$$+\frac{\lambda_{i-1}}{i} \sum_{j=1}^{i} q_{i-1}(x_1, \ldots, x_{j-1}, x_{j+1}, \ldots, x_i) b(x_j), i \geq 1.$$

An immediate substitution shows that the solution of this system has the form

$$q_i(x_1, \dots, x_i) = \frac{\lambda_0 \lambda_1 \dots \lambda_{i-1}}{i!} \prod_{j=1}^{i} (1 - B(x_j)) q_0$$

$$= \frac{\lambda^i}{i} \prod_{j=1}^{i} (1 - B(x_j)) q_0, i \geq 1.$$

Stationary state probabilities $q_i = \lim_{t \to \infty} P\{i_t = i\}$ are defined as

$$q_i = \int_0^\infty \dots \int_0^\infty q_i(x_1, \dots, x_i) dx_1 \dots dx_i = \frac{\rho^i}{i} q_0, i \geq 1, \tag{1.111}$$

$\rho = \lambda b_1$, and the probability q_0 is derived from the normalization condition.

1.7.4 $M/G/1$ with Processor Sharing and Preemptive LIFO Disciplines

So far we have considered only the FIFO and LIFO service disciplines assuming that the server takes the customer that entered the system first or last, respectively.

An important service discipline also used in real systems is the discipline of the uniform distribution of the processor resource (PS, Processor Sharing), in which the server simultaneously serves all customers in the system at the same rate (inversely proportional to the number of customers).

Since the customer service rate can change when using the PS discipline, it is useful to introduce the concept of the length and the residual length of a customer. In this case, the function $B(x)$ is considered the distribution function of an arbitrary customer length. When the system has exactly i customers during time interval Δt, the residual length of each of them decreases by the value $\frac{\Delta t}{i}$. When the residual length reaches 0 the corresponding customer leaves the system.

Let i_t be the number of customers in the system at time t. To study this process, we use the method of time random replacement. In addition to the real time t, we will consider a fictitious time τ related to the real time by the ratio $i_t d\tau = dt$. This means that if the system has i customers then the fictitious time passes i times faster than the real time. If the system is empty or there is a single customer then the fictitious time coincides with the real time.

It is easy to see that the initial system in fictitious time is equivalent to an infinite-server queue $M/G/\infty$ where the customers are served at a constant rate $i\frac{1}{i} = 1$ by each of the i servers busy at the present moment. In this case, the function $B(t)$ of the customer length distribution coincides with the service time distribution function.

When switching to fictitious time, the input intensity should also be recalculated. Since fictitious time flows faster, the input intensity λ_i to the system $M/G/\infty$ with i customers being served is defined as $\lambda_0 = \lambda, \lambda_i = i\lambda, i \geq 1$.

Obviously, the resulting infinite-server queuing system with the fictitious time completely coincides with the queue considered at the end of the previous subsection. Therefore, the stationary distribution of the number of customers is given by formula (1.111).

Since the stationary state probability of the process can be interpreted as the average fraction of time the process spends in a given state, and the real time flows i times slower than the fictitious time, then the stationary probability p_i of i customers in an $M/G/1$ queue with PS discipline is defined as $p_0 = q_0$, $p_i = iq_i, i \geq 1$ where the probabilities q_i are given by formula (1.111).

Hence, we finally obtain that

$$p_i = (1-\rho)\rho^i, i \geq 0, \rho = \lambda b_1,$$

i.e., the stationary distribution of the number of customers in an $M/G/1$ queue with PS is invariant with respect to the service time distribution and is geometric. The condition for the existence of this distribution is the fulfillment of inequality $\rho < 1$.

Another well-known service discipline in an $M/G/1$ queue is the preemptive LIFO discipline. For this discipline, a customer arriving at a nonempty system displaces a customer being service to the head of a queue similar to a stack, where the interrupted customers are placed to wait for further service.

Using the method of supplementary variables, it can be shown that in the case of such a discipline, the distribution of the number of customers in the system is geometric with the parameter ρ. For both disciplines, the output flow of the system is Poisson regardless of the type of distribution $B(x)$ (see, e.g., [121]).

With the preemptive LIFO discipline, the time to wait for service is zero. Therefore, we are interested in finding the customer sojourn time distribution. An analysis of the system's behavior makes it easy to understand that the LST $v(s)$ of this distribution coincides with the LST $\pi(s)$ of the $M/G/1$ queue busy period, which is independent of the service discipline and given by Eq. (1.79).

1.8 Priority Queues

In all the queues considered above, it was assumed that all customers arriving at a system are homogeneous, that is, they have the same service time distribution and are served in the system according to the same service discipline. However, in many real systems, the customers arriving at the system may be different both in their service time distribution and in their value for the system. Therefore, some customers may have a right to obtain priority service. Such models are investigated within the framework of the theory of priority queuing systems. This theory is rather well developed and many monographs have been devoted to its presentation (see, for example, [12, 46, 55], etc.). Here we limit the subsection to a brief description of priority systems by considering just one of them.

Consider a single-server queue system with infinite buffer to which r independent Poisson flows arrive. The kth flow has intensity λ_k, $k = 1, \dots, r$. Denote by $\Lambda_k = \sum\limits_{i=1}^{k} \lambda_i$.

The service time of the customers from the kth flow has the distribution function $B_k(t)$ with the LST $\beta_k(s)$ and the finite initial moments $b_m^{(k)} = \int\limits_0^\infty t^m dB_k(t)$, $m = 1, 2$, $k = 1, \dots, r$. The customers from the kth flow are referred to as priority k customers.

We suppose that customers from the ith flow have higher priority than the jth flow customers if $i < j$. Priority plays a role when the server completes the service. The next service will be provided to the customer with maximal priority among the customers waiting in the queue. Customers with the same priority are chosen for service according to the prescribed service discipline, for example, the FIFO discipline.

Different variants of the system behavior can be considered in the situation when a higher-priority customer arrives at the system while a lower priority customer is receiving service. The system is called a queuing system with a *relative* or *non-preemptive* priority if the arrival of such a customer does not cause the interruption of the service process. If such an interruption occurs, the system is called a queuing system with an *absolute* or *preemptive* priority. In this case, however, we need to specify the further behavior of the customer whose service was interrupted. We distinguish the following options: the interrupted customer leaves the system and is lost; the interrupted customer returns to the queue and continues its service from the interruption point after all higher-priority customers have left the system; the interrupted customer returns to the queue and starts its service over again after all higher-priority customers have left the system. The interrupted customer is served after all higher-priority customers have left the system at a time having the previous or some other distribution. It is possible that the required service time in subsequent attempts is identical to the time that was required for full service of this customer in the first attempt.

Thus, a rather large number of variants of the priority system behavior can be found in the above mentioned books. The analysis of all priority systems has in common the use of the concept of a system busy period initiated by customers of priority k and higher. The main method used in the analysis of these systems is the method of introducing complementary events briefly described in Sect. 1.6.

Let us illustrate the features of deriving the performance characteristics of priority systems by the example of the system described at the beginning of the subsection. We assume that this is a system with non-preemptive priority, and obtain the stationary waiting time distribution $w_k(t)$ of a priority k customer, $k = 1, \dots, r$ if the customer enters the system at time t (the so-called virtual waiting time).

Denote by

$$W_k(x) = \lim_{t\to\infty} P\{w_k(t) < x\}, x > 0.$$

The condition for these limits, existence is the fulfillment of inequality

$$\rho < 1, \tag{1.112}$$

where ρ is calculated as $\rho = \sum_{k=1}^{r} \lambda_k b_1^{(k)}$.

Denote also $w_k(s) = \int_0^\infty e^{-sx} dW_k(x)$.

Proposition 1.21 *The LST $w_k(s)$ of the priority k customer stationary waiting time distribution is defined by*

$$w_k(s) = \frac{(1-\rho)\mu_k(s) + \sum_{j=k+1}^{r} \lambda_j (1 - \beta_j(\mu_k(s)))}{s - \lambda_k + \lambda_k \beta_k(\mu_k(s))}, \tag{1.113}$$

where functions $\mu_k(s)$ are

$$\mu_k(s) = s + \Lambda_{k-1} - \Lambda_{k-1}\pi_{k-1}(s) \tag{1.114}$$

and functions $\pi_l(s), l = 1, \ldots, r$ are found as solutions of the functional equations

$$\Lambda_l \pi_l(s) = \sum_{i=1}^{l} \lambda_i \beta_i (s + \Lambda_l - \Lambda_l \pi_l(s)), l = 1, \ldots, r. \tag{1.115}$$

Proof Note that the function $\pi_l(s)$ is the LST of the length of the system busy period initiated by the priority l and higher customers (that is, the time interval from the moment of priority l and higher customer arrival at the empty system to the next moment when the system becomes free from l and higher priority customers). The proof that function $\pi_l(s)$ satisfies Eq. (1.115) almost literally repeats the proof of Proposition 1.13. We only note that the value of $\frac{\lambda_i}{\Lambda_l}$ is the probability that the busy period initiated by an l or higher priority customer starts with the priority $i, i = 1, \ldots, l$ customer arrival and $\beta_i(s+\Lambda_l - \Lambda_l\pi_l(s))$ is considered the probability that neither a catastrophe nor a priority l or higher customer arrives at the system during the service time of the priority i customer that initiated the current busy period.

First, instead of the process $w_k(t),\ t \geq 0$, we consider an essentially simpler auxiliary process $\bar{w}_k(t),\ t \geq 0$, which is the time the priority k customer would wait in queue $k, k = 1, \ldots, r$ if it arrives at the system at time moment t and no higher-priority customers arrive.

Let $\bar{w}_k(s,t) = \mathbf{E}e^{-s\bar{w}_k(t)}$ be the LST of the random variable $\bar{w}_k(t)$ distribution function and let us show that function $\bar{w}_k(s,t)$ is defined by

$$\bar{w}_k(s,t) = e^{\varphi_k(s)t}\Bigg[1 - s\int_0^t P_0(x)e^{-\varphi_k(s)x}dx - \sum_{j=k+1}^{r}(1-\beta_j(s))\int_0^t \bar{P}_j(x)e^{-\varphi_k(s)x}dx\Bigg] \qquad (1.116)$$

where

$$\varphi_k(s) = s - \sum_{i=1}^{k}\lambda_i(1-\beta_i(s)),$$

$P_0(x)$ is the probability that the system is empty at time x, and $\bar{P}_j(x)dx$ is the probability that a priority j customer starts its service within the time interval $(x, x+dx)$.

To prove (1.116), we apply the method of a supplementary event. Let the system have an additional Poisson flow of catastrophes of intensity s irrespective of the system's operation. Each customer is considered *bad* if a catastrophe occurs during its service, and *good* otherwise. As follows from Propositions 1.5 and 1.6, the flow of priority k and higher bad customers is the Poisson one with intensity $\sum\limits_{i=1}^{k}\lambda_i(1-\beta_i(s))$.

Introduce an event $A(s,t)$ that priority k and higher bad customers do not arrive at the system during time t. By Proposition 1.1, the probability of this event is defined as

$$P(A(s,t)) = e^{-\sum\limits_{i=1}^{k}\lambda_i(1-\beta_i(s))t}.$$

Let us calculate this probability in another way. The event $A(s,t)$ is a disjunction of three incompatible events $A_l(s,t), l = 1,2,3$.

$A_1(s,t)$ is the event that catastrophes do not occur during time t and time $\bar{w}_k(t)$. In this case, of course, only priority k and higher good customers arrive during time t. The probability of the event $A_1(s,t)$ is obviously equal to $e^{-st}\bar{w}_k(s,t)$.

$A_2(s,t)$ is the event that a catastrophe occurred in the time interval $(x, x+dx), x < t$, but the system was empty at the catastrophe's arrival time, and during the time $t-x$ no priority k and higher bad customers arrived. The probability of the event $A_2(s,t)$ is calculated as

$$P(A_2(s,t)) = \int_0^t P_0(x)e^{-\sum_{i=1}^{k}\lambda_i(1-\beta_i(s))(t-x)} se^{-sx}dx.$$

$A_3(s,t)$ is the event that a catastrophe occurred in the time interval $(x, x+dx)$, $0 \le x < t$, but a lower than k priority customer was being serviced at the catastrophe's arrival time and its service started in the interval $(u, u+du)$, $0 \le u < x$, and no priority k and higher bad customers arrived during time $t-u$. The probability of the event $A_3(s,t)$ is defined as follows:

$$P(A_3(s,t)) = \sum_{j=k+1}^{r}\int_0^t \bar{P}_j(u)e^{-\sum_{i=1}^{k}\lambda_i(1-\beta_i(s))(t-u)} du \int_u^\infty (1-B_j(x-u))se^{-sx}dx.$$

Since the event $A(s,t)$ is the disjunction of three incompatible events its probability is the sum of these events' probabilities. Thus,

$$P(A(s,t)) = e^{-st}\bar{w}_k(s,t) + \int_0^t P_0(x)e^{-\sum_{i=1}^{k}\lambda_i(1-\beta_i(s))(t-x)} se^{-sx}dx$$

$$+\sum_{j=k+1}^{r}\int_0^t \bar{P}_j(u)e^{-\sum_{i=1}^{k}\lambda_i(1-\beta_i(s))(t-u)} du \int_u^\infty (1-B_j(x-u))se^{-sx}dx.$$

Equating the two expressions obtained for probability $P(A(s,t))$ and multiplying both sides of the equality by e^{st}, after simple transformations we get (1.116).

Obviously, no catastrophe arrives during the waiting time $w_k(t)$ of a customer arriving at moment t if and only if no catastrophe and priority $k-1$ or higher customers (such that a catastrophe occurs during the busy periods initiated by them) arrive during time $\bar{w}_k(t)$. This reasoning and probabilistic sense of the LST yield the formula giving the connection between the transforms $\bar{w}_k(s,t)$ and $w_k(s,t) = \mathbf{E}e^{-sw_k(t)}$ in the obvious form

$$w_k(s,t) = \bar{w}_k(s + \Lambda_{k-1} + \Lambda_{k-1}\pi_{k-1}(s), t) = \bar{w}_k(\mu_k(s), t). \tag{1.117}$$

Taking into account (1.116), we take the limit in (1.117) as $t \to \infty$.

It can be shown that the probability $P_0(x)$ satisfies the relation

$$\int_0^\infty e^{-sx}P_0(x)dx = [s + \Lambda_1 - \Lambda_1\pi_1(s)]^{-1},$$

which, by virtue of the LST Property 1.4, results in

$$p_0 = \lim_{t\to\infty} P_0(t) = 1 - \rho.$$

In addition, we can obtain a recursive procedure to calculate the integrals $\int_0^\infty e^{-sx}\bar{P}_j(x)dx$ from (1.116) by tending t to ∞ and taking into account the boundedness of function $\bar{w}_k(s,t)$ for all $s > 0$ and $t \geq 0$ in the form

$$1 = s\int_0^\infty P_0(x)e^{-\varphi_k(s)x}dx$$

$$-\sum_{j=k+1}^{r}(1-\beta_j(s))\int_0^\infty \bar{P}_j(x)e^{-\varphi_k(s)x}dx,\ k = 0,\dots,r-1.$$

From this procedure one can obtain recursive formulas for the values

$$\bar{p}_j = \lim_{t\to\infty}\bar{P}_j(t) = \lim_{s\to 0} s\int_0^\infty \bar{P}_j(x)e^{-sx}dx,\ j = 2,\dots,r.$$

Taking into account the obtained expressions for $p_0, \bar{p}_j, j = 2,\dots,r$, as a result of the limiting transition in (1.117), we obtain the relation (1.113). Proposition 1.21 is proved. □

Denote by $W_1^{(k)}$ the expectation of the virtual waiting time of a priority k customer.

Corollary 1.9 *The values* $W_1^{(k)}, k = 1,\dots,r$, *are calculated as*

$$W_1^{(k)} = \frac{\sum_{i=1}^{r}\lambda_i b_2^{(i)}}{2\left(1-\sum_{i=1}^{k}\lambda_i b_1^{(i)}\right)\left(1-\sum_{i=1}^{k-1}\lambda_i b_1^{(i)}\right)}. \quad (1.118)$$

The proof follows from (1.113) using the LST Property 1.5.

Note that the system with two input flows and continuation of the service of interrupted customers is most easily investigated among systems with absolute priority. In this system, the characteristics of the priority flow service process are completely independent of the second flow and are calculated using the usual formulas for an $M/G/1$ system with a Poisson input with rate λ_1 and service time distribution function $B_1(x)$. In turn, the characteristics of the service process of the non-priority flow are calculated by the formulas for an unreliable system $M/G/1$

with the intensity of the Poisson input λ_2 and the service time distribution function $B_2(x)$. The arrival of a priority customer here means a breakdown of the server, and the server repair time is treated like the busy period when a server serves higher-priority customers.

1.9 Multiphase Queues

In the sections above, we considered queuing models where the customers are served by a single server or by one of several parallel identical servers. At the same time, in many real systems the service process may consist of sequential service of the customer on several servers. Queuing systems of this kind were a prototype of queuing networks and received the name *multiphase queues*.

Multiphase queues are conventionally encoded as a sequence of characters

$$A_1/B_1/n_1/m_1 \to /B_2/n_2/m_2 \to \cdots \to /B_r/n_r/m_r.$$

Here A_1 describes the input flow to the first server (phase) in a sequence of r servers,$B_k, n_k, m_k, k = 1, \ldots, r$ accordingly describe the service time distribution, the number of parallel servers, and the number of waiting places in the kth phase of the sequence. Symbols $A_1, B_k, k = 1, \ldots, r$ take their values in the same sets as the corresponding symbols in the description of the single-phase systems considered above. The symbol $\to$ means the customer transition to the input of the next server after it has been served in the previous one. In a certain sense, this symbol is a substitute for a character specifying the type of the input flow to the corresponding queuing system since the input flow to this queuing system is determined by the output flow of customers from the previous queuing system. For this reason, sometimes the coding of multiphase queues is a bit modified:

$$A_1/B_1/n_1/m_1 \to \bullet/B_2/n_2/m_2 \to \cdots \to \bullet/B_r/n_r/m_r.$$

If any character m_k takes a finite value then the question arises of the behavior of a multiphase queuing system in a situation when the buffer of the corresponding single-phase queuing system is already full, and the next customer arrives at this queue. Usually, two options are considered: the incoming customer is lost or the incoming customer stays at the server where its service has finished and temporarily blocks the further operation of this server.

Known results for a wide range of multiphase queuing systems are described quite exhaustively in [44]. Here we briefly consider three simple multiphase queuing systems.

The first of them is a multiphase queuing system with an infinite buffer before the first phase with a Poisson input of customers of intensity λ, where the service time in each of the r phases has an exponential distribution with parameter μ. It is assumed that just one customer can be served in the system at any time. Only after a customer has passed through all servers in the sequence (all phases) can the next customer start its service.

Recalling that an Erlang random variable with parameters (μ, r) is the sum of r independent random variables exponentially distributed with parameter μ, we come to the conclusion that all performance characteristics of the system considered can be easily obtained from the characteristics of the corresponding $M/E_r/1$-type single-phase queuing system. Such a system can be analyzed using the so-called Erlang phase method. The corresponding results can also be obtained from the formulas for the system $M/G/1$.

Thus, for generating function $P(z) = \sum\limits_{i=0}^{\infty} p_i z^i$ of the probabilities $p_i, i \geq 0$ of the number of customers in the system, the Pollaczek-Khinchine formula (see, for example (1.42)) taking into account the equality $\beta(s) = \left(\frac{\mu}{\mu+s}\right)^r$ yields

$$P(z) = (1-\rho)\frac{(1-z)}{1-z(1+\frac{\lambda}{\mu}(1-z))^r}, \tag{1.119}$$

where the system load ρ is defined as $\rho = \frac{r\lambda}{\mu}$.

Denote by $z_i, i = 1, \ldots, r$ the roots of equation

$$1 - z(1 + \frac{\lambda}{\mu}(1-z))^r = 0 \tag{1.120}$$

that have modulus greater than one.

Expanding the right-hand side of (1.119) by simple fractions and using the properties of the generating function, we get

$$p_0 = 1 - \rho, \tag{1.121}$$

$$p_i = (1-\rho)\sum_{l=1}^{r} K_l z_l^{-ir}\frac{1-z_l^r}{1-z_l}, \qquad i \geq 1, \tag{1.122}$$

where coefficients $K_m, m = 1, \ldots, r$ are determined as follows:

$$K_m = \prod_{l=1, l\neq m}^{r} \frac{1}{1-\frac{z_m}{z_l}}, m = 1, \ldots, r. \tag{1.123}$$

Now consider the following multiphase queuing system:

$$M/M/n_1/\infty \to /M/n_2/\infty \to \cdots \to /M/n_r/\infty,$$

i.e., the multiphase system with r consecutive multi-server queuing systems with unlimited waiting space. Let λ be the input flow intensity and μ_k be the service rate at each server in the kth system, $k = 1, \dots, r$.

Denote by $i_k(t)$ the number of customers in the kth system at time t, $t \geq 0, k = 1, \dots, r$ and by $p(i_1, \dots, i_r)$ the stationary probability of state $\{i_1, \dots, i_r\}$ of the system considered, namely,

$$p(i_1, \dots, i_r) = \lim_{t \to \infty} P\{i_1(t) = i_1, \dots, i_r(t) = i_r\}, i_k \geq 0, k = 1, \dots, r. \tag{1.124}$$

It can be shown that the condition for the limits' (1.124) existence is

$$\rho_i = \frac{\lambda}{n_i \mu_i} < 1, i = 1, \dots, r. \tag{1.125}$$

Further, we assume that these inequalities are valid. Using the Δt-method, we can obtain a system of differential equations for probabilities $P\{i_1(t) = i_1, \dots, i_r(t) = i_r\}$ that results in the following system of equations for the stationary state probabilities $p(i_1, \dots, i_r)$ by taking the limit as $t \to \infty$:

$$\left(\lambda + \sum_{k=1}^{r} \alpha_k(i_k)\mu_k\right) p(i_1, \dots, i_r) = \lambda p(i_1 - 1, i_2, \dots, i_r)(1 - \delta_{i_1,0}) \tag{1.126}$$

$$+ \sum_{k=1}^{r-1} p(i_1, \dots, i_k + 1, i_{k+1} - 1, i_{k+2}, \dots, i_r)\alpha_k(i_k + 1)(1 - \delta_{i_{k+1},0})\mu_k$$

$$+ p(i_1, \dots, i_{r-1}, i_r + 1)\alpha_r(i_r + 1)\mu_r, i_k \geq 0, k = 1, \dots, r,$$

where

$$\alpha_k(i) = \begin{cases} i, 0 \leq i \leq n_k, \\ n_k, \quad i > n_k, \end{cases}$$

$$\delta_{i,j} = \begin{cases} 1, i = j, \\ 0, i \neq j. \end{cases}$$

It is easily verified by a direct substitution that the solution of system (1.126) has the following form:

$$p(i_1, \dots, i_r) = p(0, \dots, 0) \prod_{k=1}^{r} c_k(i_k), \tag{1.127}$$

where

$$c_k(i) = \begin{cases} \frac{(n_k \rho_k)^i}{i!}, & 0 \le i \le n_k, \\ \frac{n_k^{n_k}}{n_k!}\rho_k^i, & i > n_k. \end{cases}$$

Probability $p(0, \ldots, 0)$ is given by a normalization condition

$$\sum_{i_1=0}^{\infty} \cdots \sum_{i_r=0}^{\infty} p(i_1, \ldots, i_r) = 1 \tag{1.128}$$

and has the form

$$p(0, \ldots, 0) = \prod_{k=1}^{r} (\sum_{i_k=0}^{\infty} c_k(i_k))^{-1}. \tag{1.129}$$

It can be seen from (1.127), (1.128), and formulas (1.93), (1.94) that the stationary state probabilities $p(i_1, \ldots, i_r)$ of the considered queuing system can be represented in a product form

$$p(i_1, \ldots, i_r) = \prod_{k=1}^{r} \lim_{t \to \infty} P\{i_k(t) = i_k\}, \tag{1.130}$$

that is, the joint probability that i_k customers are at phase k at an arbitrary time, $k = 1, \ldots, r$, is equal to the product of the probabilities that i_k customers are on phase k regardless of the number of customers on other phases, $k = 1, \ldots, r$.

This fact allows us to calculate the stationary state distribution of a multiphase queuing system as the product of the state probabilities of the single-phase queuing systems forming the multiphase queue follows from Burke's theorem [16], which is formulated as follows.

Theorem 1.1 *For a system with c parallel servers, Poisson input with parameter λ and identical exponential service time distribution with parameter μ for each server, the output flow of customers is Poisson with parameter λ.*

Thus, serving a simple flow at each phase of a multiphase queue with an exponential distribution of the service time does not change the nature of the flow, and as a result the joint probability distribution of the number of customers in the corresponding multiphase system has a product form.

To conclude the subsection, let us briefly consider a multiphase queuing system of type $M/G/1/\infty \to M/n/0$. The input intensity is denoted by λ, the service time distribution function is $B(t)$, and the service intensity at any server in the second phase is denoted μ.

We assume that if all servers in the second phase are busy, a customer finishing its service in the first phase leaves the system unserved (lost) at the second phase with probability $\theta, 0 \leq \theta \leq 1$, and with the complementary probability the first server is blocked and does not serve the next customer until the second-phase server becomes free. As a limiting case for $\theta = 0$, we get a system with server blocking, and a system with losses for $\theta = 1$.

Consider the two-dimensional process $\{i_{t_k}, \nu_{t_k}, k \geq 1\}$, where t_k is the kth customer service completion moment of the first phase, $i_{t_k}, i_{t_k} \geq 0$ is the number of customers in the first phase, $\nu_{t_k}, \nu_{t_k} = 0, \ldots, n$ is the number of customers in the second phase at time $t_k + 0$. It is easy to see that this process is a two-dimensional discrete-time MC.

Consider the one-step transition probabilities of this chain $P\{i_{t_{k+1}} = l, \nu_{t_{k+1}} = \nu^{'} | i_{t_k} = i, \nu_{t_k} = \nu\} = P\{(i, \nu) \rightarrow (l, \nu^{'})\}, i > 0$ and their PGF

$$R_{\nu,\nu'}(z) = \sum_{l=i-1}^{\infty} P\{(i, \nu) \rightarrow (l, \nu^{'})\} z^{l-i+1}.$$

Similarly to an embedded MC for the system $M/G/1$ considered in Sect. 1.6, the transition probabilities $P\{(i, \nu) \rightarrow (l, \nu^{'})\}$ depend on $l - i$ and are independent of i and l separately. This makes the PGF $R_{\nu,\nu'}(z)$ definition correct.

By analyzing the transitions of the two-dimensional MC, we can make sure that PGFs $R_{\nu,\nu'}(z)$ are defined as follows:

$$R_{\nu,\nu'}(z) = \begin{cases} \Gamma(z, n, n)\left(\theta + (1-\theta)\frac{n\mu}{n\mu+\lambda(1-z)}\right), & \nu = n, \nu^{'} = n, \\ \Gamma(z, \nu, \nu^{'} - 1), & \nu^{'} \leq \nu \leq n, \nu^{'} \neq n, \end{cases}$$

where

$$\Gamma(z, \nu, \nu^{'}) = \int_0^{\infty} e^{-\lambda(1-z)t} C_{\nu}^{\nu^{'}} e^{-\mu\nu^{'}t} (1 - e^{-\mu t})^{\nu-\nu^{'}} dB(t),$$

$$0 \leq \nu^{'} \leq \nu \leq n.$$

We compose the generating functions $R_{\nu,\nu'}(z)$ into a matrix PGF $R(z)$ and consider the matrix Δ with entries

$$\Delta_{\nu,\nu'} = \int_0^{\infty} \lambda e^{-\lambda t} C_{\nu}^{\nu^{'}} e^{-\mu\nu^{'}t} (1 - e^{-\mu t})^{\nu-\nu^{'}} dt,$$

$$0 \leq \nu^{'} \leq \nu \leq n.$$

The matrix Δ characterizes the transition probabilities for the number of busy servers in the second phase of the system during the time when a server is idle in the first phase waiting for a customer arrival.

Denote by

$$\pi(i, \nu) = \lim_{k\to\infty} P\{i_{t_k} = i, \nu_{t_k} = \nu\}, \tag{1.131}$$

$$\vec{\pi}(i) = (\pi(i, 0), \pi(i, 1), \ldots, \pi(i, n)),$$

$$\vec{\Pi}(z) = \sum_{i=0}^{\infty} \vec{\pi}(i) z^i, \; |z| < 1.$$

Proposition 1.22 *The vector PGF $\vec{\Pi}(z)$ satisfies the following matrix equation:*

$$\vec{\Pi}(z)(R(z) - I) = \vec{\Pi}(0)(I - \Delta z)R(z). \tag{1.132}$$

Here I is an identity matrix.

The condition for the limits (1.131) to exist and the algorithms to solve Eq. (1.132) can be found, for example, in [26, 94].

The stationary state probability distribution at an arbitrary time can be found using the theory of Markov renewal processes in a similar way as in Sect. 1.6.1.

Chapter 2
Methods to Study Queuing Systems with Correlated Arrivals

A popular model of correlated arrival processes is the Batch Markovian Arrival Process. In this chapter, we define this process and consider its basic properties, particular cases and some its generalizations. Analysis of performance characteristics of queuing systems with this arrival process requires consideration of multidimensional (at least two-dimensional) Markov processes. Therefore, for the use in the following chapters here we briefly present the known in the literature results for multidimensional Markov chains with several kinds of special structure of the generator or transition probability matrix. Some of these results were originally obtained in the papers of authors of this book.

2.1 Batch Markovian Arrival Process (*BMAP*): Phase-Type Distribution

2.1.1 Definition of the Batch Markovian Arrival Process

In a $BMAP$, arrivals of customers are governed by an irreducible continuous-time MC $\nu_t,\ t \geq 0$ with a finite state space $\{0, \ldots, W\}$. The sojourn time of the chain ν_t in state ν is exponentially distributed with parameter $\lambda_\nu,\ \nu = \overline{0, W}$. After the process sojourn time in the state expires, the process transits (makes a jump) to state ν' with probability $p_k(\nu, \nu')$ and a group of k customers is generated, $k \geq 0$. From now on, $p_0(\nu, \nu) = 0$ and

$$\sum_{\nu'=0\, \nu'\neq\nu}^{W} p_0(\nu, \nu') + \sum_{k=1}^{\infty}\sum_{\nu'=0}^{W} p_k(\nu, \nu') = 1, \ \ \nu = \overline{0, W}.$$

A. N. Dudin et al., *The Theory of Queuing Systems with Correlated Flows*,
https://doi.org/10.1007/978-3-030-32072-0_2

2.1.2 The Flow Matrix Counting Function

The main characteristic of any stationary random flow of arrivals is a flow counting function $p(n, t)$ that is the probability that n customers arrive during time t. In the case of a stationary Poisson input, the flow counting function has the form

$$p(n, t) = \frac{(\lambda t)^n}{n!} e^{-\lambda t}, n \geq 0.$$

It is impossible to specify a counting function as an unconditional probability for a $BMAP$ in a similar way since the number of customers arriving during a time interval of length t depends on the state of the process $\nu_t, t \geq 0$ at the beginning of the time interval. But we are able to calculate probabilities

$$P_{\nu,\nu'}(n, t)$$

$$= P\{n \text{ customers arrive during time } t \text{ and } \nu_t = \nu' \text{ given that } \nu_0 = \nu\}.$$

Below, we derive a system of equations for these probabilities. To this aim, we use the well-known Δt-method, which consists in deriving the equations for the state probability distribution of a random process at time t by calculating the state probability distribution of the process at time $t + \Delta t$ from the state probability distribution of this process at time t and the transition probabilities in a short time interval $(t, t + \Delta t)$. As a result, we obtain the following system of difference equations in an obvious way:

$$P_{\nu,\nu'}(n, t + \Delta t) = P_{\nu,\nu'}(n, t)e^{-\lambda_{\nu'}\Delta t} + \sum_{k=0}^{n}\sum_{r=0}^{W} P_{\nu,r}(k, t)(1 - e^{-\lambda_r \Delta t})p_{n-k}(r, \nu')$$

$$+o(\Delta t), \; n \geq 0, \nu, \nu' = \overline{0, W}.$$

In a standard way, this system results in the system of differential equations

$$P'_{\nu,\nu'}(n, t) = -\lambda_{\nu'} P_{\nu,\nu'}(n, t)$$

$$+\sum_{k=0}^{n}\sum_{r=0}^{W} P_{\nu,r}(k, t)\lambda_r p_{n-k}(r, \nu'), \; n \geq 0, \nu, \nu' = \overline{0, W}. \tag{2.1}$$

Consider the square matrix $P(n, t) = (P_{\nu,\nu'}(n, t))_{\nu,\nu'=\overline{0,W}}$ of order $\bar{W}$, where $\bar{W} = W + 1$. It is easy to see that the system of equations (2.1) can be rewritten in the form

$$P'(n, t) = \sum_{k=0}^{n} P(k, t)D_{n-k}, \; n \geq 0. \tag{2.2}$$

Here the D_k are the square matrices of order $\bar{W}$ whose entries are defined as

$$(D_0)_{\nu,\nu'} = \begin{cases} -\lambda_\nu, \ \nu = \nu', \\ \lambda_\nu p_0(\nu, \nu'), \ \nu \neq \nu', \end{cases}$$

$$(D_k)_{\nu,\nu'} = \lambda_\nu p_k(\nu, \nu'), k \geq 1,$$

and can be considered as an instantaneous intensity of the $BMAP$ since $\lambda_\nu p_k(\nu, \nu')\Delta t + o(\Delta t)$ is the probability that the $BMAP$ governing process ν_t transits from state ν to state ν' and a group of k arrivals is generated, and $1 - \lambda_\nu \Delta t + o(\Delta t)$ is the probability that the process ν_t does not make a jump and no arrivals happen during time Δt given that process ν_t was in state ν at the beginning of the Δt-interval.

The system of differential equations (2.2) contains an infinite number of matrix equations. To solve it, we introduce the matrix PGF

$$P(z, t) = \sum_{n=0}^{\infty} P(n, t) z^n, \ |z| \leq 1.$$

Multiplying the equations of system (2.2) by the corresponding powers of z and summing them up, we get the matrix differential equation

$$P'(z, t) = P(z, t) D(z), \tag{2.3}$$

where

$$D(z) = \sum_{k=0}^{\infty} D_k z^k.$$

It is obvious that the solution of the linear matrix differential equation (2.3) has the following form:

$$P(z, t) = C(z) e^{D(z)t},$$

where $C(z)$ is some matrix independent of t.

Taking into account the obvious initial condition

$$P(z, 0) = P(0, 0) = I,$$

we obtain that $C(z) = I$ for any z, where I is an identity matrix. Hence,

$$P(z, t) = e^{D(z)t}. \tag{2.4}$$

Note that for a Poisson arrival process, the PGF $P(z,t)$ has the form $P(z,t) = e^{\lambda(z-1)t}$, which agrees with the fact that $D(z) = \lambda(z-1)$ in this case.

It follows from (2.4), in particular, that the matrix $P(0,t) = (P_{\nu,\nu'}(0,t))_{\nu,\nu'=\overline{0,W}}$ of the probabilities that no customers arrive during time t has the form

$$P(0,t) = e^{D_0 t}. \tag{2.5}$$

In principle, the matrices $P(n,t),\ n \geq 1$ can be calculated from a matrix PGF $P(z,t)$ having the form (2.4) as follows:

$$P(n,t) = \frac{1}{n!}\frac{\partial^n P(z,t)}{\partial z^n}\bigg|_{z=0},\ n \geq 0.$$

However, the calculation of the matrices $P(n,t)$ in the analytical form using this formula is possible only in rare cases.

In the general case, they are calculated using a procedure based on the idea of uniformization of a Markov process, which is the following.

If H is an infinitesimal generator of a continuous-time MC then the following representation holds:

$$e^{Ht} = e^{ht(L-I)} = e^{-ht}e^{htL} = \sum_{j=0}^{\infty} e^{-ht}\frac{(ht)^j}{j!}L^j, \tag{2.6}$$

where

$$h = \max_{i=\overline{0,W}} (-H)_{ii},\ L = I + h^{-1}H.$$

The matrix L is the matrix of one-step transition probabilities of some discrete-time MC. A very useful property of representation (2.6) is the multiplicativity of the sum terms, their representation in the form of a product of a scalar function of the argument t and a matrix that does not depend on t.

It is easy to see that a similar decomposition (taking into account that L is a sub-stochastic matrix) can be applied to the matrix exponent if H is a matrix in which the off-diagonal elements are non-negative, the diagonal elements are negative, and the sum of all entries in each row is negative or zero. In particular, such an expansion is valid for the matrix exponent $e^{D_0 t}$.

Denote by $\tilde{\theta} = \max_{i=\overline{0,W}} (-D_0)_{ii}$. Then $P(0,t) = e^{D_0 t}$ can be presented in the form

$$P(0,t) = \sum_{j=0}^{\infty} e^{-\tilde{\theta}t}\frac{(\tilde{\theta}t)^j}{j!}K_0^{(j)},$$

where

$$K_0^{(j)} = (I + \tilde{\theta}^{-1} D_0)^j.$$

By analogy, the other matrices $P(n, t),\ n \geq 1$, can be found in the form

$$P(n, t) = \sum_{j=0}^{\infty} e^{-\tilde{\theta} t} \frac{(\tilde{\theta} t)^j}{j!} K_n^{(j)},\ n \geq 1, \tag{2.7}$$

where $K_n^{(j)},\ n \geq 1, j \geq 0$ are some matrices.

Substituting the expression for $P(0, t)$ and expression (2.7) into the system of differential equations (2.2) and equating the coefficients for the same powers of t, we see that representation (2.7) is valid if the matrices $K_n^{(j)},\ n \geq 0, j \geq 0$ satisfy the following system of recursive relations:

$$K_0^{(0)} = I,\ K_n^{(0)} = O,\ n \geq 1,$$

$$K_0^{(j+1)} = K_0^{(j)} (I + \tilde{\theta}^{-1} D_0), \tag{2.8}$$

$$K_n^{(j+1)} = \tilde{\theta}^{-1} \sum_{i=0}^{n-1} K_i^{(j)} D_{n-i} + K_n^{(j)} (I + \tilde{\theta}^{-1} D_0),\ n \geq 1,\ j \geq 0.$$

Formulas (2.7) and (2.8) define a numerically stable procedure for computing matrices $P(n, t),\ \ n \geq 1$. The truncation of the infinite sum in (2.7) can be performed after a sum term becomes less in a norm than a prescribed small positive number. The formula

$$K(z, y) = \sum_{n=0}^{\infty} \sum_{j=0}^{\infty} K_n^{(j)} y^j z^n = \left(I - y(I + \tilde{\theta}^{-1} D(z)) \right)^{-1}$$

can potentially be useful in choosing the truncation threshold.

2.1.3 Some Properties and Integral Characteristics of *BMAP*

From the definition of *BMAP*, it follows that:

- the matrix whose entry (ν, ν') is the probability that the first customer arrives in time interval $(t, t + dt)$ as a member of a size k batch of customers, and the governing process is in state ν' at time $t + dt$ given that the process was in state

ν at time t, is defined as follows:

$$D_k dt, \ k \geq 1, \ t \geq 0.$$

- the matrix whose entry (ν, ν') is the probability that the first customer arrives in time interval $(t, t+dt)$ as a member of a size k batch of customers, and the governing process is in state ν' exactly after their arrival given that the process was in state ν at time 0, is defined as follows:

$$e^{D_0 t} D_k dt, \ k \geq 1.$$

- the matrix whose entry (ν, ν') is the probability that the first batch of customers after time 0 is a batch of size k, and the governing process is in state ν' exactly after their arrival given that the process was in state ν at time 0, is defined as

$$\int_0^\infty e^{D_0 t} D_k dt = (-D_0)^{-1} D_k, \ k \geq 1.$$

  The existence of the integrals in the formula above and the inverse matrix follows from the definition of function of matrix and Corollary A.1 in Appendix A.
- matrix $D(1)$ is an infinitesimal generator of the process $\nu_t, t \geq 0$.
- row vector $\boldsymbol{\theta}$ of the stationary state distribution of the governing process $\nu_t, t \geq 0$ is the unique solution of the system of algebraic linear equations

$$\boldsymbol{\theta} D(1) = \mathbf{0}, \ \boldsymbol{\theta}\mathbf{e} = 1. \tag{2.9}$$

  The value λ given by the formula

$$\lambda = \boldsymbol{\theta} D'(1)\mathbf{e} \tag{2.10}$$

is called the average rate (intensity) of arrivals for the $BMAP$, where $D'(1)$ is the derivative of the matrix generating function $D(z)$ at point $z = 1$.

The notion of the mean (fundamental) rate of arrivals introduced by M. Neuts can be considered to be the average number of customers arriving per time unit. The fact that λ is the mean number of arrivals per time unit can be proved by the following formula:

$$\sum_{k=1}^\infty k P(k,t)\mathbf{e} = D'(1)t\mathbf{e} + (e^{D(1)t} - I - D(1)t)(D(1) - \mathbf{e}\boldsymbol{\theta})^{-1} D'(1)\mathbf{e}. \tag{2.11}$$

Formula (2.11) is derived as follows:

$$\sum_{k=1}^{\infty} kP(k,t)\mathbf{e} = \left(e^{D(z)t}\right)'\Big|_{z=1}\mathbf{e} = \left(\sum_{k=0}^{\infty}\frac{(D(z)t)^k}{k!}\right)'\Big|_{z=1}\mathbf{e}$$

$$= \sum_{k=1}^{\infty}\frac{t^k}{k!}(D(1))^{k-1}D'(1)\mathbf{e}.$$

Here we used the fact that matrix $D(1)$ is the generator, hence $D(1)\mathbf{e} = \mathbf{0}^T$.

Next, we need to express the resulting series again in terms of a matrix exponent but this cannot be done directly since matrix $D(1)$ is singular. To this aim, we use Lemma A.9 of Appendix A formulated for stochastic matrices. By rephrasing the lemma in terms of generators, we conclude that matrix $\mathbf{e}\boldsymbol{\theta} - D(1)$ is non-degenerate since vector $\boldsymbol{\theta}$ is a solution of the system (2.9). Using this fact and system (2.9), we can show that relation

$$(D(1))^{k-1} = -(\mathbf{e}\boldsymbol{\theta} - D(1))^{-1}(D(1))^k$$

holds for $k \geq 2$.

Now we can eliminate the series and, as a result, obtain formula (2.11).

Let us multiply Eq. (2.11) by vector $\boldsymbol{\theta}$. Taking into account the probabilistic sense of matrices $P(k,t)$, the left-hand side of the equation presents the mean number of arrivals during time t and the right-hand side is λt.

By dividing both sides of the equation by t and tending t to infinity, we obtain that the mean number of arrivals per time unit equals λ.

It can be shown in a similar way that the mean rate λ_g of batch arrivals is calculated as

$$\lambda_g = \boldsymbol{\theta}(D(1) - D_0)\mathbf{e} = -\boldsymbol{\theta} D_0\mathbf{e}.$$

Using the concept of the mean rate λ_g of batch arrivals and the mean rate λ of arrivals, we can give some more useful properties of the $BMAP$:

- the state probability distribution of the governing process $\nu_t,\ t \geq 0$ at the moment of arrival of a batch of k customers is given by the vector

$$\frac{\boldsymbol{\theta} D_k}{\lambda_g},\ k \geq 1.$$

- the probability P_k that an arbitrary customer arrives in a group of size k is calculated as follows:

$$P_k = \frac{\boldsymbol{\theta} k D_k\mathbf{e}}{\boldsymbol{\theta}\sum_{l=1}^{\infty} lD_l\mathbf{e}} = k\frac{\boldsymbol{\theta} D_k\mathbf{e}}{\boldsymbol{\theta} D'(1)\mathbf{e}} = k\frac{\boldsymbol{\theta} D_k\mathbf{e}}{\lambda},\ k \geq 1.$$

- the probability Q_k that an arbitrary arriving batch in $BMAP$ has size k is calculated as follows:

$$Q_k = \frac{\boldsymbol{\theta} D_k \mathbf{e}}{\boldsymbol{\theta} \sum_{l=1}^{\infty} D_l \mathbf{e}} = -\frac{\boldsymbol{\theta} D_k \mathbf{e}}{\boldsymbol{\theta} D_0 \mathbf{e}} = \frac{\boldsymbol{\theta} D_k \mathbf{e}}{\lambda_g}, \ k \geq 1.$$

- the probability distribution of the state of the governing process right after a batch arrival is given by the vector

$$\boldsymbol{\theta} \frac{\sum_{k=1}^{\infty} D_k}{\lambda_g} = \boldsymbol{\theta} \frac{(D(1) - D_0)}{\lambda_g} = -\boldsymbol{\theta} \frac{D_0}{\lambda_g}.$$

The average length T_g of the interval between moments of batch arrivals is calculated as

$$T_g = \frac{\boldsymbol{\theta}(-D_0) \int_0^{\infty} e^{D_0 t} dt \mathbf{e}}{\lambda_g} = \lambda_g{}^{-1} \tag{2.12}$$

or

$$T_g = \frac{\boldsymbol{\theta}(-D_0) \int_0^{\infty} t e^{D_0 t} \sum_{k=1}^{\infty} D_k dt \mathbf{e}}{\lambda_g} = \frac{\boldsymbol{\theta}(-D_0)^{-1}(D(1) - D_0)\mathbf{e}}{\lambda_g} = \lambda_g{}^{-1}.$$

To derive the two formulas above, we used two different formulas to calculate the expectation $\mathbf{E}\xi$ of a non-negative random variable ξ with a distribution function $F_\xi(x)$:

$$\mathbf{E}\xi = \int_0^{\infty} (1 - F_\xi(x)) dx$$

and

$$\mathbf{E}\xi = \int_0^{\infty} x dF_\xi(x),$$

as well as the generator properties

$$D(1)\mathbf{e} = (D_0 + \sum_{k=1}^{\infty} D_k)\mathbf{e} = \mathbf{0}^T, \ \boldsymbol{\theta} D(1) = \mathbf{0}.$$

The order m initial moment $T_g^{(m)}$ of the distribution of the intervals between the moments of batch arrivals is calculated as

$$T_g^{(m)} = \frac{\boldsymbol{\theta}(D(1) - D_0) \int\limits_0^\infty t^m e^{D_0 t} \sum\limits_{k=1}^\infty D_k dt\mathbf{e}}{\lambda_g} = m! \frac{\boldsymbol{\theta}(-D_0)^{-m+1}\mathbf{e}}{\lambda_g}, \quad m \geq 1.$$

The variance v of the time intervals between arrivals of customer groups is calculated as

$$v = T_g^{(2)} - T_g^2 = \frac{2\lambda_g \boldsymbol{\theta}(-D_0)^{-1}\mathbf{e} - 1}{{\lambda_g}^2}. \tag{2.13}$$

The correlation coefficient c_{cor} of the lengths of two consecutive intervals between arrivals of groups is given by the formula

$$c_{cor} = ({\lambda_g}^{-1} \boldsymbol{\theta}(-D_0)^{-1}(D(1) - D_0)(-D_0)^{-1}\mathbf{e} - {\lambda_g}^{-2})/v. \tag{2.14}$$

Let us explain the derivation of (2.14). Denote by ξ_1, ξ_2 the lengths of two consecutive intervals between batch arrivals in the $BMAP$. Since the random variables ξ_1, ξ_2 have the same expectation T_g and variance v, the correlation coefficient c_{cor} of these random variables is $\rho(\xi_1, \xi_2)$, given by the formula

$$\rho(\xi_1, \xi_2) = \frac{\mathbf{E}\xi_1\xi_2 - T_g^2}{v}. \tag{2.15}$$

From the above properties of the $BMAP$ it follows that the joint density $f_{\xi_1,\xi_2}(x, y)$ of random variables ξ_1, ξ_2 has the form

$$f_{\xi_1,\xi_2}(x, y) = \frac{\boldsymbol{\theta}(D(1) - D_0)e^{D_0 x}(D(1) - D_0)e^{D_0 y}(D(1) - D_0)\mathbf{e}}{\lambda_g}.$$

Consider the random variable $\eta = \xi_1\xi_2$. It is obvious that

$$P\{\eta > z\} = \int\limits_0^\infty \int\limits_{\frac{z}{x}}^\infty f_{\xi_1,\xi_2}(x, y)dxdy$$

$$= \frac{\boldsymbol{\theta}(D(1) - D_0) \int\limits_0^\infty \int\limits_{\frac{z}{x}}^\infty e^{D_0 x}(D(1) - D_0)e^{D_0 y}(D(1) - D_0)dxdy\mathbf{e}}{\lambda_g}$$

$$= \frac{\boldsymbol{\theta}(D(1) - D_0) \int\limits_0^\infty e^{D_0 x}(D(1) - D_0)e^{D_0 \frac{z}{x}}dx\mathbf{e}}{\lambda_g}.$$

Hence

$$\mathbf{E}\xi_1\xi_2 = \mathbf{E}\eta = \int_0^\infty P\{\eta > z\}dz$$

$$= \frac{\boldsymbol{\theta}(D(1) - D_0)\int_0^\infty e^{D_0 x}(D(1) - D_0)\int_0^\infty e^{D_0 \frac{z}{x}}dzdx\mathbf{e}}{\lambda_g}$$

$$= \frac{\boldsymbol{\theta}(D(1) - D_0)\int_0^\infty xe^{D_0 x}(D(1) - D_0)(-D_0)^{-1}dx\mathbf{e}}{\lambda_g}$$

$$= \frac{\boldsymbol{\theta}(-D_0)^{-1}(D(1) - D_0)(-D_0)^{-1}\mathbf{e}}{\lambda_g}.$$

This and (2.15) result in formula (2.14).

When analyzing queuing systems with a $BMAP$ and especially when deriving stability conditions, it becomes necessary to calculate the value

$$\boldsymbol{\theta}\frac{d\int_0^\infty e^{D(z)t}dB(t)}{dz}\Bigg|_{z=1}\mathbf{e},$$

where $B(t)$ is a distribution function with expectation $b_1 = \int_0^\infty tdB(t)$.

Let us show that it is equal to λb_1. Using the matrix exponent definition as a power series, we can write

$$\int_0^\infty e^{D(z)t}dB(t) = \int_0^\infty \sum_{k=0}^\infty \frac{(D(z)t)^k}{k!}dB(t).$$

The first-order derivative of the matrix $(D(z))^j$ is calculated as

$$\frac{d(D(z))^j}{dz} = \sum_{l=0}^{j-1}(D(z))^{j-l-1}\frac{dD(z)}{dz}(D(z))^l.$$

Taking into account that $\boldsymbol{\theta}D(1) = \mathbf{0}$ and $D(1)\mathbf{e} = \mathbf{0}^T$, we conclude that the first-order derivative of the matrix $(D(z))^j,\ j \geq 1$ at point $z = 1$ multiplied by the row vector $\boldsymbol{\theta}$ on the left and by the column vector $\mathbf{e}$ on the right is 0 for all $j,\ j > 1$ and equals λ for $j = 1$, which results in

$$\theta \frac{d \int_0^\infty e^{D(z)t} dB(t)}{dz}|_{z=1}\mathbf{e} = \lambda b_1. \tag{2.16}$$

2.1.4 Special Cases of $BMAP$

The most famous particular case of $BMAP$ is the stationary Poisson flow widely used in queuing theory pioneered by the work of A.K. Erlang and denoted by M in D. Kendall's classification. This flow is obtained from the $BMAP$ if the dimension of the governing process is 1, the matrix D_0 is equal to the scalar $-\lambda$, matrix D_1 is equal to the scalar λ, and matrices $D_k,\ k \geq 2$, are equal to zero.

Other flows listed below in order of their introduction into consideration can be considered as milestones on the evolutionary path from stationary Poisson flows to $BMAPs$ as models of information traffic in modern telecommunications networks.

- IPP is an Interrupted Poisson Process. Such a flow is described as follows. No arrivals happen during time having an exponential distribution with parameter φ_0. Then customers arrive from a Poisson flow with parameter λ_1 during time exponentially distributed with parameter φ_1. Further, the described arrival scenario is repeated.

  This flow is a special case of a $BMAP$ described by two non-zero matrices:

$$D_1 = \text{diag}\{0, \lambda_1\}, \quad D_0 = -D_1 + \begin{pmatrix} -\varphi_0 & \varphi_0 \\ \varphi_1 & -\varphi_1 \end{pmatrix}.$$

- SPP is a Switched Poisson Process, described as follows. Customers arrive from a Poisson flow with parameter λ_0 during time exponentially distributed with parameter φ_0. Then customers arrive from a Poisson flow with parameter λ_1 during time exponentially distributed with parameter φ_1. The described arrival scenario is repeated.

  This flow presents a special case of the $BMAP$ described by two non-zero matrices

$$D_1 = \text{diag}\{\lambda_0, \lambda_1\}, \quad D_0 = -D_1 + \begin{pmatrix} -\varphi_0 & \varphi_0 \\ \varphi_1 & -\varphi_1 \end{pmatrix}.$$

- $MMPP$ is a Markovian Modulated Poisson Process, which is a natural extension of an SPP for the case of an arbitrary number of states of a governing process. From the practical point of view, an $MMPP$ is the most important special case of a $BMAP$ and it has matrices $D_k,\ k \geq 0$, defined as follows:

$$D_0 = \Phi(P - I) - \Lambda, \quad D_1 = \Lambda, \quad D_k = 0, \ k \geq 2,$$

where $\Lambda = \text{diag}\{\lambda_0, \ldots, \lambda_W\}$, $\Phi = \text{diag}\{\varphi_0, \ldots, \varphi_W\}$, and P is a stochastic matrix.

This special case of a $BMAP$ has a transparent physical interpretation, which is as follows. There are $W + 1$ possible levels of the flow intensity. At level ν, the flow behaves like a stationary Poisson flow with arrivals of intensity λ_ν, $\nu = \overline{0, W}$. Level ν of arrivals continues for time having exponential distribution with parameter φ_ν. Then the process jumps to level ν' with probability $p_{\nu,\nu'}$. Here $p_{\nu,\nu'} = (P)_{\nu,\nu'}$ is the (ν, ν')th entry of matrix P, $\nu, \nu' = \overline{0, W}$.

- $BMMPP$ is a Batch Markovian Modulated Poisson Process. This process is a batch analogy of an $MMPP$. To describe it, we need to additionally define a set of probabilities $\delta_k^{(\nu)}$, $k \geq 1$, $\nu = \overline{0, W}$, where $\delta_k^{(\nu)}$ is the probability that a batch of customers arriving from the level ν stationary Poisson input has exactly k customers. Thus, this flow is given by a set of matrices

$$D_0 = \Phi(P - I) - \Lambda, \quad D_k = \text{diag}\{\delta_k^{(0)}, \ldots, \delta_k^{(W)}\}\Lambda, \ k \geq 1.$$

- MAP is a Markovian Arrival Process, which is an ordinary analog of a $BMAP$. It is given by two matrices D_0 and D_1.
- PH is a Phase-Type flow. It is a renewal flow, which is a special case of a MAP. Inter-arrival times in a PH flow are independent and identically distributed random variables having PH distribution, which is presented in the next subsection.

2.1.5 Phase-Type Distribution

2.1.5.1 Definition of a PH Distribution

To define a PH distribution, let us introduce some notations.

Let $\boldsymbol{\beta}$ be a stochastic vector of order M, $\boldsymbol{\beta} = (\beta_1, \ldots, \beta_M)$, where $\beta_m \geq 0$, $m = \overline{1, M}$, $\boldsymbol{\beta}\mathbf{e} = 1$, and S be a subgenerator of order M, i.e., an order M square matrix having negative diagonal entries $S_{m,m}$ and non-negative non-diagonal entries $S_{m,m'}$, $m' \neq m$, $m = \overline{1, M}$ whereby $\sum\limits_{m'=1}^{M} S_{m,m'} \leq 0$ for any m and the sum is strictly less than zero for at least one m. The matrix $S + \mathbf{S}_0\boldsymbol{\beta}$ is assumed to be irreducible.

The random variable ξ is said to have a phase-type (PH) distribution with an irreducible representation $(\boldsymbol{\beta}, S)$ if it is defined as follows.

Let η_t, $t \geq 0$ be a continuous-time MC. This chain has state space $\{1, \ldots, M, M + 1\}$, where $M + 1$ is a unique absorbing state. At initial time $t = 0$, the process η_t, $t \geq 0$ takes value m, $m \in \{1, \ldots, M\}$ with probability β_m. The time the process η_t, $t \geq 0$ stays in state m has exponential distribution with parameter $(-S)_{m,m}$. Then, the process transits to the state m', $m' \in \{1, \ldots, M\}$, $m' \neq m$ with intensity $S_{m,m'}$. In addition, the value $-\sum\limits_{m'=1}^{M} S_{m,m'}$ is the intensity of the process η_t transition from state m into absorbing state $M+1$. And the duration of a random variable ξ having PH distribution expires at the moment when process $\eta_t,\ t \geq 0$ transits to the absorbing state. Denote by $\mathbf{S}_0 = -S\mathbf{e}$. Vector $\mathbf{S}_0$ is non-negative and has at least one positive entry. The value $(\mathbf{S}_0)_m$, which is the mth entry of vector $\mathbf{S}_0$, denotes the intensity of the transition from state $m,\ m \in \{1, \ldots, M\}$ to the absorbing one.

Time with a PH distribution with irreducible representation $(\boldsymbol{\beta}, S)$ can be interpreted as follows. A virtual customer arrives at a network consisting of M nodes and passes through them. An initial node is chosen from the set $\{1, \ldots, M\}$ randomly in accordance with the probability distribution given by vector $\boldsymbol{\beta}$. Upon transition to node m, the virtual customer stays there for time exponentially distributed with parameter $-S_{m,m}$, and then it transits to some other state $m' \neq m,\ m' = \overline{1, M}$ with probability $-\frac{S_{m,m'}}{S_{m,m}}$ and makes further transitions, or it leaves the network with probability $1 + \sum\limits_{m'=1,\ m'\neq m}^{M} \frac{S_{m,m'}}{S_{m,m}}$. The time from the virtual customer arrival to departure has a phase-type distribution with irreducible representation $(\boldsymbol{\beta}, S)$.

The distribution function $A(x)$ of a random variable with a phase-type distribution has the form

$$A(x) = 1 - \boldsymbol{\beta} e^{Sx}\mathbf{e}.$$

It is obvious that if the irreducible representation $(\boldsymbol{\beta}, S)$ is known then we can easily determine the distribution function $A(x)$. The inverse problem (the definition of an irreducible representation $(\boldsymbol{\beta}, S)$ from the distribution function $A(x)$) can have numerous solutions.

The Laplace-Stieltjes transform (LST) $\alpha(s) = \int\limits_0^\infty e^{-st} dA(t)$, Re $s \geq 0$, of a phase-type distribution has the form

$$\alpha(s) = \boldsymbol{\beta}(sI - S)^{-1}\mathbf{S}_0.$$

The average inter-arrival time (the first moment) for the PH flow (the average time between two successive transitions of the process $\eta_t,\ t \geq 0$ to the absorbing state) is defined as

$$a_1 = \boldsymbol{\beta}(-S)^{-1}\mathbf{e}.$$

The order m initial moment of this distribution is defined as

$$a_m = \int_0^\infty t^m dA(t) = m!\boldsymbol{\beta}(-S)^{-m}\mathbf{e},\ m \geq 2.$$

It is known that the set of PH distributions is everywhere dense, that is, for any distribution function $A(t)$ of a non-negative random variable we can find a phase-type distribution close enough to the distribution $A(t)$ in the sense of weak convergence.

2.1.5.2 Partial Cases of PH Distribution

Many of the distributions traditionally used in queuing theory are particular cases of a PH distribution. For example,

- Erlang distribution of order k (E_k) with distribution function

$$A(t) = \int_0^t \lambda \frac{(\lambda u)^{k-1}}{(k-1)!} e^{-\lambda u} du,\ t \geq 0.$$

This distribution is defined by the vector $\boldsymbol{\beta} = (1, 0, 0, \dots, 0)$ of size K and the subgenerator of the form

$$S = \begin{pmatrix} -\lambda & \lambda & 0 & \dots & 0 \\ 0 & -\lambda & \lambda & \dots & 0 \\ \vdots & \vdots & \vdots & \ddots & \vdots \\ 0 & 0 & 0 & \dots & -\lambda \end{pmatrix}.$$

- Hyperexponential distribution of order k (HM_k). Its distribution function has the form

$$A(t) = \sum_{l=1}^{k} q_l(1 - e^{-\lambda_l t})$$

where $q_l \geq 0$, $l = \overline{1, k}$, and $\sum\limits_{l=1}^{k} q_l = 1$.

In this case,

$$\boldsymbol{\beta} = (q_1, \dots, q_k),\ S = \text{diag}\{-\lambda_1, \dots, -\lambda_k\}.$$

2.1.5.3 The Distribution of the Minimum and Maximum of Random Variables with a PH Distribution

Lemma 2.1 *Let ξ_k be independent random variables with a PH distribution with an irreducible representation $(\boldsymbol{\beta}_k, S_k)$, $k = \overline{1, K}$, and let*

$$\xi = \min_{k=\overline{1,K}} \xi_k.$$

Then the random variable ξ has a PH distribution with the irreducible representation $(\boldsymbol{\beta}_1 \otimes \cdots \otimes \boldsymbol{\beta}_K, S_1 \oplus \cdots \oplus S_K)$.

For the sake of simplicity, we give a proof for the case $K = 2$. It can be easily generalised for any K. It's obvious that

$$P\{\xi > t\} = P\{\xi_1 > t\}P\{\xi_2 > t\} = \boldsymbol{\beta}_1 e^{S_1 t}\mathbf{e}\boldsymbol{\beta}_2 e^{S_2 t}\mathbf{e}.$$

Since the product and the Kronecker product operations are equivalent for scalars, and taking into account the rule for the mixed product of matrices and the determination of the Kronecker sum of matrices (see Appendix A), we obtain

$$P\{\xi > t\} = \boldsymbol{\beta}_1 e^{S_1 t}\mathbf{e} \otimes \boldsymbol{\beta}_2 e^{S_2 t}\mathbf{e} = (\boldsymbol{\beta}_1 \otimes \boldsymbol{\beta}_2)(e^{S_1 t} \otimes e^{S_2 t})(\mathbf{e} \otimes \mathbf{e})$$
$$= (\boldsymbol{\beta}_1 \otimes \boldsymbol{\beta}_2)e^{(S_1 \oplus S_2)t}\mathbf{e}.$$

This means that random variable ξ has a PH distribution with irreducible representation $(\boldsymbol{\beta}_1 \otimes \boldsymbol{\beta}_2, S_1 \oplus S_2)$.

Lemma 2.2 *Let ξ_k, $k = \overline{1, n}$ be independent random variables with a PH distribution with the irreducible representation $(\boldsymbol{\beta}, S)$, where the size of the vector and the matrix is M, and let*

$$\eta_n = \max_{k=\overline{1,n}} \xi_k, \ n \geq 2.$$

Then the random variable η_n has a PH distribution with the irreducible representation $(\boldsymbol{\beta}^{(n)}, S^{(n)})$, which is determined recursively

$$\boldsymbol{\beta}^{(n)} = \left(\boldsymbol{\beta} \otimes \boldsymbol{\beta}^{(n-1)} \mid \mathbf{0}_{M_{n-1}} \mid \mathbf{0}_M\right),$$

$$S^{(n)} = \begin{pmatrix} S \oplus S^{(n-1)} & \mathbf{S}_0 \otimes I_{M_{n-1}} & I_M \otimes \mathbf{S}_0^{(n-1)} \\ O & S^{(n-1)} & O \\ O & O & S \end{pmatrix}$$

with initial condition

$$\boldsymbol{\beta}^{(1)} = \boldsymbol{\beta}, \ \ S^{(1)} = S,$$

where the order M_n of vector $\boldsymbol{\beta}^{(n)}$ is defined as $M_n = (M+1)^n - 1, \ n \geq 1$.

Proof The interpretation of the phase-type distribution as the virtual customer's walking time through the network before it leaves the network makes it clear that the maximum of two independent random variables with a phase-type distribution with the irreducible representations $(\boldsymbol{\gamma}^{(k)}, \Gamma^{(k)}), \ k = 1, 2$, where the size of vector $\boldsymbol{\gamma}^{(k)}$ is $M^{(k)}$ has a PH distribution with irreducible representation $(\boldsymbol{\gamma}, \Gamma)$ defined as follows:

$$\boldsymbol{\gamma} = \left(\boldsymbol{\gamma}^{(1)} \otimes \boldsymbol{\gamma}^{(2)} \mid \mathbf{0}_{M^{(2)}} \mid \mathbf{0}_{M^{(1)}}\right),$$

$$\Gamma = \begin{pmatrix} \Gamma^{(1)} \oplus \Gamma^{(2)} & | & \boldsymbol{\Gamma}_0^{(1)} \otimes I_{M^{(2)}} & | & I_{M^{(1)}} \otimes \boldsymbol{\Gamma}_0^{(2)} \\ ------ & | & ---- & | & ---- \\ O & | & \Gamma^{(2)} & | & O \\ ------ & | & ---- & | & ---- \\ O & | & O & | & \Gamma^{(1)} \end{pmatrix},$$

where $\boldsymbol{\Gamma}_0^{(k)} = -\Gamma^{(k)}\mathbf{e}$. The statement of the lemma now follows from this result in an obvious way. □

From this proof, it is not so difficult to conclude that the Lemma statement can be generalized in an obvious way to the case when the random variables ξ_k have irreducible representations depending on k.

Lemma 2.3 *Let $\xi_k, \ k = \overline{1, K}$ be independent and identically distributed random variables with a PH distribution with the irreducible representation $(\boldsymbol{\beta}, S)$ and let*

$$\eta = \max_{k=\overline{1,K}} \xi_k.$$

Then the distribution function of random variable η has the form

$$P\{\eta < t\} = \sum_{k=0}^{K} C_K^k (-1)^k \boldsymbol{\beta}^{\otimes k} e^{S^{\oplus k} t} \mathbf{e},$$

where

$$\boldsymbol{\beta}^{\otimes k} \stackrel{def}{=} \underbrace{\boldsymbol{\beta} \otimes \ldots \otimes \boldsymbol{\beta}}_{k}, \quad k \geq 1,$$

$$S^{\oplus k} \stackrel{def}{=} \underbrace{S \oplus \ldots \oplus S}_{k}, \quad k \geq 1, \quad S^{\oplus 0} \stackrel{def}{=} 0.$$

The lemma is proved by induction taking into account the experience of proving the lemma for a minimum of random variables based on the formula

$$P\{\eta < t\} = \left(1 - \boldsymbol{\beta} e^{St}\mathbf{e}\right)^K.$$

2.1.6 Calculating the Probabilities of a Fixed Number of $BMAP$ Arrivals During a Random Time

When analyzing many systems, for example, a $BMAP/G/1$ queue, there arises the following auxiliary problem. Let there be a time interval whose length is a random variable ξ with distribution function $B(t)$. Suppose we have a $BMAP$ defined by a sequence of matrices $D_k,\ k \geq 0$. Matrices $P(n,t) = (P_{\nu,\nu'}(n,t))_{\nu,\nu'=\overline{0,W}},\ n \geq 0$ specify the conditional probabilities of n arrivals during time t conditioned on the corresponding transitions of a governing process during this time. We need to calculate matrices $Y_l,\ l \geq 0$ specifying the conditional probabilities of l arrivals from the $BMAP$ during time ξ given the corresponding transitions of a governing process during this time.

The law of total probability gives the obvious formula to calculate matrices $Y_l,\ l \geq 0$:

$$Y_l = \int_0^\infty P(l,t)dB(t).$$

Since in the general case, matrices $P(n,t),\ n \geq 0$ cannot be calculated explicitly, the direct calculation of matrices $Y_l,\ l \geq 0$ is a rather tough task.

In the general case, we can use relation (2.7), which results in the following formula:

$$Y_l = \sum_{j=0}^\infty \gamma_j K_l^{(j)},\ l \geq 0,$$

where

$$\gamma_j = \int_0^\infty e^{-\tilde{\theta}t}\frac{(\tilde{\theta}t)^j}{j!}dB(t),\ j \geq 0$$

and $K_l^{(j)},\ l \geq 0,\ j \geq 0$ are matrices defined by recursion (2.8).

Let us emphasize the merit of this formula, which is as follows. The matrices $K_l^{(j)}$ depend on the input flow of customers and are independent of the service time distribution. In turn, the probabilities γ_j depend on the service time distribution and depend on the input flow only through the scalar $\tilde{\theta}$.

The calculation of matrices $Y_l,\ l \geq 0$ is considerably simplified if the distribution $B(t)$ is of PH-type.

Lemma 2.4 *If the distribution function $B(t)$ is of PH-type with irreducible representations $(\boldsymbol{\beta}, S)$ then matrices $Y_l,\ l \geq 0$ are calculated in the following way:*

$$Y_l = Z_l(I_{\bar{W}} \otimes \mathbf{S}_0),\ l \geq 0, \tag{2.17}$$

where

$$Z_0 = -(I_{\bar{W}} \otimes \boldsymbol{\beta})(D_0 \oplus S)^{-1}, \tag{2.18}$$

$$Z_l = -\sum_{i=0}^{l-1} Z_i(D_{l-i} \otimes I_M)(D_0 \oplus S)^{-1},\ l \geq 1. \tag{2.19}$$

Proof The following sequence of equalities holds:

$$Y_l = \int_0^\infty P(l,t)dB(t) = \int_0^\infty P(l,t)\boldsymbol{\beta}e^{St}\mathbf{S}_0 dt = \int_0^\infty P(l,t)I_{\bar{W}} \otimes \boldsymbol{\beta}e^{St}\mathbf{S}_0 dt$$

$$= \int_0^\infty (P(l,t) \otimes \boldsymbol{\beta}e^{St})(I_{\bar{W}} \otimes \mathbf{S}_0)dt = Z_l(I_{\bar{W}} \otimes \mathbf{S}_0),$$

where

$$Z_l = \int_0^\infty P(l,t) \otimes \boldsymbol{\beta}e^{St}dt,\ l \geq 0.$$

Here we used the form of the PH-type distribution function $B(t)$, the artificial replacement of the ordinary product by the scalar $\boldsymbol{\beta}e^{St}\mathbf{S}_0$ in the Kronecker product, and the mixed-product rule for Kronecker products, see Appendix A. Thus, formula (2.17) is obtained, and it remains to derive formulas (2.18) and (2.19) for matrices $Z_l,\ l \geq 0$. Integrating by parts, we obtain

$$Z_l = [P(l,t) \otimes (\boldsymbol{\beta}e^{St}S^{-1})]\Big|_0^\infty - \int_0^\infty P'(l,t) \otimes (\boldsymbol{\beta}e^{St}S^{-1})dt.$$

We can verify that the following relations hold: $P(l,t) \to O$ as $t \to \infty$ and $P(n,0) = \delta_{n,0} I_{\bar{W}}$. Taking these relations and the matrix differential equation (2.2) into account, we get

$$Z_l = -\delta_{l,0}(I_{\bar{W}} \otimes \boldsymbol{\beta} S^{-1}) - \int_0^\infty \sum_{i=0}^{l} (P(i,t) D_{l-i}) \otimes (\boldsymbol{\beta} e^{St} S^{-1}) dt$$

$$= -\delta_{l,0}(I_{\bar{W}} \otimes \boldsymbol{\beta} S^{-1}) - \sum_{i=0}^{l} Z_i (D_{l-i} \otimes S^{-1}).$$

This implies the formula

$$Z_l(I_{\bar{W}M} + D_0 \otimes S^{-1}) = -\delta_{l,0}(I_{\bar{W}} \otimes \boldsymbol{\beta} S^{-1}) - \sum_{i=0}^{l-1} Z_i (D_{l-i} \otimes S^{-1}). \qquad (2.20)$$

It is easy to see that the inverse of the matrix $I_{\bar{W}M} + D_0 \otimes S^{-1}$ exists and is defined as

$$(I_{\bar{W}M} + D_0 \otimes S^{-1})^{-1} = (I_{\bar{W}} \otimes S)(D_0 \oplus S)^{-1}.$$

Taking this formula into account, (2.20) results in formulas (2.18) and (2.19). □

2.1.7 *Superposition and Sifting of BMAPs*

The class of $BMAP$s has some similarities with the stationary Poisson flow with respect to the superposition and sifting operations as well.

It is known that the superposition (imposition) of two independent stationary Poisson flows with intensities λ_1 and λ_2 is again a Poisson flow with intensity $\lambda_1 + \lambda_2$.

Similarly, the superposition of two independent $BMAP$s characterized by matrix generating functions $D_1(z)$ and $D_2(z)$ is again a $BMAP$ with the matrix generating function $D(z) = D_1(z) \oplus D_2(z)$. The average rate of the superposition of two $BMAP$s equals the sum of the average rates of two overlapping $BMAP$s.

It is also known that if we apply a simple recurrent sifting procedure to a stationary Poisson flow of intensity λ where a customer (an arrival) is accepted to the sifted flow with the given probability p and is rejected with the complementary probability then the sifted flow is also a stationary Poisson one, and its intensity is $p\lambda$. Since $BMAP$ is a batch flow, two direct analogues of this procedure are possible. The first one (option a): the arriving batch is accepted to the sifted flow with probability q_a and is rejected with the complementary probability. The second one (option b): each customer from an arriving batch is accepted to the sifted flow with probability q_b independently of other customers in the batch and is rejected with the complementary probability $1 - q_b$.

Lemma 2.5 *If a $BMAP$ with a governing process $\nu_t, \ t \geq 0$ with state space $\{0, \ldots, W\}$ and matrix generating function $D(z)$ is applied to one of the sifting procedures (a or b) described above then the resulting sifted flow is also $BMAP$. The governing process of the sifted flow has the same state space as $\nu_t, \ t \geq 0$ of the initial $BMAP$. The matrix generating function of this flow has the form*

$$q_a D(z) + (1 - q_a) D(1) \textit{ for option a }, \ D(zq_b + (1 - q_b)) \textit{ for option b}.$$

The average rates of flows are λq_a or λq_b, correspondingly, where λ is the average rate of the initial $BMAP$.

Thus, the family of $BMAP$s is closed with respect to the superposition operation and the simplest sifting procedure, which is analogous to the property of stationary Poisson flows.

2.1.8 Batch Marked Markovian Arrival Process (BMMAP)

The $BMAP$ described in the previous section is a flow of homogeneous arrivals. Many flows in real systems are heterogeneous. For example, information traffic in modern telecommunications networks is a mixture (superposition) of voice streams transmitted digitally, interactive data streams, data streams that do not require on-line transmission but require high transmission reliability, video, audio, and multimedia data. To simulate such heterogeneous flows when they are correlated, the so-called Marked Markovian Arrival Process ($MMAP$) or its batch modification, the Batch Marked Markovian Arrival Process ($BMMAP$) is used.

The arrival process in a $BMMAP$, which is a superposition of L arriving flows of customers that differ in type, is controlled by an irreducible continuous-time MC $\nu_t, t \geq 0$ with a finite state space $\{0, \ldots, W\}$. The time the chain ν_t stays in some state ν has an exponential distribution with parameter $\lambda_\nu, \ \nu = \overline{0, W}$. After the time the process stays in the state expires, the process transits to state $\nu', \ \nu' \neq \nu$ and no arrival happens with probability $p_0(\nu, \nu')$ or it transits to state ν' and a size k batch of type l arrives with probability $p_k^{(l)}(\nu, \nu'), k \geq 1, \ l = \overline{1, L}$. That is,

$$p_0(\nu, \nu) = 0, \ \sum_{l=1}^{L} \sum_{k=1}^{\infty} \sum_{\nu'=0}^{W} p_k^{(l)}(\nu, \nu') + \sum_{\nu'=0}^{W} p_0(\nu, \nu') = 1, \ \nu = \overline{0, W}.$$

It is easy to keep the parameters characterizing the $BMMAP$ as square matrices $D_0,\ D_k^{(l)},\ k \geq 1\ , l = \overline{1, L}$ of order $\bar{W} = W + 1$ whose entries are defined as follows:

$$(D_0)_{\nu,\nu} = -\lambda_\nu,\ (D_0)_{\nu,\nu'} = \lambda_\nu p_0(\nu, \nu'),\ \nu \neq \nu',$$

$$(D_k^{(l)})_{\nu,\nu'} = \lambda_\nu p_k^{(l)}(\nu, \nu'),\ \nu, \nu' = \overline{0, W},\ k \geq 1,\ l = \overline{1, L}.$$

Denote

$$D(1) = D_0 + \sum_{l=1}^{L} \sum_{k=1}^{\infty} D_k^{(l)},\ \hat{D}_k^{(l)} = \sum_{i=k}^{\infty} D_i^{(l)},$$

$$\mathcal{D}_k^{(l)} = \sum_{i=k}^{\infty} \sum_{\bar{l}=1, \bar{l}\neq l}^{L} D_i^{(\bar{l})},\ k \geq 1,\ l = \overline{1, L}.$$

Vector $\boldsymbol{\theta}$ of the stationary state probability distribution of the MC $\nu_t,\ t \geq 0$ is the unique solution of the system of linear algebraic equations

$$\boldsymbol{\theta} D(1) = \mathbf{0}, \boldsymbol{\theta}\mathbf{e} = 1.$$

The average rate λ_l of the type l arrivals is calculated as

$$\lambda_l = \boldsymbol{\theta} \sum_{k=1}^{\infty} k D_k^{(l)} \mathbf{e}, l = \overline{1, L}.$$

The average rate $\lambda_l^{(b)}$ of arrivals of type l batches is calculated by the formula

$$\lambda_l^{(b)} = \boldsymbol{\theta} \hat{D}_1^{(l)} \mathbf{e},\ \ l = \overline{1, L}.$$

The variance v_l of the time intervals between the arrivals of type l batches is calculated by the formula

$$v_l = \frac{2\boldsymbol{\theta}(-D_0 - \mathcal{D}_1^{(l)})^{-1}\mathbf{e}}{\lambda_l^{(b)}} - \left(\frac{1}{\lambda_l^{(b)}}\right)^2,\ l = \overline{1, L}.$$

The correlation coefficient $C_{cor}^{(l)}$ of the lengths of two adjoining intervals between the arrivals of type l batches is calculated by the formula

$$C_{cor}^{(l)} = \left[\frac{\boldsymbol{\theta}(-D_0 - \mathcal{D}_1^{(l)})^{-1}}{\lambda_l^{(b)}} \hat{D}_1^{(l)} (-D_0 - \mathcal{D}_1^{(l)})^{-1}\mathbf{e} - \left(\frac{1}{\lambda_l^{(b)}}\right)^2\right] v_l^{-1},\ l = \overline{1, L}.$$

2.1.9 Semi-Markovian Arrival Process (SM)

This arrival process is more general than a MAP and is defined as follows. Consider a regular ergodic semi-Markovian random process $m_t,\ t \geq 0$ characterized by its state space $\{1, 2, \dots, M\}$ and a semi-Markovian kernel $B(x)$ with entries $B_{m,m'}(x)$. Function $B_{m,m'}(x)$ is considered to be the probability that the sojourn time of the process in the current state does not exceed x, and the process will make the next transition to state m' given that the current state of the process is $m,\ \ m, m' = \overline{1, M}$. Matrix $B(\infty)$ is the transition probability matrix for the MC embedded at all jump moments of the process $m_t,\ t \geq 0$. We suppose that the embedded MC is irreducible, and hence vector $\boldsymbol{\delta} = (\delta_1, \dots, \delta_M)$ of its stationary state probability distribution satisfying the relations

$$\boldsymbol{\delta} B(\infty) = \boldsymbol{\delta},\ \ \boldsymbol{\delta}\mathbf{e} = 1$$

exists. It is also assumed that the integral $\int\limits_0^\infty t dB(t)$ exists.

A Semi-Markovian (SM) arrival process is a random flow of arrivals where the inter-arrival time intervals are the consequent sojourn times of the semi-Markovian process $m_t,\ t \geq 0$ in its states. The kth initial moment of the inter-arrival time intervals is calculated as $b_k = \boldsymbol{\delta} \int\limits_0^\infty t^k dB(t)\mathbf{e},\ k \geq 1$.

An important special case of the semi-Markovian arrival process is one with a semi-Markovian kernel $B(t)$ presented in the form

$$B(t) = \mathrm{diag}\{B_1(t), \dots, B_M(t)\}P,$$

where $B_m(t)$ is the distribution function of the sojourn time of the process $m_t,\ t \geq 0$ in state $m,\ m = \overline{1, M}$, and $P = (p_{i,j})_{i,j=\overline{1,M}} = B(\infty)$.

From the definition of a semi-Markovian arrival process, it follows that, generally speaking, the inter-arrival times can be correlated. In the special case of a semi-Markovian arrival process described above, the formula for calculating the coefficient of correlation of the successive inter-arrival intervals for the SM arrival process is

$$c_{cor} = \frac{\sum\limits_{i=1}^{M}\sum\limits_{j=1}^{M} \delta_i b_1^{(i)} p_{i,j} b_1^{(j)} - (\sum\limits_{i=1}^{M} \delta_i b_1^{(i)})^2}{\sum\limits_{i=1}^{M} \delta_i b_2^{(i)} - (\sum\limits_{i=1}^{M} \delta_i b_1^{(i)})^2},$$

where $b_k^{(i)} = \int\limits_0^\infty t^k dB_i(t),\ k = 1, 2,\ i = \overline{1, M}$.

2.2 Multidimensional Birth-and-Death Processes

When studying classical Markovian queuing systems, it is useful to apply known results for the so-called birth-and-death processes.

A birth-and-death process $i_t, \ t \geq 0$ is a special case of a homogeneous continuous-time MC with a finite or countable state space. Let us consider a birth-and-death process with a countable state space $\{0, 1, 2, \dots\}$. The matrix Q of the infinitesimal coefficients (a generator) of the process has a three-diagonal structure

$$Q = \begin{pmatrix} -\lambda_0 & \lambda_0 & 0 & 0 \dots \\ \mu_1 & -(\lambda_1 + \mu_1) & \lambda_1 & 0 \dots \\ 0 & \mu_2 & -(\lambda_2 + \mu_2) & \lambda_2 \dots \\ \vdots & \vdots & \vdots & \vdots \ddots \end{pmatrix}.$$

Parameter λ_i is called the birth intensity and μ_i is the death intensity of state i.

Probabilities $p_i(t) = P\{i_t = i\}, \ i \geq 0$ arranged in a row vector $\mathbf{p}(t) = (p_0(t), p_1(t), p_2(t), \dots)$ satisfy an infinite system of equations

$$\frac{d\mathbf{p}(t)}{dt} = \mathbf{p}(t)Q$$

with the normalization condition $\mathbf{p}(t)\mathbf{e} = 1$.

If the initial probability distribution $\mathbf{p}(0)$ is known then the solution of this system of equations has the form

$$\mathbf{p}(t) = \mathbf{p}(0)e^{Qt}.$$

Denote $\rho_k = \frac{\prod\limits_{l=0}^{k-1} \lambda_l}{\prod\limits_{l=1}^{k} \mu_l}, \ k \geq 0.$

Given that the series $\sum\limits_{k=0}^{\infty} \rho_k$ converges and the series $\sum\limits_{k=1}^{\infty} \prod\limits_{i=1}^{k} \frac{\mu_i}{\lambda_i}$ diverges, the stationary state probabilities of the birth-and-death process $i_t, \ t \geq 0$

$$p_i = \lim_{t \to \infty} p_i(t), \ i \geq 0$$

exist and they are defined by relations

$$p_0 = (\sum_{k=0}^{\infty} \rho_k)^{-1}, \ p_i = p_0 \rho_i, \ i \geq 1.$$

2.2.1 Definition of the Quasi-Birth-and-Death Process and Its Stationary State Distribution

In the study of many queuing systems, for example, systems with MAP and (or) service processes like PH, systems operating in a random environment, tandem queues, etc., it becomes necessary to analyze the stationary behavior of a two- or multidimensional Markov process $\xi_t = \{i_t, \mathbf{v}_t\}, t \geq 0$ with state space

$$\mathcal{S} = \{(i, \mathbf{v}), \mathbf{v} \in \mathcal{V}_i,\ i \geq 0\},$$

where $\mathcal{V}_i$ is a finite set if the Markov process $\xi_t = \{i_t, \mathbf{v}_t\}, t \geq 0$ is two-dimensional, or some finite set of finite-dimensional vectors if the Markov process $\xi_t = \{i_t, \mathbf{v}_t\}, t \geq 0$ is multidimensional.

Enumerate the states of the process $\xi_t = \{i_t, \mathbf{v}_t\}, t \geq 0$ in lexicographic order and combine the states for which the component i_t has value i into the macro-state i, sometimes also called the level of process $\xi_t = \{i_t, \mathbf{v}_t\}, t \geq 0$. According to such an enumeration, the generator Q of this process can be presented in a block form $Q = (Q_{i,l})_{i,l \geq 0}$, where $Q_{i,l}$ is a matrix formed by the intensities $q_{(i,\mathbf{r});(l,\mathbf{v})}$ of the process $\xi_t, t \geq 0$ transitions from state $(i, \mathbf{r}), \mathbf{r} \in \mathcal{V}_i$ to state $(l, \mathbf{v}), \mathbf{v} \in \mathcal{V}_l$. The diagonal elements of matrix $Q_{i,i}$ are defined as $q_{(i,\mathbf{r});(i,\mathbf{v})} = -\sum\limits_{(j,\mathbf{v}) \in \mathcal{S} \setminus (i,\mathbf{r})} q_{(i,\mathbf{r});(j,\mathbf{v})}$.

The process $\xi_t = \{i_t, \mathbf{v}_t\}, t \geq 0$ is called a multidimensional (vector) Quasi-Birth-and-Death process (QBD) if its irreducible block generator $Q = (Q_{i,l})_{i,l \geq 0}$ can be presented in the form

$$Q = \begin{pmatrix} Q_{0,0} & Q_0 & O & O & \cdots \\ Q_2 & Q_1 & Q_0 & O & \cdots \\ O & Q_2 & Q_1 & Q_0 & \cdots \\ \vdots & \vdots & \vdots & \vdots & \ddots \end{pmatrix}. \tag{2.21}$$

It follows from the form (2.21) of generator Q that the sets $\mathcal{V}_i$ for different values of $i,\ i > 0$ here coincide.

Theorem 2.1 *A necessary and sufficient condition for the existence of a stationary probability distribution of the MC* $\xi_t,\ t \geq 0$ *is the validity of inequalities*

$$\mathbf{y}Q_2\mathbf{e} > \mathbf{y}Q_0\mathbf{e}, \tag{2.22}$$

where the row vector $\mathbf{y}$ *is the unique solution of the system of linear algebraic equations*

$$\mathbf{y}(Q_0 + Q_1 + Q_2) = \mathbf{0},\ \mathbf{y}\mathbf{e} = 1. \tag{2.23}$$

The proof of this condition is given in [93]. This condition is also automatically obtained from conditions (2.66)–(2.67) proved below in Sect. 2.4 for MCs of $M/G/1$-type.

Below we assume that condition (2.22) holds. Then there exist limits

$$\pi(i, \mathbf{v}) = \lim_{t\to\infty} P\{i_t = i,\ \mathbf{v}_t = \mathbf{v}\},\ \mathbf{v} \in \mathcal{V}_i,\ i \geq 0,$$

which are called stationary state probabilities of the MC.

Corresponding to a lexicographic ordering of the MC states, we group these probabilities into row vectors

$$\boldsymbol{\pi}_i = (\pi(i, \mathbf{v}),\ \mathbf{v} \in \mathcal{V}_i),\ i \geq 0.$$

Theorem 2.2 *The vectors of stationary state probabilities of the MC $\xi_t,\ t \geq 0$ are calculated as follows:*

$$\boldsymbol{\pi}_i = \boldsymbol{\pi}_0 \mathcal{R}^i,\ i \geq 0, \tag{2.24}$$

where matrix $\mathcal{R}$ is a minimal non-negative solution of the matrix equation

$$\mathcal{R}^2 Q_2 + \mathcal{R} Q_1 + Q_0 = O \tag{2.25}$$

and vector $\boldsymbol{\pi}_0$ is the unique solution of the following system of linear algebraic equations:

$$\boldsymbol{\pi}_0(Q_{0,0} + \mathcal{R} Q_2) = \mathbf{0}, \tag{2.26}$$

$$\boldsymbol{\pi}_0(I - \mathcal{R})^{-1}\mathbf{e} = 1. \tag{2.27}$$

Proof It is known that vector $\boldsymbol{\pi} = (\boldsymbol{\pi}_0, \boldsymbol{\pi}_1, \boldsymbol{\pi}_2, \ldots)$ is a solution of the system of equilibrium equations

$$\boldsymbol{\pi} Q = \mathbf{0} \tag{2.28}$$

with the normalization condition

$$\boldsymbol{\pi}\mathbf{e} = 1. \tag{2.29}$$

Suppose that the solution of system (2.28) has the form (2.24). Substituting vectors $\boldsymbol{\pi}_i,\ i \geq 1$ of form (2.24) into equations

$$\boldsymbol{\pi}_{i-1} Q_0 + \boldsymbol{\pi}_i Q_1 + \boldsymbol{\pi}_{i+1} Q_2 = \mathbf{0},\ i \geq 1$$

of system (2.28), it is easy to see that these equations result in the identities if matrix $\mathcal{R}$ satisfies matrix equation (2.25). Substituting vectors $\boldsymbol{\pi}_i,\ i \geq 1$ of form (2.24) into equations

$$\boldsymbol{\pi}_0 Q_{0,0} + \boldsymbol{\pi}_1 Q_2 = \mathbf{0}$$

of system (2.28), we obtain that vector $\boldsymbol{\pi}_0$ satisfies Eq. (2.26).

It is known, see [93], that the ergodicity condition (2.22) holds if and only if the spectral radius $\varrho(\mathcal{R})$ of matrix $\mathcal{R}$ is strictly less than 1. Under this condition, the series $\sum\limits_{l=0}^{\infty} \mathcal{R}^l$ converges and is equal to $(I - \mathcal{R})^{-1}$. Equation (2.27) now follows from (2.24) and the normalization condition $\sum\limits_{i=0}^{\infty} \boldsymbol{\pi}_i \mathbf{e} = 1$. □

Note that matrix $\mathcal{R}$ can be represented in the form $\mathcal{R} = Q_0 \mathcal{N}$, where entries $\mathcal{N}_{\mathbf{v},\mathbf{v}'}$ of matrices $\mathcal{N}$ characterize the average time the MC $\xi_t = \{i_t, \mathbf{v}_t\}, t \geq 0$ is in some macro-state i before it enters level $i-1$ the first time given that the component $\mathbf{v}_t$ transits from state $\mathbf{v}$ to state $\mathbf{v}'$ during this time.

Various methods are used to solve Eq. (2.25). The simplest of them is the method of successive approximations. The zero matrix is taken as an initial approximation $\mathcal{R}_0$ for matrix $\mathcal{R}$. The formula for calculating successive approximations follows in an obvious manner from (2.25) in the form

$$\mathcal{R}_{k+1} = (\mathcal{R}_k^2 Q_2 + Q_0)(-Q_1)^{-1},\ k \geq 0. \tag{2.30}$$

We note that the inverse matrix in (2.30) exists since matrix Q_1 is an irreducible subgenerator. It is shown in [93] that the sequence of matrices $\{\mathcal{R}_k,\ k \geq 0\}$ given by recursion (2.30) converges to the minimal non-negative solution of matrix equation (2.25).

Another algorithm, the so-called algorithm with logarithmic reduction, uses the relation

$$\mathcal{R} = Q_0(-Q_1 - Q_0 \mathcal{G})^{-1}$$

between the matrix $\mathcal{R}$, which is a solution of Eq. (2.25), and the matrix $\mathcal{G}$, which is a solution of the matrix equation

$$Q_0 \mathcal{G}^2 + Q_1 \mathcal{G} + Q_2 = O.$$

In turn, matrix $\mathcal{G}$ is found by the method of successive approximations on the basis of the relation

$$\mathcal{G} = (-Q_1)^{-1} Q_0 \mathcal{G}^2 + (-Q_1)^{-1} Q_2.$$

The above results for QBD processes are easily extended to the case when the block generator of the process $\xi_t,\ t \geq 0$ has the form

$$Q = \begin{pmatrix} \tilde{Q}_{0,0} & \tilde{Q}_0 & O & O & \cdots \\ \tilde{Q}_2 & Q_1 & Q_0 & O & \cdots \\ O & Q_2 & Q_1 & Q_0 & \cdots \\ \vdots & \vdots & \vdots & \vdots & \ddots \end{pmatrix}$$

where matrices $\tilde{Q}_{0,0},\ \tilde{Q}_0,\ \tilde{Q}_2$ can have size different from that of blocks $Q_0,\ Q_1,\ Q_2$, and blocks $\tilde{Q}_0,\ \tilde{Q}_2$ are not square matrices unlike (2.21).

In this case, a necessary and sufficient condition for the existence of a stationary state distribution for the process $\xi_t,\ t \geq 0$ is also given by inequality (2.22) where vector $\mathbf{y}$ is the solution of system (2.23). The formulas for calculating the vectors of stationary state probabilities have the following form here:

$$\boldsymbol{\pi}_i = \boldsymbol{\pi}_1 \mathcal{R}^{i-1},\ i \geq 1,$$

where matrix $\mathcal{R}$ is the minimal non-negative solution of matrix equation (2.25), and vectors $\boldsymbol{\pi}_0$ and $\boldsymbol{\pi}_1$ are the unique solution of the following system of linear algebraic equations:

$$\boldsymbol{\pi}_0 \tilde{Q}_{0,0} + \boldsymbol{\pi}_1 \tilde{Q}_2 = \mathbf{0},$$

$$\boldsymbol{\pi}_0 \tilde{Q}_0 + \boldsymbol{\pi}_1 (Q_1 + \mathcal{R} Q_2) = \mathbf{0},$$

$$\boldsymbol{\pi}_0 \mathbf{e} + \boldsymbol{\pi}_1 (I - \mathcal{R})^{-1} \mathbf{e} = 1.$$

Similarly, the above results can be generalized to the case of QBD processes for which not one but some finite number J of block rows of the process block generator have a form different from the one in (2.21).

For illustration, we apply the above information for QBD processes to the analysis of a $MAP/PH/1$-type queue.

2.2.2 Application of the Results Obtained for QBD Processes to the $MAP/PH/1$ Queue Analysis

Consider a single-server queuing system with unlimited waiting space. The input of customers is a MAP defined by a governing process, which is an irreducible continuous-time MC $\nu_t,\ t \geq 0$ with a finite state space $\{0, 1, \ldots, W\}$ and matrices D_0 and D_1. The PH-type service process is defined by a continuous-time MC $\eta_t,\ t \geq 0$ with non-absorbing states $\{1, \ldots, M\}$ and an irreducible representation $(\boldsymbol{\beta}, S)$, where $\boldsymbol{\beta}$ is a stochastic row vector and S is a subgenerator.

The system behavior is described by a three-dimensional MC $\xi_t = \{i_t, \mathbf{v}_t\},\ t \geq 0$, where i_t is the number of customers in the system at time t and the two-dimensional process $\mathbf{v}_t = (\nu_t, \eta_t),\ t \geq 0$ describes the states of both the governing process of arrivals and the service process, $i_t \geq 0,\ \nu_t = \overline{0, W},\ \eta_t = \overline{1, M}$.

Note that at moments when $i_t = 0$, that is, there are no customers in the system, in general, the state of the component η_t is undefined. However, in order to avoid the ensuing complexity of notation and results, we can agree that a state of the process $\eta_t,\ t \geq 0$ is randomly chosen according to probability vector $\boldsymbol{\beta}$ when the system becomes empty, and this state is considered to be *frozen* until the next customer arrives and its service starts.

It is easy to verify that the MC $\xi_t = \{i_t, \mathbf{v}_t\} = \{i_t, \nu_t, \eta_t\},\ t \geq 0$ is a QBD process and its block generator has the form (2.21) with the size $\bar{W} \times M$ blocks having the form

$$Q_{0,0} = D_0 \otimes I_M,\ Q_0 = D_1 \otimes I_M,\ Q_1 = D_0 \oplus S,\ Q_2 = I_{\bar{W}} \otimes (\mathbf{S}_0\boldsymbol{\beta}).$$

Here $\otimes$ and $\oplus$ are the so-called Kronecker product and Kronecker sum of matrices defined as follows. For two matrices $A = (a_{i,j})$ and $C = (c_{i,j})$, their Kronecker product is a matrix denoted by $A \otimes C$ and consisting of blocks $(a_{i,j}C)$. The Kronecker sum of these matrices is the matrix denoted by $A \oplus C$ and given by the formula $A \oplus C = A \otimes I_C + I_A \otimes C$, where I_A and I_C are identity matrices of size the same as matrices A and C, respectively. The usefulness of these operations when considering an MC with several finite components is explained by the following fact. If A and C are matrices of the one-step transition probabilities of the independent MCs $\xi_n^{(1)}$ and $\xi_n^{(2)}$, respectively, then the transition probability matrix of the two-dimensional discrete-time MC $\{\xi_n^{(1)}, \xi_n^{(2)}\}$ is matrix $A \otimes C$. A similar property holds for the generator of a two-dimensional continuous-time MC. We recommend the books [9] and [49] for detailed information on the properties of the Kronecker product and sum of matrices. Useful information about them is also given in the Appendix. As one of the most valuable properties of the Kronecker product, we note the so-called mixed-product rule

$$(AB) \otimes (CD) = (A \otimes C)(B \otimes D).$$

Since we will often need to write out more complex generators in the following sections, let us briefly explain the form of blocks of this generator for the sake of easy understanding. The generator block $Q_{0,0}$ sets the intensity of the process $\{\nu_t, \eta_t\}$ transitions without changing the state 0 of the process i_t. Such transitions are possible only for process ν_t transitions with no arrivals. Intensities of such transitions are given by off-diagonal entries of the matrix D_0. The diagonal entries of matrix D_0 are negative and give the process ν_t exit intensity from the corresponding states (up to a sign). Since only one of the two processes $\{\nu_t, \eta_t\}$ can make a transition during an infinitesimal time, the transition of process ν_t implies that process η_t does not make a transition, that is, its transition probability matrix is

I_M. Thus, we obtain the formula $Q_{0,0} = D_0 \otimes I_M$. The block form $Q_0 = D_1 \otimes I_M$ is explained similarly.

This block describes the transition intensities of the processes $\{\nu_t, \eta_t\}$ when process i_t transits from state i to $i+1$. Such transitions are possible only when the process ν_t makes a transition with a customer arrival. Intensities of such transitions are described by the matrix D_1 entries. And here again, process η_t keeps its state, thus its transition probability matrix is I_M. The form $Q_2 = I_{\bar{W}} \otimes (\mathbf{S}_0\boldsymbol{\beta})$ of the block describing the transitions of the processes $\{\nu_t, \eta_t\}$ when process i_t changes its state from i to $i-1$ is easily explained as follows. Such a transition is only possible at the end of a customer service. The intensities of such transitions for process η_t are defined by the entries of the column vector $\mathbf{S}_0$. After this transition, the initial state of the process η_t is instantaneously determined for the next customer service. The process η_t initial state is chosen according to the probabilities given by the entries of vector $\boldsymbol{\beta}$. The governing process ν_t cannot make any transitions, so its transition probability matrix is $I_{\bar{W}}$. The block Q_2 form is now explained. Finally, the generator block Q_1 describes the transition intensities of the processes $\{\nu_t, \eta_t\}$ without changing state i of the process i_t. Such transitions are possible only when process ν_t makes a transition without a customer arrival or process η_t transits to a non-absorbing state. So these arguments result in the formula $Q_1 = D_0 \otimes I_M + I_{\bar{W}} \otimes S = D_0 \oplus S$.

Let us derive the stability condition for the MC $\xi_t = \{i_t, \mathbf{v}_t\}, t \geq 0$.

Here, matrix $Q_0 + Q_1 + Q_2$ has the form

$$(D_0 + D_1) \oplus (S + \mathbf{S}_0\boldsymbol{\beta}) = (D_0 + D_1) \otimes I_M + I_{\bar{W}} \otimes (S + \mathbf{S}_0\boldsymbol{\beta}).$$

Therefore, taking into account the rule of the mixed product for the matrix Kronecker product, it is easy to verify that vector $\mathbf{y}$, which is a solution of the system of equations (2.23), has the form

$$\mathbf{y} = \boldsymbol{\theta} \otimes \boldsymbol{\psi}, \tag{2.31}$$

where $\boldsymbol{\theta}$ is the vector of stationary state probabilities of the MAP governing process satisfying the conditions $\boldsymbol{\theta}(D_0 + D_1) = \mathbf{0}$, $\boldsymbol{\theta}\mathbf{e} = 1$, and vector $\boldsymbol{\psi}$ satisfies the equations

$$\boldsymbol{\psi}(S + \mathbf{S}_0\boldsymbol{\beta}) = \mathbf{0}, \ \boldsymbol{\psi}\mathbf{e} = 1. \tag{2.32}$$

It is easy to verify that the solution of the system of equations (2.32) is the vector

$$\boldsymbol{\psi} = \mu\boldsymbol{\beta}(-S)^{-1}, \tag{2.33}$$

where μ is the average rate of service (the inverse of the average service time):

$$\mu^{-1} = b_1 = \boldsymbol{\beta}(-S)^{-1}\mathbf{e}.$$

Substituting the vector $\mathbf{y}$ of form (2.31) into inequality (2.22) and using expression (2.33), we obtain the inequality

$$\mu > \lambda. \tag{2.34}$$

Thus, we have proved the following statement.

Theorem 2.3 *The stationary state distribution of the MC* $\xi_t = \{i_t, \mathbf{v}_t\},\ t \geq 0$ *exists if and only if inequality (2.34) holds.*

We note that inequality (2.34) is intuitively obvious. The stationary state distribution for the MC describing the system under consideration exists if and only if the average service rate is greater than the average rate of arrivals.

Assuming the inequality (2.34) to hold, the stationary state probability distribution for the considered queuing system is found from (2.24), where matrix $\mathcal{R}$ is the solution of Eq. (2.25) and vector $\boldsymbol{\pi}_0$ is the solution of the system of linear algebraic Eqs. (2.26)–(2.27).

Next, we find the stationary distribution of the virtual waiting time for an arbitrary customer in the system under consideration in terms of the Laplace-Stieltjes transform (LST). Let $W(x)$ be the distribution function of the stationary distribution of the waiting time and $w(s) = \int\limits_0^\infty e^{-sx} dW(x)$, Re $s \geq 0$ be its LST. It is known that given a real s, the value $w(s)$ can be interpreted as the probability that no catastrophe from an imaginary stationary Poisson input of catastrophes with a parameter s happens during a waiting time. Accordingly, the probability that no catastrophe appears during a service time with a phase-type distribution with an irreducible representation $(\boldsymbol{\beta}, S)$ is $\boldsymbol{\beta}(sI - S)^{-1}\mathbf{S}_0$. The mth entry $((sI - S)^{-1}\mathbf{S}_0)_m$ of the vector function $(sI - S)^{-1}\mathbf{S}_0$ gives the probability that no catastrophe arrives during a residual service time given that the governing process of service is in state $m,\ m = \overline{1, M}$ at the present time.

Taking into account this probabilistic interpretation of the LST, the formula of total probability results in

$$w(s) = \boldsymbol{\pi}_0\mathbf{e} + \sum_{i=1}^{\infty} \boldsymbol{\pi}_i(\mathbf{e}_{\bar{W}} \otimes I_M)(sI - S)^{-1}\mathbf{S}_0(\boldsymbol{\beta}(sI - S)^{-1}\mathbf{S}_0)^{i-1}$$

$$= \boldsymbol{\pi}_0\mathbf{e} + \boldsymbol{\pi}_0\mathcal{R}(I - \mathcal{R}\beta(s))^{-1}(\mathbf{e}_{\bar{W}} \otimes I_M)(sI - S)^{-1}\mathbf{S}_0,$$

where

$$\beta(s) = \boldsymbol{\beta}(sI - S)^{-1}\mathbf{S}_0.$$

If the service time distribution is exponential with parameter μ, this formula is simplified to

$$w(s) = \boldsymbol{\pi}_0 \sum_{i=0}^{\infty} \mathcal{R}^i \mathbf{e}(\frac{\mu}{\mu + s})^i = \boldsymbol{\pi}_0(I - \mathcal{R}\frac{\mu}{\mu + s})^{-1}\mathbf{e}.$$

Note that there is another approach (the so-called spectral approach) to the solution of the problem of calculating vectors $\boldsymbol{\pi}_i,\ i \geq 0$ that are the stationary state probabilities for a QBD process $\xi_t,\ t \geq 0$. Below, we give a short description of this approach.

2.2.3 *Spectral Approach for the Analysis of Quasi-Birth-and-Death Processes*

Consider the matrix polynomial

$$Q(\lambda) = Q_0 + Q_1\lambda + Q_2\lambda^2,$$

where the square matrices $Q_0,\ Q_1,\ Q_2$ of order N are the elements of the block generator (2.21). Let λ_k be the eigenvalues of the polynomial, i.e., λ_k are the roots of the equation

$$\det Q(\lambda) = 0. \tag{2.35}$$

Under the ergodicity condition for the process under consideration, there are exactly N roots of Eq. (2.35) in the unit disk of the complex plane. Below we consider only these roots. For the sake of simplicity, we assume that all these roots are simple. Denote by $\boldsymbol{\psi}_k$ the left eigenvector of polynomial $Q(\lambda)$ corresponding to the eigenvalue λ_k, i.e., vector $\boldsymbol{\psi}_k$ satisfies the equation

$$\boldsymbol{\psi}_k Q(\lambda_k) = \mathbf{0}. \tag{2.36}$$

The equilibrium equations for the stationary state probability vectors $\boldsymbol{\pi}_i,\ i \geq 0$ for the QBD process have the form

$$\boldsymbol{\pi}_0 Q_{0,0} + \boldsymbol{\pi}_1 Q_2 = \mathbf{0}, \tag{2.37}$$

$$\boldsymbol{\pi}_{j-1} Q_0 + \boldsymbol{\pi}_j Q_1 + \boldsymbol{\pi}_{j+1} Q_2 = \mathbf{0},\ \ j \geq 1. \tag{2.38}$$

By direct substitution into the equations of system (2.38) by means of (2.36), we can verify that the solution is the set of vectors

$$\boldsymbol{\pi}_j = \sum_{k=1}^{N} x_k \boldsymbol{\psi}_k \lambda_k^j,\ \ j \geq 0, \tag{2.39}$$

where $x_k, k = \overline{1, N}$ are constants. Note that some roots λ_k can be complex, not real. The conjugate to the root λ_k of (2.35) is also a root of this equation. The corresponding eigenvectors $\boldsymbol{\psi}_k$ are also complex and conjugate. In this case, the constants x_k are also complex but the vectors $\boldsymbol{\pi}_j$ calculated by formulas (2.39) are real.

To obtain constants $x_k, k = \overline{1, N}$, we use the system of equations (2.39) and the normalization condition.

Thus, the following statement is valid.

Theorem 2.4 *The stationary state probability vectors of the MC* $\xi_t, \ t \geq 0$ *are calculated by formula (2.39) where constants* $x_k, k = \overline{1, N}$ *are the unique solution of the system of linear algebraic equations*

$$\sum_{k=1}^{N} x_k \boldsymbol{\psi}_k (Q_{0,0} + Q_2 \lambda_k) = \mathbf{0},$$

$$\sum_{k=1}^{N} x_k \boldsymbol{\psi}_k \mathbf{e} \frac{1}{1 - \lambda_k} = 1.$$

A description and examples of the use of the spectral approach for analysis of queuing systems and a discussion of its advantages and disadvantages compared to the approach given in Theorem 2.2 can be found in [89].

2.3 $G/M/1$-Type Markov Chains

The multidimensional birth-and-death processes described in the previous section are the simplest and most well-studied case of multidimensional $G/M/1$- and $M/G/1$-type MCs. Such MCs are considered in the present and subsequent sections. By contrast with QBD processes, which are continuous-time MCs, a description of $G/M/1$- and $M/G/1$-type MCs and the results of their investigation are given below for the case of discrete time.

2.3.1 Definition of a $G/M/1$-Type MC and its Stationary State Distribution

Consider the multidimensional Markov process $\xi_n = \{i_n, \mathbf{v}_n\}, n \geq 1$ with state space

$$\mathcal{S} = \{(i, \mathbf{v}), \mathbf{v} \in \mathcal{V}_i, \ i \geq 0\},$$

where $\mathcal{V}_i$ is a finite set if the Markov process $\xi_n, n \geq 1$ is two-dimensional and is a finite set of finite-dimensional vectors if the process $\xi_n, n \geq 1$ is multidimensional.

Enumerate the states of the process $\xi_n = \{i_n, \mathbf{v}_n\}, n \geq 1$ in lexicographical order and combine the states with the component $i_n = i$ into the macro-state i, sometimes called the level of the process $\xi_n = \{i_n, \mathbf{v}_n\}, n \geq 1$. Corresponding to this enumeration, the matrix P of one-step transition probabilities of this process can be written in the block form $P = (P_{i,l})_{i,l \geq 0}$, where $P_{i,l}$ is the matrix formed by probabilities $p_{(i,\mathbf{r});(l,\boldsymbol{\nu})}$ of the one-step transition of the MC $\xi_n, n \geq 1$ from state $(i, \mathbf{r}),\ \mathbf{r} \in \mathcal{V}_i$ to state $(l, \boldsymbol{\nu}),\ \boldsymbol{\nu} \in \mathcal{V}_l$.

The process $\xi_n = \{i_n, \mathbf{v}_n\}, n \geq 1$ is called a $G/M/1$-type MC if its matrix P of one-step transition probabilities has the following structure:

$$P = \begin{pmatrix} B_0 & A_0 & O & O & O & \dots \\ B_1 & A_1 & A_0 & O & O & \dots \\ B_2 & A_2 & A_1 & A_0 & O & \dots \\ B_3 & A_3 & A_2 & A_1 & A_0 & \dots \\ \vdots & \vdots & \vdots & \vdots & \vdots & \ddots \end{pmatrix}. \tag{2.40}$$

It follows from the form (2.40) of matrix P that here the sets $\mathcal{V}_i$ for different values $i,\ i > 0$ coincide. Below we use the notation $\mathcal{V}_i = \mathcal{V},\ i \geq 0$.

This denotation for this type of Markov chains is explained by the fact that the matrix of transition probabilities of the MC embedded at the arrival moments in a $G/M/1$ queue has a structure similar to (2.40). The only difference of (2.40) from the described matrix is that the entries of matrix (2.40) are not scalars but matrices.

A matrix having the structure (2.40) is called a block lower Hessenberg matrix. The component i_n is skip-free to the left, i.e., its one-step jump to the left does not exceed 1.

Suppose that the MC $\xi_n = \{i_n, \mathbf{v}_n\}, n \geq 1$ with one-step transition probability matrix P is irreducible and aperiodic, and matrices $A = \sum\limits_{i=0}^{\infty} A_i$ and $B = \sum\limits_{i=0}^{\infty} B_i$ are stochastic and irreducible.

Denote by $\boldsymbol{\delta}$ the row vector that is the unique solution of the system

$$\boldsymbol{\delta} = \boldsymbol{\delta} A, \ \ \boldsymbol{\delta}\mathbf{e} = 1$$

and by $\boldsymbol{\gamma}$ the column vector $\boldsymbol{\gamma} = \sum\limits_{i=1}^{\infty} i A_i \mathbf{e}$.

Theorem 2.5 *The MC $\xi_n = \{i_n, \mathbf{v}_n\}, n \geq 1$ has a stationary state probability distribution if and only if the inequality*

$$\boldsymbol{\delta}\boldsymbol{\gamma} > 1 \tag{2.41}$$

holds.

The theorem is proved in [93]. Below we suppose the condition (2.41) to hold. Then the stationary state probability distribution of the MC $\xi_n = \{i_n, \mathbf{v}_n\}, n \geq 1$ exists:

$$\pi(i, \mathbf{v}) = \lim_{n \to \infty} P\{i_n = i, \mathbf{v}_n = \mathbf{v}\}, \ \mathbf{v} \in \mathcal{V}_i, \ i \geq 0.$$

Let us order the stationary probabilities for each i in lexicographic order and introduce vectors

$$\boldsymbol{\pi}_i = (\pi(i, \mathbf{v}), \ \mathbf{v} \in \mathcal{V}), \ i \geq 0.$$

Theorem 2.6 *The stationary state probability vectors of the MC* $\xi_n = \{i_n, \mathbf{v}_n\}, n \geq 1$ *are calculated as follows:*

$$\boldsymbol{\pi}_i = \boldsymbol{\pi}_0 \mathcal{R}^i, \ i \geq 0, \tag{2.42}$$

where matrix $\mathcal{R}$ *is the minimal non-negative solution of the matrix equation*

$$\mathcal{R} = \sum_{j=0}^{\infty} \mathcal{R}^j A_j, \tag{2.43}$$

and vector $\boldsymbol{\pi}_0$ *is the unique solution of the following system of linear algebraic equations:*

$$\boldsymbol{\pi}_0 \sum_{j=0}^{\infty} \mathcal{R}^j B_j = \boldsymbol{\pi}_0, \tag{2.44}$$

$$\boldsymbol{\pi}_0 (I - \mathcal{R})^{-1} \mathbf{e} = 1. \tag{2.45}$$

Proof The vector $\boldsymbol{\pi} = (\boldsymbol{\pi}_0, \ \boldsymbol{\pi}_1, \ \boldsymbol{\pi}_2, \ldots)$ of the stationary distribution is a solution of the system of equilibrium equations

$$\boldsymbol{\pi} P = \boldsymbol{\pi} \tag{2.46}$$

supplemented by the normalization condition

$$\boldsymbol{\pi} \mathbf{e} = 1. \tag{2.47}$$

Suppose that the solution of system (2.46) has the form (2.42). Substituting the vectors $\boldsymbol{\pi}_i, \ i \geq 1$ of form (2.42) into equations

$$\boldsymbol{\pi}_j = \sum_{i=j-1}^{\infty} \boldsymbol{\pi}_i A_{i-j+1}, \ j \geq 1$$

of system (2.46), it is easy to verify that these equations become identities if the matrix $\mathcal{R}$ satisfies the matrix equation (2.43). Substituting the vectors $\boldsymbol{\pi}_i, \ i \geq 1$ of form (2.42) into equation

$$\boldsymbol{\pi}_0 = \sum_{i=0}^{\infty} \boldsymbol{\pi}_i B_i \tag{2.48}$$

of system (2.46), we obtain that vector $\boldsymbol{\pi}_0$ satisfies (2.44).

It is known, see, e.g., [93], that if the ergodicity condition for (2.40) holds then the spectral radius $\varrho(\mathcal{R})$ of matrix $\mathcal{R}$ is strictly less than 1. Given that this condition holds, the series $\sum\limits_{l=0}^{\infty} \mathcal{R}^l$ converges to $(I - \mathcal{R})^{-1}$. Equation (2.45) now follows from (2.42) and the normalization condition $\sum\limits_{i=0}^{\infty} \boldsymbol{\pi}_i \mathbf{e} = 1$. □

The nonlinear matrix equation (2.43) for matrix $\mathcal{R}$ can be solved using the method of successive approximations. Taking the matrix $\mathcal{R}_0 = O$ as the initial approximation, the subsequent approximations can be calculated from the formula

$$\mathcal{R}_{k+1} = \sum_{j=0}^{\infty} \mathcal{R}_k^j A_j, \ k \geq 0$$

or

$$\mathcal{R}_{k+1} = \sum_{j=0,\ j \neq 1}^{\infty} \mathcal{R}_k^j A_j (I - A_1)^{-1}, \ k \geq 0.$$

It is proven in [93] that the successive approximations calculated by these iterative procedures form a sequence of matrices with monotonically increasing elements converging to the minimal non-negative solution of Eq. (2.43).

The above results for a $G/M/1$-type MC can be easily generalized to the case when the block transition probability matrix has the form

$$P = \begin{pmatrix} B_0 & C_0 & O & O & O & \cdots \\ B_1 & A_1 & A_0 & O & O & \cdots \\ B_2 & A_2 & A_1 & A_0 & O & \cdots \\ B_3 & A_3 & A_2 & A_1 & A_0 & \cdots \\ \vdots & \vdots & \vdots & \vdots & \vdots & \ddots \end{pmatrix}.$$

Note that the sets $\mathcal{V}_i$ coincide for values $i \geq 1$ here. But the set $\mathcal{V}_0$ can differ both in composition and in dimension.

In this case the necessary and sufficient condition for the existence of the stationary state probability distribution of the process $\xi_n, \ n \geq 1$ also has the form of inequality (2.41) supplemented by the condition that the stochastic matrix

$$\mathcal{T} = \begin{pmatrix} B_0 & C_0 \\ \sum_{j=1}^{\infty} \mathcal{R}^{j-1} B_j & \sum_{j=1}^{\infty} \mathcal{R}^{j-1} A_j \end{pmatrix}$$

has a positive left eigenvector.

The formulas for calculating the vectors of stationary probabilities here have the following form:

$$\boldsymbol{\pi}_i = \boldsymbol{\pi}_1 \mathcal{R}^{i-1}, \ i \geq 1,$$

where matrix $\mathcal{R}$ is minimal non-negative solution of matrix equation (2.43) and the vectors $\boldsymbol{\pi}_0$ and $\boldsymbol{\pi}_1$ are the unique solution of the following system of linear algebraic equations:

$$(\boldsymbol{\pi}_0, \boldsymbol{\pi}_1)\mathcal{T} = (\boldsymbol{\pi}_0, \boldsymbol{\pi}_1),$$

$$\boldsymbol{\pi}_0 \mathbf{e} + \boldsymbol{\pi}_1 (I - \mathcal{R})^{-1} \mathbf{e} = 1.$$

The above results for $G/M/1$-type MCs can be similarly generalized to the case when the matrix of transition probabilities of the process has the form

$$P = \begin{pmatrix} B_0^{(0)} & \dots & B_0^{(J)} & C_0 & O & O & O & \dots \\ B_1^{(0)} & \dots & B_1^{(J)} & A_1 & A_0 & O & O & \dots \\ B_2^{(0)} & \dots & B_2^{(J)} & A_2 & A_1 & A_0 & O & \dots \\ B_3^{(0)} & \dots & B_3^{(J)} & A_3 & A_2 & A_1 & A_0 & \dots \\ \vdots & \ddots & \vdots & \vdots & \vdots & \vdots & \vdots & \ddots \end{pmatrix},$$

where J is a finite integer.

To illustrate this fact, we apply the above results for a $G/M/1$-type MC to investigation of $G/PH/1$-type queue.

2.3.2 Application of Results to Analysis of a $G/PH/1$-Type Queue

Consider a single-server queuing system with unlimited waiting space. The input flow is renewal with distribution function of inter-arrival times $H(t)$. Denote the flow intensity by $\lambda = (\int_0^{\infty} t dH(t))^{-1}$.

The service times have a PH-type distribution given by a continuous-time MC $\eta_t,\ t \geq 0$ with a space of non-absorbing states $\{1, \ldots, M\}$ and irreducible representation $(\boldsymbol{\beta}, S)$, where $\boldsymbol{\beta}$ is a stochastic row vector and S is a subgenerator. Denote by μ the mean intensity of service $\mu^{-1} = b_1 = \boldsymbol{\beta}(-S)^{-1}\mathbf{e}$.

The process $i_t,\ t \geq 0$, which is the number of customers in the system at time t, is not Markovian. For its analysis, we apply the method of embedded MCs, see, e.g. [30]. To this aim, first we consider the queue behavior at the arrival moments.

Let t_n be the nth moment of customer arrival at the system, $n \geq 1$. It is easy to see that the two-dimensional process $\xi_n = \{i_{t_n-0}, \eta_{t_n+0}\},\ n \geq 1$, where η_{t_n+0} is the state of the service governing process at time $t_n + 0$ is an MC, $i_{t_n-0} \geq 0,\ \eta_{t_n+0} = \overline{1, M}$.

Let $P(k, u),\ k \geq 0$ be a matrix whose entry $(P(k, u))_{\eta,\eta'}$ is the probability that during time interval $[0, u)$ exactly k customers arrive from the renewal flow with phase-type distribution of inter-arrival times with an irreducible representation $(\boldsymbol{\beta}, S)$ and the state of the governing process $\eta_t,\ t \geq 0$ of arrivals is η' at time u given that it was η at time 0. In other words, the entries of matrix $P(k, u)$ give the probabilities of the service governing process transitions during time u with exactly k customers served given that the queue was not empty and at each service completion moment the queue had a customer for service so that the service process was not interrupted in the time interval $[0, u)$.

Since the arrivals from a renewal flow with a PH distribution of inter-arrival times with an irreducible representation $(\boldsymbol{\beta}, S)$ is a special case of MAP, the matrices $P(k, u),\ k \geq 0$ can be derived from the expansion

$$\sum_{k=0}^{\infty} P(k, u) z^k = e^{(S+\mathbf{S}_0\boldsymbol{\beta} z)u},\quad u \geq 0,\ |z| \leq 1.$$

Then, the entries of matrix $\Omega_k = \int\limits_0^\infty P(k, u) dH(u),\ k \geq 0$ give the probability that k customers get service and the service governing process makes the corresponding transitions during an inter-arrival interval given that the service process was not interrupted due to the absence of customers in the queue. Denote by

$$\Omega(z) = \sum_{k=0}^{\infty} \Omega_k z^k = \int\limits_0^\infty e^{(S+\mathbf{S}_0\boldsymbol{\beta} z)u} dH(u).$$

Enumerate the states of MC $\xi_n, n \geq 1$ in lexicographic order and combine the states with $i_n = i$ into macro-state i. Then, the matrix P of one-step transition probabilities of the MC can be written in the block form $P = (P_{i,l})_{i,l\geq 0}$, where $P_{i,l}$ is a matrix formed by the probabilities of the one-step transitions from state (i, η) to state (l, η') of the MC $\xi_n, n \geq 1,\ \eta, \eta' = \overline{1, M}$.

It is easy to see that matrices $P_{i,l},\ i,\ l \geq 0$ are given as follows:

$$P_{i,l} = \begin{cases} O, & l > i+1, \\ \Omega_{i+1-l}, & 0 < l \leq i+1, \\ \sum\limits_{k=i+1}^{\infty} \Omega_k \mathbf{e}\boldsymbol{\beta}, & l = 0. \end{cases}$$

Therefore, the MC $\xi_n = \{i_{t_n-0}, \eta_{t_n+0}\},\ n \geq 1$ is of $G/M/1$-type with the matrix of one-step transition probabilities (2.40) having blocks

$$A_j = \Omega_j,\ B_j = \sum_{k=j+1}^{\infty} \Omega_k \mathbf{e}\boldsymbol{\beta},\ j \geq 0.$$

Theorem 2.7 *The system under consideration is stable, i.e., its stationary state probability distribution exists, if and only if the following inequality holds:*

$$\lambda < \mu. \tag{2.49}$$

Proof It follows from Theorem 2.1 that the stationary state probability distribution of the system under consideration exists if and only if the inequality of type (2.41) holds where vector $\boldsymbol{\delta}$ is the unique solution of equation

$$\boldsymbol{\delta} = \boldsymbol{\delta}\Omega(1),\ \boldsymbol{\delta}\mathbf{e} = 1$$

and $\boldsymbol{\gamma} = \frac{d\Omega(z)}{dz}|_{z=1}\mathbf{e}$. A direct substitution shows that for the MC describing the considered queuing system the vector $\boldsymbol{\delta}$ has the form

$$\boldsymbol{\delta} = \mu\boldsymbol{\beta}(-S)^{-1}.$$

Taking into account that $\boldsymbol{\delta}(S + \mathbf{S}_0\boldsymbol{\beta}) = \mathbf{0}$ and $(S + \mathbf{S}_0\boldsymbol{\beta})\mathbf{e} = \mathbf{0}$, it can be shown in the same way as was used for the derivation of (2.16) that inequality (2.41) results in $\mu\lambda^{-1} > 1$, which is equivalent to (2.49). □

Under the assumption that the stability condition (2.49) holds, the stationary distribution of the system states can be found in a matrix-geometric form by the algorithm given by Theorem 2.2. The relation

$$B_j = \sum_{k=j+1}^{\infty} A_k \mathbf{e}\boldsymbol{\beta},\ j \geq 0 \tag{2.50}$$

between matrices B_j and A_j allows us to get simpler formulas for stationary state probability vectors than those for an arbitrary $G/M/1$-type MC. Namely, vector $\boldsymbol{\pi}_0$, which is numerically calculated as a solution of a system of linear algebraic equations (2.44)–(2.45) in the general case, here has the explicit form

$$\boldsymbol{\pi}_0 = c\boldsymbol{\beta}, \tag{2.51}$$

$$c = [\boldsymbol{\beta}(I - \mathcal{R})^{-1}\mathbf{e}]^{-1}. \tag{2.52}$$

Indeed, by substituting expression (2.50) into Eq. (2.48), we obtain

$$\boldsymbol{\pi}_0 = \boldsymbol{\pi}_0 \sum_{j=0}^{\infty} \mathcal{R}^j B_j = \boldsymbol{\pi}_0 \sum_{j=0}^{\infty} \mathcal{R}^j \sum_{k=j+1}^{\infty} \Omega_k \mathbf{e}\boldsymbol{\beta} = \boldsymbol{\pi}_0 \sum_{k=1}^{\infty} \sum_{j=0}^{k-1} \mathcal{R}^j \Omega_k \mathbf{e}\boldsymbol{\beta}$$

$$= \boldsymbol{\pi}_0 \sum_{k=1}^{\infty} (I - \mathcal{R})^{-1}(I - \mathcal{R}^k)\Omega_k \mathbf{e}\boldsymbol{\beta} = \boldsymbol{\pi}_0 (I - \mathcal{R})^{-1}(\Omega(1) - \sum_{k=0}^{\infty} \mathcal{R}^k \Omega_k)\mathbf{e}\boldsymbol{\beta}$$

$$= \boldsymbol{\pi}_0 (I - \mathcal{R})^{-1}(\Omega(1) - \mathcal{R})\mathbf{e}\boldsymbol{\beta} = \boldsymbol{\pi}_0 \mathbf{e}\boldsymbol{\beta}.$$

The solution of the obtained equation $\boldsymbol{\pi}_0 = \boldsymbol{\pi}_0\mathbf{e}\boldsymbol{\beta}$ with the normalization condition (2.45) results in formulas (2.51), (2.52).

Corollary 2.1 *The mean number L_1 of customers in the system immediately before an arrival moment is calculated by the formula*

$$L_1 = c\boldsymbol{\beta}(I - \mathcal{R})^{-2}\mathbf{e} - 1.$$

We now consider the problem of finding the stationary probability distribution of the system states at an arbitrary moment of time.

Let us study the process $\{i_t, \eta_t\}$, $t \geq 0$, where i_t is the number of customers in the system and η_t is the state of the service governing process at time t, $i_t \geq 0$, $\eta_t = \overline{1, M}$.

Denote

$$p(0) = \lim_{t\to\infty} P\{i_t = 0\},\ p(i, \eta) = \lim_{t\to\infty} P\{i_t = i, \eta_t = \eta\},\ i \geq 1,\ \eta = \overline{1, M},$$

$$\mathbf{p}_i = (p(i, 1), p(i, 2), \ldots, p(i, M)),\ i \geq 1.$$

It can be shown that the stationary probability $p(0)$ and the stationary state probability vectors $\mathbf{p}_i$, $i \geq 1$ exist if and only if condition (2.49) for the existence of the stationary state probability distribution for MC $\xi_n, n \geq 1$ is fulfilled.

Theorem 2.8 *The stationary probability $p(0)$ and the vectors of stationary state probabilities $\mathbf{p}_i$, $i \geq 1$ are calculated as follows:*

$$p(0) = 1 - \rho, \tag{2.53}$$

$$\mathbf{p}_i = c\lambda\boldsymbol{\beta}\mathcal{R}^{i-1}(\mathcal{R} - I - \mathcal{R}\mathbf{e}\boldsymbol{\beta})S^{-1},\ i \geq 1, \tag{2.54}$$

where $\rho = \frac{\lambda}{\mu}$ *is the system load and the constant c is given by formula (2.52).*

Proof The technique for deriving the stationary probability distribution of the process states at an arbitrary time based on the probability distribution of the process states at some embedded moments is briefly described in Sect. 3.1 and can be applied to prove this theorem. However, here we can do it in a simpler way. From the theory of renewal flows, it is known that if there is a stationary renewal flow of customers with distribution function of inter-arrival times $H(t)$, and we choose an arbitrary moment of time, the time passed from the previous arrival moment to an arbitrary moment has the distribution function $\tilde{H}(t) = \lambda \int\limits_0^t (1 - H(u))du$ where λ is the arrival intensity calculated as $\lambda = (\int\limits_0^\infty (1 - H(u))du)^{-1}$.

Using this fact and the formula of total probability, we obtain the following relation between the probability distributions of the considered system states at an arbitrary time and the embedded moments:

$$\mathbf{p}_i = \sum_{j=i-1}^{\infty} \boldsymbol{\pi}_j \lambda \int_0^\infty P(j+1-i, t)(1 - H(t))dt, \ i \geq 1.$$

Taking into account formula (2.42), this relation can be rewritten in the form

$$\mathbf{p}_i = \boldsymbol{\pi}_0 \mathcal{R}^{i-1} \lambda \sum_{m=0}^{\infty} \mathcal{R}^m \int_0^\infty P(m, t)(1 - H(t))dt, \ i \geq 1. \tag{2.55}$$

Denote

$$\Psi(m) = \lambda \int_0^\infty P(m, t)(1 - H(t))dt, \ m \geq 0,$$

$$\Psi(\mathcal{R}) = \sum_{m=0}^{\infty} \mathcal{R}^m \Psi(m) = \lambda \sum_{m=0}^{\infty} \mathcal{R}^m \int_0^\infty P(m, t)(1 - H(t))dt$$

and show that

$$\Psi(\mathcal{R}) = \lambda(\mathcal{R} - \mathcal{R}\mathbf{e}\boldsymbol{\beta} - I)S^{-1}. \tag{2.56}$$

To do this, we use the matrices introduced earlier, $\Omega_m = \int\limits_0^\infty P(m, t)dH(t), \ m \geq 0,$ and the system (2.2) of matrix differential equations for matrices $P(n, t), \ n \geq 0$. Since there is no idle period for the server, the service process becomes a special case of a MAP with matrices $D_0 = S,\ D_1 = \mathbf{S}_0\boldsymbol{\beta}$, and it follows from system (2.2) that the matrix functions $P(n,t),\ n \geq 0$ satisfy the system of differential equations

$$\frac{dP(0,t)}{dt} = P(0,t)S,\ \frac{dP(n,t)}{dt} = P(n,t)S + P(n-1,t)\mathbf{S}_0\boldsymbol{\beta},\ n \geq 1. \quad (2.57)$$

Using the integration by parts formula, we have

$$\Omega_m = \int_0^\infty P(m,t)dH(t) = -\int_0^\infty P(m,t)d(1-H(t))$$

$$= -P(m,t)(1-H(t))|_0^\infty + \int_0^\infty \frac{dP(m,t)}{dt}(1-H(t)).$$

Taking into account the notations introduced and the system (2.57), we obtain

$$\Omega_m = \delta_{m,0}I + \lambda^{-1}\Psi(m)S + (1-\delta_{m,0})\lambda^{-1}\Psi(m-1)\mathbf{S}_0\boldsymbol{\beta},\ m \geq 0.$$

Transforming the above into the matrix generating functions (by multiplying by matrix $\mathcal{R}^m$ from the left and summing over m) and taking into account that (2.43) results in $\sum_{m=0}^{\infty} \mathcal{R}^m\Omega_m = \mathcal{R}$, we get

$$\mathcal{R} = I + \lambda^{-1}\Psi(\mathcal{R})S + \lambda^{-1}\mathcal{R}\Psi(\mathcal{R})\mathbf{S}_0\boldsymbol{\beta}. \quad (2.58)$$

Multiplying relation (2.58) by vector $\mathbf{e}$ and taking into account both the fact that vector $\boldsymbol{\beta}$ is stochastic and the relation between matrix S and the column vector $\mathbf{S}_0$, we obtain

$$(I-\mathcal{R})\mathbf{e} = \lambda^{-1}(I-\mathcal{R})\Psi(\mathcal{R})\mathbf{S}_0.$$

Since the spectral radius $\varrho(\mathcal{R})$ of matrix $\mathcal{R}$ is strictly less than 1, the matrix $(I-\mathcal{R})^{-1}$ exists, so the following relations hold:

$$\mathbf{e} = \lambda^{-1}\Psi(\mathcal{R})\mathbf{S}_0,$$

$$\mathbf{e}\boldsymbol{\beta} = \lambda^{-1}\Psi(\mathcal{R})\mathbf{S}_0\boldsymbol{\beta}.$$

Substituting the last expression into (2.58), we obtain

$$\lambda^{-1}\Psi(\mathcal{R})S = \mathcal{R} - I - \mathcal{R}\mathbf{e}\boldsymbol{\beta},$$

which results in (2.56) by taking into account the non-singularity of the subgenerator S. Formula (2.54) follows from (2.55) and (2.56) in an obvious way.

To prove (2.53), we use the normalization condition. According to it,

$$p(0) = 1 - \sum_{i=1}^{\infty} \mathbf{p}_i \mathbf{e}.$$

Then, formulas (2.51) and (2.54) yield

$$p(0) = 1 - c\lambda \boldsymbol{\beta}(I - \mathcal{R})^{-1}(\mathcal{R} - \mathcal{R}\mathbf{e}\boldsymbol{\beta} - I)S^{-1}\mathbf{e}.$$

Given that the average service time b_1 is defined by formula $b_1 = \boldsymbol{\beta}(-S)^{-1}\mathbf{e}$ and using (2.52), we obtain

$$p(0) = 1 + c\lambda b_1(-1 - \boldsymbol{\beta}(I - \mathcal{R})^{-1}\mathcal{R}\mathbf{e})$$
$$= 1 - \lambda b_1(\boldsymbol{\beta}(I - \mathcal{R})^{-1}\mathbf{e})^{-1}(1 + \boldsymbol{\beta}(I - \mathcal{R})^{-1}\mathcal{R}\mathbf{e}) = 1 - \lambda b_1.$$

Here we used the relation

$$1 + \boldsymbol{\beta}(I - \mathcal{R})^{-1}\mathcal{R}\mathbf{e} = 1 + \boldsymbol{\beta}(I - \mathcal{R})^{-1}(\mathcal{R} - I + I)\mathbf{e} = \boldsymbol{\beta}(I - \mathcal{R})^{-1}\mathbf{e}.$$

□

Corollary 2.2 *The mean number L_2 of customers in the system at an arbitrary time is calculated by the formula*

$$L_2 = \rho L_1 - c\lambda \boldsymbol{\beta}(I - \mathcal{R})^{-1}S^{-1}\mathbf{e}.$$

Consider now the waiting time distribution. Let $W(t)$ be the waiting time distribution function of an arbitrary customer and $w(s) = \int_0^{\infty} e^{-st}dW(t),\ Re\ s \geq 0$ be its LST.

Taking into account the probabilistic sense of the LST, the formula of total probability leads to the formula

$$w(s) = \boldsymbol{\pi}_0\mathbf{e} + \sum_{i=1}^{\infty} \boldsymbol{\pi}_i(sI - S)^{-1}\mathbf{S}_0(\boldsymbol{\beta}(sI - S)^{-1}\mathbf{S}_0)^{i-1}.$$

Substituting here the expressions for $\boldsymbol{\pi}_i$ from (2.53) and (2.54), we get the expression

$$w(s) = c\boldsymbol{\beta}\left(\mathbf{e} + \sum_{i=1}^{\infty} \mathcal{R}^i(sI - S)^{-1}\mathbf{S}_0(\boldsymbol{\beta}(sI - S)^{-1}\mathbf{S}_0)^{i-1}\right).$$

Corollary 2.3 *The mean waiting time V_{wait} of an arbitrary customer is calculated by the formula*

$$V_{wait} = \lambda^{-1} L_2 - b_1.$$

2.4 $M/G/1$-Type Markov Chains

2.4.1 Definition of an $M/G/1$-Type Markov Chain and Its Ergodicity Criteria

Consider a multidimensional Markov process $\xi_n = \{i_n, \mathbf{v}_n\}, n \geq 1$ with state space

$$\mathcal{S} = \{(i, \mathbf{v}), \mathbf{v} \in \mathcal{V}_i, \ i \geq 0\},$$

where $\mathcal{V}_i$ is finite set if the Markov process $\xi_n, n \geq 1$ is two-dimensional and a finite set of finite-dimensional vectors if the Markov process $\xi_n, n \geq 1$ is multidimensional.

Combine all states of the MC $\xi_n = \{i_n, \mathbf{v}_n\}, n \geq 1$, with value i of the component i_n into the macro-state i and enumerate the levels in a lexicographical order. The states from the same level can be enumerated in any order. Without loss of generality, we assume that this is also the lexicographic order. Corresponding to this enumeration, the matrix P of one-step transition probabilities of the process is written in a block form $P = (P_{i,l})_{i,l \geq 0}$, where $P_{i,l}$ is a matrix whose entry $p_{(i,\mathbf{r});(l,\boldsymbol{\nu})}$ is the one-step transition probability of the MC $\xi_n, n \geq 1$ from state $(i, \mathbf{r}), \mathbf{r} \in \mathcal{V}_i$ to state $(l, \boldsymbol{\nu}), \boldsymbol{\nu} \in \mathcal{V}_l$.

The process $\xi_n = \{i_n, \mathbf{v}_n\}, n \geq 1$ is called an $M/G/1$-type MC, or a multidimensional quasi-Toeplitz MC (QTMC), if its matrix P of one-step transition probabilities has the following structure:

$$P = \begin{pmatrix} V_0 & V_1 & V_2 & V_3 & \ldots \\ Y_0 & Y_1 & Y_2 & Y_3 & \ldots \\ O & Y_0 & Y_1 & Y_2 & \ldots \\ O & O & Y_0 & Y_1 & \ldots \\ \vdots & \vdots & \vdots & \vdots & \ddots \end{pmatrix}. \tag{2.59}$$

From the structure of the matrix P it is seen that all sets $\mathcal{V}_i, i \geq 1$ coincide. In the following, we assume that the cardinality of these sets is K.

The denotation of the chain is explained by the fact that a quasi-Toeplitz matrix of one-step transition probabilities of the MC embedded at service completion moments in an $M/G/1$ queue has a structure similar to (2.59). The difference between (2.59) and such a matrix is that the entries of matrix (2.59) are not scalars but matrices.

A matrix with the structure (2.59) is called a block upper-Hessenberg matrix. The component i_n is skip-free to the left, i.e., its one-step jump to the left does not exceed 1.

Consider the matrix PGFs $Y(z) = \sum_{i=0}^{\infty} Y_i z^i$ and $V(z) = \sum_{i=0}^{\infty} V_i z^i, \ |z| \le 1$.

Theorem 2.9 *Let the MC $\xi_n = \{i_n, \mathbf{v}_n\}, n \ge 1$ be irreducible and aperiodic with one-step transition probability matrix P, matrices $Y(1)$ and $V(1)$ be stochastic and irreducible, and the inequalities $Y'(1) < \infty$ and $V'(1) < \infty$ hold. The MC ξ_n, $n \ge 1$ is ergodic if and only if the inequality*

$$[\det(zI - Y(z))]'_{z=1} > 0 \tag{2.60}$$

holds.

Proof To derive the ergodicity condition for an MC, usually the notion of a mean one-step drift is used. For a one-dimensional MC $i_n, \ i_n \ge 0, \ n \ge 1$ and an arbitrary non-negative test function x_i defined on the state space of the MC $i_n, \ n \ge 1$ the mean one-step x-drift γ_i^x of the chain i_n in state i is defined as follows:

$$\gamma_i^x = \mathbf{E}(x_{i_{n+1}} - x_{i_n} | i_n = i), i \ge 0,$$

where $\mathbf{E}$ is the expectation, or if $p_{ij}, \ i, \ j \ge 0$ are the transition probabilities then

$$\gamma_i^x = \sum_{j=0}^{\infty} p_{ij} x_j - x_i, \ i \ge 0.$$

Then vectors $\mathbf{x}_i, \ \mathbf{\Gamma}_i^X, \ i \ge 1$, formed by $x_{i,\mathbf{v}}$ and $\gamma_{i,\mathbf{v}}^x$, respectively, ordered in lexicographic order of increasing indices $\mathbf{v}$ can be considered as the analogues of $x_i, \ \gamma_i^x$ for the MC $\xi_n = \{i_n, \mathbf{v}_n\}, n \ge 1$. It is not difficult to show that the vectors $\mathbf{\Gamma}_i^X, \ i \ge 0$ have the form

$$\mathbf{\Gamma}_i^X = \sum_{j=\max\{0,i-1\}} P_{ij} \mathbf{x}_j - \mathbf{x}_i, \ i \ge 0,$$

where

$$P_{ij} = \begin{cases} V_j, \ i = 0, \ j \ge 0, \\ Y_{j-i+1}, \ j \ge i - 1, \ i > 0. \end{cases}$$

Let vectors $\mathbf{\Gamma}_i^X, \ i \ge 0$ be called the vectors of the mean one-step X-drifts of the MC $\xi_n, \ n \ge 1$, in its states corresponding to level i of its denumerable component.

Let the vector test function $\mathbf{x}_i$, $i \geq 0$, defined on the MC ξ_n, $n \geq 1$ state space have the form

$$\mathbf{x}_i = \begin{cases} \mathbf{0}, \text{ if } i < J, \\ (i+1)\mathbf{e} + \boldsymbol{\alpha}, \text{ if } i \geq J, \end{cases} \tag{2.61}$$

where $\boldsymbol{\alpha}$ is some real-valued vector of order K, $J = [\alpha_{\max}]$, and $\alpha_{\max}$ is the maximum of the modules of the vector $\boldsymbol{\alpha}$ entries. For a test function of the form (2.61), the vectors $\boldsymbol{\Gamma}_i^X$ of the mean drifts for $i > J$ are as follows:

$$\boldsymbol{\Gamma}_i^X = \sum_{j=i-1}^{\infty} Y_{j-i+1}[(j+1)\mathbf{e} + \boldsymbol{\alpha}] - [(i+1)\mathbf{e} + \boldsymbol{\alpha}], \; i > J.$$

These vectors are easily expressed in terms of the generating function $Y(z)$ as

$$\boldsymbol{\Gamma}_i^X = [(Y(z) - zI)]'_{z=1}\mathbf{e} + [Y(1) - I]\boldsymbol{\alpha}, \; i > J.$$

Then, by Lemma A.10 (see Appendix A), there exists a vector $\boldsymbol{\Delta}$ such that

$$\boldsymbol{\Gamma}_i^X = -\boldsymbol{\Delta}, \; i > J, \tag{2.62}$$

where the sign of each component of vector $\boldsymbol{\Delta}$ coincides with the sign of the derivative $[\det(zI - Y(z))]'_{z=1}$ (or $\boldsymbol{\Delta} = 0$ if the derivative is 0).

Let us prove that inequality (2.60) is sufficient for the MC ξ_n, $n \geq 1$ to be ergodic. To this end, we use Moustafa's theorem and formulate an analogue of this theorem for our MC.

The irreducible non-periodic MC ξ_n, $n \geq 1$ is ergodic if there exist a real value $\varepsilon > 0$, a positive integer J, and a set of non-negative vectors $\mathbf{x}_i$, $i \geq 0$ of size K such that the following inequalities hold:

$$\sum_{j=0}^{\infty} P_{ij}\mathbf{x}_j - \mathbf{x}_i < -\varepsilon\,\mathbf{e}, \; i > J, \tag{2.63}$$

$$\sum_{j=0}^{\infty} P_{ij}\mathbf{x}_j < +\infty, \; i = \overline{0, J}. \tag{2.64}$$

By definition, the left-hand side of (2.63) is vector $\boldsymbol{\Gamma}_i^X$ (taking into account that $P_{ij} = O$, $j < i-1$). Given the form (2.61) of the test function, the vectors $\boldsymbol{\Gamma}_i^X$, $i > J$ have the form (2.62). Then if inequality (2.60) holds, we are guaranteed to find the vector $\boldsymbol{\alpha}$ defining the test function with $\boldsymbol{\Gamma}_i^X < 0$, $i > J$, that is, conditions (2.63) of Moustafa's theorem hold. It is also easy to show that the inequalities $Y'(1) < \infty$

and $V'(1) < \infty$ ensure that condition (2.64) of Moustafa's theorem holds. It follows from the foregoing that the fact that inequality (2.60) is valid is sufficient for the considered MC to be ergodic.

The proof that (2.60) is necessary for the considered MC to be ergodic is omitted here.

□

The ergodicity condition (2.60) has an abstract mathematical form. In addition, the calculation of the left-hand side of (2.60) requires considerable computational effort. Therefore, we give another, more easily verified, form of the condition (2.60).

Corollary 2.4 *Inequality (2.60) is equivalent to the following inequality:*

$$\mathbf{x}Y'(1)\mathbf{e} < 1, \tag{2.65}$$

where the vector $\mathbf{x}$ *satisfies the system of linear algebraic equations*

$$\begin{cases} \mathbf{x}Y(1) = \mathbf{x}, \\ \mathbf{x}\mathbf{e} = 1. \end{cases} \tag{2.66}$$

Proof Let us calculate the left-hand side of (2.60) as follows. We expand the determinant $\det(zI - Y(z))$ over some column (without loss of generality, we can take the first column) and, by differentiating the corresponding expression, we obtain that

$$(\det(zI - Y(z)))'|_{z=1} = \nabla(zI - Y(z))'|_{z=1}\mathbf{e}, \tag{2.67}$$

where the vector ∇ is the row vector of the algebraic adjuncts of the first column of $\det(I - Y(1))$. As mentioned in the proof of Lemma A.10 for a stochastic irreducible matrix $Y(1)$, the vector ∇ is positive. Besides, it is known that the vector ∇ presents the solution $\mathbf{x}$ of the system of linear algebraic equations

$$\mathbf{x}Y(1) = \mathbf{x}$$

up to a constant multiplier, that is,

$$\mathbf{x} = c\nabla.$$

Then the unique solution of system (2.66) has the form

$$\mathbf{x} = (\nabla\mathbf{e})^{-1}\nabla, \ \nabla > 0.$$

From the last equality, (2.60), and (2.67) it follows that (2.66) holds. □

Further we assume that condition (2.60) holds and investigate the stationary probability distribution of the MC $\xi_n,\ n \geq 1$

$$\pi(i, \mathbf{v}) = \lim_{n\to\infty} P\{i_n = i, \mathbf{v}_n = \mathbf{v}\}, \ \ \mathbf{v} \in \mathcal{V}_i, \ i \geq 0.$$

To simplify the explanations, we will further assume that $\mathcal{V}_i = \mathcal{V},\ i \geq 0$ and the set $\mathcal{V}$ has cardinality K, $K \geq 1$.

Then we enumerate the stationary probabilities in lexicographic order and introduce the vectors

$$\boldsymbol{\pi}_i = (\pi(i, \mathbf{v}), \ \mathbf{v} \in \mathcal{V}), \ i \geq 0$$

of size K.

In the literature, there are three different approaches to calculate the stationary state probability vectors $\boldsymbol{\pi}_i,\ i \geq 0$. Let us describe them below.

2.4.2 *The Method of Generating Functions to Find the Stationary State Distribution of an $M/G/1$-Type MC*

The stationary state probability vectors $\boldsymbol{\pi}_l$, $l \geq 0$ of the MC ξ_n, $n \geq 1$ satisfy the system of equilibrium equations

$$\boldsymbol{\pi}_l = \boldsymbol{\pi}_0 V_l + \sum_{i=1}^{l+1} \boldsymbol{\pi}_i Y_{l-i+1}, \ l \geq 0. \tag{2.68}$$

System (2.68) is derived by using the formula of total probability and the matrix of the MC transition probabilities of the form (2.59).

The stationary state distribution $\boldsymbol{\pi}_l, l \geq 0$ is completely determined by the vector PGF

$$\boldsymbol{\Pi}(z) = \sum_{l=0}^{\infty} \boldsymbol{\pi}_l z^l,$$

which is analytic in the domain $|z| < 1$ and is continuous on its boundary $|z| = 1$. For this function the following statement holds.

Theorem 2.10 *The vector PGF $\boldsymbol{\Pi}(z)$ of the stationary state distribution $\boldsymbol{\pi}_l$, $l \geq 0$ of MC ξ_n, $n \geq 1$ satisfies the matrix functional equation*

$$\boldsymbol{\Pi}(z)(zI - Y(z)) = \boldsymbol{\pi}_0 H(z), \ |z| \leq 1 \tag{2.69}$$

with

$$H(z) = (zV(z) - Y(z)).$$

Moreover, $\mathbf{\Pi}(z)$ *is the unique solution of (2.69) that is analytic in the disk* $|z| < 1$ *and continuous on its boundary* $|z| = 1$ *such that* $\mathbf{\Pi}(1)\mathbf{e} = 1$.

Proof Formula (2.69) is obtained by multiplying the lth equation of system (2.68) by z^l and further summing all equations over all $l \geq 0$.

We prove the uniqueness of the solution by contradiction. Let $\mathbf{\Pi}(z)$ be a non-unique solution of equation (2.69) satisfying the theorem conditions. Then another function $\mathbf{P}(z)$ exists that is analytic in the domain $|z| < 1$ and continuous on its boundary $|z| = 1$ such that it satisfies the equation $\mathbf{P}(1)\mathbf{e} = 1$ and is a solution of equation (2.69), that is

$$\mathbf{P}(z)(zI - Y(z)) - \mathbf{P}(0)H(z) = 0, \ |z| \leq 1. \tag{2.70}$$

Expanding the left-hand side of (2.70) into the power series of z^l, $l \geq 0$, and equating the coefficients of z^l, $l > 0$ to 0, we obtain the system of Eqs. (2.68) for vectors $\mathbf{p}_l$, $l \geq 0$ defined by the following expansion:

$$\mathbf{P}(z) = \sum_{l=0}^{\infty} \mathbf{p}_l z^l. \tag{2.71}$$

Since the solution of the equilibrium equations is determined up to a constant, the vectors $\mathbf{p}_l$, $l \geq 0$ are related to the stationary state distribution $\boldsymbol{\pi}_l$, $l \geq 0$ as follows:

$$\mathbf{p}_l = c\boldsymbol{\pi}_l, \ l \geq 0. \tag{2.72}$$

Taking into account the form (2.72) of coefficients $\mathbf{p}_l$, $l \geq 0$, we can state that the series (2.71) converges on the boundary of domain $|z| < 1$, and therefore the representation (2.72) at the boundary points remains valid for the function $\mathbf{P}(z)$ continuous on the boundary $|z| = 1$.

In particular,

$$\mathbf{P}(1) = \sum_{i=0}^{\infty} \mathbf{p}_i = c \sum_{i=0}^{\infty} \boldsymbol{\pi}_i$$

and $c = 1$ follows from the fact that $\mathbf{P}(1)\mathbf{e} = 1$.

Then, due to (2.72), the coefficients $\boldsymbol{\pi}_l$, $\mathbf{p}_l$, $l \geq 0$ of the expansion of the functions $\mathbf{\Pi}(z)$ and $\mathbf{P}(z)$ into series in powers of z coincide. But these series uniquely determine the functions $\mathbf{\Pi}(z)$ and $\mathbf{P}(z)$ in the disk $|z| \leq 1$, which yields $\mathbf{P}(z) = \mathbf{\Pi}(z)$ at $|z| \leq 1$. □

It follows from Theorem 2.10 that if we manage to find a solution of the functional equation (2.69) satisfying the theorem conditions then we solve the problem of finding the required stationary state distribution.

Note that in the case when the multidimensional QTMC $\xi_n,\ n \geq 1$ describes the behavior of a queuing system, Eq. (2.69) is actually a matrix analogue of the Pollaczek-Khinchine formula for the stationary state distribution of the $M/G/1$-type queue length.

Indeed, for such a queue with intensity of arrivals λ and service time distribution function $B(t)$, we have $Y(z) = V(z) = \beta(\lambda(1-z))$, where $\beta(s) = \int\limits_0^\infty e^{-st}dB(t),\ \text{Re}\, s \geq 0$, so Eq. (2.69) results in

$$\Pi(z) = \pi_0 \frac{(z-1)\beta(\lambda(1-z))}{z - \beta(\lambda(1-z))}.$$

If in the case of an $M/G/1$ queue the derivation of such a formula almost completely solves the problem of finding the stationary state distribution since the probability π_0 is easily found from the normalization condition, in the vector case this problem is much more complicated. Here, the equation can also be considered solved if the vector $\boldsymbol{\pi}_0$ is defined. Finding this vector is a key problem in solving equation (2.69). For the sake of simplicity, we make the two following assumptions.

Assumption 2.1 *Let matrix $Y(1)$ be irreducible.*

Assumption 2.2 *Among the entries of the matrix $Y(z)$, there is at least one entry $Y_{k,k'}(z)$ such that the coefficients $Y_l(k,k')$ of its expansion $Y_{k,k'}(z) = \sum\limits_{l=0}^{\infty} Y_l(k,k')z^l$ in powers of z satisfy the condition: there exists $l,\ l \geq 0$ such that*

$$Y_l(k,k')Y_{l+1}(k,k') > 0.$$

If these assumptions hold and the MC considered is ergodic, we can prove that equation

$$\det(zI - Y(z)) = 0 \tag{2.73}$$

has exactly K roots (taking into account their multiplicities) in a disk $|z| \leq 1$, one of which is the simple root $z = 1$, and all other roots lie inside the unit circle. Denote by $\tilde{K}$ the number of different roots z_k such that $|z_k| < 1$ and by n_k their multiplicities, $n_k \geq 1, k = \overline{1, \tilde{K}}, \sum\limits_{k=1}^{\tilde{K}} n_k = K - 1$.

Rewrite equation (2.69) in the form

$$\boldsymbol{\Pi}(z) = \frac{\boldsymbol{\Pi}(0)H(z)\text{Adj}(zI - Y(z))}{\det(zI - Y(z))}. \tag{2.74}$$

Corollary 2.5 *The PGF* $\boldsymbol{\Pi}(z)$ *of the stationary state distribution* $\pi_i,\ i \geq 0$ *of the MC* $\xi_n,\ n \geq 1$ *is the solution of the functional equation (2.69) if and only if the system of linear algebraic equations for the vector* $\boldsymbol{\Pi}(0)$ *entries*

$$\boldsymbol{\Pi}(0)\frac{d^n}{dz^n}[H(z)Adj(zI - Y(z))]_{z=z_k} = 0,\ n = \overline{0, n_k - 1},\ k = \overline{1, \tilde{K}}, \tag{2.75}$$

$$\boldsymbol{\Pi}(0)\frac{d}{dz}[H(z)Adj(zI - Y(z))]_{z=1}\mathbf{e} = \frac{d}{dz}[\det(zI - Y(z))]_{z=1} \tag{2.76}$$

has a unique solution.

Moreover, $\boldsymbol{\Pi}(z)$ *is the unique solution of (2.74) analytic in the disk* $|z| < 1$ *and continuous on the boundary* $|z| = 1$ *such that* $\boldsymbol{\Pi}(1)\mathbf{e} = 1$.

Proof Let $\boldsymbol{\Pi}(z)$ be the solution of (2.69). As stated above, it is the unique solution in the class of functions under consideration. Since $\boldsymbol{\Pi}(z)$ is analytic in the disk $|z| < 1$ and continuous on its boundary, the numerator of (2.74) and its $n_k - 1$ derivatives vanish at each of the points $z_k,\ k = \overline{1, \tilde{K}}$, so that relations (2.75) hold. The value $\boldsymbol{\Pi}(z)$ at $z = 1$ can be calculated by applying L'Hôpitals rule to the right-hand side of (2.74). Multiplying the resulting relation by the vector $\mathbf{e}$, we obtain Eq. (2.76).

The inhomogeneous system of linear equations (2.75), (2.76) has a unique solution. Suppose the contrary. Then either there is no vector $\boldsymbol{\Pi}(0)$ such that PGF $\boldsymbol{\Pi}(z)$ is the solution of equation (2.69) (the conditions of the solution analyticity in the domain $|z| < 1$, the continuity on the boundary $|z| = 1$, and the normalization condition do not hold, all simultaneously or some of them) or there exist two or more linearly independent vectors $\boldsymbol{\Pi}(0)$, and Eq. (2.69) has more than one solution in the class of functions under consideration. All this contradicts the assumption made at the beginning of the proof.

Now let the system of Eqs. (2.75), (2.76) have a unique solution. Then it follows from the construction of this system that substitution of its solution $\boldsymbol{\Pi}(0)$ into (2.74) gives us a function $\boldsymbol{\Pi}(z)$ that is analytic in the domain $|z| < 1$, continuous on the boundary $|z| = 1$, and satisfies the normalization condition $\boldsymbol{\Pi}(1)\mathbf{e} = 1$. This function is also a solution of (2.69). Then, by virtue of the uniqueness of the solution of (2.69) in the class of functions under consideration, the defined function is exactly the PGF of the stationary state distribution of the MC $\xi_n,\ n \geq 1$. □

The formulated results justify the algorithm to find the stationary state distribution of the multidimensional QTMC $\xi_n,\ n \geq 1$ consisting of the following steps.

Step 1. Checking the Condition for the Existence of the Stationary State Distribution

Conditions $Y'(1) < \infty,\ V'(1) < \infty$ hold almost always for the MCs describing the operation of a queuing system since their violation means that with a non-zero probability the denumerable component of the MC can make an infinitely large jump to the right.

Then we solve the system of linear algebraic equations (2.66) and verify whether the inequality (2.65) holds. If the inequality does not hold, then the considered MC does not have a stationary state distribution and we stop the algorithm. Otherwise, we go to step 2.

It is worth noting that quite often when investigating specific queuing systems and having sufficient experience, the solution of system (2.66) can be guessed (with subsequent verification by substituting the guessed solution into the system). In such cases, the ergodicity condition can be obtained in a more or less simple analytical form.

We also note that by finding the parameters of the queuing system for which the left-hand side of (2.65) becomes equal to 1, we can find some practically important characteristics of the system's operation such as the throughput (i.e., the limiting value of the average intensity of arrivals), the minimum acceptable rate of customer service with which the system does not accumulate an infinite queue, etc.

Step 2. The Vector $\mathbf{\Pi}(\mathbf{0})$ Computation

The vector $\mathbf{\Pi}(0)$ is the unique solution of the system of linear algebraic equations (2.75), (2.76). Therefore, the first step in calculating $\mathbf{\Pi}(0)$ is the composition of a coefficient matrix for unknowns in the system (2.75), (2.76).

Then we solve Eq. (2.73) and, as above, denote by $z_1, \dots, z_{\tilde{K}}$ the roots of multiplicities $n_1, \dots, n_{\tilde{K}}$ inside the unit circle and $z_K = 1$. When forming the matrix of the system (2.75), (2.76), the following information about this system is useful.

The rank of system (2.75) is $K - 1$ while the number of equations is $K(K - 1)$ (each root generates K equations). In the case of simple roots $z_1, \dots, z_{K-1}$ $(\tilde{K} = K - 1)$, we can show that the rank of the subsystem generated by each of the roots is 1. In this case, the system (2.75), (2.76) yields

$$\mathbf{\Pi}(0)[H(z)\mathrm{Adj}(zI - Y(z))]_{z=z_k}\mathbf{e}^{(1)} = 0, \ k = \overline{1, K - 1}, \tag{2.77}$$

$$\mathbf{\Pi}(0)\frac{d}{dz}[H(z)\mathrm{Adj}(zI - Y(z))]_{z=1}\mathbf{e} = [\det(zI - Y(z))]'_{z=1}, \tag{2.78}$$

where $\mathbf{e}^{(1)}$ is a column vector with the first entry equal to 1, and the others are zero.

In the case of multiple roots, it is not possible to analytically isolate $K-1$ linearly independent equations from $K(K - 1)$ equations (2.75). In this case, the system is obviously redundant, and corresponding software is required to find linearly independent equations. Forming the matrix of system (2.75), (2.76), we solve the system of linear algebraic equations and find the vector $\mathbf{\Pi}(0)$.

Step 3. Calculation of the Stationary State Probabilities

With the vector $\mathbf{\Pi}(0) = \boldsymbol{\pi}_0$ calculated, the problem of finding the stationary state probabilities of the MC ξ_n, $n \geq 1$ can be considered solved. Using the equilibrium equation (2.68), the values of any number $\tilde{L}$ of vectors $\boldsymbol{\pi}_l$, $l = \overline{0, \tilde{L}}$ can be calculated by the recurrent formulas

$$\boldsymbol{\pi}_{l+1} = [\boldsymbol{\pi}_l - \boldsymbol{\pi}_0 V_l - \sum_{i=1}^{l} \boldsymbol{\pi}_i Y_{l-i+1}] Y_0^{-1}, \ l \geq 0. \tag{2.79}$$

An obstacle to this can be the singularity of matrix Y_0. We note however that the assumption of the non-singularity of matrix Y_0 is rather non-restrictive when an MC describes the operation of a single-server queuing system with a BMAP. Another obstacle may be the numerical instability of the recursion (2.79) due to the loss of accuracy (the disappearance of the order) caused by the repeated use of the subtraction operation. If this happens, or if the matrix Y_0 is singular, then we should use a different recursion derived from the matrix-analytical approach of M. Neuts described below in Sect. 3.2 to find the vectors $\boldsymbol{\pi}_l, \ l \geq 1$.

2.4.3 Calculation of Factorial Moments

Introduce the notation for factorial moments

$$\boldsymbol{\Pi}^{(m)} = \frac{d^m}{dz^m} \boldsymbol{\Pi}(z)|_{z=1}, \ m \geq 1.$$

Let $\boldsymbol{\Pi}^{(0)} = \boldsymbol{\Pi}(1)$.

In the case when the distribution $\boldsymbol{\pi}_l, \ l \geq 0$ has not too heavy a tail (that is, the probability mass of the distribution is not concentrated in states with a large value of the first component), the value $\sum_{l=0}^{\tilde{L}} \boldsymbol{\pi}_l \mathbf{e}$ is close to 1 for not very large $\tilde{L}$, so the factorial moments of the distribution can be calculated directly from

$$\boldsymbol{\Pi}^{(m)} = \sum_{i=m}^{\infty} \frac{i!}{(i-m)!} \boldsymbol{\pi}_i, \ m \geq 1.$$

If the tail is heavy, then the sum $\sum_{l=0}^{\tilde{L}} \boldsymbol{\pi}_l \mathbf{e}$ tends to 1 slowly as $\tilde{L}$ grows, and the sums $\sum_{i=m}^{\tilde{L}} \frac{i!}{(i-m)!} \boldsymbol{\pi}_i$ converge much more slowly. Then, to calculate the factorial moments, we use the following statement, whose proof is based on the use of functional equation (2.69).

Corollary 2.6 *Let $Y^{(m)}(1) < \infty$, $V^{(m)}(1) < \infty$, $m = \overline{1, M+1}$, where M is some integer. Then vectors $\mathbf{\Pi}^{(m)}$, $m = \overline{0, M}$ are calculated recursively from*

$$\mathbf{\Pi}^{(m)} = \Bigg[\Bigg(\mathbf{\Pi}(0)H^{(m)}(1) - \sum_{l=0}^{m-1}\binom{m}{l}\mathbf{\Pi}^{(l)}(zI - Y(z))^{(m-l)}|_{z=1}\Bigg)\tilde{I}$$

$$+\frac{1}{m+1}\Bigg(\mathbf{\Pi}(0)H^{(m+1)}(1) - \sum_{l=0}^{m-1}\binom{m+1}{l}\mathbf{\Pi}^{(l)}(zI - Y(z))^{(m+1-l)}|_{z=1}\Bigg)\mathbf{e}\hat{\mathbf{e}}\Bigg]\tilde{\mathbf{A}}^{-1}, \tag{2.80}$$

where

$$\tilde{\mathbf{A}} = (I - Y(1))\tilde{I} + (I - Y'(1))\mathbf{e}\hat{\mathbf{e}},$$

$$\hat{\mathbf{e}} = (1, 0, \ldots, 0),$$

$\tilde{I} = \text{diag}\{0, 1, \ldots, 1\}$.

Proof For the sake of brevity, we denote $\mathbf{A}(z) = zI - Y(z)$, $\mathbf{b}(z) = \boldsymbol{\pi}_0(zV(z) - Y(z)) = \boldsymbol{\pi}_0 H(z)$. Successively differentiating equation (2.69), we obtain the relations

$$\mathbf{\Pi}^{(m)}(z)\mathbf{A}(z) = \mathbf{b}^{(m)}(z) - \sum_{l=0}^{m-1}\binom{m}{l}\mathbf{\Pi}^{(l)}(z)\mathbf{A}^{(m-l)}(z), \ m \geq 0. \tag{2.81}$$

Matrix $\mathbf{Y}(1)$ is stochastic, so matrix $\mathbf{A}(1)$ is singular and it is not possible to construct the recursion to calculate the vectors of the factorial moments $\mathbf{\Pi}^{(m)}$, $m \geq 0$ directly from (2.81). So we modify Eq. (2.81) as follows. By setting $z = 1$ in (2.81), substituting $m+1$ instead of m, multiplying both sides of the obtained equation by $\mathbf{e}$, and taking into account the obvious relation $\mathbf{A}(1)\mathbf{e} = \mathbf{0}^T$, we finally get the equation

$$(m+1)\mathbf{\Pi}^{(m)}\mathbf{A}^{(1)}(1)\mathbf{e} = \mathbf{b}^{(m+1)}(1)\mathbf{e} - \sum_{l=0}^{m-1}\binom{m+1}{l}\mathbf{\Pi}^{(l)}\mathbf{A}^{(m+1-l)}(1)\mathbf{e}. \tag{2.82}$$

It can be shown that the right-hand side of (2.82) is not zero. Replacing one of the equations (2.81) at point $z = 1$ (without loss of generality we will replace the first equation) by Eq. (2.82), we obtain the following system of equations with respect to the components of vector $\mathbf{\Pi}^{(m)}$:

$$\mathbf{\Pi}^{(m)}\tilde{\mathbf{A}} = \left[\left(\mathbf{b}^{(m)}(1) - \sum_{l=0}^{m-1}\binom{m}{l}\mathbf{\Pi}^{(l)}\mathbf{A}^{(m-l)}(1)\right)\tilde{I}\right.$$

$$\left.+\frac{1}{m+1}\left(\boldsymbol{b}^{(m+1)}(1) - \sum_{l=0}^{m-1}\binom{m+1}{l}\mathbf{\Pi}^{(l)}\mathbf{A}^{(m+1-l)}(1)\right)\mathbf{e}\hat{\mathbf{e}}\right]. \tag{2.83}$$

Equation (2.80) can be directly obtained from (2.83) if we show that matrix $\tilde{\mathbf{A}}$ is non-singular. To this aim, we calculate the determinant of this matrix. It can be seen that $\det\tilde{\mathbf{A}} = c[\det\mathbf{A}(z)]'_{z=1} = c[\det(zI - Y(z)]'_{z=1}$, where $c \neq 0$. By virtue of inequality (2.60), we conclude that the required determinant is not zero.

□

The algorithms described above solve the problem of finding the vector $\mathbf{\Pi}(0) = \boldsymbol{\pi}_0$, the required number of vectors $\boldsymbol{\pi}_l, l = \overline{1, \tilde{L}}$, or the required number of factorial moments $\mathbf{\Pi}^{(k)},\ k = \overline{0, M}$ using a given form of matrix PGFs $Y(z),\ V(z)$ describing the transitions of the multidimensional QTMC.

Note that Eq. (2.80) can be used to obtain an inhomogeneous equation in the system of linear algebraic equations (2.77), (2.78) for the entries of vector $\mathbf{\Pi}(0)$ that has a more constructive form than Eq. (2.78).

From Eq. (2.80), we have for $m = 0$

$$\mathbf{\Pi}(1) = \mathbf{\Pi}(0)S, \tag{2.84}$$

where

$$S = (H(1)\tilde{I} + H^{(1)}(1)\mathbf{e}\hat{\mathbf{e}})\tilde{\mathbf{A}}^{-1}. \tag{2.85}$$

Therefore, it follows from the normalization condition that Eq. (2.78) can be replaced by

$$\mathbf{\Pi}(0)S\mathbf{e} = 1.$$

Another possibility for obtaining a relation between vectors $\mathbf{\Pi}(1)$ and $\mathbf{\Pi}(0)$ and further derivation of an equation similar to the inhomogeneous equation (2.78) for vector $\mathbf{\Pi}(0)$ from the normalization condition is as follows.

It follows from Lemma A.9 that if $\boldsymbol{\pi}$ is a stochastic left eigenvector of the irreducible matrix $Y(1)$, that is

$$\boldsymbol{\pi}Y(1) = \boldsymbol{\pi},\ \ \boldsymbol{\pi}\mathbf{e} = 1, \tag{2.86}$$

then matrix $I - Y(1) + \mathbf{e}\boldsymbol{\pi}$ is non-degenerate. Taking into account that the vector $\mathbf{\Pi}(1)$ is stochastic and $\mathbf{\Pi}(1)\mathbf{e}\boldsymbol{\pi} = \boldsymbol{\pi}$, adding the vector $\mathbf{\Pi}(1)\mathbf{e}\boldsymbol{\pi}$ to both sides of Eq. (2.69) at $z = 1$ and further multiplying them by the matrix $(I - Y(1) + \mathbf{e}\boldsymbol{\pi})^{-1}$, we obtain

$$\mathbf{\Pi}(1) = (\mathbf{\Pi}(1)\mathbf{e}\boldsymbol{\pi} + \mathbf{\Pi}(0)H(1))\,(I - Y(1) + \mathbf{e}\boldsymbol{\pi})^{-1},$$

which along with (2.86) yields

$$\mathbf{\Pi}(1) = \boldsymbol{\pi} + \mathbf{\Pi}(0)H(1)(I - Y(1) + \mathbf{e}\boldsymbol{\pi})^{-1}. \qquad (2.87)$$

Equation (2.87) presents the relation between vectors $\mathbf{\Pi}(1)$ and $\mathbf{\Pi}(0)$ (with participation of the known vector $\boldsymbol{\pi}$) in an alternative way to the derived relation (2.84).

2.4.4 *A Matrix-Analytic Method to Find the Stationary State Probability Distribution of an $M/G/1$-Type MC (the Method of M. Neuts)*

An essential step in the method of solution of functional equation (2.69) using the property of analyticity of the vector PGF is the procedure for computing the zeroes of the determinant of the matrix $Y(z) - zI$ in the unit disk of the complex plane. Theoretically, this procedure is quite simple, and there are many methods to solve it. However, in practice during the implementation of this procedure there can arise problems related to poor root separation and the insufficient precision of a personal computer.

The necessity to find an unknown vector $\mathbf{\Pi}(0)$ while avoiding searching for the complex-variable function roots was the driving factor behind the algorithm of Neuts [94]. Note that the equilibrium equations (2.68) contain all the available information on the behavior of an $M/G/1$-type MC. However, formula (2.69) for the PGF $\mathbf{\Pi}(z)$ obtained on the basis of the equilibrium equations defines this function only up to an unknown constant vector $\mathbf{\Pi}(0)$. Based on the equilibrium equations, it is not possible to obtain any additional information about this vector. The way out is to use additional observations on the vector $\mathbf{\Pi}(z)$ or the vector $\mathbf{\Pi}(0)$. In Sect. 2.4.2 we used additional information about the fact that a PGF is analytic in the unit disk of the complex plane. M. Neuts used another fact known from the theory of MCs: the stationary probability of a chain state is inversely proportional to the expectation of the time before the first return to this state. For a homogeneous discrete-time MC, this time is measured by the number of transitions before returning to the state. A detailed exposition of M. Neuts's approach requires the reader to know the theory of Markov renewal processes. Therefore, we restrict ourselves to only a schematic description of this approach.

We are interested in the value of the vector $\mathbf{\Pi}(0)$ whose νth component $\pi(0, \nu)$ is inversely proportional to the average number of transitions the chain makes after leaving the state $(0, \nu)$ before it first returns to this state. Therefore, we have to determine the distribution of the number of transitions of the MC between two neighboring transitions into the state with a zero value of the denumerable component. In this case, however, we have to keep track of the transitions of the finite component during this time as well.

The cornerstone concept in the approach of M. Neuts is the concept of a fundamental period. The fundamental period is the time interval from the moment when the value of the denumerable component is equal to i until the first moment when the value of this component becomes $i-1, i \geq 1$. From the definition of an $M/G/1$-type MC (QTMC) it follows that the length of the fundamental period does not depend on the value i.

Denote by $G(z)$ a matrix PGF whose (ν, ν')th entry is the PGF of the number of transitions of the MC during a fundamental period that ends when the value of the finite component is ν', given that at the beginning of the fundamental period the value of the finite component was ν; $\nu, \nu' = \overline{0, W}$. Using facts from the theory of Markov renewal processes, we can show that matrix $G(z)$ satisfies the equation

$$G(z) = z \sum_{j=0}^{\infty} Y_j G^j(z). \tag{2.88}$$

This equation is also easily derived by using the known probabilistic interpretation of the PGF in terms of coloring the customers, see, for example, [26, 30], and the previous chapter.

The structure of the time between neighboring moments when the denumerable component hits the zero state is clear: at the first step, the denumerable component passes from the zero state to state j, $j \geq 0$ (the transition probabilities of the finite component during this time are determined by the matrix V_j) and then j fundamental periods (if $j \geq 1$) follow during which the denumerable component successively decreases by one until it reaches the value 0.

Denote by $K(z)$ the matrix PGF whose (ν, ν')th entry is the PGF of the number of transitions of the MC during the time before the denumerable component returns to state 0 with the state of the finite component ν' at the moment of return, given that this component was in state ν when the denumerable component exited state 0, $\nu, \nu' = \overline{0, W}$.

Analyzing the time structure described above between two moments the denumerable component enters the state 0, it is easy to verify that

$$K(z) = z \sum_{j=0}^{\infty} V_j G^j(z).$$

Denote by $\boldsymbol{\kappa}$ and $\boldsymbol{\kappa}^*$ the row vector and column vector given by the relations

$$\boldsymbol{\kappa} K(1) = \boldsymbol{\kappa}, \ \boldsymbol{\kappa}\mathbf{e} = 1,$$

$$\boldsymbol{\kappa}^* = K'(1)\mathbf{e}.$$

The component κ_ν of vector $\boldsymbol{\kappa}$ is the stationary state probability that the process ν_t is in state ν at the moment the denumerable component exits the state 0, $\nu = \overline{0, W}$. The component κ^*_ν of vector $\boldsymbol{\kappa}^*$ is the mean number of transitions of the QTMC during the time period until the denumerable component enters the state 0 again, given that the finite component was in state ν at the beginning of the period, $\nu = \overline{0, W}$.

Using the abovementioned interpretation of the stationary probability in terms of the average number of transitions until the system returns to a state and information from the theory of Markov renewal processes, we obtain the final formula specifying the value of the vector $\boldsymbol{\Pi}(0)$:

$$\boldsymbol{\Pi}(0) = \frac{\boldsymbol{\kappa}}{\boldsymbol{\kappa}\boldsymbol{\kappa}^*}. \tag{2.89}$$

In addition to the possibility of obtaining the vector $\boldsymbol{\Pi}(0)$ without using the roots of Eq. (2.73), the matrix-analytic approach allows us to obtain a recursive procedure to calculate the vectors of stationary probabilities $\boldsymbol{\pi}_l, \ l \geq 1$ in an alternative way to recursion (2.79).

It was noted above that recursion (2.79) is inapplicable when the matrix Y_0 is singular. In addition, recursion (2.79) has the following significant drawback. If the distribution has a heavy tail then to compute a sufficient number of vectors $\boldsymbol{\pi}_l, \ l \geq 0$ we have to make many steps in the recursive procedure (2.79), which contains subtraction operations. As practice shows, a recursive procedure containing a subtraction operation has the ability to accumulate computational errors which (with current computer technologies) can lead to unacceptably inaccurate results. In such a situation, the development of an alternative algorithm that does not have these drawbacks is of immediate interest.

To develop an alternative algorithm, we will use the idea of censored MCs, see, for example, [56]. For the sake of simplicity, we first describe the concept of a censored MC for the case of a one-dimensional chain.

Let $i_n, \ i_n \geq 0, \ n \geq 1$ be an MC with one-step transition probabilities $P_{i,l}$. Let us fix an arbitrary integer $N, \ N \geq 0$. A random process, paths of which are obtained from those of the MC $i_n, \ n \geq 1$ by means of removal of path sections where $i_n > N$ and further coupling of the remaining sections is called a censored (or censoring) MC $i_n^{(N)}, \ n \geq 1$, with a censoring level N with respect to the MC $i_n, \ n \geq 1$.

It is easy to see that the random process obtained as a result of a censoring procedure is also an MC. This chain $i_n^{(N)}, n \geq 1$ has state space $\{0, 1, 2, \ldots, N\}$

and one-step transition probabilities $P_{i,l}^{(N)}$ calculated as $P_{i,l}^{(N)} = P_{i,l},\ l < N$ and $P_{i,N}^{(N)} = \sum_{l=N}^{\infty} P_{i,l}$.

It is known that the stationary state probabilities of the initial MC and those of the censored MC coincide on the censored MC state space up to a multiplicative constant.

Now consider a multidimensional MC $\xi_n = \{i_n, \mathbf{v}_n\},\ n \geq 1$ with state space $\mathcal{S} = \{(i, \mathbf{v}), \mathbf{v} \in \mathcal{V}\}$. We construct a multidimensional censored MC $\xi_n^{(N)} = \{i_n^{(N)}, \mathbf{v}_n^{(N)}\}, n \geq 1$ with respect to $\xi_n,\ n \geq 1$ with the censoring level $N,\ N \geq 0$ by means of removal of path sections where $i_n > N$.

Denote the matrix of transition probabilities of the censored Markov chain from states $(i, \mathbf{v})$ to states $(l, \mathbf{v}')$ by $P_{i,l}^{(N)},\ i, l = \overline{0, N}$. It is easy to see that for $l < N$ $P_{i,l}^{(N)} = P_{i,l}$ as in the case of a one-dimensional MC. However, calculation of the matrices $P_{i,N}^{(N)}$ in the case of a multidimensional MC is much more complicated than in the case of a one-dimensional MC since it is required to keep track of all transitions of the finite component $\mathbf{v}_n^{(N)},\ n \geq 1$ of the MC $\xi_n^{(N)} = \{i_n^{(N)}, \mathbf{v}_n\}, n \geq 1$ over the removed sections of the path.

In the case of a multidimensional MC with a block matrix of the transition probabilities with an arbitrary structure, calculation of the transition probabilities of the finite component $\mathbf{v}_n,\ n \geq 1$ over the removed sections of the path is a tough task. Therefore, below we will use the specifics of the block matrix (2.59) of the transition probabilities of the $M/G/1$-type MCs considered in this section. First, this specificity includes the fact that it is impossible to reduce the denumerable component i_n of the MC $\xi_n,\ n \geq 1$ by more than 1 at one step: $P_{i,l} = O,\ l < i - 1$. Second, the matrix of form (2.59) is a block quasi-Toeplitz one, i.e., $P_{i,i+l-1} = Y_l,\ l \geq 0$. In other words, the probability of transition from a state that belongs to level i to a state from level k depends only on the difference $i - k$ and is independent of i and k separately. Usage of these specifics greatly simplifies the problem of calculating the matrices $P_{i,N}^{(N)}$.

Let us analyze the behavior of the MC $\xi_n = \{i_n, \mathbf{v}_n\},\ n \geq 1$ on a removed section of the path from the moment when the first component is in state i, $i \leq N$ and the next state of this component is larger than N till the moment when the MC first enters the state on level N.

This behavior is as follows. Either at the first step the MC $\xi_n = \{i_n, \mathbf{v}_n\}, n \geq 1$ enters a state with the first component from level N, or at this first step the MC enters a state of the first component with value $l, l > N$. Then the MC can make one or any finite number of transitions until the first component reaches the level $l - 1$. The number of such transitions is finite since the original MC is assumed to be ergodic. Then the MC can make one or more transitions until the first component reaches level $l - 2$ and so on until this component reaches level N.

Denote by G_i the transition probability matrix of the component $\mathbf{v}_n$ until the component i_n reaches the value i for the first time starting from level $i + 1$.

Using the formula of total probability, it is not difficult to see that matrices $G_i,\ i \geq 0$, satisfy the equations

$$G_i = P_{i+1,i} + \sum_{l=i+1}^{\infty} P_{i+1,l} G_{l-1} G_{l-2} \dots G_i,\ i \geq 0. \tag{2.90}$$

The fact that the MC is quasi-Toeplitz suggests that matrices $G_i,\ i \geq 0$ are independent of i. Let all of them be equal to some matrix G. Then from (2.90) it follows that matrix G satisfies the equation

$$G = \sum_{j=0}^{\infty} Y_j G^j. \tag{2.91}$$

Note that Eq. (2.91) also follows directly from (2.88) since it follows from the definition of matrices G and $G(z)$ that $G(1) = G$.

The nonlinear matrix equation (2.91) for matrix G can be solved using the method of successive approximations in the same way as the nonlinear matrix equation (2.43) for matrix $\mathcal{R}$ in Sect. 2.3. Taking as an initial approximation of the matrix G the matrix $G^{(0)} = O$ or $G^{(0)} = I$, or $G^{(0)}$ equal to an arbitrary stochastic matrix, the following approximations can be calculated from the formula:

$$G^{(k+1)} = \sum_{j=0}^{\infty} Y_j (G^{(k)})^j,\ k \geq 0$$

or

$$G^{(k+1)} = \sum_{j=0,\ j \neq 1}^{\infty} (I - Y_1)^{-1} Y_j (G^{(k)})^j,\ k \geq 0.$$

The sequence of matrices $G^{(k)}$, $k \geq 0$ converges to the solution of equation (2.91).

Having calculated matrices $G_i = G,\ i \geq 0$, and taking into account the above-described behavior of the MC $\xi_n = \{i_n, \mathbf{v}_n\},\ n \geq 1$ from the moment when the first component of the censored MC $\xi_n^{(N)} = \{i_n^{(N)}, \mathbf{v}_n^{(N)}\},\ n \geq 1$ was $i,\ i \leq N$ to the moment when it first enters the state N, it is easy to verify the correctness of the relation

$$P_{i,N}^{(N)} = P_{i,N} + \sum_{l=N+1}^{\infty} P_{i,l} G_{l-1} G_{l-2} \dots G_N = P_{i,N} + \sum_{l=N+1}^{\infty} P_{i,l} G^{l-N}.$$

Denote by $\boldsymbol{\pi}_i,\ i \geq 0$ the stationary state probability vectors of the initial MC ξ_n, $n \geq 1$ and by $\boldsymbol{\pi}_i^{(N)}, i = \overline{0, N}$, the stationary state probability vectors of the censored MC $\xi_n^{(N)}$, $n \geq 1$.

The vectors $\boldsymbol{\pi}_i^{(N)}, \ i = \overline{0, N}$ satisfy the Chapman-Kolmogorov equations (equilibrium equations)

$$\boldsymbol{\pi}_i^{(N)} = \sum_{l=0}^{i+1} \boldsymbol{\pi}_l^{(N)} P_{l,i}, \ i = \overline{0, N-1},$$

$$\boldsymbol{\pi}_N^{(N)} = \sum_{l=0}^{N} \boldsymbol{\pi}_l^{(N)} P_{l,N}^{(N)}. \tag{2.92}$$

A known fact, which was mentioned above, is that the stationary state probabilities of the initial MC and those of the censored MC coincide up to a multiplicative constant on the state space of the censored MC:

$$\boldsymbol{\pi}_i^{(N)} = c\boldsymbol{\pi}_i, \ i = \overline{0, N}.$$

Therefore, it follows from (2.92) that vectors $\boldsymbol{\pi}_i, \ i \geq 0$ of the stationary state probabilities of the initial MC satisfy the equations

$$\boldsymbol{\pi}_N = \sum_{l=0}^{N} \boldsymbol{\pi}_l P_{l,N}^{(N)}. \tag{2.93}$$

Since the censoring level N is chosen arbitrarily, $N \geq 0$, it follows from (2.93) that vectors $\boldsymbol{\pi}_i, \ i \geq 0$ of the stationary state probabilities of the initial MC satisfy the equations

$$\boldsymbol{\pi}_j = \sum_{i=0}^{j} \boldsymbol{\pi}_i \bar{P}_{i,j}, \ i \geq 0, \tag{2.94}$$

where $\bar{P}_{i,j} \stackrel{def}{=} P_{i,j}^{(j)}$ and matrices $\bar{P}_{i,j}$ are given by formulas

$$\bar{P}_{i,j} = P_{i,j} + \sum_{l=j+1}^{\infty} P_{i,l} G^{l-j}, \ j \geq i. \tag{2.95}$$

From (2.94) it follows that vectors $\boldsymbol{\pi}_i, \ i \geq 0$ of the stationary state probabilities can be presented in the form

$$\boldsymbol{\pi}_i = \boldsymbol{\pi}_0 \Phi_i, \ i \geq 0, \tag{2.96}$$

where matrices $\Phi_i,\ i \geq 0$ satisfy the recursive relations

$$\Phi_0 = I, \tag{2.97}$$

$$\Phi_l = \left(\bar{P}_{0,l} + \sum_{i=1}^{l-1} \Phi_i \bar{P}_{i,l}\right)(I - \bar{P}_{l,l})^{-1},\ l \geq 1. \tag{2.98}$$

The non-singularity of the matrix $I - \bar{P}_{l,l}$ and the non-negativity of the matrix $(I - \bar{P}_{l,l})^{-1},\ l \geq 1$ follow from the easily verified fact that matrix $\bar{P}_{l,l}$ is strictly sub-stochastic as well as from O. Taussky's theorem, see [43], and Lemma A.6. Non-negativity of matrices $(I - \bar{P}_{l,l})^{-1},\ l \geq 1$ implies that the subtraction operation is not used in recursion (2.98), which ensures the computational stability of this recursion.

The relations (2.96) define vectors $\boldsymbol{\pi}_i,\ i \geq 0$ of the stationary state probabilities up to a yet unknown vector $\boldsymbol{\pi}_0$. This vector can be calculated using the vector PGF method or formula (2.89). However, the above results suggest another procedure for calculating the vector $\boldsymbol{\pi}_0$.

For $j = 0$, Eq. (2.94) result in

$$\boldsymbol{\pi}_0(I - \bar{P}_{0,0}) = \mathbf{0}. \tag{2.99}$$

By multiplying relation (2.95) by vector $\mathbf{e}$ from the right, it is easy to verify that matrix $\bar{P}_{0,0}$ is stochastic. By construction, this matrix is irreducible. Therefore, the rank of the system (2.99) is one unit less than the dimension of vector $\boldsymbol{\pi}_0$. Hence, the system (2.99) defines the vector $\boldsymbol{\pi}_0$ up to a constant. Consequently, if we succeed in obtaining another, non-homogeneous equation for the components of vector $\boldsymbol{\pi}_0$ then the resulting system of equations will have a unique solution. Such an equation is easily obtained from (2.96) and the normalization condition and finally has the form

$$\boldsymbol{\pi}_0 \sum_{i=0}^{\infty} \Phi_i \mathbf{e} = 1. \tag{2.100}$$

Note that given that the ergodicity condition for the stationary state distribution of the considered MC holds, the following marginal relations for the norms of matrices $\Phi_l,\ l \geq 1$ are valid: $||\Phi_l|| \to 0$ as $l \to \infty$. Therefore, the calculation of the infinite sum on the left-hand side of Eq. (2.100) can be stopped when $||\Phi_l|| < \varepsilon$, where ε is the required accuracy of calculation.

Summing up, we formulate the following algorithm to find vectors $\boldsymbol{\pi}_i,\ i \geq 0$ of the stationary state probabilities:

Step 1. Find the matrix G as a solution of the nonlinear matrix equation (2.91).
Step 2. Calculate the matrices $\bar{P}_{i,l}$ by formulas (2.96).
Step 3. Calculate the matrices Φ_l by recurrent formulas (2.98).

Step 4. Calculate the vector $\boldsymbol{\pi}_0$ as the unique solution of system (2.99), (2.100).
Step 5. Find the required number of vectors $\boldsymbol{\pi}_i$ by formulas (2.96).

The difference between this algorithm for calculating the vectors $\boldsymbol{\pi}_i,\ i \geq 0$ of the stationary state probabilities and the one based on recursion (2.79) is the former's numerical stability in addition to the possibility of computing vector $\boldsymbol{\pi}_0$. Note that V. Ramaswami proposed such an algorithm to find the vectors $\boldsymbol{\pi}_i,\ i \geq 0$ of the stationary state probabilities under the assumption that vector $\boldsymbol{\pi}_0$ is known and given by formula (2.89).

Comparing the approach to find the stationary state probability distribution of the MC states based on the vector PGFs (analytical approach) with M. Neuts's approach, we can state the following: both approaches have their strengths and weaknesses.

The weaknesses of the analytical approach are:

- the necessity to search for a required number of roots of a complex-variable function in the unit disc of the complex plane. Theoretically, the search problem is not difficult. In practice, it is rather complicated especially when the roots are poorly separated or the roots are multiple;
- the necessity of performing differentiation (analytically or by computer) of the matrix functions in (2.75) in the case of multiple roots of Eq. (2.73). Note that the case of complex roots does not complicate the problem and does not require the use of complex arithmetic in calculations by a computer since each such root gives two equations with real coefficients obtained by separately equating the real and imaginary parts of the equations to zero;
- the necessity to search for a required number of linearly independent equations in the system (2.75), (2.76);
- the use of numerically unstable recursion (2.79) to calculate the vectors $\boldsymbol{\pi}_i, i \geq 1$.

The weaknesses of M. Neuts's approach are the following:

- the necessity of solving nonlinear matrix equation (2.91);
- the necessity to calculate infinite sums (see formulas (2.95), (2.100));
- the occurrence of problems with the calculation of factorial distribution moments in the case of a heavy-tailed distribution.

The strength of the analytical approach is the existence of simple recursive formulas for calculating the factorial distribution moments.

The strengths of M. Neuts's approach are the following:

- the numerically stable recursion for the calculation of the vectors $\boldsymbol{\pi}_i, i \geq 1$ of stationary state probabilities;
- the possibility of its efficient extention to the case of spatially inhomogeneous MCs (see Sect. 2.6 below).

Note that when solving the problem of finding factorial moments, it is sometimes rational to use a combination of two approaches: the analytic approach and M. Neuts's approach. By means of M. Neuts's approach, we find the vector $\boldsymbol{\pi}_0$ and calculate the factorial moments by formula (2.80). Thus, it is possible to overcome difficulties in calculating factorial moments for the probabilities $\boldsymbol{\pi}_i,\ i \geq 1$ in the case of a heavy tail.

2.5 $M/G/1$-Type Markov Chains with Finite State Space

If the first component i_n of an $M/G/1$-type MC $\xi_n = \{i_n, \mathbf{v}_n\}, n \geq 1$ takes only a finite set of values, for example, $i_n = \overline{0, N}$, then its stationary state distribution can be obtained by using an algorithm that is a modification of the algorithm presented in Sect. 2.4.4 on the basis of constructing a family of censored MCs. This algorithm to calculate vectors $\boldsymbol{\pi}_i,\ i = \overline{0, N}$ of the stationary state probabilities consists of the following steps:

Step 1. Calculate the matrices $G_i,\ i = \overline{0, N-1}$ as a solution of the following matrix recursion:

$$G_{N-1} = (I - P_{N,N})^{-1} P_{N,N-1},$$

$$G_i = (I - \sum_{l=i+1}^{N} P_{i+1,l} G_{l-1} G_{l-2} \cdot \ldots \cdot G_{i+1})^{-1} P_{i+1,i},\ i = \overline{0, N-2}.$$

Step 2. Calculate matrices $\bar{P}_{i,l}$ by formulas

$$\bar{P}_{i,l} = P_{i,l} + \bar{P}_{i,l+1} G_l,\ i = \overline{0, N},\ l = \overline{i, N},\ \bar{P}_{i,N+1} = O.$$

Step 3. Calculate matrices Φ_l by formulas

$$\Phi_0 = I,\ \Phi_l = \sum_{i=0}^{l-1} \Phi_i \bar{P}_{i,l} (I - \bar{P}_{l,l})^{-1},\ l = \overline{1, N}.$$

Step 4. Calculate vector $\boldsymbol{\pi}_0$ as the unique solution of the system of equations

$$\boldsymbol{\pi}_0 (I - \bar{P}_{0,0}) = \mathbf{0},\ \boldsymbol{\pi}_0 \sum_{l=0}^{N} \Phi_l \mathbf{e} = 1.$$

Step 5. Calculate vectors $\boldsymbol{\pi}_i,\ i = \overline{0, N}$ by formulas

$$\boldsymbol{\pi}_i = \boldsymbol{\pi}_0 \Phi_i,\ i = \overline{0, N}.$$

2.6 Asymptotically Quasi-Toeplitz Discrete-Time Markov Chains

Consider a multidimensional Markovian process $\xi_n = \{i_n, \mathbf{v}_n\}, n \geq 1$ with state space

$$\mathcal{S} = \{(i, \mathbf{v}), \mathbf{v} \in \mathcal{V}_i, \ i \geq 0\},$$

where $\mathcal{V}_i$ is a finite set if the Markov process $\xi_n, n \geq 1$ is two-dimensional and a finite set of finite-dimensional vectors if the Markov process $\xi_n, n \geq 1$ is multidimensional.

Let's combine the states of the MC $\xi_n = \{i_n, \mathbf{v}_n\}, n \geq 1$, having the state i of component i_n into a macro-state $\mathbf{i}$ sometimes referred to as a level of the process $\xi_n = \{i_n, \mathbf{v}_n\}, n \geq 1$, and let's renumber the levels in lexicographical order. States with the same level can be renumbered in any order. Without loss of generality, we assume that this is also a lexicographic order. In accordance with this numbering, matrix P of the one-step transition probabilities of the process can be written in the block form $P = (P_{i,l})_{i,l \geq 0}$, where $P_{i,l}$ is a matrix formed by the probabilities $p_{(i,\mathbf{r});(l,\boldsymbol{\nu})}$ of the one-step transitions from state $(i, \mathbf{r}), \mathbf{r} \in \mathcal{V}_i$ into state $(l, \boldsymbol{\nu}), \boldsymbol{\nu} \in \mathcal{V}_l$. The number of states included in the level i is the cardinality of the set $\mathcal{V}_i$ and is denoted by $K_i, \ i \geq 0$.

Further assume that there exists a non-negative integer i^* and a set $\mathcal{V}$ such that $\mathcal{V}_i = \mathcal{V}$ if $i > i^*$. The cardinality of the set $\mathcal{V}_i$ is denoted by K.

Definition 2.1 An irreducible non-periodic MC $\xi_n = \{i_n, \mathbf{v}_n\}, n \geq 1$ is called an asymptotically quasi-Toeplitz MC (AQTMC) if the blocks $P_{i,l}$ of the matrix P of one-step transition probabilities satisfy the following conditions:

(a) $P_{i,l} = O$ if $l < i - 1, i > 1$.
(b) There exist matrices $Y_k, k \geq 0$ such that

$$\lim_{i \to \infty} P_{i,i+k-1} = Y_k, \ k \geq 0 \tag{2.101}$$

and the matrix $\sum\limits_{k=0}^{\infty} Y_k$ is stochastic.

Remark 2.1 Here the convergence of a sequence of matrices is assumed to be elementwise convergence. Since only sequences of non-negative matrices of finite order are considered this is equivalent to the normal convergence.

Remark 2.2 Matrices $Y_k, \ k \geq 0$ contain information about the limiting behavior of the MC $\xi_n = \{i_n, \mathbf{v}_n\}, n \geq 1$. They can be considered as matrices of one-step transition probabilities of some quasi-Toeplitz chain limiting with respect to the AQTMC.

Further, we give conditions for the ergodicity and nonergodicity of such AQTMCs and develop an algorithm to find their ergodic (stationary) distribution.

2.6.1 The Ergodicity and Nonergodicity Conditions for an Asymptotically Quasi-Toeplitz MC

To obtain ergodicity and nonergodicity conditions of a multidimensional AQTMC, we use an approach based on the analysis of the mean shifts (the approach of A.A. Lyapunov), see, for example, [80, 90], and the results of [102], where sufficient nonergodicity conditions for one-dimensional MCs are obtained.

Below we formulate the direct analogues of Theorem 4 from [102] and Moustafa's theorem from [90] for a multidimensional AQTMC $\xi_n = \{i_n, \mathbf{v}_n\}, n \geq 1$.

Let

- $x_{i,\mathbf{r}}$ be a non-negative function defined on the MC $\xi_n, n \geq 1$ state space $\mathcal{S}$;
- $\gamma^x_{i,\mathbf{r}} = \sum\limits_{(j,\boldsymbol{\nu})\in\mathcal{S}} p_{(i,\mathbf{r})(j,\boldsymbol{\nu})}(x_{j,\boldsymbol{\nu}} - x_{i,\mathbf{r}})$ be an x-shift (see [102]) in state $(i, \mathbf{r})$;
- $\mathbf{X}_i$, $\boldsymbol{\Gamma}^x_i$ be vectors composed of components $x_{i,\mathbf{r}}$ and $\gamma^x_{i,\mathbf{r}}$, respectively, renumbered in lexicographical order, $i \geq 0$.

It is obvious that

$$\boldsymbol{\Gamma}^x_i = \sum_{j=\max\{0,i-1\}}^{\infty} P_{i,j}\mathbf{X}_j - \mathbf{X}_i, \, i \geq 0. \tag{2.102}$$

Following [102], we say that the MC $\xi_n, n \geq 1$ satisfies the generalized Kaplan condition if there exists a positive constant B, an integer $L > 0$, and a constant c, $0 \leq c < 1$ such that $\Psi_{i,\mathbf{r}}(z) \geq -B$ for $(i, \mathbf{r}) \in \mathcal{S}$, $i > L$, and $z \in [c, 1)$, where the generalized Kaplan function $\Psi_{i,\mathbf{r}}(z)$ is given as follows:

$$\Psi_{i,\mathbf{r}}(z) = -\sum_{(j,\boldsymbol{\nu})\in\mathcal{S}} p_{(i,\mathbf{r})(j,\boldsymbol{\nu})}(z^{x_{j,\boldsymbol{\nu}}} - z^{x_{i,\mathbf{r}}})/(1-z), \ (i, \mathbf{r}) \in \mathcal{S}.$$

The following statement is a direct analogue of Theorem 4 from [102].

Proposition 2.1 *Suppose that the MC $\xi_n, n \geq 1$ satisfies the generalized Kaplan condition, $\boldsymbol{\Gamma}^x_i < \infty, i \geq 0$, and the state space $\mathcal{S} = \{(i, \mathbf{r}), \mathbf{r} \in \mathcal{R}_i, i \geq 0\}$ is divided into two sets $\mathcal{S}_1, \mathcal{S}_2$ such that $\mathcal{S}_1$ is a finite set, $\mathcal{S} = \mathcal{S}_1 \cup \mathcal{S}_2, \mathcal{S}_1 \cap \mathcal{S}_2 = \emptyset$, and the following inequalities hold:*

$$\min_{(i,\mathbf{r})\in\mathcal{S}_2} x_{i,\mathbf{r}} > \max_{(i,\mathbf{r})\in\mathcal{S}_1} x_{i,\mathbf{r}}, \quad \gamma^x_{i,\mathbf{r}} \geq 0, \ (i, \mathbf{r}) \in \mathcal{S}_2.$$

Then the MC $\xi_n, n \geq 1$ is nonergodic.

Proposition 2.2 (The Analogue of Moustafa's Theorem from [90]) *For ergodicity of the MC $\xi_n, n \geq 1$, it is sufficient that there exist a positive number ε, a positive integer J_0, $J_0 \geq i^*$, and a set of non-negative vectors $\mathbf{X}_i, i \geq 0$ of order K such that the following inequalities hold:*

$$\sum_{j=0}^{\infty} P_{i,j}\mathbf{X}_j - \mathbf{X}_i < -\varepsilon\mathbf{e}, \quad i > J_0, \tag{2.103}$$

$$\sum_{j=0}^{\infty} P_{i,j}\mathbf{X}_j < \infty, \quad i = \overline{0, J_0}. \tag{2.104}$$

It is known that the ergodicity or nonergodicity of an MC is determined by the chain's behavior at large values of the denumerable component. Therefore, it is intuitively clear that ergodicity or nonergodicity of an AQTMC is determined by the corresponding ergodicity or nonergodicity of the limiting QTMC. Below this will be proved formally.

Let $Y(z) = \sum\limits_{k=0}^{\infty} Y_k z^k$, $|z| \leq 1$. We distinguish cases where the $Y(1)$ matrix is irreducible and reducible.

In the case of an irreducible matrix $Y(1)$, the ergodicity condition is determined by the sign of the value $\beta = [\det(zI - Y(z))]'_{z=1}$.

To prove the ergodicity condition, we need the following Lemma. The symbol *sign* there applied to a vector means that all components of the vector have the same sign.

Lemma 2.6 *Let $Y'(1) < \infty$. Then there exists a column vector $\mathbf{\Delta}$ such that* $\text{sign}\mathbf{\Delta} = \text{sign}\beta$ *(or $\mathbf{\Delta} = \mathbf{0}$ if $\beta = 0$), and the system of linear algebraic equations with respect to the vector $\boldsymbol{\alpha}$ components*

$$(I - Y(1))\boldsymbol{\alpha} = (Y(z) - zI)'|_{z=1}\mathbf{e} + \mathbf{\Delta} \tag{2.105}$$

has an infinite set of solutions.

Proof Since matrix $Y(1)$ is irreducible and stochastic, the rank of the matrix $I - Y(1)$ is $K - 1$ (see, e.g., [43]). Let $D_n(\mathbf{\Delta}), n = \overline{1, K}$ be the determinant of matrix Y_n^* derived from matrix $I - Y(1)$ by replacing its nth column with the vector on the right side of system (2.105). From linear algebra it is known that the system of Eqs. (2.105) has an infinite set of solutions if there is a vector $\mathbf{\Delta}$ satisfying the system of equations

$$D_n(\mathbf{\Delta}) = 0, n = \overline{1, K}. \tag{2.106}$$

Let us establish the existence of such a vector $\mathbf{\Delta}$.

Decomposing the determinant $D_n(\mathbf{\Delta})$ of matrix Y_n^* by the nth column elements, we obtain the following system of linear algebraic equations for the vector $\mathbf{\Delta}$ components:

$$\mathbf{u}_n \mathbf{\Delta} = \mathbf{u}_n (zI - Y(z))'|_{z=1}\mathbf{e}, \ n = \overline{1, K}. \tag{2.107}$$

Here $\mathbf{u}_n$ is a row vector of the algebraical adjuncts to the elements of the nth column of matrix $I - Y(1)$.

From the theory of matrices it is well known (see, e.g., [6, 43]) that all vectors $\mathbf{u}_n, \ n = \overline{1, K}$ coincide and are positive for an irreducible stochastic matrix $Y(1)$:

$$\mathbf{u}_n = \mathbf{u} > 0, \ n = \overline{1, K}. \tag{2.108}$$

Thus, the system (2.107) is equivalent to the equation

$$\mathbf{u}\mathbf{\Delta} = \mathbf{u}(zI - Y(z))'|_{z=1}\mathbf{e}. \tag{2.109}$$

The right side of Eq. (2.109) is equal to $(\det(zI - Y(z)))'|_{z=1} = \beta$. To verify this, we add all columns of matrix $zI - Y(z)$ to a randomly selected column, expand the determinant of the resulting matrix by the elements of this column and differentiate it at $z = 1$. As a result, we obtain the required relation

$$\beta = \mathbf{u}(zI - Y(z))'|_{z=1}\mathbf{e}. \tag{2.110}$$

Thus, Eq. (2.109) is equivalent to $\mathbf{u}\mathbf{\Delta} = \beta$.

Since according to (2.108) $\mathbf{u} > 0$, there exists a vector $\mathbf{\Delta}$ such that $\mathrm{sign}\mathbf{\Delta} = \mathrm{sign}\beta$ (or $\mathbf{\Delta} = \mathbf{0}$ if $\beta = 0$), satisfying this equation. By construction, the same vector is the solution of the system (2.106). Substituting this solution into (2.105), we get a system of linear algebraic equations for the components of vector $\boldsymbol{\alpha}$ that has an infinite set of solutions. □

Now we can formulate and prove the ergodicity and nonergodicity conditions of an AQTMC.

Theorem 2.11 *Let the matrix $Y(1)$ be irreducible. Assume that the series $\sum_{k=1}^{\infty} kP_{i,i+k-1}\mathbf{e}$ converges for $i = \overline{0, i^*}$ and the series $\sum_{k=1}^{\infty} kP_{i,i+k-1}$ converges for any $i > i^*$, and there exists an integer $j_1, \ j_1 > i^*$ such that the second series converges uniformly in the region $i \geq j_1$.*

Then the following statements are valid:

- *If $\beta > 0$ then AQTMC $\xi_n, \ n \geq 1$ is ergodic.*
- *If $\beta < 0$ then AQTMC $\xi_n, \ n \geq 1$ is non-ergodic.*

Proof Suppose that $\beta \neq 0$. Validity of the conditions of Theorem 2.11 implies the validity of the condition $Y'(1) < \infty$ of Lemma 2.6. Then we can find the vector $\mathbf{\Delta}$ such that $\mathrm{sign}\mathbf{\Delta} = \mathrm{sign}\beta$, and Eq. (2.105) has a solution $\boldsymbol{\alpha} = (\alpha_r)_{r\in\mathcal{R}}$. Let us take the vector test function $\boldsymbol{X}_i, i \geq 0$ of the form

$$\boldsymbol{X}_i = \begin{cases} \mathbf{0}, & i < J, \\ (i+1)\mathbf{e} + \boldsymbol{\alpha}, & i \geq J \end{cases} \tag{2.111}$$

where $J = \max_{\mathbf{r} \in \mathcal{R}} \{j_1, [|\alpha_{\mathbf{r}}|] + 1\}$.

It is obvious that this function is non-negative. It can be shown that for $i > J$, the vectors $\boldsymbol{\Gamma}_i^x$ defined by formula (2.102) are

$$\boldsymbol{\Gamma}_i^x = (\sum_{k=1}^{\infty} k P_{i,i+k-1} - I)\mathbf{e} + (\sum_{k=0}^{\infty} P_{i,\,i+k-1} - I)\boldsymbol{\alpha},\ i > J. \tag{2.112}$$

From Lemma 2.6 it follows that

$$0 = (Y(z) - zI)'|_{z=1}\mathbf{e} + (Y(1) - I)\boldsymbol{\alpha} + \boldsymbol{\Delta}. \tag{2.113}$$

Subtraction of (2.113) from (2.112) yields

$$\boldsymbol{\Gamma}_i^x = -\boldsymbol{\Delta} + \boldsymbol{\epsilon}_i,\ i > J, \tag{2.114}$$

where $\boldsymbol{\epsilon}_i = [\sum_{k=1}^{\infty} k P_{i,i+k-1} - Y'(1)]\mathbf{e} + [\sum_{k=0}^{\infty} P_{i,i+k-1} - Y(1)]\boldsymbol{\alpha},\ i > J.$

From Definition 2.1 (b) for the AQTMC and the uniform convergence of series $\sum_{k=1}^{\infty} k P_{I,I+k-1}$ for large values I it follows that $\boldsymbol{\epsilon}_i$ tends to $\mathbf{0}$ as $i \to \infty$. Therefore, there is a positive integer $J_0 \geq \max\{j_1, J\}$ such that

$$\text{sign}\, \boldsymbol{\Gamma}_i^x = \text{sign}\,(-\boldsymbol{\Delta}),\ i > J_0. \tag{2.115}$$

Let $\beta > 0$.

From equality (2.102) and Definition 2.1 (a) for the AQTMC, it follows that the left side of inequality (2.103) is $\boldsymbol{\Gamma}_i^x$. Vectors $\boldsymbol{\Gamma}_i^x,\ i > J$, have the form (2.114) and satisfy the relation (2.115) for $i > J_0 \geq J$. In the case under consideration, we have $\boldsymbol{\Delta} > 0$ since $\beta > 0$. Hence from (2.115) it follows that $\boldsymbol{\Gamma}_i^x < 0,\ i > J_0$. This means that condition (2.103) of Proposition 2.2 holds.

It is easy to verify that the convergence of the series $\sum_{k=1}^{\infty} k P_{i,i+k-1}\mathbf{e}, i = \overline{0, i^*}$ and series $\sum_{k=1}^{\infty} k P_{i,i+k-1}, i > i^*$ implies the validity of condition (2.104) of Proposition 2.2.

Thus, we have shown that if $\beta > 0$ then the conditions of Proposition 2.2 are valid. Therefore, the MC $\xi_n,\ n \geq 1$ is ergodic. In other words, the validity of inequality $\beta > 0$ is a sufficient condition for the ergodicity of the MC $\xi_n,\ n \geq 1$.

Now, let $\beta < 0$.

By means of simple transformations, we can make sure that for a vector test function of the form (2.111), the MC $\xi_n,\ n \geq 1$ satisfies the generalized Kaplan condition. Taking into account the convergence of the series $\sum\limits_{k=1}^{\infty} kP_{i,i+k-1}\mathbf{e},\ i = \overline{0, i^*}$ and $\sum\limits_{k=1}^{\infty} kP_{i,i+k-1},\ i > i^*$, it is easy to see that the mean shift vectors $\mathbf{\Gamma}_i^x$ satisfy the inequalities $\mathbf{\Gamma}_i^x < \infty,\ i \geq 0$.

Decompose the state space $\mathcal{S} = \{(i, \mathbf{r}), \mathbf{r} \in \mathcal{R}_i, i \geq 0\}$ of the MC into two sets $\mathcal{S}_1, \mathcal{S}_2$ in the following way. The set $\mathcal{S}_1$ includes the states $(i, \mathbf{r}), i \leq J_0$. The rest of the states form the set $\mathcal{S}_2$. Now move all states $(j, \boldsymbol{\nu}) \in \mathcal{S}_2$ such that $x_{j,\boldsymbol{\nu}} \leq \max\limits_{(i,\mathbf{r})\in\mathcal{S}_1} x_{i,\mathbf{r}}$ from the set $\mathcal{S}_2$ into the set $\mathcal{S}_1$. As a result, we get a finite set $\mathcal{S}_1$ and a denumerable set $\mathcal{S}_2$ such that $\min\limits_{(i,\mathbf{r})\in\mathcal{S}_2} x_{i,\mathbf{r}} > \max\limits_{(i,\mathbf{r})\in\mathcal{S}_1} x_{i,\mathbf{r}}$.

Because the set $\mathcal{S}_2$ includes only states $(i, \mathbf{r}), i > J_0$, and $\mathbf{\Delta} < 0$, fulfillment of inequality $\beta < 0$ implies that inequalities $\gamma_{i,\mathbf{r}}^x > 0,\ (i, \mathbf{r}) \in \mathcal{S}_2$ stem from (2.115).

Thus, we have shown that if $\beta < 0$ then all the conditions of Proposition 2.1 hold. Therefore, the MC $\xi_n,\ n \geq 1$ is non-ergodic, i.e., the inequality $\beta < 0$ is a sufficient condition for nonergodicity of the MC $\xi_n,\ n \geq 1$. □

The conditions $\beta = [\det(zI - Y(z))]'_{z=1} = \mathbf{u}(zI - Y(z))'|_{z=1}\mathbf{e} > 0 (< 0)$ are not very convenient to verify. Let us get more easy-to-use conditions.

Corollary 2.7 *The AQTMC $\xi_n,\ n \geq 1$ is ergodic if*

$$\mathbf{y}Y'(1)\mathbf{e} < 1 \tag{2.116}$$

and non-ergodic if

$$\mathbf{y}Y'(1)\mathbf{e} > 1,$$

where the row vector $\mathbf{y}$ *is the unique solution of the system of equations*

$$\mathbf{y}Y(1) = \mathbf{y},\ \mathbf{y}\mathbf{e} = 1. \tag{2.117}$$

Proof It is known from the theory of matrices (see, e.g., [43]) that all solutions of the equation $\mathbf{y}Y(1) = \mathbf{y}$ have the form $\mathbf{y} = c\mathbf{u}$, where $\mathbf{u}$ is a column vector of the algebraical adjuncts introduced in the proof of Lemma 2.6, and c is a constant. Then the unique solution of system (2.117) has the form

$$\mathbf{y} = (\mathbf{u}\mathbf{e})^{-1}\mathbf{u}. \tag{2.118}$$

Note that $(\mathbf{u}\mathbf{e})^{-1} > 0$ by virtue of (2.108). It follows from relations (2.110) and (2.118) that the inequality $\beta > 0\ (\beta < 0)$ is equivalent to the inequality $\mathbf{y}Y'(1)\mathbf{e} < 1\ \ (\mathbf{y}Y'(1)\mathbf{e} > 1)$. Therefore, the Corollary statement to be proved is equivalent to the statement of Theorem 2.11. □

Now let the matrix $Y(1)$ be reducible. Then the matrix PGF $Y(z)$ is also reducible and has the same structure as the matrix $Y(1)$. Without loss of generality, we suppose that the matrix $Y(z)$ is already presented in a canonical normal form (see, e.g., Appendix A, or [43, 94] for the definition), that is, it has the following structure:

$$Y(z) = \begin{pmatrix} Y^{(1)}(z) & 0 & \cdots & 0 & 0 & \cdots & 0 \\ 0 & Y^{(2)}(z) & \cdots & 0 & 0 & \cdots & 0 \\ \vdots & \vdots & \ddots & \vdots & \vdots & \ddots & \vdots \\ 0 & 0 & \cdots & Y^{(m)}(z) & 0 & \cdots & 0 \\ Y^{(m+1,1)}(z) & Y^{(m+1,2)}(z) & \cdots & Y^{(m+1,m)}(z) & Y^{(m+1)}(z) & \cdots & 0 \\ \vdots & \vdots & \ddots & \vdots & \vdots & \ddots & \vdots \\ Y^{(s,1)}(z) & Y^{(s,2)}(z) & \cdots & Y^{(s,m)}(z) & Y^{(s,m+1)}(z) & \cdots & Y^{(s)}(z) \end{pmatrix}, \tag{2.119}$$

where $Y^{(1)}(z), \ldots, Y^{(m)}(z)$ are irreducible square matrices of order $r_1, \ldots, r_m$, respectively, $\sum_{n=1}^{m} r_n \le K$, and each row $Y^{(n,1)}(z), \ldots, Y^{(n,n-1)}(z), n = m+1, \ldots, s$ has at least one non-zero matrix.

Let $\beta_l = [\det(zI - Y^{(l)}(z))]'_{z=1}, \ l = \overline{1,m}$.

Lemma 2.7 *There exists a column vector* $\mathbf{\Delta} = (\mathbf{\Delta}_1, \ldots, \mathbf{\Delta}_m, \mathbf{\Delta}_{m+1})$, *where* $\mathbf{\Delta}_l$ *is a vector of order* r_l, *satisfying the conditions* $\text{sign}\, \mathbf{\Delta}_l = \text{sign}\, \beta_l$ *for* $\beta_l \neq 0$ *and* $\mathbf{\Delta}_l = \mathbf{0}$ *for* $\beta_l = 0, \ l = \overline{1,m}$ *such that the system of linear equations (2.105) has an infinite set of solutions.*

Lemma 2.7 is proved in a similar way to Lemma 2.6.

Theorem 2.12 *Let matrix* $Y(1)$ *be reducible with the canonical normal form (2.119). If the series* $\sum_{k=1}^{\infty} kP_{i,i+k-1}\mathbf{e}, \ i = \overline{0,i^*}$ *and* $\sum_{k=1}^{\infty} kP_{i,i+k-1}, \ i > i^*$ *satisfy the conditions of Theorem 2.1 then the following propositions hold:*

- *If* $\beta_l > 0, \ l = \overline{1,m}$ *then the AQTMC* $\xi_n, \ n \ge 1$ *is ergodic.*
- *If* $\beta_l < 0, \ l = \overline{1,m}$ *then the AQTMC* $\xi_n, \ n \ge 1$ *is nonergodic.*

The proof of Theorem 2.12 is similar to the proof of Theorem 2.11 but using Lemma 2.7 instead of Lemma 2.6.

Corollary 2.8 *The AQTMC* $\xi_n, \ n \ge 1$ *is ergodic if*

$$\mathbf{y}_l \frac{dY^{(l)}(z)}{dz}\Big|_{z=1}\mathbf{e} < 1, \ l = \overline{1,m}, \tag{2.120}$$

and it is non-ergodic if

$$\mathbf{y}_l \frac{dY^{(l)}(z)}{dz}\Big|_{z=1}\mathbf{e} > 1, \ l = \overline{1,m}.$$

Here the row vectors $\mathbf{y}_l,\ l = \overline{1,m}$ *are the unique solutions of the systems of equations*

$$\mathbf{y}_l Y^{(l)}(1) = \mathbf{y}_l,\ \mathbf{y}_l \mathbf{e} = 1. \tag{2.121}$$

Remark 2.3 If condition (b) of Definition 2.1 holds in a stronger form, i.e., there exists an integer N, $N \geq 1$ such that $P_{i,i+k-1} = Y_k, k \geq 0$ for any $i \geq N$ then the AQTMC is a multidimensional QTMC (or an $M/G/1$-type MC) with N boundary states.

In this case, inequalities (2.116) and (2.120) specify the ergodicity criteria for the MC.

2.6.2 Algorithm to Calculate the Stationary State Probabilities of an Asymptotically Quasi-Toeplitz Markov Chain

Suppose that the ergodicity conditions for the AQTMC $\xi_n,\ n \geq 1$ hold and consider the stationary state probabilities of the AQTMC

$$\pi(i, \mathbf{r}) = \lim_{n\to\infty} P\{i_n = i, \mathbf{r}_n = \mathbf{r}\}.$$

Combine the states with the value i of the denumerable component i_t numbered in lexicographic order into level i and consider the row vectors $\boldsymbol{\pi}_i$ of the stationary probabilities of the states belonging to level i, $i \geq 0$.

Chapman-Kolmogorov equations for these vectors have the form

$$\boldsymbol{\pi}_j = \sum_{i=0}^{j+1} \boldsymbol{\pi}_i P_{i,j},\ j \geq 0. \tag{2.122}$$

Note that we have already faced the problem of solving a similar system of equations (the system (2.68)) in our analysis of multidimensional MCs in Sect. 2.4. There, we applied two approaches to solve the system: an approach based on vector PGFs and one going back to M. Neuts. In the case of an AQTMC, the approach based on vector PGFs is applicable in rare cases only. One such case will be discussed below in the study of $BMAP/SM/1$ queues with retrials and a constant retrial rate. Here, we focus on the second approach.

As was noted in Sect. 2.4, that approach is based on the idea of censored MCs, see [56]. In Sect. 2.4, this idea was clarified on the example of one-dimensional MCs and further was applied for multidimensional MCs. Here we start immediately with the case of multidimensional MCs.

Let us fix an arbitrary integer $N,\ N \geq 0$. Denote by $\xi_n^{(N)},\ n \geq 1$ a censored MC with state space $(i, \mathbf{r}),\ \mathbf{r} \in \mathcal{V}_i,\ i = \overline{0, N}$. Denote by $P^{(N)} = (P_{i,l}^{(N)})_{i,l=\overline{0,N}}$ the matrix of one-step transition probabilities of the censored MC $\xi_n^{(N)}$, $n \geq 1$. In the general case, if the matrix of one-step transition probabilities of the original MC is presented in a block form

$$P = \begin{pmatrix} T & U \\ R & Q \end{pmatrix},$$

where T, U, R, Q are matrices of corresponding size then the matrix of one-step transition probabilities of the censored MC is given as follows:

$$P^{(N)} = T + U\Psi R,$$

where $\Psi = \sum\limits_{i=0}^{\infty} Q^i$.

The useful property of the censored MC $\xi_n^{(N)}$, $n \geq 1$ is the fact that its stationary state probabilities coincide with the stationary state probabilities of the corresponding states of the original MC up to a normalizing constant.

Since the block one-step transition probability matrix of the AQTMC is a block upper-Hessenberg matrix according to Property a) in Definition 2.1, then the matrix $P^{(N)} = (P_{i,l}^{(N)})_{i,l=\overline{0,N}}$ of one-step transition probabilities of the censored MC is determined more simply than in the general case. In particular, it is easy to see that

$$P_{i,l}^{(N)} = P_{i,l}, i = \overline{0, N}, l = \overline{0, N-1}.$$

The expression for the matrices $P_{i,N}^{(N)}$ is somewhat more difficult to derive because their calculation requires taking into account the possible transitions of the original MC on the removed sections of its path. Thus, here we take into account that any path of transitions of the original MC from a state having value i of the denumerable component to a state having value j of this component, $i > j \geq 0$, has to visit all intermediate states with values $i-1, i-2, \ldots, j+1$ of denumerable component at least once. Due to Property (a), this feature of the AQTMC opens the following way to calculate the matrices $P_{i,N}^{(N)}$.

Introduce into consideration matrices

$$G_i = (g_i(\mathbf{r}; \boldsymbol{\nu}))_{\mathbf{r}\in\mathcal{V}_{i+1},\, \boldsymbol{\nu}\in\mathcal{V}_i}, i \geq N,$$

whose entry $g_i(\mathbf{r}; \boldsymbol{\nu})$ is the conditional probability that the MC ξ_n, $n \geq 1$ enters the set of states $\{(i, \mathbf{s}), \mathbf{s} \in \mathcal{V}_i\}$ by entering the state $(i, \boldsymbol{\nu})$ given that it starts from the state $(i + 1, \mathbf{r})$. Matrices G_i were introduced into consideration by M. Neuts, see [94], and we have already used them in Sect. 2.4.

By analyzing the behavior of the MC $\xi_n,\ n \geq 1$ and using the formula of total probability, one can make sure that matrices $G_i, i \geq N$ satisfy the following recurrent relations:

$$G_i = P_{i+1,i} + \sum_{l=i+1}^{\infty} P_{i+1,l} G_{l-1} G_{l-2} \dots G_i, i \geq N.$$

Now, given that the MC has to visit all the intermediate levels, it is easy to see that

$$P_{i,N}^{(N)} = P_{i,N} + \sum_{l=1}^{\infty} P_{i,N+l} G_{N+l-1} G_{N+l-2} \dots G_N, i = \overline{0, N}.$$

Thus, the problem of calculating the blocks of matrix $P^{(N)}$ is solved completely. The vectors of stationary state probabilities $\boldsymbol{\pi}_i^{(N)},\ i = \overline{0, N}$ satisfy the system of linear algebraic equations

$$\boldsymbol{\pi}_l^{(N)} = \sum_{i=0}^{l+1} \boldsymbol{\pi}_i^{(N)} P_{i,l}^{(N)}, l = \overline{0, N-1},$$

$$\boldsymbol{\pi}_N^{(N)} = \sum_{i=0}^{N} \boldsymbol{\pi}_i^{(N)} P_{i,N}^{(N)}. \tag{2.123}$$

The abovementioned fact that the stationary state probabilities of the censored MC coincide with the corresponding stationary state probabilities of the original MC up to a normalizing constant implies that the vectors of the stationary state probabilities $\boldsymbol{\pi}_i,\ i = \overline{0, N}$ of the initial MC satisfy Eq. (2.123) as well, that is

$$\boldsymbol{\pi}_N = \sum_{i=0}^{N} \boldsymbol{\pi}_i P_{i,N}^{(N)},$$

or

$$\boldsymbol{\pi}_N = \sum_{i=0}^{N-1} \boldsymbol{\pi}_i P_{i,N}^{(N)} (I - P_{N,N}^{(N)})^{-1}. \tag{2.124}$$

The non-singularity of the matrix $I - P_{N,N}^{(N)}$ for $N \geq 1$ follows from the fact that the matrix $P_{N,N}^{(N)}$ is irreducible and sub-stochastic.

Since Eqs. (2.124) hold for any number $N,\ N \geq 0$ they give an alternative in relation (2.122) system of equations to calculate the vectors $\boldsymbol{\pi}_j,\ j \geq 1$.

It follows from (2.124) that all vectors $\boldsymbol{\pi}_l, l \geq 1$ can be expressed in terms of vector $\boldsymbol{\pi}_0$ as follows:

$$\boldsymbol{\pi}_l = \boldsymbol{\pi}_0 \Phi_l,\ l \geq 1, \tag{2.125}$$

where the matrices $\Phi_l,\ l \geq 1$ are calculated by the recurrence equations

$$\Phi_0 = I,\ \ \Phi_l = \sum_{i=0}^{l-1} \Phi_i P_{i,l}^{(l)} (I - P_{l,l}^{(l)})^{-1},\ l \geq 1.$$

To find an unknown vector $\boldsymbol{\pi}_0$ we set $N = 0$ in (2.123), which results in the following homogeneous system of linear algebraic equations:

$$\boldsymbol{\pi}_0 (I - P_{0,0}^{(0)}) = \mathbf{0}. \tag{2.126}$$

It can be shown that the matrix $P_{0,0}^{(0)}$ is irreducible and stochastic. Therefore, system (2.126) defines the vector $\boldsymbol{\pi}_0$ up to a constant that is determined from the normalization condition.

Thus, from the theoretical point of view, the problem of finding the stationary state distribution of the AQTMC $\xi_n, n \geq 1$ is solved. The solution algorithm consists of the following fundamental steps.

- Calculate the matrices G_i from the recurrence relations

$$G_i = P_{i+1,i} + \sum_{l=i+1}^{\infty} P_{i+1,l} G_{l-1} G_{l-2} \dots G_i,\ i \geq 0 \tag{2.127}$$

as

$$G_i = (I - \sum_{l=i+1}^{\infty} P_{i+1,l} G_{l-1} G_{l-2} \dots G_{i+1})^{-1} P_{i+1,i},\ \ i \geq 0.$$

- Calculate the matrices $\bar{P}_{i,l}$ by formulas

$$\bar{P}_{i,l} = P_{i,l} + \sum_{n=l+1}^{\infty} P_{i,n} G_{n-1} G_{n-2} \dots G_l,\ \ l \geq i, i \geq 0. \tag{2.128}$$

- Calculate the matrices $\Phi_l, l \geq 1$ from recursion

$$\Phi_0 = I,\ \ \Phi_l = \sum_{i=0}^{l-1} \Phi_i \bar{P}_{i,l} (I - \bar{P}_{l,l})^{-1},\ \ l \geq 1. \tag{2.129}$$

- Calculate the vector $\boldsymbol{\pi}_0$ as the unique solution of the system of equations

$$\boldsymbol{\pi}_0(I - \bar{P}_{0,0}) = \mathbf{0}, \quad (2.130)$$

$$\boldsymbol{\pi}_0 \sum_{l=0}^{\infty} \Phi_l \mathbf{e} = 1. \quad (2.131)$$

- Calculate vectors $\boldsymbol{\pi}_l,\ l \geq 1$ by formula (2.125).

Remark 2.4 Note that all inverse matrices presented in the algorithm exist and are non-negative. This ensures the numerical stability of the proposed algorithm.

However, the practical implementation of the described algorithm encounters some difficulties. The most significant of them consists in organizing the computation of the matrices G_i from recursion (2.127). The point is that this is an infinite backward (reverse) recursion, and the matrix G_i can be computed from it only if all the matrices $G_m, m > i$ are known. To overcome this difficulty, we exploit the probabilistic meaning of these matrices and the asymptotic properties of an AQTMC. Recall that an AQTMC for large values of the component i_n behaves similarly to an AQTMC with the transition probability matrices $P_{i,i+k-1} = Y_k,\ k \geq 0$. In Sect. 2.4, it was noted that for a QTMC the matrices G_i coincide for all $i,\ i \geq 0$ and are equal to the matrix G, which is the minimal non-negative solution of the nonlinear matrix equation

$$G = \sum_{k=0}^{\infty} Y_k G^k. \quad (2.132)$$

Section 2.4 discussed ways to solve this equation.

It follows that in a computer implementation of the proposed algorithm, one can use recursion (2.127) where the G_i are supposed to be G for all $i \geq \tilde{i}$ where $\tilde{i}$ is a sufficiently large number. We give a more rigorous justification for such a solution of the computation problem of the reverse recursion.

Denote by $R_i = G - \sum\limits_{k=i}^{\infty} P_{i+1,k} G^{k-i}$ a residual in (2.127) where all matrices $G_m, m \geq i$, are replaced by matrix G. The norm of matrix R_i determines the degree of proximity of the solution $\{G_i, G_{i+1}, \ldots\}$ of recursion (2.127) to the set of matrices $\{G, G, \ldots\}$.

Theorem 2.13 *The norm of matrix R_i vanishes as $i \to \infty$.*

Proof The norm $||A||$ of a matrix $A = (a_{i,j})$ is considered to be the value $\max\limits_i \sum\limits_j |a_{i,j}|$. It is easy to see that for non-negative matrices $P_{i,j}$ and the stochastic matrix G the following relation holds:

$$||\sum_{k=k_0}^{\infty} P_{i+1,i+k} G^k|| = ||\sum_{k=k_0}^{\infty} P_{i+1,i+k}||, \; i \geq 0, k_0 \geq 0. \tag{2.133}$$

It was assumed above that the ergodicity conditions of the AQTMC $\xi_n, n \geq 1$ hold. One of these conditions is the uniform convergence of the series $\sum_{k=1}^{\infty} k P_{i+1,i+k}$ for large i. Then the series $\sum_{k=0}^{\infty} P_{i+1,i+k}$ uniformly converges for large i as well. Taking into account the relation (2.133) and Remark 2.1, we conclude that the series $\sum_{k=0}^{\infty} P_{i+1,i+k} G^k$ uniformly converges for large i as well.

From this and (2.101) it follows that the following relations hold:

$$\lim_{i\to\infty} \sum_{k=0}^{\infty} P_{i+1,i+k} G^k = \sum_{k=0}^{\infty} \lim_{i\to\infty} P_{i+1,i+k} G^k = \sum_{k=0}^{\infty} Y_k G^k.$$

These relations imply that the sequence of matrices

$$R_i = G - \sum_{k=i}^{\infty} P_{i+1,k} G^{k-i} = \sum_{k=0}^{\infty} Y_k G^k - \sum_{k=0}^{\infty} P_{i+1,i+k} G^k, \; i \geq 0$$

converges elementwise to the zero matrix. This yields

$$\lim_{i\to\infty} ||R_i|| = 0.$$

□

Thus, the numerical realization of the infinite reverse recursion (2.127) can be performed as follows. We fix a large number $\tilde{i}$, $\tilde{i} \geq i^*$. This number depends on the rate of convergence of the matrices $P_{i,i+k-1}$ to the corresponding matrices $Y_k, k \geq 0$. The number $\tilde{i}$ must be chosen as the minimum of the numbers i for which the norm $||R_i||$ of the residual is less than the prescribed small number ε. We set $G_i = G$ for $i \geq \tilde{i} \geq i^*$ and calculate the matrices $G_i, i = \overline{0, \tilde{i}-1}$ from recursion (2.127).

Note that the value $\tilde{i}$ can also be found by comparing the norm of matrix $G_{\tilde{i}-1} - G$ with ε where $G_{\tilde{i}-1}$ is the matrix calculated in the first step of the described recursion. If this norm is less than ε then we assume that the number $\tilde{i}$ is large enough. Otherwise, we increase the number $\tilde{i}$ and make one step of the reverse recursion again. We repeat this procedure until the norm of matrix $G_{\tilde{i}-1} - G$ becomes less than ε. The fact that we will find a required number $\tilde{i}$ sooner or later follows from the statement of Theorem 2.12.

With the described method of the recursion (2.127) calculations, formula (2.128) can be rewritten as

$$\bar{P}_{i,l} = P_{i,l} + \sum_{n=l+1}^{\infty} P_{i,n} G^{\max\{0,n-\max\{\tilde{i},l\}\}} G_{\min\{\tilde{i},n\}-1} \cdots G_l, \; l \geq i, \; i \geq 0.$$

The described procedure to select the value $\tilde{i}$ can be considered to be a heuristic one. But this procedure is unimprovable in the sense that a more precise computer procedure cannot be constructed in principle if we take the so-called computer *epsilon* as ε, that is the minimum modulo number with which a computer can operate.

Note that at the last step of the algorithm described above, it is required to truncate the infinite series in (2.131). This can be done by stopping the computations in (2.131) after the norm of the next matrix Φ_l becomes smaller than some prescribed small number ε_1. The fact that this happens sooner or later follows from the ergodicity of the considered MC. For the abovementioned multidimensional AQTMC with N boundary states, the described algorithm is simplified since here all matrices $G_i, i \geq N$ are equal to the matrix G calculated from Eq. (2.132).

2.7 Asymptotically Quasi-Toeplitz Continuous-Time Markov Chains

2.7.1 Definition of a Continuous-Time AQTMC

In this subsection, we introduce continuous-time AQTMCs. The investigation of these MCs is based essentially on the results of Sect. 2.6, where discrete-time AQTMCs were studied. In this case, we use the fact that a discrete-time process describing the continuous-time MC transitions at the moments of change of its states (embedded MC, or jump MC) is a discrete-time MC, and the connection between these MCs can be described as follows. Let A be the generator of a continuous-time MC, P be the one-step transition probability matrix of its embedded MC, and T be the diagonal matrix whose diagonal elements coincide with the diagonal elements of the generator A taken with the opposite sign. Then the following relation holds:

$$P = T^{-1}A + I.$$

Thus, let $\xi_t = \{i_t, \mathbf{r}_t\}, t \geq 0$ be a regular irreducible continuous-time MC. We suppose that this MC has the same state space $\mathcal{S}$ as a discrete-time AQTMC considered in Sect. 2.6, namely,

$$\mathcal{S} = \{(i, \mathbf{r}), \mathbf{r} \in \mathcal{R}_i, i = 0, 1, \ldots, i^*; (i, \mathbf{r}), \mathbf{r} \in \mathcal{R}, i > i^*\}.$$

Concerning the numbering of states, we make the same assumptions as in Sect. 2.6: the states $(i, \mathbf{r})$ are renumbered in increasing order of the component i, and for a fixed value i the states $(i, \mathbf{r})$, $\mathbf{r} \in \mathcal{R}_i$, are renumbered in lexicographic order, $i \geq 0$.

Let us represent the generator A of the MC $\xi_t, t \geq 0$ in a block form $A = (A_{i,l})_{i,l \geq 0}$, where $A_{i,l}$ is the matrix of size $K_i \times K_l$ formed by the intensities $a_{(i,\mathbf{r});(l,\boldsymbol{\nu})}$ of the MC transitions from state $(i, \mathbf{r})$, $\mathbf{r} \in \mathcal{R}_i$ into state $(l, \boldsymbol{\nu})$, $\boldsymbol{\nu} \in \mathcal{R}_l$. Note that for $i, l \geq i^*$ the matrices $A_{i,l}$ are square matrices of size K. The diagonal entries of matrix $A_{i,i}$ are defined as $a_{(i,\mathbf{r});(i,\mathbf{r})} = -\sum\limits_{(j,\boldsymbol{\nu}) \in \mathcal{S} \setminus (i,\mathbf{r})} a_{(i,\mathbf{r});(j,\boldsymbol{\nu})}$.

Denote by $T_i, i \geq 0$ a diagonal matrix with diagonal entries $-a_{(i,\mathbf{r});(i,\mathbf{r})}$.

Definition 2.2 A regular irreducible continuous-time MC $\xi_t, t \geq 0$ is called asymptotically quasi-Toeplitz if

1^0. $A_{i,l} = O$ for $l < i - 1$, $i > 0$.
2^0. There exist matrices $Y_k, k \geq 0$ such that

$$Y_k = \lim_{i \to \infty} T_i^{-1} A_{i,i+k-1}, k = 0, 2, 3, \ldots, \tag{2.134}$$

$$Y_1 = \lim_{i \to \infty} T_i^{-1} A_{i,i} + I \tag{2.135}$$

and the matrix $\sum\limits_{k=0}^{\infty} Y_k$ is stochastic.

3^0. The embedded MC $\xi_n, n \geq 1$ is aperiodic.

Remark 2.5 Conditions $[2^0]$ hold automatically if the following conditions are valid:

2^*. There exists a matrix T such that $\lim\limits_{i \to \infty} T_i^{-1} = T$.
3^*. There exist integers $i_1, k_1 \geq 0$ such that matrices $A_{i,i+k}$ are independent of i for $i \geq i_1, k \geq k_1$.
4^*. The limits $\lim\limits_{i \to \infty} T_i^{-1} A_{i,i+k}$, $k = -1, 0, \ldots, k_1 - 1$ exist.

Conditions $[2^*]$–$[4^*]$ are more restrictive than conditions $[2^0]$. We describe them since they are valid for the MCs describing many multi-server queuing systems with retrials (see, for example, the retrial $BMAP/PH/N$ system discussed in Sect. 4.2 below), while they are much simpler to verify than conditions $[2^0]$.

Remark 2.6 If there exists an integer N, $N \geq 1$ such that $A_{i,i+k-1} = \tilde{A}_k, k \geq 0$, for all $i \geq N$ then condition $[2^0]$ in Definition 2.2 necessarily holds. A continuous-time MC of such a type is called a multidimensional continuous-time QTMC with N boundary levels.

2.7.2 *Ergodicity Conditions for a Continuous-Time Asymptotically Quasi-Toeplitz MC*

For analysis of a continuous-time AQTMC, we consider its embedded MC $\xi_n = \{i_n, \mathbf{r}_n\}, \quad n \geq 1$ with state space $\mathcal{S}$ and the matrices of one-step transition probabilities $P_{i,l}, \ i, l \geq 0$, given as follows:

$$P_{i,l} = \begin{cases} O, & l < i-1, \ i > 0; \\ T_i^{-1} A_{i,l}, & l \geq \max\{0, i-1\}, \ l \neq i; \\ T_i^{-1} A_{i,i} + I, & l = i, \ i \geq 0. \end{cases} \quad (2.136)$$

The following statement holds.

Lemma 2.8 *An MC $\xi_n, \ n \geq 1$ with matrices of one-step transition probabilities $P_{i,l}, \ i, l \geq 0$ of form (2.136) belongs to the class of discrete-time AQTMCs.*

Proof We show that the MC $\xi_n, n \geq 1$ satisfies Definition 2.1.

The MC $\xi_n, \ n \geq 1$ is irreducible, that is, all its states are communicating, since the same is true for the MC $\xi_t, \ t \geq 0$, which is aperiodic due to item $[3^0]$ of Definition 2.2.

Conditions (a) and (b) of Definition 2.1 hold due to (2.134)–(2.136).

Thus the MC $\xi_n, \ n \geq 1$, satisfies all the conditions of Definition 2.1 and, hence, it is an AQTMC. □

Denote by $Y(z)$ the PGF of the matrices $Y_k, \ k \geq 0$ given by formulas (2.134) and (2.135),

$$Y(z) = \sum_{k=0}^{\infty} Y_k z^k, \ |z| \leq 1.$$

A sufficient condition for ergodicity of the AQTMC $\xi_t, \ t \geq 0$ is given in terms of this matrix PGF. Just as in the case of a discrete-time AQTMC, we distinguish between cases when the matrix $Y(1)$ is irreducible or reducible.

Theorem 2.14 *Let the matrix $Y(1)$ be irreducible. Suppose that*

(a) the series $\sum_{k=1}^{\infty} k A_{i,i+k-1}\mathbf{e}$ converges for $i = \overline{0, i^*}$;
(b) the series $\sum_{k=1}^{\infty} k A_{i,i+k-1}$ converges for all $i > i^*$, and there exists an integer $j_1 > i^*$ such that these series uniformly converge for $i \geq j_1$;
(c) the sequence $T_i^{-1}, \ i \geq 0$, is upper-bounded.

Then a sufficient condition for the ergodicity of the AQTMC $\xi_t,\ t \geq 0$ is the fulfilment of the inequality

$$(\det(zI - Y(z)))'|_{z=1} > 0. \tag{2.137}$$

Proof First of all, we show that if the conditions of the theorem hold then inequality (2.137) determines a sufficient ergodicity condition for the embedded MC $\xi_n,\ n \geq 1$. To this aim, we use Theorem 2.11.

According to this Theorem, it is sufficient to prove that the series $\sum_{k=1}^{\infty} kP_{i,i+k-1}\mathbf{e}$ converges for $i = \overline{0, i^*}$, the series $\sum_{k=1}^{\infty} kP_{i,i+k-1}$ converges for $i > i^*$ and converges uniformly for large values of i.

The convergence of the series $\sum_{k=1}^{\infty} kP_{i,i+k-1}\mathbf{e}$ follows from (2.136) and condition (a) of the theorem to be proved.

For $i > i^*$, the series $\sum_{k=1}^{\infty} kP_{i,i+k-1}$ can be presented in the form

$$\sum_{k=1}^{\infty} kP_{i,i+k-1} = T_i^{-1} \sum_{k=1}^{\infty} kA_{i,i+k-1} + I. \tag{2.138}$$

The convergence of series (2.138) for $i > i^*$ follows from the convergence of the series $\sum_{k=1}^{\infty} kA_{i,i+k-1}$ for $i > i^*$. The uniform convergence of series (2.138) follows from the uniform convergence of the series $\sum_{k=1}^{\infty} kA_{i,i+k-1}$ and the uniform boundedness of the matrices T_i^{-l} for large values i.

Thus, it is shown that validity of (2.137) is a sufficient condition for ergodicity of the MC $\xi_n, n \geq 1$.

We can write the ergodic (stationary) distribution of the MC $\xi_n,\ n \geq 1$ in the form $(\boldsymbol{\pi}_0, \boldsymbol{\pi}_1, \boldsymbol{\pi}_2, \ldots)$, where the row vector $\boldsymbol{\pi}_i$ consists of the probabilities of states having value i of the denumerable (first) component. It is easy to see that for any constant c, the row vector

$$(\mathbf{p}_0, \mathbf{p}_1, \mathbf{p}_2, \ldots)$$

with

$$\mathbf{p}_i = c\boldsymbol{\pi}_i T_i^{-1},\ i \geq 0$$

satisfies the system of Chapman-Kolmogorov equations (equilibrium equations) for the stationary state distribution of the initial continuous-time MC $\xi_t,\ t \geq 0$. Since this MC is regular and irreducible then according to Foster's theorem, see, e.g, [80], a sufficient condition for its ergodicity is the difference from zero of the constant c having the form

$$c = (\sum_{i=0}^{\infty} \boldsymbol{\pi}_i T_i^{-1} \mathbf{e})^{-1}. \tag{2.139}$$

Taking into account the uniform boundedness of the matrices T_i^{-1} for large values of i, it is easy to see that the series on the right-hand side of (2.139) converges to a positive number. Therefore, the constant defined by (2.139) is finite and positive. Consequently, the row vector $(\boldsymbol{p}_0, \boldsymbol{p}_1, \boldsymbol{p}_2, \ldots)$ with a constant of the form (2.139) gives the stationary state distribution of the MC $\xi_t, t \geq 0$. □

Corollary 2.9 *If the AQTMC satisfies conditions* [2*]–[3*] *in Remark 2.5 then condition (c) of Theorem 2.14 can be omitted and condition (b) reduces to the condition*

(b′) the series $\sum\limits_{k=1}^{\infty} kA_{i,i+k-1}$ *converges for* $i = \overline{i^* + 1, i_1 - 1}$.

Corollary 2.10 *For the AQTMC defined in Remark 2.6, condition (c) of Theorem 2.14 can be omitted and condition (b) reduces to the condition:*

(b″) the series $\sum\limits_{k=1}^{\infty} kA_{i,i+k-1}$ *converges for* $i = \overline{i^* + 1, N}$.

Corollary 2.11 *Inequality (2.137) is equivalent to the inequality*

$$\mathbf{y}Y'(1)\mathbf{e} < 1, \tag{2.140}$$

where the row vector $\mathbf{y}$ *is the unique solution of the system of equations*

$$\mathbf{y}Y(1) = \mathbf{y},\ \mathbf{ye} = 1.$$

Now let the matrix $Y(1)$ be reducible.

Theorem 2.15 *Suppose that the matrix* $Y(1)$ *is reducible and reduced to a normal form of the form (2.119). If conditions (a)–(c) of Theorem 2.14 hold then a sufficient condition for the ergodicity of the AQTMC* $\xi_t,\ t \geq 0$ *is the validity of the inequalities*

$$(\det(zI - Y^{(l)}(z)))'|_{z=1} > 0,\ l = \overline{1, m}. \tag{2.141}$$

The theorem is proved in a similar way to Theorems 2.14 and 2.11.

Corollary 2.12 *Inequalities (2.141) are equivalent to inequalities of the form (2.120).*

Verification of the ergodicity condition for the reducible matrix $Y(1)$ requires preliminary reduction of this matrix to the normal form. This reduction can be done by a consistent permutation of the rows and columns of the matrix. We do not know the formal procedures for such a reduction; therefore, when considering the MCs describing a specific queuing system, certain experience and art are required. The case of a reducible matrix $Y(1)$ often occurs when considering multi-server queueing systems. In this case, the reducible matrix $Y(1)$ quite often has a special form that allows us to consider only part of the matrix PGF $Y(z)$ when checking the ergodicity condition. Usually, this part is a diagonal block of the matrix $Y(z)$ with a much smaller dimension than the entire matrix $Y(z)$. In this case, the results of the following lemma and theorem can be useful.

Lemma 2.9 *Let $Y(1)$ be a reducible stochastic matrix of size K that can be represented as*

$$Y(1) = \begin{pmatrix} Y_{11}(1) & Y_{12}(1) \\ O & Y_{22}(1) \end{pmatrix},$$

where $Y_{11}(1)$ and $Y_{22}(1)$ are square matrices of size L_1 and L_2, respectively, $0 \le L_1 \le K-1,\ L_1+L_2 = K$. Suppose that all the diagonal and subdiagonal elements of matrix $Y_{11}(1)$ are zero.

Let $Y_{22}^{(l)}(1),\ l = \overline{1,m}$ be irreducible stochastic blocks of a normal form $Y_{22}^{\{N\}}(1)$ of matrix $Y_{22}(1)$.

Then these blocks are also irreducible stochastic blocks of the normal form $Y^{\{N\}}(1)$ of matrix $Y(1)$, and matrix $Y^{\{N\}}(1)$ does not have other stochastic blocks.

Proof By a consistent permutation of rows and columns, the matrix $Y(1)$ can be reduced to the form

$$\hat{Y} = \begin{pmatrix} Y_{22}(1) & O \\ \hat{Y}_{12} & \hat{Y}_{11} \end{pmatrix},$$

where all diagonal and off-diagonal elements of the matrix $\hat{Y}_{11}$ are zero.

Let us reduce the block $Y_{22}(1)$ to its normal form $Y_{22}^{\{N\}}(1)$ by a consistent permutation of the rows and columns of matrix $\hat{Y}$. As a result, we obtain the normal form $Y^{\{N\}}(1)$ of matrix $Y(1)$

$$Y^{\{N\}}(1) = \begin{pmatrix} Y_{22}^{\{N\}}(1) & O \\ \tilde{Y}_{12} & \hat{Y}_{11} \end{pmatrix}.$$

All irreducible stochastic diagonal blocks of the matrix $Y^{\{N\}}(1)$ are included in matrix $Y_{22}^{\{N\}}(1)$ since the irreducible diagonal blocks of matrix $\hat{Y}_{11}$ are sub-stochastic matrices of size 1×1; more precisely, each of these blocks is a scalar equal to zero. □

Theorem 2.16 *Suppose that all conditions of Theorem 2.15 hold and the matrix $Y(1)$ satisfies the conditions of Lemma 2.9. Then a sufficient condition for the*

AQTMC $\xi_t,\ t \geq 0$ *to be ergodic is the validity of the inequalities*

$$(\det(zI - Y_{22}^{(l)}(z)))|'_{z=1} > 0,\ l = \overline{1,m}.$$

The proof of the theorem follows from Lemma 2.9 and Theorem 2.15.

Remark 2.7 In the case of a continuous-time QTMC with N boundary levels (see Remark 2.6), the matrix PGF $Y(z)$ is calculated as $Y(z) = zI + T^{-1}\tilde{A}(z)$, where $\tilde{A}(z) = \sum\limits_{k=0}^{\infty} \tilde{A}_k z^k$, and T is a diagonal matrix with diagonal entries that coincide with the modules of the corresponding diagonal entries of the matrix $\tilde{A}_1$. In this case, inequalities (2.140) in the case of an irreducible matrix $Y(1)$ or inequality (2.120) in the case of a reducible matrix $Y(1)$ give not only a sufficient but also a necessary condition for the ergodicity of the corresponding MCs.

2.7.3 Algorithm to Calculate the Stationary State Distribution

The algorithm to calculate the stationary state distribution of a continuous-time AQTMC can be developed on the basis of the corresponding algorithm for a discrete-time AQTMC described in Sect. 2.6. In this case, we use relation (2.136) between the blocks of a continuous-time MC generator and the blocks of the one-step transition probability matrix of the embedded MC, and also the relation

$$\boldsymbol{\pi}_i = c^{-1}\mathbf{p}_i T_i,\ i \geq 0$$

between the vectors $\boldsymbol{\pi}_i$ of the stationary state probabilities of the embedded MC and the vectors $\mathbf{p}_i$ of the continuous-time MC under consideration. Here, the constant c is determined by formula (2.139).

The result is the following procedure for calculating the vectors of stationary probabilities $\mathbf{p}_i,\ i \geq 0$:

- Calculate the matrices G_i from the reverse recursion

$$G_i = (-\sum_{n=i+1}^{\infty} A_{i+1,n}G_{n-1}G_{n-2}\dots G_{i+1})^{-1}A_{i+1,i},\ i \geq 0. \tag{2.142}$$

- Calculate the matrices $\bar{A}_{i,l},\ l \geq i,\ i \geq 0$ by formulas

$$\bar{A}_{i,l} = A_{i,l} + \sum_{n=l+1}^{\infty} A_{i,n}G_{n-1}G_{n-2}\dots G_l,\ \ l \geq i,\ i \geq 0. \tag{2.143}$$

- Calculate the matrices $F_l,\ l \geq 0$ using the recursive relations

$$F_0 = I,\ F_l = \sum_{i=0}^{l-1} F_i \bar{A}_{i,l}(-\bar{A}_{l,l})^{-1},\ l \geq 1. \tag{2.144}$$

- Calculate the vector $\mathbf{p}_0$ as the unique solution of the system of linear algebraic equations

$$\mathbf{p}_0(-\bar{A}_{0,0}) = \mathbf{0},\ \mathbf{p}_0 \sum_{l=0}^{\infty} F_l \mathbf{e} = 1.$$

- Calculate the vectors $\mathbf{p}_l,\ l \geq 1$ by formulas

$$\mathbf{p}_l = \mathbf{p}_0 F_l,\ l \geq 1. \tag{2.145}$$

Note that the inverse matrices in (2.142), (2.144) exist and are non-negative. The problem of finding the terminal condition for recursion (2.142) can be solved in the same way as was done in Sect. 2.6 for a discrete-time MC.

Chapter 3
Queuing Systems with Waiting Space and Correlated Arrivals and Their Application to Evaluation of Network Structure Performance

Queuing systems with waiting space, i.e., systems that have an input buffer so that a customer arriving when all servers are busy is placed in the buffer and served later when one of the servers becomes free, are adequate mathematical models of many real systems including wired and wireless telecommunication systems. This chapter presents the results of analyzing various systems under the assumption that the input flow is a $BMAP$.

3.1 $BMAP/G/1$ Queue

A $BMAP/G/1$ queue is a single-server queuing system with unlimited waiting space. Customers arrive at the system according to a $BMAP$. As was described above, a $BMAP$ is characterized by an underlying process $\nu_t, t \geq 0$, with state space $\{0, 1, \ldots, W\}$ and a matrix generating function $D(z)$. Customer service times are mutually independent and characterized by a distribution function $B(t)$ with LST $\beta(s)$ and finite initial moments $b_k = \int\limits_0^\infty t^k dB(t),\ k = 1, 2.$

We are interested in investigation of the processes i_t (the number of customers in the system at time t) and w_t (the virtual waiting time in the system at time $t,\ t \geq 0$). The subjects of the study are the stationary distributions of these processes and the system performance characteristics calculated by means of these distributions.

The processes i_t and w_t are not Markovian, therefore it is not possible to determine their stationary distributions directly. To solve the problem, we need to apply a Markovization procedure.

For a Markovization, we use a combination of two methods: the embedded-MC method and the extension of the state space by introducing complementary variables.

A. N. Dudin et al., *The Theory of Queuing Systems with Correlated Flows*,
https://doi.org/10.1007/978-3-030-32072-0_3

First, we consider the problem of finding the stationary distribution of a specially constructed embedded MC. And then we describe a technique of transition from the stationary state distribution of this MC to the stationary state distribution of the processes $i_t, t \geq 0$ and $w_t, t \geq 0$ at arbitrary time moments.

3.1.1 Stationary Distribution of an Embedded Markov Chain

Since the distribution of the number of customers arriving in a $BMAP$ depends on the states of the $BMAP$ underlying process $\nu_t,\ t \geq 0$, it is natural to conclude that a Markov process cannot be constructed without taking into account the states of this process. First, we consider a two-dimensional process $\{i_t, \nu_t\},\ t \geq 0$. Since this two-dimensional process is not Markovian, we apply the embedded-MC method for its analysis.

Let $t_k,\ k \geq 1$ be the kth service completion moment, i_{t_k} be the number of customers in the system at time $t_k + 0$, and ν_{t_k} be the state of the underlying process at time t_k. It is easy to see that the two-dimensional process $\xi_k = \{i_{t_k}, \nu_{t_k}\},\ k \geq 1$ is an MC.

We order its one-step transition probabilities

$$P\left\{(i,\nu) \to (l,\nu')\right\} = P\left\{i_{t_{k+1}} = l, \nu_{t_{k+1}} = \nu' | i_{t_k} = i, \nu_{t_k} = \nu\right\},$$

$$i > 0,\ l \geq i-1,\ \nu, \nu' = \overline{0, W}$$

in lexicographical order and combine them into the matrices

$$P_{i,l} = \Omega_{l-i+1} = \int\limits_0^\infty P(l-i+1,t)dB(t),\ i > 0,\ l \geq i-1$$

of size $\bar{W} \times \bar{W}$.

The matrices $\Omega_m,\ m \geq 0$ play an important role in the analysis of queues with $BMAP$. The problem of calculation of the matrices $P(n,t),\ n \geq 0$ was discussed above (see, e.g., formula (2.7) in Sect. 2.1). The generating function of these matrices is defined by formula (2.4). It follows from this formula that the PGF of matrices $\Omega_m, m \geq 0$ is the matrix function

$$\sum_{m=0}^\infty \Omega_m z^m = \beta(-D(z)) = \int\limits_0^\infty e^{D(z)t} dB(t).$$

Comparing the one-step transition probabilities of the MC under consideration with the definition of an $M/G/1$-type MC, we see that this chain belongs to the class of $M/G/1$-type MCs. Thus, the problem of the calculation of its stationary distribution (and its stability condition) can be effectively solved using the results presented in the previous chapter if we give the explicit forms of the matrix PGFs $Y(z)$ and $V(z)$.

The matrix $Y(z)$ is the PGF of the probabilities of the MC's transitions from the states with nonzero value of the denumerable component i. For the MC $\{i_{t_k}, \nu_{t_k}\}$, $k \geq 1$, these transition probabilities are described by the matrices Ω_m, $m \geq 0$. Thus we get

$$Y(z) = \sum_{m=0}^{\infty} \Omega_m z^m = \beta(-D(z)) = \int_0^{\infty} e^{D(z)t} dB(t). \tag{3.1}$$

In this case, of course, the the matrices Y_m of expansion $Y(z) = \sum_{m=0}^{\infty} Y_m z^m$ coincide with the matrices Ω_m, $m \geq 0$.

To find the form of the matrix PGF $V(z)$ of the probabilities of the MC transitions from states with a zero value of the denumerable component, we analyze the behavior of the queuing system after it enters the states with the value $i = 0$, which correspond to the empty queue at a service completion moment. In this case, the system remains empty until the arrival of a new customer or batch of customers. The BMAP underlying process transitions during the waiting time are characterized, as noted in Sect. 2.1, by the matrix $e^{D_0 t} D_k dt$, given that this waiting time is equal to t and the arriving batch consists of k customers. Hence, taking into account that the MC's further transitions are performed in the same way as transitions that start at service completion moments with exactly k customers in the system, $k \geq 1$, and using the total probability formula, we obtain the following expression for the matrices V_m:

$$V_m = \sum_{k=1}^{m+1} \int_0^{\infty} e^{D_0 t} D_k dt Y_{m+1-k} = -D_0^{-1} \sum_{k=1}^{m+1} D_k Y_{m+1-k}. \tag{3.2}$$

Accordingly, the PGF $V(z) = \sum_{m=0}^{\infty} V_m z^m$ has the form

$$V(z) = (-D_0)^{-1} \frac{1}{z} (D(z) - D_0) \beta(-D(z)). \tag{3.3}$$

Substituting the obtained expressions for the matrix PGF in Eq. (2.70) for the vector PGF $\mathbf{\Pi}(z)$ of the stationary distribution of an $M/G/1$-type MC, we obtain the equation for the vector PGF of the stationary distribution of an embedded MC:

$$\mathbf{\Pi}(z)\,(zI - \beta(-D(z))) = \mathbf{\Pi}(0)(-D_0)^{-1} D(z) \beta(-D(z)). \tag{3.4}$$

Below we obtain a condition for the existence of the stationary distribution of the MC $\xi_k,\ k \geq 1$, based on Corollary 2.3. Since the matrix PGF $Y(z)$ has the form (3.1), it is easy to verify that the vector $\mathbf{x}$, which is a solution of Eq. (2.67), is equal to the vector $\boldsymbol{\theta}$. Now, taking into account the formula for calculation of the value

$$\boldsymbol{\theta}\frac{d\int\limits_0^\infty e^{D(z)t}dB(t)}{dz}|_{z=1}\mathbf{e}$$

presented in Sect. 2.1, it is not difficult to verify that the criterion for the existence of the stationary distribution of the embedded MC is the fulfillment of the inequality

$$\rho = \lambda b_1 < 1, \tag{3.5}$$

where λ is the $BMAP$ intensity defined in Sect. 2.1, and $b_1 = \int\limits_0^\infty (1 - B(t))dt$ is the mean service time. Below, this condition is assumed to hold.

To apply the algorithm for calculating the stationary distribution of the embedded MC, which was described above, it is required to specify a way of calculating the values of $Y^{(m)}(1)$ and $V^{(m)}(1)$, which are the mth derivatives of the matrix PGF $Y(z)$ and $V(z)$ at the point $z = 1,\ m \geq 0$. Despite the relatively simple form (3.1), (3.3) of these functions, the problem of finding the matrices $Y^{(m)}(1)$ and $V^{(m)}(1)$, $m \geq 0$ is rather complicated even in relatively simple cases.

Let us present two examples.

Example 3.1 If the input flow is an $MMPP$ given by the matrices $D_1 = \Lambda$, $D_0 = -\Lambda + H$, where $H = \Phi(P - I)$, $D_k = 0,\ k \geq 2$, $\Lambda = \text{diag}\{\lambda_0, \ldots, \lambda_W\}$ is a diagonal matrix with diagonal entries $\lambda_\nu,\ \nu = \overline{0, W}$, $\Phi = \text{diag}\{\varphi_0, \ldots, \varphi_W\}$, P is a stochastic matrix, and the service time has a hyper-Erlang distribution, i.e.,

$$B(t) = \sum_{i=1}^{k} q_i \int_0^t \frac{\gamma_i(\gamma_i\tau)^{h_i-1}}{(h_i - 1)!}e^{-\gamma_i\tau}d\tau,$$

$$q_i \geq 0,\ \sum_{i=1}^{k} q_i = 1,\ \gamma_i > 0,\ h_i \geq 1,\ i = \overline{1,k},\ k \geq 1$$

then

$$\int_0^\infty e^{D(z)t}dB(t) = \sum_{i=1}^{k} q_i(\gamma_i)^{h_i}(\gamma_i I - H + \Lambda - \Lambda z)^{-h_i}.$$

Thus

$$\Omega_l = \sum_{i=1}^{k} q_i \gamma_i^{h_i} D_l^{(i)},\ l \geq 0,$$

where

$$D_l^{(i)} = \sum_{(n_1,\ldots,n_{h_i}) \in N_{h_i}{}^{(l)}} \Gamma_{n_{h_i}}^{(i)} \prod_{r=1}^{h_i-1} \Gamma_{n_{h_i}-n_{h_i-r+1}}^{(i)},$$

$$\Gamma_l^{(i)} = (S_i \Lambda)^l S_i,\ S_i = (\gamma_i I - H + \Lambda)^{-1},$$

$$N_{h_i}^{(l)} = \{(n_1, \ldots, n_{h_i}) : l = n_1 \geq n_2 \geq \ldots \geq n_{h_i} \geq 0\}.$$

In this case, the formulas for calculation of the matrices $Y^{(m)}(1)$, $m = \overline{1,2}$, are as follows:

$$Y^{(1)}(1) = \sum_{i=1}^{k} q_i \left[\sum_{r=1}^{h_i} \gamma_i^{h_i - r} \Lambda \bar{S}_i^{h_i+1+r} - I \right],$$

$$Y^{(2)}(1) = \sum_{i=1}^{k} q_i \left[\sum_{r=1}^{h_i} \sum_{l=0}^{h_i-r} \gamma_i^{r+l-1} \Lambda \bar{S}_i^{l+1} \Lambda \bar{S}_i^r \right],$$

where

$$\bar{S}_i = (\gamma_i I - H)^{-1}.$$

Example 3.2 If the service time is exponentially distributed with parameter γ and the input flow is a $BMMPP$ with a geometrical distribution of the batch size with the parameter q_ν during the sojourn time of the underlying process ν_t in state ν, $\nu = \overline{0, W}$ then

$$D_0 = -\Lambda + H,\ D(z) = D_0 + \Lambda \theta(z),$$

where

$$\theta(z) = \text{diag}\{\theta_0(z), \ldots, \theta_W(z)\},$$

$$\theta_\nu(z) = z \frac{1 - q_\nu}{1 - q_\nu z},\ 0 < q_\nu < 1,\ \nu = \overline{0, W}.$$

In addition,

$$\int_0^\infty e^{D(z)t} dB(t) = \gamma(\gamma I - H + \Lambda - \Lambda\theta(z))^{-1}$$

and

$$\Omega_0 = \gamma S,\ S = (\gamma I - H + \Lambda)^{-1},$$

$$\Omega_l = \gamma((S(\Lambda + Q\tilde{S}^{-1}))^l - Q(S(\Lambda + Q\tilde{S}^{-1}))^{l-1}),\ l \geq 1,$$

$$\tilde{S} = (\gamma I - H)^{-1},\ Q = \text{diag}\{q_0, \ldots, q_W\}.$$

In this case the formulas for calculation of the matrices $Y^{(m)}(1)$, $m = \overline{1,2}$, are as follows:

$$Y^{(1)}(1) = \tilde{Q}\Lambda\tilde{S} - I,$$

$$Y^{(2)}(1) = \tilde{Q}\Lambda\tilde{S}\tilde{Q}\Lambda\tilde{S} + \tilde{Q}^2\Lambda\tilde{S}$$

where $\tilde{Q} = \text{diag}\left\{\frac{1}{1-q_0}, \ldots, \frac{1}{1-q_W}\right\}$, $\tilde{S} = (\gamma I - H)^{-1}$.

In the case of an arbitrary $BMAP$ and arbitrary service time distribution function $B(t)$, the values $Y^{(m)}(1)$, $m \geq 1$ can be calculated by means of (2.7) as follows:

$$\Omega_m = \int_0^\infty \sum_{j=0}^\infty \frac{(\tilde{\theta}t)^j}{j!} e^{-\tilde{\theta}t} K_m^{(j)} dB(t),\ m \geq 0,$$

$$Y^{(m)}(1) = \sum_{l=m}^\infty l(l-1)\ldots(l-m+1)\Omega_l,\ m \geq 1.$$

The matrices $V^{(m)}(1)$ are derived via $Y^{(m)}(1)$, $m \geq 0$ based on relations (3.3).

To illustrate the implementation of the above-described algorithm for calculation of the stationary distribution of a QTMC based on the PGF method, we consider a numerical example of calculating the stationary distribution of the queuing system under consideration.

Suppose the underlying process ν_t, $t \geq 0$ of the $BMAP$ has state space $\{0, 1, 2, 3\}$, i.e., $W = 3$, and its behavior is described by the matrices D_0, D_1,

D_2 of the form

$$D_0 = \begin{pmatrix} -1.45 & 0.2 & 0.15 & 0.1 \\ 0.2 & -2.6 & 0.1 & 0.3 \\ 0.2 & 0.1 & -3.7 & 0.4 \\ 0.1 & 0.05 & 0.15 & -4.3 \end{pmatrix},$$

$$D_1 = D_2 = \text{diag}\{0.5;\ 1;\ 1.5;\ 2\}.$$

The sojourn times of the underlying process $\nu_t,\ t \geq 0$ in the states $\{0, 1, 2, 3\}$ are exponentially distributed with parameters $\{1.45;\ 2.6;\ 3.7;\ 4.3\}$, respectively. At the moments of the underlying process jump either no arrivals occur (and transition to another state happens) or one or two arrivals occur with equal probability and the process $\nu_t,\ t \geq 0$ remains in its state until the next jump.

An infinitesimal generator of the process $\nu_t,\ t \geq 0$ has the form

$$D(1) = D_0 + D_1 + D_2 = \begin{pmatrix} -0.45 & 0.2 & 0.15 & 0.1 \\ 0.2 & -0.6 & 0.1 & 0.3 \\ 0.2 & 0.1 & -0.7 & 0.4 \\ 0.1 & 0.05 & 0.15 & -0.3 \end{pmatrix},$$

and the row vector $\boldsymbol{\theta}$ of the stationary distribution of this process is as follows:

$$\boldsymbol{\theta} = (0.238704;\ 0.144928;\ 0.167945;\ 0.448423).$$

The intensity λ of the $BMAP$ is $\lambda \approx 4.23913$, that is, approximately 4.239 customers arrive at the queue per time unit.

Suppose that the customer service time has an Erlang distribution with parameters (3, 15), that is,

$$B(t) = \int_0^t \frac{15(15\tau)^2}{2!} e^{-15\tau} d\tau,\ \beta(s) = \left(\frac{15}{15+s}\right)^3,\ b_1 = \frac{3}{15} = 0.2.$$

The matrix LST $\beta(-D(z))$ has the form

$$\beta(-D(z)) = 15^3(15I - D_0 - D_1 z - D_2 z^2)^{-3}.$$

As already noted above, for the MC under consideration the condition for the existence of the stationary distribution has the form (3.5) and there is no need to perform step 1 of the algorithm. Here, the system load ρ is 0.8478, therefore the MC has a stationary distribution.

Let's describe the algorithm steps that are implemented to calculate the stationary distribution.

Equation (2.74) has three simple roots in the unit circle of the complex plane:

$$z_1 = 0.737125,\ z_2 = 0.787756,\ z_3 = 0.063899$$

and a simple root $z = 1$.

Matrix $A = Y(1) - I$ has the form

$$A = \begin{pmatrix} -0.082987 & 0.035370 & 0.026676 & 0.020941 \\ 0.036077 & -0.109504 & 0.018607 & 0.054820 \\ 0.035864 & 0.018279 & -0.125627 & 0.071493 \\ 0.019084 & 0.009720 & 0.026836 & -0.055641 \end{pmatrix}.$$

Matrix $\tilde{\mathbf{A}}$ has the form

$$\tilde{\mathbf{A}} = \begin{pmatrix} -0.669415 & 0.035370 & 0.026676 & 0.020941 \\ -0.381314 & -0.109504 & 0.018607 & 0.054820 \\ -0.103749 & 0.018270 & -0.125627 & 0.071493 \\ 0.179084 & 0.009720 & 0.026836 & -0.055641 \end{pmatrix}.$$

Matrix $H(1) = Y(1) - V(1)$ has the form

$$\begin{pmatrix} 0.263048 & -0.091349 & -0.077829 & -0.093871 \\ -0.048899 & 0.191608 & -0.037651 & -0.105058 \\ -0.034798 & -0.024618 & 0.152498 & -0.093082 \\ -0.017330 & -0.011891 & -0.026394 & 0.055615 \end{pmatrix}.$$

Matrix $H(1)\tilde{I} + H^{(1)}(1)\mathbf{e}\hat{\mathbf{e}}$ has the form

$$\begin{pmatrix} -1.663108 & -0.091349 & -0.077829 & -0.093871 \\ -1.562058 & 0.191608 & -0.037651 & -0.105058 \\ -1.499574 & -0.024618 & 0.152498 & -0.093082 \\ -1.468385 & -0.011891 & -0.026394 & 0.055615 \end{pmatrix}.$$

Here $\tilde{I} = \text{diag}\{0, 1, \ldots, 1\}$, $\hat{\mathbf{e}} = (1, 0, \ldots, 0)$.

Matrix S has the form

$$\begin{pmatrix} 2.956759 & 3.557284 & 4.301839 & 11.832143 \\ 3.551012 & 0.462866 & 2.630622 & 7.060740 \\ 2.576184 & 1.617995 & 0.658583 & 5.082831 \\ 2.016273 & 1.248130 & 1.425717 & 2.820937 \end{pmatrix}.$$

The matrix of the system of linear algebraic equations (2.76), (2.77) has the form

$$\begin{pmatrix} 22.648025 & -0.000001 & 0.000026 & 0.000000 \\ 13.705240 & -0.000015 & -0.000037 & -0.000000 \\ 9.935593 & 0.000011 & -0.000003 & -0.000001 \\ 7.511057 & 0.000006 & -0.000002 & 0.000001 \end{pmatrix}.$$

The vector $\boldsymbol{\Pi}(0)$, which is the solution of system (2.76), (2.77), has the form

$$\boldsymbol{\Pi}(0) = (0.0239881;\ 0.0150405;\ 0.0113508;\ 0.0183472).$$

The vectors $\boldsymbol{\pi}_l$ calculated by (2.80) have the form

$$\boldsymbol{\pi}_1 = (0.0205056;\ 0.0150091;\ 0.0133608;\ 0.0244394),$$

$$\boldsymbol{\pi}_2 = (0.0129573;\ 0.0110476;\ 0.0121506;\ 0.0256954),$$

$$\boldsymbol{\pi}_3 = (0.010361;\ 0.00883173;\ 0.0110887;\ 0.0259778), \ldots$$

As the system load $\rho = 0.8478$ is rather large, the tail of the probability distribution $\boldsymbol{\pi}_l,\ l \geq 0$ is heavy, that is the specific gravity of probabilities for large numbers l is quite large. Thus, the sum of all entries of vectors $\boldsymbol{\pi}_l,\ l = \overline{0, 14}$, is 0.668468, i.e., the number of customers in the system exceeds 14 with probability 0.331532. In this situation, the calculation of the factorial moments directly by their definition $\boldsymbol{\Pi}^{(m)} = \sum\limits_{i=m}^{\infty} \frac{i!}{(i-m)!}\boldsymbol{\pi}_i,\ m \geq 1$ is ineffective. To calculate them, we have to use formula (2.81). Aiming to calculate the factorial moments $\boldsymbol{\Pi}^{(m)}$ for $m = \overline{0, 3}$, we successively get

$$\boldsymbol{\Pi}^{(0)} = \boldsymbol{\Pi}(1) = (0.190571;\ 0.133559;\ 0.176392;\ 0.499478),$$

$$\boldsymbol{\Pi}^{(1)} = (2.34708;\ 1.56858;\ 2.3965;\ 7.76206),$$

$$\boldsymbol{\Pi}^{(2)} = (67.9937;\ 44.6851;\ 69.0566;\ 229.087),$$

$$\boldsymbol{\Pi}^{(3)} = (2994.5;\ 1963.99;\ 3034.92;\ 10103.2).$$

Note that if we change the value 15 of the Erlang distribution intensity to 25 then the load ρ decreases to 0.5087, and the sum of all entries of the vectors $\boldsymbol{\pi}_l$, $l = \overline{0, 14}$, becomes equal to 0.9979135, that is, the tail of the distribution of the number of customers in the system becomes light. In this case, the values of the factorial moments calculated directly by their definition in terms of the stationary state probability vectors are very close to those calculated by formula (2.81).

3.1.2 The Stationary Distribution of the System at an Arbitrary Time

In the previous subsection, we investigated the embedded MC $\xi_k = \left\{i_{t_k}, \nu_{t_k}\right\}, k \geq 1$ by means of vector PGFs. But, as was already noted above, this MC has an auxiliary character. The process $\zeta_t = \{i_t, \nu_t\}, i \geq 0$ is more interesting from the practical point of view. It is easy to see that in contrast to the embedded MC $\{i_{t_k}, \nu_{t_k}\}, k \geq 1$, the process $\{i_t, \nu_t\}, t \geq 0$ is not Markovian. At the same time, it is intuitively clear that the distribution of the MC $\{i_{t_k}, \nu_{t_k}\}$ keeps the information about the distribution of the process $\{i_t, \nu_t\}$. An approach based on embedded Markov renewal processes (MRP)[22] allows us to use this information to find the stationary distribution of the process $\{i_t, \nu_t\}$.

It is natural to consider the process $\{i_{t_k}, \nu_{t_k}, t_k\}, k \geq 1$ as an embedded MRP for our process $\zeta_t, t \geq 0$. The semi-Markovian kernel $Q(t)$ of the process $\{i_{t_k}, \nu_{t_k}, t_k\}$, $k \geq 1$, has a block structure of the form

$$Q(t) = \begin{pmatrix} \tilde{V}_0(t) & \tilde{V}_1(t) & \tilde{V}_2(t) & \cdots \\ \tilde{Y}_0(t) & \tilde{Y}_1(t) & \tilde{Y}_2(t) & \cdots \\ O & \tilde{Y}_0(t) & \tilde{Y}_1(t) & \cdots \\ \vdots & \vdots & \vdots & \ddots \end{pmatrix},$$

where blocks $\tilde{V}_l(t), \tilde{Y}_l(t), l \geq 0$ of order $(W+1) \times (W+1)$ are defined as follows:

$$\tilde{Y}_l(t) = \int_0^t P(l, u)dB(u),$$

$$\tilde{V}_l(t) = \sum_{k=1}^{l+1} \int_0^t e^{D_0(t-u)} D_k \tilde{Y}_{l+1-k}(u)du.$$

Denote by $H(t)$ the matrix renewal function of the process $\{i_{t_k}, \nu_{t_k}, t_k\}, k \geq 1$. The infinite-size matrix $H(t)$ has the block structure $H(t) = \left(H_{i,j}(t)\right)_{i,j\geq 0}$, where block $H_{i,j}(t)$ has the form $H_{i,j}(t) = \left(h_{i,j}^{\nu,r}(t)\right)_{\nu,r=\overline{0,W}}$. The entry $h_{i,j}^{\nu,r}(t)$ of matrix $H(t)$ is the expectation of the number of visits to the state (j, r) by the MC $\{i_{t_k}, \nu_{t_k}\}$ within the time interval $(0, t)$ given that the MC was in state (i, ν) at time 0. It can be shown that the value $dh_{i,j}^{\nu,r}(u)$ is an element of a conditional probability that the MC $\{i_{t_k}, \nu_{t_k}\}$ visits the state (j, r) in the interval $(u, u+du)$. Let us fix a state (i_0, ν_0) of the process $\zeta_t,\ t \geq 0$. For convenience of the following calculations, we denote by $\mathbf{h}_j(t),\ j \geq 0$ the vector that is the ν_0th row of the matrix $H_{i_0,j}(t)$, i.e., $\mathbf{h}_j(t) = \left(h_{i_0,j}^{\nu_0,0}(t), \ldots, h_{i_0,j}^{\nu_0,W}(t)\right)$.

Let $p(i, \nu, t)$, $i \geq 0$, $\nu = \overline{0, W}$ be the state probability distribution of the process $\{i_t, \nu_t\}$, $t \geq 0$ at time t. Introduce the vectors $\mathbf{p}(i, t) = (p(i, 0, t), \dots, p(i, W, t))$, $i \geq 0$. Without loss of generality, we assume that the time moment $t = 0$ coincides with a service completion moment, and at that moment the MC $\{i_{t_k}, \nu_{t_k}\}$ is in state (i_0, ν_0). Using the probabilistic interpretation of the values $dh_{i_0,j}^{\nu_0,r}(t)$ and the total probability formula, we obtain the following expressions for the vectors $\mathbf{p}(i, t)$:

$$\mathbf{p}(0, t) = \int_0^t d\mathbf{h}_0(u) e^{D_0(t-u)},$$

$$\mathbf{p}(i, t) = \int_0^t d\mathbf{h}_0(u) \int_0^{t-u} e^{D_0 v} \sum_{k=1}^{i} D_k dv P(i-k, t-u-v)(1 - B(t-u-v))$$

$$+ \sum_{k=1}^{i} \int_0^t d\mathbf{h}_k(u) P(i-k, t-u)(1 - B(t-u)), \ i > 0. \tag{3.6}$$

Next, we need the notion of the *fundamental mean* of the MRP. The value τ of the fundamental mean is equal to the expectation of the length of the interval between neighboring renewal points. It can be calculated as the scalar product of the invariant vector of the matrix $Q(\infty)$ and the column $\int_0^\infty t dQ(t)\mathbf{e}$ of the expectations of the sums of the row elements of the matrix $Q(t)$. For the system under consideration, the value τ is defined as the expectation of the length of the interval between neighboring service completion moments. Thus, the following intuitive expression is obtained:

$$\tau = b_1 + \boldsymbol{\pi}_0(-D_0)^{-1}\mathbf{e}. \tag{3.7}$$

The first term is the mean service time and the second one defines the mean idle time after an arbitrary service completion. Note that when the system is operating in a stationary regime, the value τ is equal to the inverse of the $BMAP$ intensity λ, that is,

$$\tau = \lambda^{-1}. \tag{3.8}$$

Now we find the stationary distribution of the process $\{i_t, \nu_t\}$, $t \geq 0$. Note that the stability condition is given by (3.5). To find the form of the steady state probabilities, we pass (3.6) to the limit as $t \to \infty$ and then we can apply to the right-hand sides of (3.6) the key Markov renewal theorem [22], which has in this case the form

$$\lim_{t\to\infty}\sum_{k=0}^{\infty}\int_0^t d\mathbf{h}_k(u)G_i(k,t-u)=\frac{1}{\tau}\sum_{k=0}^{\infty}\boldsymbol{\pi}_k\int_0^{\infty}G_i(k,u)du,\ i\geq 0. \tag{3.9}$$

Here $G_i(k,t-u)$ denotes the integrands of the right-hand sides of relations (3.6), that is,

$$G_0(0,t-u)=e^{D_0 t-u},$$

$$G_i(0,t-u)=\int_0^{t-u}e^{D_0 v}\sum_{k=1}^{i}D_k dv P(i-k,t-u-v)(1-B(t-u-v)),$$

$$G_i(k,t-u)=P(i-k,t-u)(1-B(t-u)),\ 1\leq k\leq i,\ i\geq 1,$$

$$G_i(k,u)=O,\ k>i\geq 0,\ u\geq 0.$$

After passing to the limit in (3.6), we use formula (3.9) and obtain the expressions for the vectors $\mathbf{p}(i)=\lim_{t\to\infty}\mathbf{p}(i,t),\ i\geq 0$, of the stationary distribution of the process $\{i_t,\nu_t\},\ t\geq 0$ as follows:

$$\mathbf{p}(0)=\lambda\boldsymbol{\pi}_0\int_0^{\infty}e^{D_0 u}du=\lambda\boldsymbol{\pi}_0(-D_0)^{-1},$$

$$\mathbf{p}(i)=\lambda\boldsymbol{\pi}_0\int_0^{\infty}e^{D_0 v}\sum_{k=1}^{i}D_k dv\int_0^{\infty}P(i-k,u)(1-B(u))du$$

$$+\lambda\sum_{k=1}^{i}\boldsymbol{\pi}_k\int_0^{\infty}P(i-k,u)(1-B(u))du$$

$$=\lambda\sum_{k=1}^{i}(\boldsymbol{\pi}_0(-D_0)^{-1}D_k+\boldsymbol{\pi}_k)\tilde{\Omega}_{i-k},\ i>0, \tag{3.10}$$

where $\tilde{\Omega}_m=\int_0^{\infty}P(m,u)(1-B(u))du,\ m\geq 0.$

Note that the matrices $\tilde{\Omega}_m$ can be easily calculated recursively through the matrices Ω_m, as follows:

$$\tilde{\Omega}_0 = (\Omega_0 - I)D_0^{-1},$$

$$\tilde{\Omega}_m = (\Omega_m - \sum_{k=0}^{m-1} \tilde{\Omega}_k D_{m-k})D_0^{-1}, \ m \geq 1.$$

Multiplying Eq. (3.10) by the corresponding powers of z, summing up and taking into account the equations

$$\int_0^\infty e^{D(z)t}(1 - B(t))dt = (-D(z))^{-1}(I - \beta(-D(z)))$$

and (3.4), after algebraic transformations we conclude that the following statement holds.

Theorem 3.1 *The vector PGF* $\mathbf{p}(z) = \sum_{i=0}^{\infty} \mathbf{p}(i)z^i$ *of the stationary distribution* $\mathbf{p}(i)$, $i \geq 0$ *of the system states at an arbitrary moment is expressed in terms of the PGF* $\mathbf{\Pi}(z)$ *of the stationary distribution of the system states at service completion moments as follows:*

$$\mathbf{p}(z)D(z) = \lambda(z - 1)\mathbf{\Pi}(z). \tag{3.11}$$

Expanding (3.11) in the Maclaurin series and equating the coefficients for the same powers of z, we obtain recursive formulas that allow us to calculate sequentially a predetermined number of the vectors $\mathbf{p}(i)$:

$$\mathbf{p}(0) = -\lambda\boldsymbol{\pi}_0 D_0^{-1}, \tag{3.12}$$

$$\mathbf{p}(i + 1) = \left[\sum_{j=0}^{i} \mathbf{p}(j)D_{i+1-j} - \lambda(\boldsymbol{\pi}_i - \boldsymbol{\pi}_{i+1})\right](-D_0)^{-1}, \ i \geq 0. \tag{3.13}$$

From (3.7), (3.8), and (3.12) we get

$$\mathbf{p}(0)\mathbf{e} = 1 - \rho. \tag{3.14}$$

Thus, the probability that a $BMAP/G/1$-type queuing system is empty at an arbitrary time equals $1 - \rho$, which is the same as for an $M/G/1$ queue.

The factorial moments $\mathbf{P}^{(m)}$, $m \geq 0$ of the distribution $\mathbf{p}(i)$, $i \geq 0$ are calculated similarly to the factorial moments $\mathbf{\Pi}^{(m)}$. The moment $\mathbf{P}^{(0)}$ of order 0 coincides with the vector $\boldsymbol{\theta}$ of stationary probabilities of the $BMAP$ underlying process calculated as a solution of the system of Eq. (2.9).

The following moments $\mathbf{P}^{(m)},\ m \geq 1$ are recursively calculated as

$$\mathbf{P}^{(m)} = \left[\left(m\lambda\boldsymbol{\Pi}^{(m-1)} - \sum_{l=0}^{m-1} C_m^l \mathbf{P}^{(l)} D^{(m-l)}\right)\tilde{I}\right.$$

$$\left. + \left(\lambda\boldsymbol{\Pi}^{(m)}\mathbf{e} - \frac{1}{m+1}\sum_{l=0}^{m-1} C_{m+1}^l \mathbf{P}^{(l)} D^{(m+1-l)}\mathbf{e}\right)\hat{\mathbf{e}}\right]\hat{A}^{-1}, \tag{3.15}$$

where

$$\hat{A} = D(1)\tilde{I} + D^{(1)}\mathbf{e}\hat{\mathbf{e}}.$$

Let us continue our consideration of the numerical example described in the previous subsection. Having calculated the values of the stationary probabilities $\boldsymbol{\pi}_i,\ i \geq 0$ and the factorial moments $\boldsymbol{\Pi}^{(m)}$ of the embedded MC, using formulas (3.12)–(3.15) we can easily get the probabilities $\mathbf{p}(i)$, $i \geq 0$ and factorial moments $\mathbf{P}^{(m)}$.

Thus, the values of vectors $\mathbf{p}(i)$, $i = \overline{0, 2}$ are as follows:

$$\mathbf{p}(0) = (0.0786320; 0.0317218; 0.0180150; 0.0238051),$$

$$\mathbf{p}(1) = (0.0219269; 0.0146647; 0.0116901; 0.0196986),$$

$$\mathbf{p}(2) = (0.0179116; 0.0137114; 0.0127269; 0.0240296).$$

The values of the factorial moments $\mathbf{P}^{(m)}$, $m = \overline{0, 3}$ are as follows:

$$\mathbf{P}^{(0)} = (0.238704; 0.144928; 0.167945; 0.448423),$$

$$\mathbf{P}^{(1)} = (2.10506; 1.41165; 2.13705; 6.87336),$$

$$\mathbf{P}^{(2)} = (60.2861; 39.6502; 61.2431; 202.954),$$

$$\mathbf{P}^{(3)} = (2653.26; 1740.33; 2689.32; 8951.38).$$

The probability p_0 that the system is empty at an arbitrary time is calculated as $p_0 = \mathbf{p}(0)\mathbf{e} = 1 - \rho = 0.1521739$.

The mean number L of customers in the system is calculated as follows:

$$L = \mathbf{P}^{(1)}\mathbf{e} = 12.52712.$$

Note that when the $BMAP$ is approximated by a stationary Poisson input with the same intensity $\lambda = 4.239$, the average number $\hat{L}$ of customers in the corresponding $M/G/1$ system would be calculated as

$$\hat{L} = \rho + \frac{\lambda^2 b_2}{2(1-\rho)} = 3.99613$$

where $b_2 = \int_0^\infty t^2 dB(t)$.

Comparing the value of L calculated exactly and its approximate value $\hat{L}$ we see that the approximation error exceeds 300%.

It is also interesting to note that if we use a more inaccurate approximation of the $BMAP/G/1$ queue by an $M/M/1$ queue, that is, we additionally assume that the service time has not an Erlang distribution but an exponential distribution with the same mean, we get an approximate value $\tilde{L}$ as

$$\tilde{L} = \frac{\rho}{1-\rho} = 5.5703.$$

That is, the approximate formula gives an error of about 2.24 times. The poorer approximation gives a lesser error than the initial approximation. This happens, apparently, due to the mutual offset of two errors in the approximation.

3.1.3 The Virtual and Real Waiting Time Distributions

Recall that w_t is the virtual waiting time at the moment t, $t \geq 0$. Let $\mathbf{W}(x)$ be the row vector whose νth entry denotes the stationary probability that the $BMAP$ underlying process is in state ν, $\nu = \overline{0, W}$ at an arbitrary time and the virtual waiting time does not exceed x:

$$\mathbf{W}(x) = \lim_{t\to\infty} (P\{w_t < x, \nu_t = 0\}, \ldots, P\{w_t < x, \nu_t = W\}).$$

Denote by $\mathbf{w}(s) = \int\limits_0^\infty e^{-sx} d\mathbf{W}(x)$ the row vector consisting of the LSTs of the entries of the vector $\mathbf{W}(x)$.

Theorem 3.2 *The vector LST* $\mathbf{w}(s)$ *is defined as follows:*

$$\mathbf{w}(s)\,(sI + D(\beta(s))) = s\mathbf{p}(0), \;\; Re\ s > 0, \tag{3.16}$$

where $D(\beta(s)) = \sum\limits_{k=0}^{\infty} D_k(\beta(s))^k$, $\beta(s) = \int\limits_0^\infty e^{-st} dB(t)$.

Proof It is obvious that $\mathbf{W}(x) = \mathbf{W}(+0) + \mathbf{T}_0(x) + \mathbf{T}_1(x)$ where $\mathbf{W}(+0)$ is the vector of probabilities that a virtual customer arrives at a moment when the system is empty, $\mathbf{T}_0(x)$ is the vector of probabilities that such a customer arrives during the first service time of a customer in a busy period and will wait no longer than x, and $\mathbf{T}_1(x)$ is the vector of probabilities that a virtual customer arrives during the service time of the second, third, or other customer in a busy period and will wait no longer than time x.

By analyzing the system behavior and using our experience of proving the theorem in the previous subsection, it is not too difficult to obtain the following expression for the vector function $\mathbf{T}_0(x)$:

$$\mathbf{T}_0(x) = \lambda\boldsymbol{\pi}_0 \sum_{i=0}^{\infty} \int_0^{\infty} dt \sum_{k=1}^{\infty} \int_0^{t} e^{D_0 v} D_k P(i, t - v) dv \tag{3.17}$$

$$\times \int_0^{x} dB(t + u - v) B^{(i+k-1)}(x - u).$$

Here $B^{(m)}(y)$ is the order m convolution of the distribution function $B(y)$, which is the distribution function of the sum of m independent random variables each having the same distribution function $B(y)$.

In deriving formula (3.17), the variable t is associated with an arbitrary moment of time (the moment of a virtual customer arrival), v is the preceding moment when a group of k customers arrived at the empty system and one of these customers started its service, and u is this customer's residual service time at time t. The time countdown starts from the moment of the end of the previous busy period (the time when the system was empty at the service completion moment).

Changing the order of integration over t and v, we rewrite formula (3.17) in the form

$$\mathbf{T}_0(x) = \mathbf{p}(0) \sum_{i=0}^{\infty} \sum_{k=1}^{\infty} \int_0^{\infty} dy D_k P(i, y) \int_0^{x} dB(y + u) B^{(i+k-1)}(x - u). \tag{3.18}$$

Similarly, we write out the expression for the vector function $\mathbf{T}_1(x)$:

$$\mathbf{T}_1(x) = \lambda \sum_{i=1}^{\infty} \sum_{k=1}^{i} \boldsymbol{\pi}_k \int_0^{\infty} dt P(i - k, t) \int_0^{x} dB(t + u) B^{(i-1)}(x - u). \tag{3.19}$$

Then we pass from the vectors $\mathbf{T}_m(x)$ to their LSTs:

$$\mathbf{T}_m^*(s) = \int_0^\infty e^{-sx} d\mathbf{T}_m(x), \ m = 0, 1.$$

As a result, we obtain formulas

$$\mathbf{T}_0^*(s) = \mathbf{p}(0) \sum_{i=0}^\infty \sum_{k=1}^\infty \int_0^\infty \int_0^\infty e^{-su} D_k P(i, t) dB(t + u) dt \beta^{i+k-1}(s), \quad (3.20)$$

$$\mathbf{T}_1^*(s) = \lambda \sum_{i=1}^\infty \sum_{k=1}^i \boldsymbol{\pi}_k \int_0^\infty \int_0^\infty e^{-su} P(i - k, t) dB(t + u) dt \beta^{i-1}(s).$$

Multiplying relations (3.20) by the matrix $(sI + D(\beta(s)))$ from the right, we get

$$T_0^*(s)\,(sI + D(\beta(s))) = sT_0^*(s)$$

$$+\mathbf{p}(0) \sum_{i=0}^\infty \sum_{k=1}^\infty \int_0^\infty \int_0^\infty e^{-su} D_k P(i, y) dB(y + u) dy (\beta(s))^{i+k-1} \sum_{r=0}^\infty D_r (\beta(s))^r$$

$$= sT_0^*(s) + \mathbf{p}(0) \sum_{k=1}^\infty D_k \mathcal{A}_k(s),$$

where $\mathcal{A}_k(s)$ denotes

$$\mathcal{A}_k(s) = \int_0^\infty \int_0^\infty e^{-su} \sum_{i=0}^\infty P(i, y) dB(y + u) dy (\beta(s))^{i+k-1} \sum_{r=0}^\infty D_r (\beta(s))^r.$$

Making the change of the summation variable $j = i + r$ in the expression above, we obtain

$$\mathcal{A}_k(s) = \sum_{i=0}^\infty \sum_{r=0}^\infty \int_0^\infty \int_0^\infty P(i, y) D_r e^{-su} dB(y + u) dy (\beta(s))^{i+k+r-1}$$

$$= \sum_{j=0}^\infty \sum_{r=0}^j \int_0^\infty \int_0^\infty P(j - r, y) D_r e^{-su} dB(y + u) dy (\beta(s))^{j+k-1}.$$

Taking into account the matrix differential Eq. (2.2) written in the form

$$P'(i,y) = \sum_{r=0}^{i} P(i-r,y)D_r$$

yields

$$\mathcal{A}_k(s) = \sum_{i=0}^{\infty} \int_0^{\infty} \int_0^{\infty} P'(i,y)e^{-su} dB(y+u)dy(\beta(s))^{i+k-1}.$$

This implies the fulfillment of the formula

$$T_0^*(s)D(\beta(s)) = \mathbf{p}(0) \sum_{k=1}^{\infty} D_k \int_0^{\infty} \sum_{i=0}^{\infty} \int_0^{\infty} P'(i,y)e^{-su} dB(y+u)dy(\beta(s))^{i+k-1}.$$

Let us simplify this formula. The inner integral on the right-hand side is calculated by parts

$$\int_0^{\infty}\int_0^{\infty} P'(i,y)e^{-su} dB(y+u)dy = \begin{bmatrix} v = y+u \\ u = v-y \end{bmatrix}$$

$$= \int_0^{\infty} e^{-sv} \int_0^{v} P'(i,y)e^{sy}\, dy\, dB(v) = \begin{bmatrix} u = e^{sy}, & du = se^{sy} \\ dv = P'(i,y)dy, & v = P(i,y) \end{bmatrix}$$

$$= \int_0^{\infty} e^{-sv} \left(\left(P(i,y)e^{sy}\right)\Big|_0^v - s\int_0^{v} P(i,y)e^{sy}\, dy \right) dB(v)$$

$$= \int_0^{\infty} \left(P(i,v) - P(i,0)e^{-sv} - e^{-sv}s \int_0^{v} P(i,y)e^{sy}\, dy \right) dB(v).$$

Making the change of integration variables in (3.20), we can verify that

$$sT_0^*(s) = s\mathbf{p}(0) \sum_{i=0}^{\infty}\sum_{k=1}^{\infty} D_k \int_0^{\infty} e^{-sv} \int_0^{v} P(i,y)e^{sy}\, dy\, dB(v)(\beta(s))^{i+k-1}.$$

As a result of our calculations, we obtain the formula

$$T_0^*(s)\,(sI + D(\beta(s)))$$

$$= \mathbf{p}(0)\sum_{k=1}^{\infty} D_k \int_0^{\infty} \sum_{i=0}^{\infty} \left(P(i,v) - P(i,0)e^{-sv}\right) dB(v)(\beta(s))^{i+k-1}.$$

Taking into account the denotation $Y_l = \int\limits_0^{\infty} P(l,v)dB(v)$ and the obvious formula

$$P(i,0) = \begin{cases} I, & i = 0, \\ 0, & i \neq 0, \end{cases}$$

the formula above can be rewritten in the form

$$T_0^*(s)\,(sI + D(\beta(s))) = \mathbf{p}(0)\sum_{k=1}^{\infty} D_k \sum_{i=0}^{\infty} Y_i(\beta(s))^{i+k-1} - \mathbf{p}(0)\sum_{k=1}^{\infty} D_k(\beta(s))^k.$$

Making the change of the summation variable $l = i + k - 1$ in the double sum, taking into account the formulas (3.2) and (3.12), this formula can be rewritten in the form

$$T_0^*(s)\,(sI + D(\beta(s))) = -\mathbf{p}(0)\sum_{k=1}^{\infty} D_k(\beta(s))^k + \lambda\boldsymbol{\pi}_0 \sum_{k=0}^{\infty} V_k(\beta(s))^k.$$

Performing similar manipulations with the function $T_1^*(s)$, relation (3.20) results in the following:

$$T_1^*(s)\,(sI + D(\beta(s))) = sT_1^*(s)$$

$$+\lambda \sum_{l=1}^{\infty} \boldsymbol{\pi}_l \sum_{i=0}^{\infty} \int_0^{\infty} P(i,y) \int_0^{\infty} e^{-su}\, dB(y+u)dy(\beta(s))^{i+l-1} \sum_{r=0}^{\infty} D_r(\beta(s))^r$$

$$= sT_1^*(s) + \lambda \sum_{l=1}^{\infty} \boldsymbol{\pi}_l \sum_{i=0}^{\infty} \int_0^{\infty}\int_0^{\infty} P'(i,y)e^{-su} dB(y+u)dy(\beta(s))^{i+l-1}$$

$$= \lambda \sum_{l=1}^{\infty} \boldsymbol{\pi}_l \sum_{i=0}^{\infty} \int_0^{\infty} (P(i,v) - P(i,0)e^{-sv}) dB(v)(\beta(s))^{i+l-1}$$

$$= -\lambda \sum_{l=1}^{\infty} \boldsymbol{\pi}_l (\beta(s))^l + \lambda \sum_{l=1}^{\infty} \boldsymbol{\pi}_l \sum_{i=0}^{\infty} Y_i (\beta(s))^{i+l-1}.$$

From the abovementioned formula $\mathbf{W}(x) = \mathbf{W}(+0) + \mathbf{T}_0(x) + \mathbf{T}_1(x)$ it follows that

$$\mathbf{w}(s)(sI + D(\beta(s))) = (W(+0) + T_0^*(s) + T_1^*(s))(sI + D(\beta(s)))$$

$$= \mathbf{p}(0)sI + \mathbf{p}(0) \sum_{k=0}^{\infty} D_k (\beta(s))^k - \mathbf{p}(0) \sum_{k=1}^{\infty} D_k (\beta(s))^k + \lambda \boldsymbol{\pi}_0 \sum_{k=0}^{\infty} V_k (\beta(s))^k$$

$$-\lambda \sum_{l=1}^{\infty} \boldsymbol{\pi}_l (\beta(s))^l + \lambda \sum_{l=1}^{\infty} \boldsymbol{\pi}_l \sum_{i=0}^{\infty} Y_i (\beta(s))^{i+l-1}.$$

Taking into account the equation

$$\sum_{i=0}^{\infty} \boldsymbol{\pi}_i (\beta(s))^i = \boldsymbol{\pi}_0 \sum_{i=0}^{\infty} V_i (\beta(s))^i + \sum_{i=0}^{\infty} \sum_{k=1}^{i+1} \boldsymbol{\pi}_k Y_{i-k+1} (\beta(s))^i$$

$$= \boldsymbol{\pi}_0 \sum_{i=0}^{\infty} V_i (\beta(s))^i + \sum_{l=1}^{\infty} \boldsymbol{\pi}_l \sum_{i=0}^{\infty} Y_i (\beta(s))^{i+l-1},$$

which follows from the equilibrium equations (2.69) for the vectors $\boldsymbol{\pi}_i,\ i \geq 0$, we simplify the right-hand side of the equation for the vector function $\mathbf{w}(s)$ and write it in the form

$$\mathbf{w}(s)(sI + D(\beta(s))) = s\mathbf{p}(0)$$

which was to be proved.

□

Corollary 3.1 *The vector $\mathbf{W}_1$, whose νth component is the mean virtual waiting time at the moment when the BMAP governing process is in state ν, $\nu = \overline{0, W}$, is calculated as follows:*

$$\mathbf{W}_1 = (-(\mathbf{p}(0) + \boldsymbol{\theta}(D'(1)b_1 - I))\tilde{I} - \boldsymbol{\theta}\frac{1}{2}(D''(1)b_1^2 + D'(1)b_2)\mathbf{e}\hat{\mathbf{e}})$$

$$\times (D(1)\tilde{I} + (D'(1)b_1 - I)\mathbf{e}\hat{\mathbf{e}})^{-1}. \tag{3.21}$$

Proof Introduce the expansions

$$\mathbf{w}(s) = \mathbf{w}(0) + \mathbf{w}'(0)s + \mathbf{w}''(0)\frac{s^2}{2} + o(s^2),$$

$$D(\beta(s)) = D(1) - D'(1)b_1 s + (D''(1)b_1^2 + D'(1)b_2)\frac{s^2}{2} + o(s^2).$$

Substituting them into formula (3.16) and equating the coefficients in the left and right parts for the same powers of s, we obtain the following equations:

$$\mathbf{w}(0)D(1) = \mathbf{0}, \tag{3.22}$$

$$\mathbf{w}'(0)D(1) + \mathbf{w}(0)(I - D'(1)b_1) = \mathbf{p}(0), \tag{3.23}$$

$$\frac{1}{2}\mathbf{w}''(0)D(1) + \mathbf{w}'(0)(I - D'(1)b_1) + \mathbf{w}(0)(D''(1)b_1^2 + D'(1)b_2)\frac{1}{2} = \mathbf{0}. \tag{3.24}$$

Equation (3.22) obviously implies the relation $\mathbf{w}(0) = \boldsymbol{\theta}$. Relation (3.21) is derived from formulas (3.23) and (3.24) similarly to the proof of Corollary 2.6, taking into account that $\mathbf{W}_1 = -\mathbf{w}'(0)$. □

Corollary 3.2 *Let $\mathbf{V}(x)$ be the row vector whose νth entry is the probability that the $BMAP$ governing process is in state ν at an arbitrary time and the virtual sojourn time of a customer in the system does not exceed x, and $\mathbf{v}(s) = \int_0^\infty e^{-sx}d\mathbf{V}(x)$. Then*

$$\mathbf{v}(s)\,(sI + D(\beta(s))) = s\mathbf{p}(0)\beta(s). \tag{3.25}$$

Formula (3.25) follows in an elementary way from (3.21) by taking into account how the sojourn time is related to the waiting time and the service time.

Next, consider the distribution of the real waiting time in the system. Let $w_t^{(a)}$ denote the actual waiting time for a customer arriving at time t, $t \geq 0$. Let $W^{(a)}(x)$ be the stationary probability that the actual waiting time does not exceed x, $W^{(a)}(x) = \lim_{t\to\infty} P\{w_t^{(a)} < x\}$, and $w^{(a)}(s) = \int_0^\infty e^{-sx}dW^{(a)}(x)$, $\operatorname{Re} s \geq 0$ be the LST of the distribution function $W^{(a)}(x)$. We suppose that an arbitrary customer arriving at the system within a batch of size k will be served with probability $\frac{1}{k}$ as the jth customer of the batch, $j = \overline{1,k}$.

Theorem 3.3 *The LST $w^{(a)}(s)$ of the real customer waiting time distribution function has the form*

$$w^{(a)}(s) = -\lambda^{-1}\mathbf{w}(s)(1 - \beta(s))^{-1}D(\beta(s))\mathbf{e}. \tag{3.26}$$

Proof The real waiting time of an arbitrary tagged customer consists of the virtual waiting time that starts with the arrival of the batch to which the customer belongs, and the service time of the customers within the batch that are served before the tagged customer. The vector $\mathbf{w}(s)$ gives the probability that no catastrophe happens from a stationary Poisson input of catastrophes with parameter s during a virtual waiting time conditioned on the corresponding states of the $BMAP$ governing process ν_t at the batch arrival moment. Under a fixed state probability distribution of the governing process at an arbitrary time, the probabilities that a tagged customer arrives within a batch of size k are specified by the vector $\frac{kD_k\mathbf{e}}{\lambda}$ (see Sect. 2.1). As was assumed above, this customer will be served as the jth one within the batch with probability $\frac{1}{k}$, $j = \overline{1,k}$. In this case the probability that no catastrophe happens during the service time of $j-1$ customers (to be served before the tagged one) is $(\beta(s))^{j-1}$. Given the reasoning above, the following formula is derived from the total probability formula:

$$w^{(a)}(s) = \lambda^{-1}\mathbf{w}(s)\sum_{k=1}^{\infty} kD_k\mathbf{e}\sum_{j=1}^{k}\frac{1}{k}(\beta(s))^{j-1}.$$

From this relation, formula (3.26) follows in an obvious way. □

Corollary 3.3 *The mean waiting time $W_1^{(a)}$ of a customer in the system is given as follows:*

$$W_1^{(a)} = \lambda^{-1}[\mathbf{W}_1 D'(1) + \frac{1}{2}b_1\boldsymbol{\theta} D''(1)]\mathbf{e}. \tag{3.27}$$

Corollary 3.4 *The mean sojourn time $V_1^{(a)}$ of a customer in the system is given as follows:*

$$V_1^{(a)} = \lambda^{-1}[\mathbf{W}_1 D'(1) + \frac{1}{2}b_1\boldsymbol{\theta} D''(1)]\mathbf{e} + b_1. \tag{3.28}$$

Note that it can be shown that Little's formula holds for the given system as follows:

$$V_1^{(a)} = \lambda^{-1}L,$$

where L is the mean number of customers in the system at an arbitrary time.

3.2 $BMAP/SM/1$ Queue

For a $BMAP/G/1$ queue, it was assumed that the customer service times are independent and identically distributed (with the distribution function $B(t)$) random variables. In many real systems, the service times of consecutive customers are essentially dependent and can be distributed according to different laws. In the literature, it has been suggested to use the formalism of an SM process, described in Sect. 2.1 to model such service processes. Let us briefly recall its description.

Let $m_t,\ t \geq 0$ be a semi-Markovian random process having a stationary state distribution. It is known that this process is fully characterized by its state space and a semi-Markovian kernel. Let a set $\{1, 2, \ldots, M\}$ be the state space of the process m_t, and matrix $B(t)$ with entries $B_{m,m'}(t)$ be its kernel. The function $B_{m,m'}(t)$ is interpreted as the probability that the time the process is in the current state does not exceed t and the next state of the process is m' given that the current state is m, $m, m' = \overline{1, M}$.

The service process is SM if the service times of consecutive customers are given by the consecutive sojourn times of the process m_t is its states. The system $BMAP/SM/1$ was first considered by D. Lucantoni and M. Neuts on the basis of the analytical approach of M. Neuts. Below we briefly describe the results of the system analysis.

3.2.1 The Stationary Probability Distribution of an Embedded Markov Chain

Using the experience of studying a $BMAP/G/1$ queue, we begin our investigation of a $BMAP/SM/1$ queue by analyzing the embedded process $\{i_n, \nu_n, m_n\}$, $n \geq 1$, where i_n is the number of customers in the system at time $t_n + 0$ (where t_n is the nth customer service completion moment), ν_n is the state of the $BMAP$ governing process at time t_n, and m_n is the state of the SM governing service process at time $t_n + 0$.

Analyzing the behavior of this process, it is easy to understand that it is a three-dimensional QTMC. The matrix PGFs $Y(z)$ and $V(z)$ characterizing the one-step transition probabilities of this MC have the following form:

$$Y(z) = \int_0^\infty e^{D(z)t} \otimes dB(t) = \hat{\beta}(-D(z)), \tag{3.29}$$

$$V(z) = \frac{1}{z}(-\tilde{D}_0)^{-1}(\tilde{D}(z) - \tilde{D}_0)\hat{\beta}(-D(z)),$$

where $\tilde{D}_k = D_k \otimes I_M$ and $\tilde{D}(z) = D(z) \otimes I_M$. In form, these matrix PGFs coincide with the similar PGFs for a $BMAP/G/1$ queue but the scalar multiplication operation under the integral sign is replaced by the symbol of the Kronecker product of matrices.

Theorem 3.4 *The stationary state distribution of the process* $\{i_n, \nu_n, m_n\}$, $n \geq 1$ *exists if and only if the following inequality holds:*

$$\rho = \lambda b_1 < 1, \tag{3.30}$$

where λ *is the mean intensity of customer arrival from the* $BMAP$, $\lambda = \boldsymbol{\theta} D'(1)\mathbf{e}$, b_1 *is the mean service time,* $b_1 = \boldsymbol{\delta} \int\limits_0^\infty t dB(t)\mathbf{e}$, *and* $\boldsymbol{\delta}$ *is the invariant vector of the stochastic matrix* $B(\infty)$, *i.e., the vector satisfying the system of equations*

$$\boldsymbol{\delta} B(\infty) = \boldsymbol{\delta}, \ \boldsymbol{\delta}\mathbf{e} = 1.$$

Proof It is easy to see that the ergodicity conditions (2.66), (2.67) for the quasi-Toeplitz chain $\{i_n, \nu_n, m_n\}, n \geq 1$ are equivalent to

$$\mathbf{x}(\hat{\beta}(-D(z)))'|_{z=1}\mathbf{e} < 1, \tag{3.31}$$

where the vector $\mathbf{x}$ satisfies the system of linear algebraic equations

$$\mathbf{x}(I - \hat{\beta}(-D(1))) = \mathbf{0}, \ \mathbf{x}\mathbf{e} = 1. \tag{3.32}$$

Using the rule of a mixed product, it is not difficult to see that the unique solution of system (3.32) has the form

$$\mathbf{x} = \boldsymbol{\theta} \otimes \boldsymbol{\delta}. \tag{3.33}$$

Substituting the matrix function $\hat{\beta}(-D(z))$ of the form

$$\hat{\beta}(-D(z)) = \int_0^\infty \sum_{l=0}^\infty \frac{(D(z)t)^l}{l!} \otimes dB(t)$$

and vector $\mathbf{x}$ of the form (3.33) into (3.31) and making simple transformations, we obtain the inequality

$$(\boldsymbol{\theta} D'(1)\mathbf{e})(\boldsymbol{\delta} \int\limits_0^\infty t dB(t)\mathbf{e}) < 1,$$

which is equivalent to (3.30). □

Below we assume that condition (3.30) holds. Then there exist stationary probabilities

$$\pi(i, \nu, m) = \lim_{n\to\infty} P\{i_n = i, \nu_n = \nu, m_n = m\}, \ i \geq 0, \ \nu = \overline{0, W}, \ m = \overline{1, M}.$$

Denote

$$\boldsymbol{\pi}(i,\nu) = (\pi(i,\nu,1),\ldots,\pi(i,\nu,M)),$$

$$\boldsymbol{\pi}_i = (\boldsymbol{\pi}(i,0),\ldots,\boldsymbol{\pi}(i,W)),$$

$$\boldsymbol{\Pi}(z) = \sum_{i=0}^{\infty} \boldsymbol{\pi}_i z^i, \ |z| < 1.$$

From the general Eq. (2.70) for the vector PGF of a multidimensional QTMC, taking into account (3.29), the following statement holds.

Theorem 3.5 *The vector PGF* $\boldsymbol{\Pi}(z)$ *of the stationary state distribution of the embedded MC* $\{i_n, \nu_n, m_n\}$, $n \geq 1$ *satisfies the matrix functional equation*

$$\boldsymbol{\Pi}(z)(\hat{\beta}(-D(z)) - zI) = \boldsymbol{\Pi}(0)\tilde{D}_0^{-1}\tilde{D}(z)\hat{\beta}(-D(z)). \tag{3.34}$$

Since the form of the matrix PGFs $Y(z)$ and $V(z)$ is known, we can use the algorithms described in the previous chapter to find the unknown vectors $\boldsymbol{\Pi}(0)$ and $\boldsymbol{\Pi}(z)$. In this case, there are no additional principal difficulties compared to the implementation of these algorithms for a $BMAP/G/1$ queue except the method of calculating the matrices $Y_l, l > 0$.

We describe the procedure to calculate the matrices Y_l for the most important and practically interesting case, i.e., when the semi-Markovian kernel has the form

$$B(t) = \text{diag}\{B_1(t),\ldots,B_M(t)\}P = \text{diag}\{B_m(t),\ m = \overline{1,M}\}P, \tag{3.35}$$

where P is a stochastic matrix and $B_m(t),\ m = \overline{1,M}$ are some distribution functions.

The form (3.35) of the semi-Markovian kernel means the following. The sojourn time of the semi-Markovian process m_t in state m depends only on the number of this state. The future state is determined, according to the transition probability matrix P at the moment the process finishes its stay in its current state.

Using the mixed-product rule (see Appendix A) for the Kronecker product of matrices, we obtain the following formula for the matrix LST

$$\hat{\beta}(-D(z)) = \left(\int_0^{\infty} e^{D(z)t} \otimes \text{diag}\{dB_m(t),\ m = \overline{1,M}\}\right)(I_{\bar{W}} \otimes P).$$

It is well known that the multiplication of a matrix on the left or right by elementary matrices rearranges its rows and columns. By direct calculations, it is easy to verify that

$$\hat{\beta}(-D(z)) = \left(Q\text{diag}\{\int_0^{\infty} e^{D(z)t} dB_m(t),\ m = \overline{1,M}\}Q^T\right)(I_{\bar{W}} \otimes P). \tag{3.36}$$

Here Q is a product of elementary matrices. If $\bar{W} = M$, the explicit form of these matrices is as follows:

$$Q = \prod_{i=0}^{M-2} \prod_{j=i+2}^{M} S_{Mi+j,M(j-1)+i+1},$$

where $S_{l,k}$ is an elementary matrix obtained from an identity matrix I by moving a unit in the lth row out of the diagonal to the kth column and in the kth row out of the diagonal to the lth column, $l, k = \overline{1, M}$. In case $\bar{W} \neq M$, the matrices Q are easily constructed algorithmically.

Denote by $Y_l^{(m)}$ matrices that are coefficients of expansions

$$\sum_{l=0}^{\infty} Y_l^{(m)} z^l = \int_0^{\infty} e^{D(z)t} dB_m(t), \ m = \overline{1, M}.$$

The way we calculated these matrices was described in Chap. 2. Then it obviously follows from (3.36) that matrix Y_l, which is a term of expansion $\sum_{l=0}^{\infty} Y_l z^l = \int_0^{\infty} e^{D(z)t} \otimes dB(t)$, is defined by the formula

$$Y_l = \left(Q\text{diag}\{Y_l^{(m)}, \ m = \overline{1, M}\}Q^T\right)(I_{W+1} \otimes P), \ l \geq 0.$$

Remark 3.1 The problem with the need to rearrange the rows and columns discussed above could have been avoided if we had initially considered the MC $\{i_n, m_n, \nu_n\}, \ n \geq 1$ instead of $\{i_n, \nu_n, m_n\}, \ n \geq 1$, where i_n is the number of customers in the system at time $t_n + 0$ (t_n is the nth customer service completion moment), ν_n is the state of the $BMAP$ governing process at time t_n, and m_n is the state of the SM governing service process at time $t_n + 0$. In this case, the analysis would be carried out in a completely similar way to the above by changing the denotations. For example, the function $Y(z)$ given by (3.29) has the following form in the new denotation:

$$Y(z) = \hat{\beta}(-D(z)) = \int_0^{\infty} dB(t) \otimes e^{D(z)t}.$$

3.2.2 *The Stationary Probability Distribution of the System States at an Arbitrary Time*

We now consider the problem of finding the stationary state distribution of the process $\{i_t, \nu_t, m_t\}, t \geq 0$ at an arbitrary time.

Suppose

$$p(i, \nu, m) = \lim_{t\to\infty} P\{i_t = i, \nu_t = \nu, m_t = m\},$$

$$\mathbf{p}(i, \nu) = (p(i, \nu, 1), \dots, p(i, \nu, M)), \quad \mathbf{p}(i) = (\mathbf{p}(i, 0), \dots, \mathbf{p}(i, W)),$$

$$\mathbf{P}(z) = \sum_{i=0}^{\infty} \mathbf{p}(i)z^i, \ |z| < 1.$$

Then the following statement holds.

Theorem 3.6 *The PGF* $\mathbf{P}(z)$ *of the stationary state distribution of the process* $\{i_t, \nu_t, m_t\}$ *at an arbitrary time and the PGF* $\mathbf{\Pi}(z)$ *of the distribution at service completion moments are related as follows:*

$$\mathbf{P}(z)\tilde{D}(z) = \lambda\mathbf{\Pi}(z)(z(\hat{\beta}(-D(z)))^{-1}\nabla^*(z) - I), \tag{3.37}$$

where $\nabla^*(z) = \int\limits_0^\infty e^{D(z)t} \otimes d\nabla_B(t)$*, where* $\nabla_B(t)$ *is a diagonal matrix with diagonal entries* $B(t)\mathbf{e}_j$*,* $j = \overline{1, M}$*.*

Proof Using the approach based on the use of an embedded MRP described in the previous section, we obtain the following formulas relating vectors $\mathbf{p}(i)$, $i \geq 0$, to vectors $\boldsymbol{\pi}_i$, $i \geq 0$:

$$\mathbf{p}(0) = \lambda\boldsymbol{\pi}_0 \int\limits_0^\infty (e^{D_0 t} \otimes I_M)dt = \lambda\boldsymbol{\pi}_0(-\tilde{D}_0)^{-1},$$

$$\mathbf{p}(i) = \lambda\boldsymbol{\pi}_0 \int\limits_0^\infty \int\limits_0^t (e^{D_0 v} \otimes I_M) \sum_{k=1}^{i} (D_k \otimes I_M)dv$$

$$\times[P(i-k, t-v)\otimes(I_M - \nabla_B(t-v))]dt + \lambda\sum_{k=1}^{i} \boldsymbol{\pi}_k \int\limits_0^\infty [P(i-k, t)\otimes(I_M - \nabla_B(t))]dt$$

$$= \lambda\boldsymbol{\pi}_0(-\tilde{D}_0)^{-1} \sum_{k=1}^{i} \int\limits_0^\infty [D_k P(i-k, t) \otimes (I_M - \nabla_B(t))]dt$$

$$+ \lambda\sum_{k=1}^{i} \boldsymbol{\pi}_k \int\limits_0^\infty [P(i-k, t) \otimes (I_M - \nabla_B(t))]dt, \ i > 0. \tag{3.38}$$

Multiplying Eq. (3.38) by the corresponding powers of z and summing them up, we obtain the following expression for the vector PGF $\mathbf{P}(z)$:

$$\mathbf{P}(z)=\lambda\left(\boldsymbol{\pi}_0(-\tilde{D}_0)^{-1}+(\boldsymbol{\pi}_0(-\tilde{D}_0)^{-1}\tilde{D}(z)+\boldsymbol{\Pi}(z))\int\limits_0^\infty e^{\tilde{D}(z)t}\otimes(I-\nabla_B(t))dt\right). \tag{3.39}$$

We calculate the integral in (3.39):

$$\int\limits_0^\infty e^{\tilde{D}(z)t}\otimes(I-\nabla_B(t))dt=(\tilde{D}(z))^{-1}e^{\tilde{D}(z)t}\otimes(I-\nabla_B(t))|_0^\infty$$

$$+(\tilde{D}(z))^{-1}\int\limits_0^\infty e^{\tilde{D}(z)t}\otimes d\nabla_B(t)$$

$$=-(\tilde{D}(z))^{-1}+(\tilde{D}(z))^{-1}\int\limits_0^\infty e^{\tilde{D}(z)t}\otimes d\nabla_B(t)$$

$$=(\tilde{D}(z))^{-1}(\nabla^*(z)-I).$$

Substituting the expression obtained in (3.39) and making transformations by means of Eq. (3.34) for the PGF $\boldsymbol{\Pi}(z)$, we finally get (3.37). □

Corollary 3.5 *The scalar PGF of the queue length distribution for a $BMAP/SM/1$ queue has the form*

$$\mathbf{P}(z)\mathbf{e}=\lambda(z-1)\boldsymbol{\Pi}(z)(\tilde{D}(z))^{-1}\mathbf{e}.$$

The proof follows from the facts that the matrices $z\hat{\beta}^{-1}(-D(z))\nabla^*(z)-I$ and $(\tilde{D}(z))^{-1}$ are permutable, and equation $\nabla^*(z)\mathbf{e}=\hat{\beta}(z)\mathbf{e}$ holds.

Corollary 3.6 *If the semi-Markovian kernel $B(t)$ has the form (3.35) then formula (3.37) can be rewritten in the form*

$$\mathbf{P}(z)\tilde{D}(z)=\lambda\boldsymbol{\Pi}(z)(zI\otimes P^{-1}-I).$$

3.3 $BMAP/SM/1/N$ Queue

Both models considered in the previous subsections assume that the waiting space in the system is unlimited. At the same time, the study of systems with a finite buffer is also of immediate interest. One of the reasons for this fact is the rapid development of telecommunications networks. The growth in the number of users and the types of information transferred leads to an increasing load on the network servers and, although the buffer capacity is hundreds to hundreds of thousands of packets, situations when the buffer overflows still happen. At the same time, when the load of the channel (the ratio of the input intensity to the packet transmission rate) is in the range 0.9–1.1, the probability of packet loss is very sensitive to both load changes and buffer size change. Therefore, in order to solve the problem of choosing the optimal buffer size successfully, it is necessary to be able to accurately calculate the probability of failure with a fixed form of the input flow, the service process, and a fixed buffer size. Thus, the problem of calculating the characteristics of a $BMAP/SM/1/N$ system is very relevant.

A significant contribution to the study of finite-buffer systems was made by P.P. Bocharov, who intensively studied systems with a finite buffer and an ordinary MAP. The assumption that the input flow is a $BMAP$ leads to some complications of the analysis caused by the need to consider different scenarios of the system's behavior in a situation when there are free places at the moment a batch of customers arrives but these places are not enough to accept all customers in the batch. In the literature, the following three disciplines for accepting customers are popular:

- *Partial admission* (PA) assumes that some of the customers in a batch corresponding into the number of available places are accepted into the system, and the rest of them are lost;
- *Complete admission* (CA) assumes that all customers from an arriving batch are accepted into the queue (for example, by using extended memory);
- *Complete rejection* (CR) assumes that the whole batch is not accepted and lost. For example, this discipline is reasonable when all customers in a batch belong to the same message, and the situation when at least one customer is lost means that message transmission fails.

First, we analyze a $BMAP/SM/1/N$ queue with the partial admission discipline, which is the simplest of the three disciplines in the sense of its analysis complexity. Then we consider the other two disciplines. Let us briefly mention a hybrid discipline that provides for a randomized selection of one of the three listed disciplines each time a batch of customers arrives. We will study the distribution of the number of customers in the system. In particular, we derive one of its most important characteristics, namely, the probability of an arbitrary customer loss.

3.3.1 Analysis of the System with the Partial Admission Discipline

3.3.1.1 Transition Probabilities of the Embedded Markov Chain

Let us consider a three-dimensional process $\{i_n, \nu_n, m_n\}$, $n \geq 1$, where $i_n = i_{t_n}$ is the number of customers in the system at time $t_n + 0$, $i_n = \overline{0, N}$; $\nu_n = \nu_{t_n}$ is the state of the $BMAP$ governing process at time t_n, $\nu_n = \overline{0, W}$; $m_n = m_{t_n}$ is the state of the SM governing service process at time $t_n + 0$, $m_n = \overline{1, M}$.

Denote by $P\{(i, \nu, m) \to (j, \nu', m')\} = P\{i_{n+1} = j, \nu_{n+1} = \nu', m_{n+1} = m' \mid i_n = i, \nu_n = \nu, m_n = m\}$, $n \geq 1$ the one-step transition probabilities of the process $\{i_n, \nu_n, m_n\}$, $n \geq 1$. We assume that the states of the MC $\{i_n, \nu_n, m_n\}$, $n \geq 1$ are ordered in lexicographical order and introduce the denotation for block matrices of transition probabilities

$$P_{i,j} = \|P\{(i, \nu, m) \to (j, \nu', m')\}\|_{\nu,\nu'=\overline{0,W},\ m,m'=\overline{1,M}} \ .$$

Using standard arguments, it is easy to verify the following statement.

Theorem 3.7 *The matrices of the transition probabilities $P_{i,j}$ are defined as follows:*

$$P_{i,j} = Y_{j-i+1}, \quad 0 < i < N, \quad j \geq i - 1,$$

$$P_{0,j} = V_j, \quad j = \overline{0, N-1},$$

$$P_{0,N} = V(1) - \sum_{l=0}^{N-1} V_l, \ P_{i,N} = Y(1) - \sum_{l=0}^{N-i} Y_l, \quad i = \overline{1, N},$$

where matrices Y_l, V_l are found as the expansion coefficients

$$\sum_{l=0}^{\infty} Y_l z^l = Y(z) = \int_0^{\infty} e^{D(z)t} \otimes dB(t),$$

$$\sum_{l=0}^{\infty} V_l z^l = V(z) = \frac{1}{z}(-\tilde{D}_0)^{-1}(\tilde{D}(z) - \tilde{D}_0)Y(z).$$

Proof The probabilities $P_{i,j}$, $j < N$ coincide with the corresponding probabilities for a $BMAP/SM/1$ queue with an infinite buffer. The form of the probabilities $P_{i,N}$ obviously follows from the normalization condition and the probabilistic meaning of matrices $Y(1)$ and $V(1)$, which describe the transitions of the finite components $\{\nu_n, m_n\}$ during a customer service period (the time from the moment when the system becomes empty at a customer service completion moment to the first service completion moment in a system busy period). □

3.3.1.2 A Direct Algorithm to Find the Stationary State Distribution of the Embedded Markov Chain

Since the MC $\{i_n, \nu_n, m_n\}$, $n \geq 1$ has a finite state space, under the above assumption about the irreducibility of the governing processes of input and service, the stationary state probabilities

$$\pi(i, \nu, m) = \lim_{n\to\infty} P\{i_n = i, \nu_n = \nu, m_n = m\},\ i = \overline{0, N},\ \nu = \overline{0, W},\ m = \overline{1, M}$$

exist for any values of the system parameters.

In accordance with the above lexicographic ordering of the MC components, we introduce the vectors $\boldsymbol{\pi}_i$, $i = \overline{0, N}$ of the stationary state probabilities $\pi(i, \nu, m)$ corresponding to level i of the component i_n.

Taking into account the form of the transition probability matrices given above, it is easy to obtain a system of equilibrium equations for the probability vectors $\boldsymbol{\pi}_i$:

$$\boldsymbol{\pi}_l = \boldsymbol{\pi}_0 V_l + \sum_{i=1}^{l+1} \boldsymbol{\pi}_i Y_{l+1-i}, \quad l = \overline{0, N-1}, \tag{3.40}$$

$$\boldsymbol{\pi}_N = \boldsymbol{\pi}_0 (V(1) - \sum_{l=0}^{N-1} V_l) + \sum_{i=1}^{N} \boldsymbol{\pi}_i (Y(1) - \sum_{l=0}^{N-i} Y_l). \tag{3.41}$$

Analyzing the structure of system (3.40)–(3.41), it is easy to see that by setting $l = 0, 1, .., N-1$ in (3.40) we can consequently express all vectors $\boldsymbol{\pi}_l$, $l = \overline{1, N}$ through the vector $\boldsymbol{\pi}_0$. To find the latter, we can use Eq. (3.41) (where all the vectors $\boldsymbol{\pi}_l$ are already known up to $\boldsymbol{\pi}_0$) and the normalization condition.

As a result, we can verify the following statement.

Theorem 3.8 *The stationary probabilities of the MC under consideration are defined as follows:*

$$\boldsymbol{\pi}_l = \boldsymbol{\pi}_0 F_l, \quad l = \overline{0, N},$$

where matrices F_l are defined recurrently

$$F_0 = I, \quad F_{l+1} = (-V_l F_l - \sum_{i=1}^{l} F_i Y_{l+1-i}) Y_0^{-1}, \quad l = \overline{0, N-1}, \tag{3.42}$$

and vector $\boldsymbol{\pi}_0$ *is the unique solution of the system of linear algebraic equations*

$$\boldsymbol{\pi}_0\big[-F_N V(1) - \sum_{l=0}^{N-1} V_l \sum_{i=1}^{N} F_i(Y(1) - \sum_{l=0}^{N-i} Y_l)\big] = \mathbf{0}, \tag{3.43}$$

$$\boldsymbol{\pi}_0 \sum_{l=0}^{N} F_l \,\mathbf{e} = 1. \tag{3.44}$$

Remark 3.2 One of the important practical problems that can be solved by means of the obtained results is the choice of the buffer size N providing (with fixed characteristics of the flow and the service process) an acceptable level of arbitrary customer loss probability. While solving this problem, the procedure for calculating the stationary state distribution can be applied for several successive values of N. The advantage of the procedure to find the stationary state distribution given by this theorem is the fact that it is not necessary to repeat the procedure again as the value N increases. It is only necessary to calculate the required number of additional matrices F_l by recursive formulas (3.42) and solve the system of linear algebraic equations (3.43)–(3.44) with the updated matrix of the system.

3.3.1.3 Modified Algorithm to Find the Stationary State Distribution of the Embedded Markov Chain

The recursion given by (3.42) is not numerically stable since it contains subtractions of matrices, which may cause underflows and the accumulation of rounding errors especially in the case of large buffer capacity N. Using the general algorithm for multi-dimensional MCs with a finite state space described in Sect. 2.5, one can obtain another recursion given by the following theorem.

Theorem 3.9 *The vectors* $\boldsymbol{\pi}_l, \quad l = \overline{0, N}$ *of the stationary state probabilities of the embedded MC* $\{i_n, \nu_n, m_n\},\ n \geq 1$ *are calculated as follows:*

$$\boldsymbol{\pi}_l = \boldsymbol{\pi}_0 A_0, \quad l \geq 0, \tag{3.45}$$

where the matrices A_l *satisfy the recurrent relations*

$$A_0 = I,\ A_l = (\overline{V}_l - \sum_{i=1}^{l-1} A_i \overline{Y}^{(l)}_{l+1-i})(I - \overline{Y}^{(l)}_1)^{-1}, \quad l = \overline{1, N-1}, \tag{3.46}$$

$$A_N = (\overline{V}_N - \sum_{i=1}^{N-1} A_i \tilde{Y}_{N+1-i})(I - \tilde{Y}_1)^{-1}, \tag{3.47}$$

where

$$\tilde{Y}_l = Y(1) - \sum_{i=0}^{l-1} Y_i, \quad l = \overline{1, N}$$

and matrices $\overline{V}_k$, $\overline{Y}_l^{(k)}$ *are defined as follows:*

$$\overline{V}_N = \tilde{V}_N = V(1) - \sum_{i=0}^{N-1} V_i, \tag{3.48}$$

$$\overline{V}_l = V_l + \overline{V}_{l+1} G_{l+1}, \quad l = \overline{0, N-1},$$

$$\overline{Y}_i^{(N)} = \tilde{Y}_i, \quad i = \overline{1, N}, \tag{3.49}$$

$$\overline{Y}_i^{(k)} = Y_i + \overline{Y}_{i+1}^{(k+1)} G_{k+1}, \quad i = \overline{1, k}, \quad k = \overline{1, N-1}, \tag{3.50}$$

where the matrices G_l *are defined recursively*

$$G_N = (I - \tilde{Y}_1)^{-1} Y_0,$$

$$G_l = Y_0 \sum_{i=1}^{N-l} Y_i G_{l+i-1} \cdot \ldots \cdot G_l + \tilde{Y}_{N-l+1} G_N \cdot \ldots \cdot G_l, \quad l = \overline{1, N-1} \tag{3.51}$$

or by formula

$$G_l = (I - \overline{Y}_1^{(l)})^{-1} Y_0, \quad l = \overline{1, N}. \tag{3.52}$$

The vector $\boldsymbol{\pi}_0$ *is defined as the unique solution of the system of linear algebraic equations*

$$\boldsymbol{\pi}_0 (I - \overline{V}_0) = \mathbf{0}, \tag{3.53}$$

$$\boldsymbol{\pi}_0 \sum_{l=0}^{N} A_l \mathbf{e} = 1. \tag{3.54}$$

Proof As noted above, formulas (3.46), (3.47) are obtained by constructing a censored MC for the MC $\{i_n, \nu_n, m_n\}$, $i_n = \overline{0, N}$, $\nu_n = \overline{0, W}$, $m_n = \overline{1, M}$, $n \geq 1$. Note that the matrix G_i describes the transitions of the process $\{\nu_n, m_n\}$ at the time when the component i_n starting from the state i reaches $i - 1$ for the first time. Unlike a system with an infinite buffer, this matrix is not invariant with respect to i. Another way to prove the theorem is to use the equilibrium equations (3.40),

(3.41). Eliminating $\boldsymbol{\pi}_N, \boldsymbol{\pi}_{N-1}, .., \boldsymbol{\pi}_1$ successively from these equations, we get Eqs. (3.53), (3.54) for the probability vector $\boldsymbol{\pi}_0$, where the matrices involved in (3.53), (3.54) are given by the relations (3.46)–(3.52). Note that all matrices that have to be inverted in these relations are non-singular because the matrices subtracted from matrix I are sub-stochastic. □

Remark 3.3 The recursion (3.46), (3.47) includes only non-negative matrices, therefore it is much more stable numerically than recursion (3.42).

3.3.1.4 The Stationary Probability Distribution of the System States at an Arbitrary Time

As mentioned earlier, the process $\{i_t, \nu_t, m_t\}, t \geq 1$ of interest to us is not Markovian. One approach to find its stationary state distribution uses the probability distribution of the process states at embedded moments. In our case, we consider the moments $t = t_n, n \geq 1$ of customer service completion as embedded moments. Having found the stationary state distribution of the embedded MC $\{i_n, \nu_n, m_n\}, n \geq 1$, we are able to find the stationary state distribution of the process $\{i_t, \nu_t, m_t\}, t \geq 0$.

Denote

$$p(i, \nu, m) = \lim_{t\to\infty} P\{i_t = i, \nu_t = \nu, m_t = m\},$$

$$i = \overline{0, N+1}, \ \nu = \overline{0, W}, \ m = \overline{1, M}.$$

Ordering these probabilities in lexicographical order, we get the probability vectors $\mathbf{p}_i, i = \overline{0, N+1}$. Suppose that a semi-Markovian kernel $B(t)$ has the form (3.35).

Theorem 3.10 *The stationary state distribution $\mathbf{p}_i, i = \overline{0, N}$ of the system states at an arbitrary time is defined as follows:*

$$\mathbf{p}_0 = \Lambda \boldsymbol{\pi}_0(-\tilde{D}_0)^{-1}, \tag{3.55}$$

$$\mathbf{p}_i = \left[\sum_{j=0}^{i-1} \mathbf{p}_j \tilde{D}_{i-j} - \Lambda(\boldsymbol{\pi}_{i-1}(I_{W1} \otimes P)^{-1} - \boldsymbol{\pi}_i)\right](-\tilde{D}_0)^{-1}, i = \overline{1, N}, \tag{3.56}$$

$$\mathbf{p}_{N+1}\mathbf{e} = 1 - \sum_{i=0}^{N} \mathbf{p}_i \mathbf{e}, \tag{3.57}$$

where

$$\Lambda = \tau^{-1}, \ \tau = \boldsymbol{\pi}_0(-\tilde{D}_0)^{-1}\mathbf{e}b_1. \tag{3.58}$$

Proof The standard method to find the stationary state distribution at an arbitrary time based on the stationary state distribution of the embedded MC is to use results from the theory of Markov renewal processes [22]. Applying this theory, we obtain the following relations:

$$\mathbf{p}_0 = \Lambda \boldsymbol{\pi}_0 (-\tilde{D}_0)^{-1}, \tag{3.59}$$

$$\mathbf{p}_i = \Lambda \boldsymbol{\pi}_0 (-\tilde{D}_0)^{-1} \sum_{k=1}^{i} \int_0^\infty (\tilde{D}_k P(i-k,t)) \otimes (I - \tilde{B}(t)) dt$$

$$+ \Lambda \sum_{k=1}^{i} \boldsymbol{\pi}_k \int_0^\infty P(i-k,t) \otimes ((I - \tilde{B}(t)) dt, \tag{3.60}$$

$$\mathbf{p}_N = \Lambda \boldsymbol{\pi}_0 (-\tilde{D}_0)^{-1} \Big[(\tilde{D}(1) - \tilde{D}_0) \int_0^\infty e^{D(1)t} \otimes (I - \tilde{B}(t)) dt$$

$$- \sum_{k=1}^{N} \int_0^\infty (\tilde{D}_k \sum_{l=0}^{N-k-1} P(l,t)) \otimes (I - \tilde{B}(t)) dt \Big] \tag{3.61}$$

$$+ \Lambda \sum_{k=1}^{N} \boldsymbol{\pi}_k \Big[\int_0^\infty e^{D(1)t} \otimes (I - \tilde{B}(t)) dt - \sum_{l=0}^{N-k-1} P(l,t) \otimes (I - \tilde{B}(t)) dt \Big].$$

Here the value Λ is given by formula (3.58) and determines the departure intensity of the served customers. Matrix $P(i,t)$ defines the probabilities of the process ν_t transition and i customers' arrival in the $BMAP$ during time $(0,t)$. The matrix $\tilde{B}(t)$ is diagonal. Its diagonal elements are given by vector $B(t)\mathbf{e}$.

Equation (3.59) coincides with (3.55). Analyzing the remaining equation of the system (3.60)–(3.61) including matrices $P(i,t)$, which are explicitly unknown and are given by their generating function

$$\sum_{i=0}^{\infty} P(i,t) z^i = e^{D(z)t},$$

we do not see a simple direct way to derive formulas (3.56), (3.57) from the above. Recall that in the case of a system with an infinite buffer, the generating functions of the vectors $\boldsymbol{\pi}_i$ and $\mathbf{p}_i$ are related as follows:

$$\sum_{i=0}^{\infty} \mathbf{p}_i z^i = \lambda \sum_{i=0}^{\infty} \boldsymbol{\pi}_i z^i (z \hat{\beta}^{-1} (-D(z)) \bigtriangledown^* (z) - I)(\tilde{D}(z))^{-1}. \tag{3.62}$$

Here

$$\hat{\beta}^{-1}(-D(z)) = \int_0^\infty e^{D(z)t} \otimes dB(t), \quad \nabla^*(z) = \int_0^\infty e^{D(z)t} \otimes d\tilde{B}(t).$$

For the form (3.35) of the semi-Markovian kernel $B(t)$, the matrix PGFs $\hat{\beta}^{-1}(-D(z))$ and $\nabla^*(z)$ are associated as follows:

$$\hat{\beta}^{-1}(-D(z)) = \nabla^*(z)(I_{W1} \otimes P),$$

since formula (3.62) has the form

$$\sum_{i=0}^{\infty} \mathbf{p}_i z^i = \lambda \sum_{i=0}^{\infty} \boldsymbol{\pi}_i z^i (z(I \otimes P)^{-1} - I). \tag{3.63}$$

We now consider the system (3.59), (3.60) for $0 \le i \le \infty$. Passing to the PGFs, we obtain (3.63) up to the factor λ, which is replaced by Λ in this case. Expanding this relation in a series, we obtain (3.56) for $i = \overline{1, N}$. The value $\mathbf{p}_{N+1}\mathbf{e}$ is derived from the normalization condition. □

One of the most important probabilistic characteristics of a finite buffer system is the probability P_{loss} that an arbitrary customer is lost due to a buffer overflow.

Theorem 3.11 *The probability P_{loss} that an arbitrary customer is lost in the system under consideration is given as follows:*

$$P_{loss} = 1 - \frac{1}{\lambda} \sum_{i=0}^{N} \mathbf{p}_i \sum_{k=0}^{N+1-i} (k - (N+1) + i)\tilde{D}_k \mathbf{e} = 1 - \frac{1}{\lambda \tau}. \tag{3.64}$$

Remark 3.4 The presence of two different formulas for calculating the probability P_{loss} of an arbitrary customer loss is useful at the stage of computer realization of the obtained algorithms for calculating the stationary state probabilities and the probability P_{loss}.

Algorithms to calculate the probability distribution of the system states at an arbitrary time and at the embedded moments of time are implemented in the form of a SIRIUS software package developed at the Belarusian State University, see [29]. The results of numerical experiments conducted by means of this software, the purpose of which was to determine the regions of the stable computations of the direct and additional algorithms depending on the buffer size N, obtaining the dependence of the loss probability P_{loss} on the buffer size N and the importance of taking into account the correlation in the input flow can be found in [32]. In particular, it has been clarified that the use of a stationary Poisson input as a model of real data traffic with even a relatively small (about 0.2) correlation of the neighboring interval lengths can give a loss probability 100,000 times less than the real value of this probability.

3.3.2 *Analysis of the System with the Complete Admission and Complete Rejection Disciplines*

3.3.2.1 The Transition Probabilities of the Embedded Markov Chain

Note that the state space of the process i_t, which is the number of customers in the system at time t, $t > 0$, is the finite set $\{0, 1, \ldots, N+1\}$ for the PA and CR disciplines. In the case of CA, this state space is infinite. The process i_t is not Markovian, and we consider the embedded MC $\xi_n = \{i_n, \nu_n, m_n\}$, $n \geq 1$ for its analysis. Our consideration starts with the derivation of matrices $P_{i,l}$ consisting of transition probabilities

$$P\{i_{n+1} = l, \nu_{n+1} = \nu', m_{n+1} = m' | i_n = i, \nu_n = \nu, m_n = m\},$$

$$\nu, \nu' = \overline{0, W}, m, m' = \overline{1, M}.$$

Calculation of the matrices $P_{i,l}$ first needs the calculation of the matrices $P^{(j)}(n,t)$ whose (ν, ν')th entry is the conditional probability that n customers are accepted into the system during the time interval $(0, t]$, and the state of the $BMAP$ governing process ν_t is ν' at time t given that the initial state of the process was ν at time 0 and at most j customers were accepted into the system (due to the buffer limitations) during time interval $(0, t]$, $n = \overline{0, j}$.

In the case of the PA discipline, the matrices $P^{(j)}(n,t)$ are easily calculated by the obvious formulas

$$P^{(j)}(n,t) = \begin{cases} P(n,t), & n < j, \\ \sum\limits_{l=j}^{\infty} P(l,t), & n = j, \end{cases} \tag{3.65}$$

where the matrices $P(n,t)$ giving the probabilities of n customer arrivals during the time interval $(0, t]$ are given as coefficients of the matrix expansion

$$e^{D(z)t} = \sum_{n=0}^{\infty} P(n,t) z^n. \tag{3.66}$$

The problem of the calculation of these matrices was discussed in Sect. 2.1.6.

In the case of the CA and CR disciplines, the matrices $P^{(j)}(n,t)$ cannot be easily calculated through matrices $P(n,t)$ similarly to (3.65), (3.66), which is the main reason why these disciplines are relatively poorly studied.

Recall that the numerical procedure for calculating the matrices $P(n,t)$ given in Sect. 2.1.6 is as follows. Let ψ be defined as $\psi = \max_{\nu=\overline{0,W}} (-D_0)_{\nu,\nu}$. Then

$$P(n,t) = e^{-\psi t} \sum_{i=0}^{\infty} \frac{(\psi t)^i}{i!} U_n^{(i)}, \tag{3.67}$$

where matrices $U_n^{(i)}$ are calculated in a recursive way as

$$U_n^{(0)} = \begin{cases} I, \ n = 0, \\ 0, \ n > 0, \end{cases}$$

$$U_n^{(i+1)} = U_n^{(i)}(I + \psi^{-1} D_0) + \psi^{-1} \sum_{l=0}^{n-1} U_l^{(i)} D_{n-l}, \ i \geq 0, \ n \geq 0.$$

These formulas were obtained using the system of matrix differential equations (2.1), which, in turn, was obtained on the basis of a system of matrix differential equations for matrices $P(n,t), \ n \geq 0$.

Proceeding similarly for the CA and CR disciplines, we can verify the following statement.

Lemma 3.1 *Matrices $P^{(j)}(n,t)$ for the CA and CR disciplines are calculated by formulas*

$$P^{(j)}(n,t) = e^{-\psi t} \sum_{i=0}^{\infty} \frac{(\psi t)^i}{i!} U_n^{(i)}(j),$$

where the matrices $U_n^{(i)}(j)$ are calculated by

$$U_n^{(0)}(j) = \begin{cases} I, \ n = 0, \\ 0, \ n > 0 \end{cases}$$

for both disciplines, and

$$U_n^{(i+1)}(j) = U_n^{(i)}(j) \left(I + \psi^{-1}(D_0 + \hat{D}_{j+1-n}) \right) + \psi^{-1} \sum_{l=0}^{n-1} U_l^{(i)}(j) D_{n-l},$$
$$i \geq 0, \ j = \overline{0, N+1}, \ n = \overline{0, j},$$
$$\hat{D}_l = \sum_{m=l}^{\infty} D_m$$

in the case of CR, and

$$U_n^{(i+1)}(j) = U_n^{(i)}(j)(I + \psi^{-1} D_0) + \psi^{-1} \sum_{l=0}^{n-1} U_l^{(i)}(j) D_{n-l},$$
$$i \geq 0,\ j = \overline{0, N+1},\ n = \overline{0, j-1},$$
$$U_n^{(i+1)}(j) = U_n^{(i)}(j) \left(I + \psi^{-1} D(1)\right) + \psi^{-1} \sum_{l=0}^{j-1} U_l^{(i)}(j) D_{n-l},$$
$$i \geq 0,\ j = \overline{0, N+1},\ n \geq j$$

in the case of CA.

Lemma 3.2 *The matrices of the transition probabilities $P_{i,l}$ are calculated by formulas*

$$P_{0,l} = -(D_0 + \hat{D}_{N+2})^{-1} \sum_{k=1}^{l+1} D_k \int_0^\infty P^{(N+1-k)}(l+1-k, t) dB(t),\ l = \overline{0, N},$$

$$P_{i,l} = \int_0^\infty P^{(N+1-i)}(l+1-i, t) dB(t), i = \overline{1, N},\ l = \overline{i-1, N},$$
$$P_{i,l} = 0,\ i = \overline{i, N},\ l < i-1$$

for the CR discipline, and

$$P_{0,l} = (-D_0)^{-1} \sum_{k=1}^{l+1} D_k \int_0^\infty P^{(N+1-k)}(l+1-k, t) dB(t),\ l = \overline{0, N},$$
$$P_{0,l} = (-D_0)^{-1} \left(D_{l+1} G + \sum_{k=1}^{N} D_k \int_0^\infty P^{(N+1-k)}(l+1-k, t) dB(t) \right),\ l > N,$$

$$P_{i,l} = \int_0^\infty P^{(N+1-i)}(l+1-i, t) dB(t),\ i = \overline{1, N},\ l \geq i-1,$$

$$P_{i,l} = O,\ i > N,\ l \neq i-1 \text{ and } i > 0,\ l < i-1,$$

$$P_{i,l} = \int_0^\infty e^{D(1)t} \otimes dB(t),\ i > N,\ l = i-1$$

for the CA discipline.

The proof of this lemma is fairly obvious if we take into account the probabilistic meaning of the matrices appearing there. In particular, the elements of the matrix

$$\int_0^\infty P^{(N+1-k)}(l+1-k,t)dB(t)$$

specify the probability of accepting $l+1-k$ customers to the system and the corresponding process $\nu_t, t \geq 0$ transitions during a customer service time given that there were $N+1-k$ free places in the queue at the beginning of service.

In the case of the CR discipline, the entries of the matrix

$$-(D_0+\hat{D}_{N+2})^{-1}D_k = \int_0^\infty e^{(D_0+\hat{D}_{N+2})t}D_k dt$$

specify the probabilities that an idle period of the system finishes with a batch of k arrivals, and the process $\nu_t, t \geq 0$ makes the corresponding transitions during the time when the system is empty, $k = \overline{1, N+1}$. Note that the derivation of this expression uses the stability property of the matrix $D_0 + \hat{D}_{N+2}$.

Using the expressions obtained for the transition probabilities of the embedded MC and the algorithms to find the stationary probabilities of the MC described in Sects. 2.4 and 2.5, we can calculate the vectors of the stationary state probabilities of the embedded MC.

3.3.2.2 Calculation of the Vectors of the Stationary State Probabilities of the Embedded Markov Chain

Theorem 3.12 *In the case of the CR discipline, the vectors $\boldsymbol{\pi}_i$, $i = \overline{0, N}$ are calculated by the formulas*

$$\boldsymbol{\pi}_i = \boldsymbol{\pi}_0\Phi_i,\ i = \overline{0, N},$$

where the matrices Φ_i are calculated by the recursive formulas

$$\Phi_0 = I,\ \Phi_l = \sum_{i=0}^{l-1}\Phi_i\bar{P}_{i,l}(I-\bar{P}_{l,l})^{-1},\ l = \overline{1, N},$$

and vector $\boldsymbol{\pi}_0$ is the unique solution of the system

$$\boldsymbol{\pi}_0(I-\bar{P}_{0,0}) = 0,\ \boldsymbol{\pi}_0\sum_{l=0}^{N}\Phi_l\mathbf{e} = 1.$$

Here the matrices $\bar{P}_{i,l}$ are calculated recursively:

$$\bar{P}_{i,l} = P_{i,l} + \bar{P}_{i,l+1}G_l, \ i = \overline{0, N}, \ l = \overline{i, N}, \ \bar{P}_{i,N+1} = O,$$

and the matrices G_i are also calculated recursively

$$G_{N-1} = (I - P_{N,N})^{-1} P_{N,N-1},$$

$$G_i = (I - \sum_{l=i+1}^{N} P_{i+1,l} G_{l-1} G_{l-2} \cdot \ldots \cdot G_{i+1})^{-1} P_{i+1,i}, \ i = \overline{0, N-2}.$$

Theorem 3.13 *In the case of the CA discipline, the stationary distribution of the system states exists if and only if the inequality $\lambda b_1 < 1$ holds. Given that this condition holds, the vectors $\boldsymbol{\pi}_i$, $i \geq 0$ of the stationary state probabilities are calculated by the formulas*

$$\boldsymbol{\pi}_i = \boldsymbol{\pi}_0 \Phi_i, \ i \geq 0,$$

where the matrices Φ_i and the vector $\boldsymbol{\pi}_0$ are determined by the corresponding formulas in the previous theorem if N is supposed to be ∞ and all matrices G_i, $i \geq N$ are equal to the matrix G defined by formula (2.92).

3.3.2.3 The Stationary Distribution of the System States at an Arbitrary Time and the Loss Probability

Having calculated the stationary state distribution of the embedded MC, we are able to find the stationary distribution of the system states at an arbitrary time. Denote

$$p(i, \nu, m) = \lim_{t \to \infty} P\{i_t = i, \nu_t = \nu, m_t = m\}, \ \nu = \overline{0, W}, m = \overline{1, M}.$$

Let $\mathbf{p}_i$, $i \geq 0$ be the vectors of these probabilities renumbered in lexicographical order.

Theorem 3.14 *In the case of the CR disciplines, the vectors $\mathbf{p}_i$, $i = \overline{0, N+1}$, are calculated as follows:*

$$\mathbf{p}_0 = \tau^{-1} \boldsymbol{\pi}_0 (-1) \tilde{D}^{-1},$$

where

$$\tilde{D} = (D_0 + \hat{D}_{N+2}) \otimes I_M,$$

$$\mathbf{p}_i = \tau^{-1}\Big(\boldsymbol{\pi}_0(-1)\tilde{D}^{-1}\sum_{k=1}^{i}\int_0^\infty \tilde{D}_k P^{(N+1-k)}(i-k,t)\otimes(I-B(t))dt$$

$$+\sum_{k=1}^{\min\{i,N\}}\boldsymbol{\pi}_k\int_0^\infty P^{(N+1-k)}(i-k,t)\otimes(I-B(t))dt\Big),\ i=\overline{1,N+1}.$$

The mean time τ between customer service completions is given by the formula

$$\tau = b_1 + \boldsymbol{\pi}_0(-1)\tilde{D}^{-1}\mathbf{e}.$$

Theorem 3.15 *In the case of the CA discipline, the vectors $\mathbf{p}_i$, $i \geq 0$ are calculated as follows:*

$$\mathbf{p}_0 = \tau^{-1}\boldsymbol{\pi}_0(-\tilde{D}_0)^{-1},$$

$$\mathbf{p}_i = \tau^{-1}\big(\boldsymbol{\pi}_0(-\tilde{D}_0)^{-1}+\sum_{k=1}^{i}\int_0^\infty \tilde{D}_k P^{(N+1-k)}(i-k,t)\otimes(I-B(t))dt$$

$$+\sum_{k=1}^{i}\boldsymbol{\pi}_k\int_0^\infty P^{(N+1-k)}(i-k,t)\otimes(I-B(t))dt\big),\ i=\overline{1,N+1},$$

$$\mathbf{p}_{N+l} = \tau^{-1}\big(\boldsymbol{\pi}_0(-\tilde{D}_0)^{-1}\sum_{k=1}^{N+1}\int_0^\infty \tilde{D}_k P^{(N+1-k)}(N+l-k,t)\otimes(I-B(t))dt$$

$$+\sum_{k=1}^{N}\boldsymbol{\pi}_k\int_0^\infty P^{(N+1-k)}(N+l-k,t)\otimes(I-B(t))dt$$

$$+\big(\boldsymbol{\pi}_0(-\tilde{D}_0)^{-1}\tilde{D}_{N+l}+\boldsymbol{\pi}_{N+l}\big)\int_0^\infty e^{D(1)t}\otimes(I-B(t))dt\big),\ l>1.$$

The mean time τ between customer service completions is given by the formula

$$\tau = b_1 + \boldsymbol{\pi}_0(-\tilde{D}_0)^{-1}\mathbf{e}.$$

Theorem 3.16 *The probability P_{loss} of an arbitrary customer loss is calculated by the formula*

$$P_{loss} = 1 - \lambda^{-1} \sum_{i=0}^{N} \sum_{k=1}^{N+1-i} k\mathbf{p}_i \tilde{D}_k \mathbf{e} = 1 - (\tau\lambda)^{-1}$$

in the case of the CR discipline, and by the formula

$$P_{loss} = 1 - \lambda^{-1} \sum_{i=0}^{N} \sum_{k=1}^{\infty} k\mathbf{p}_i \tilde{D}_k \mathbf{e} = 1 - (\tau\lambda)^{-1}$$

in the case of the CA discipline.

Numerical results for these disciplines including a comparison between them and the PA discipline are given in [33]. In [28], the results are extended to systems where the access strategy for a batch of customers at each moment a batch arrives and there are not enough places in the buffer is randomly selected from the set of disciplines PA, CA, and CR.

3.4 $BMAP/PH/N$ Queue

All the systems discussed above in this chapter have a single server. Many real systems have several servers. In this section, we consider a multi-server system with an infinite buffer and a $BMAP$. For single-server systems, the analysis can be carried out for the case when the successive service times of customers are independent and identically distributed random variables (see Sect. 3.1) and even for the case when the successive service times of customers are determined by a semi-Markov process (see Sect. 3.2). This is done above by constructing a multidimensional MC embedded at the service completion moments. In the case of multiple servers, it is impossible to build such a chain. Therefore, when considering the multi-server systems, the most common models, which can be analyzed analytically, are models where the successive service times of customers are independent and identically distributed random variables with a phase-type distribution (PH distribution), see Sect. 2.1.5.

In the following, we assume that all servers of the system are identical, and the distribution of the successive customer service times is governed by a continuous-time MC η_t, $t \geq 0$. The MC has state space $\{1, \dots, M, M+1\}$ with state $M+1$ absorbing. The initial state of the MC at the time a customer service starts is selected randomly according to the probability distribution specified by the entries of vector $\boldsymbol{\beta} = (\beta_1, \dots, \beta_M)$, where $\beta_m \geq 0$, $m = \overline{1, M}$, $\boldsymbol{\beta}\mathbf{e} = 1$. The transitions between states within the set $\{1, \dots, M\}$ happen with the intensities given by the entries the of subgenerator $S = S_{m,m'}$, $m, m' = \overline{1, M}$. Transitions out of the states $\{1, \dots, M\}$

into the absorbing state $M + 1$ happen with the intensities given by the entries of the column vector $\mathbf{S}_0 - S\mathbf{e}$. Transition of the MC η_t into the absorbing state $M + 1$ causes a customer service completion. We assume that matrix $S+\mathbf{S}_0\boldsymbol{\beta}$ is irreducible.

Let us describe the number of customers in the considered $BMAP/PH/N$ queue by a multidimensional continuous-time MC. In fact, in addition to the process i_t, which is the number of customers in the system under consideration at time t, and the $BMAP$ governing process ν_t, this MC should completely describe the current state of the service processes at all servers of the system. Note that there are several ways to describe the current state of the service processes.

The simplest way is to describe the state of the MC η_t governing the service process on each server. If a server is idle, the state of this process can be 0. This method is probably the only possible way if the servers of the system are not identical. But the state space of such a process consists of $(M + 1)^N$ elements and can be very large, which can result in significant problems at the stage of numerical realization of the obtained analytical results.

Another way to specify the governing service processes that essentially exploits the identity of servers and has a smaller dimension is as follows. We specify the state of the MC η_t that governs the service process at each *busy* server at a given moment t. At the same time, it is necessary to make an agreement about a dynamic renumbering of the busy servers in the system. A good renumbering method is, for example, as follows. At each time, the minimum number is assigned to the server which has the longest service time (among all busy servers) passed to the current time moment, and so on, and finally the server with the shortest currently passed service time (the server that started its current service the latest) gets the maximum number. If a server finishes its service and there are no more waiting customers in the system, the server loses its number. The numbers of all servers larger than the number of the server that completed its service are reduced by 1. If there are waiting customers in the queue, the server that has finished the service does not change its number and starts to serve a new customer. At the beginning of the service of a customer that arrived at the system when there were idle servers, one of these servers receives a number, which is the number of servers that are currently busy plus one, and begins serving a customer.

Thus, for this method of specifying the governing service processes (only for busy servers) for $i,\ i < N$ busy servers, the state space of the governing service processes consists of M^i elements. For $i,\ i \geq N$ busy servers, the state space of the governing service processes consists of M^N elements. Therefore, this way of specifying the governing service processes is more economic than the first method. There is an even more economic way but it will be described and used somewhat later.

It is easy to see that the behavior of a $BMAP/PH/N$ queue under consideration is described by a multidimensional MC

$$\zeta_t = \{i_t, \nu_t, \eta_t^{(1)}, \ldots, \eta_t^{(\min\{i_t,N\})}\},\ t \geq 0.$$

Denote by level i the set of process ζ_t states with $i_t = i$, and the rest of the components are renumbered in lexicographical order. Denote also by $Q_{i,j}$ the matrix consisting of the intensities of the MC ζ_t transitions from level i to level j, $j \geq \min\{i-1, 0\}$. Since the customers are each served by one server, it is obvious that $Q_{i,j} = O$ if $j < i-1$, i.e., the block matrix Q consisting of blocks $Q_{i,j}$ is an upper-Hessenberg matrix.

Lemma 3.3 *The infinitesimal generator Q of the MC ζ_t has the following structure:*

$$Q = (Q_{i,j})_{i,j\geq 0} \tag{3.68}$$

$$= \begin{pmatrix} Q_{0,0} & Q_{0,1} & Q_{0,2} & \cdots & Q_{0,N} & Q_{0,N+1} & Q_{0,N+2} & Q_{0,N+3} & \cdots \\ Q_{1,0} & Q_{1,1} & Q_{1,2} & \cdots & Q_{1,N} & Q_{1,N+1} & Q_{1,N+2} & Q_{1,N+3} & \cdots \\ O & Q_{2,1} & Q_{2,2} & \cdots & Q_{2,N} & Q_{2,N+1} & Q_{2,N+2} & Q_{2,N+3} & \cdots \\ \vdots & \vdots & \vdots & \ddots & \vdots & \vdots & \vdots & \vdots & \ddots \\ O & O & O & \cdots & Q_{N-1,N} & Q_{N-1,N+1} & Q_{N-1,N+2} & Q_{N-1,N+3} & \cdots \\ O & O & O & \cdots & R^0 & R_1^+ & R_2^+ & R_3^+ & \cdots \\ O & O & O & \cdots & R^- & R^0 & R_1^+ & R_2^+ & \cdots \\ O & O & O & \cdots & O & R^- & R^0 & R_1^+ & \cdots \\ \vdots & \vdots & \vdots & \ddots & \vdots & \vdots & \vdots & \vdots & \ddots \end{pmatrix},$$

where

$$Q_{i,i} = D_0 \oplus S^{\oplus i},\ i = \overline{0, N-1}, \tag{3.69}$$

$$Q_{i,i-1} = I_{\bar{W}} \otimes \mathbf{S}_0^{\oplus i},\ i = \overline{1, N}, \tag{3.70}$$

$$Q_{i,i+k} = \begin{cases} D_k \otimes I_{M^i} \otimes \boldsymbol{\beta}^{\otimes i},\ i = \overline{0, N-1},\ i+k \leq N, \\ D_k \otimes I_{M^i} \otimes \boldsymbol{\beta}^{\otimes (i+k-N)},\ i = \overline{0, N-1},\ i+k > N, \end{cases} \tag{3.71}$$

$$R^0 = D_0 \oplus S^{\oplus N}, \tag{3.72}$$

$$R^- = I_{\bar{W}} \otimes (\mathbf{S}_0 \boldsymbol{\beta})^{\oplus N}, \tag{3.73}$$

$$R_k^+ = D_k \otimes I_{M^N}, \tag{3.74}$$

where

$$\boldsymbol{\beta}^{\otimes k} \stackrel{def}{=} \underbrace{\boldsymbol{\beta} \otimes \ldots \otimes \boldsymbol{\beta}}_{k}, k \geq 1, \boldsymbol{\beta}^{\otimes 0} \stackrel{def}{=} 1, \tag{3.75}$$

$$S^{\oplus i} \stackrel{def}{=} \underbrace{S \oplus \ldots \oplus S}_{i}, i \geq 1, S^{\oplus 0} \stackrel{def}{=} 0, \tag{3.76}$$

$$S_0^{\oplus i} \stackrel{def}{=} \sum_{m=0}^{i-1} I_{M^m} \otimes S_0 \otimes I_{M^{i-m-1}}, i \geq 1. \tag{3.77}$$

Proof The lemma is easily proved by analogy with the explanation of the form of the three-dimensional MC generator describing the behavior of a $MAP/PH/1$ queue in Sect. 2.2.2. A $MAP/PH/1$ queue is a special case of the $BMAP/PH/N$ model considered in this section with the number of servers $N = 1$ and an ordinary MAP input of customers instead of a $BMAP$. In Sect. 2.2.2, we explained the importance and effectiveness of using the Kronecker product and sum of the matrices when analyzing the multidimensional MCs. Because of the batch nature of the input flow and the presence of $N \geq 1$ servers, the role of these operations here is even greater.

Let us give a brief comment on the form (3.69)–(3.74) of the blocks of generator (3.68). As already noted in Sect. 2.2.2, all entries of the generator mean the intensity of the transitions between the corresponding states of the MC with the exception of the diagonal entries, which are negative and equal in their modulus to the intensity of the MC transition out of the corresponding state. To avoid unnecessary repetition of the text here and in future sections, we shall no longer separately mention the special role of the diagonal elements, and we shall call all the entries of the generator intensities.

A transition of the process ζ_t from level i to the same level during a time interval of infinitesimal length is possible when the governing process ν_t of arrivals makes a transition without generating a customer (the corresponding transition intensities are given by the entries of the matrix D_0) or when the service governing process η_t of one of the busy servers makes a transition not to the absorbing state. The introduced notation (3.76) specifies the intensities of such transitions of the governing process for one of i servers. Hence for $i = \overline{0, N-1}$ we obtain

$$Q_{i,i} = D_0 \otimes I_{M^i} + I_{\bar{W}} \otimes S^{\oplus i} = D_0 \oplus S^{\oplus i},$$

and formula (3.69) is proved. The matrix R^0 has the sense of matrix $Q_{i,i}$ for $i \geq N$. For $i \geq N$ only N customers are being served, therefore

$$Q_{i,i} = D_0 \oplus S^{\oplus N} = R^0, \;\; i \geq N.$$

Formula (3.72) is proved.

A transition of the process ζ_t from level i to level $i-1$ during a time interval of infinitesimal length is possible when the service governing process η_t of one of the busy servers transits to the absorbing state. The introduced notation (3.77) specifies the intensity of the transitions of the governing process of one of i servers to the absorbing state. Hence, for $i = \overline{1, N}$ we obtain

$$Q_{i,i-1} = I_{\bar{W}} \otimes \mathbf{S}_0^{\oplus i}, \; i = \overline{1, N},$$

and formula (3.70) is proved. The matrix R^- has the sense of the matrix $Q_{i,i-1}$ for $i > N$. For $i > N$, only N customers are served. In addition, since the queue is non-empty, as the service time expires at one of the busy servers, the initial phase for the next customer service process at the idle server is immediately determined with the probabilities given by the entries of the vector $\boldsymbol{\beta}$. Thus,

$$Q_{i,i-1} = R^- = I_{\bar{W}} \otimes (\mathbf{S}_0\boldsymbol{\beta})^{\oplus N}, \quad i \geq N.$$

Formula (3.73) is proven.

A transition of the process ζ_t from level i to level $i + k, \;\; k \geq 1$ during a time interval of infinitesimal length is possible when the governing process ν_t of arrivals makes a transition accompanied by a size k batch arrival (the corresponding transition intensities are given by the entries of the matrix D_k). When this happens, if $i \geq N$ then the service governing processes at the N busy servers do not make any transitions. Thus for $i \geq N$, we have

$$Q_{i,i+k} = R_k^+ = D_k \otimes I_{M^N},$$

i.e., formula (3.74) is valid. If $i < N$ and $i + k \leq N$ then the size k batch arrival makes it necessary to determine the initial phase of the next customer service process with probabilities given by the entries of the vector $\boldsymbol{\beta}$ at k servers. The vector $\boldsymbol{\beta}^{\otimes k}$ defined by (3.75) specifies the initial phases for the service process in k servers. If $i < N$ and $i + k > N$ then the size k batch arrival implies the need to determine an initial phase for the next customer service process at only $i + k - N$ servers, and the rest of the customers in the batch do not have an available server and start waiting in the queue. Formula (3.71) is proved. □

It is easy to see that the MC ζ_t belongs to the class of AQTMCs, analysis of which is given in Sect. 2.7. Moreover, starting from level N, the form of the matrices $Q_{i,i+l}$ does not depend on i but on $l, \;\; l \geq -1$ only. Thus, in the algorithm we avoid the problems of calculating the stationary state probability vectors related to the determination of the terminal condition for the backward recursion for matrices $G_i, \; i \geq 0$.

In particular, it is easy to verify that the system under consideration is stable if and only if condition

$$\lambda < Nb_1$$

holds, where b_1 is the average customer service time.

If the input to the system is an ordinary MAP then the generator (3.68) becomes block-three-diagonal, and the stationary state probability vectors of a process ζ_t can be calculated in the similar way as in Sect. 2.2.1. The analysis of the waiting time stationary distribution for an arbitrary customer is almost identical to the analysis of this distribution for a $MAP/PH/1$ queue, taking into account the fact that for all busy servers, the set of N servers has the same productivity as one server with the subgenerator $S^{\oplus N}$ of the PH service governing process.

3.5 $BMAP/PH/N/N$ Queue

In some real-life systems, there are no places to wait for customers arriving when all servers are busy, and a customer arriving at the system when all servers are busy is lost. Since the input of customers is assumed to be a batch arrival process, there are possible situations when there are idle servers in the system but the number of arriving customers exceeds the number of idle servers. As in Sect. 3.3, we consider the different disciplines to admit customers: partial admission (PA), complete admission (CA), and complete rejection (CR).

The importance of studying the $BMAP/PH/N/N$ queue is explained, in particular, by the fact that it is a natural essential generalization of the $M/M/N/N$ queue, whose analysis, pioneered by A.K. Erlang, began the history of queuing theory at the beginning of the twentieth century. Results of this queue analysis by Erlang were applied to the calculation of the performance characteristics of telephone networks, and the calculated characteristics (for example, the connection failure probability) agreed well enough with the measurement results on real networks. This was somewhat unexpected, since the duration of a telephone call is not exponentially distributed as was suggested by the Erlang model. This good match between the results of calculation and measurements was later explained by the fact of the invariance property of the stationary distribution of the number of customers in an $M/G/N/N$ queue with an arbitrary service time distribution with respect to the type of this distribution. That is, the stationary distribution of the number of customers in an $M/G/N/N$ queue is the same as in an $M/M/N/N$ queue with the same average service time. This fact was rigorously proved by B.A. Sevastyanov.

Numerical experiments conducted on the basis of the algorithmic results presented in this section indicate that the invariance property of the stationary distribution of the number of customers in a queue with $BMAP$ is not valid. Therefore, it is necessary to analyze $BMAP/PH/N/N$ queues with service time distribution close to the sample distribution obtained as a result of monitoring the system operation rather than an essentially simpler $BMAP/M/N/N$ queue with exponentially distributed service time.

3.5.1 The Stationary Distribution of the System States for the PA Discipline

Recall that the PA discipline implies that in at case when a size k batch of customers arrives at the system and there are exactly m idle servers then $min\{k, m\}$ customers of the batch immediately start their service. If $k > m$ then m customers start their service, and $k - m$ others are lost.

It is easy to see that the system behavior is described by the MC

$$\xi_t = \{i_t, \nu_t, \eta_t^{(1)}, \dots, \eta_t^{(i_t)}\}, t \geq 0,$$

where i_t is the number of customers in the system, ν_t is the state of the $BMAP$ governing process, $\eta_t^{(i)}$ is the state of the PH service governing process at the ith server, $i_t = \overline{0, N}$, $\nu_t = \overline{0, W}$, $\eta_t^{(i)} = \overline{1, M}$, $i = \overline{1, N}$, $t \geq 0$.

Lemma 3.4 *The infinitesimal generator Q of the MC $\xi_t, t \geq 0$ has the following structure:*

$$Q = (Q_{m,m'})_{m,m'=\overline{0,N}} \tag{3.78}$$

$$= \begin{pmatrix} D_0 & T_1^1(0) & T_2^2(0) & \dots & T_{N-1}^{N-1}(0) & \sum\limits_{k=0}^{\infty} T_{N+k}^{N}(0) \\ I_{\overline{W}} \otimes S_0^{\oplus 1} & D_0 \oplus S^{\oplus 1} & T_1^1(1) & \dots & T_{N-2}^{N-2}(1) & \sum\limits_{k=0}^{\infty} T_{N+k-1}^{N-1}(1) \\ 0 & I_{\overline{W}} \otimes S_0^{\oplus 2} & D_0 \oplus S^{\oplus 2} & \dots & T_{N-3}^{N-3}(2) & \sum\limits_{k=0}^{\infty} T_{N+k-2}^{N-2}(2) \\ \dots & \dots & \dots & \dots & \dots & \dots \\ 0 & 0 & 0 & \dots & I_{\overline{W}} \otimes S_0^{\oplus N} & \sum\limits_{k=0}^{\infty} D_k \oplus S^{\oplus N} \end{pmatrix},$$

where

$$T_k^r(m) = D_k \otimes I_{M^m} \otimes \boldsymbol{\beta}^{\otimes r}, \ k \geq 1, \ m = \overline{0, N}, \ r = \overline{0, N}.$$

Since the MC is irreducible and its state space is finite, for any values of the system parameters there exists the stationary state distribution

$$p(i, \nu, m^{(1)}, \dots, m^{(i)}) = \lim_{t \to \infty} P\{i_t = i, \nu_t = \nu, \eta_t^{(1)} = m^{(1)}, \dots, \eta_t^{(i)} = m^{(i)}\}.$$

By means of a lexicographic ordering of the MC ξ_t states, we form vectors $\mathbf{p}_i$, $i = \overline{0, N}$ of the probabilities of the MC states having the value i in their first component. Denote also $\mathbf{p} = (\mathbf{p}_0, .., \mathbf{p}_N)$.

The vector $\mathbf{p}$ satisfies the system of equations

$$\mathbf{p}Q = \mathbf{0}, \ \mathbf{pe} = 1, \tag{3.79}$$

where the generator Q of the MC ξ_t is given by formula (3.78).

The proof of Lemma 3.4 is similar to the proof of Lemma 3.3 above.

In the case when the dimension of the vector $\mathbf{p}$ is small, the finite system of Eq. (3.79) with matrix Q given by formula (3.78) can be easily solved by a computer. But the size of matrix Q is equal to $K = \bar{W}\frac{M^{N+1}-1}{M-1}$ and can be rather large. For example, if $\bar{W} = 2, M = 2$ the size is 126 for $N = 5$, it is already 254 for $N = 6$, and so on. Therefore, for a large number N of servers, which is typical for modern telecommunications networks, a solution of system (3.79) can be quite long or impossible at all because of a lack of computer RAM.

Fortunately, matrix Q has a special structure (it is upper Hessenberg). This allows us to develop more efficient procedures to solve system (3.79). One of the possible procedures consists of sequential elimination of the block components of vector $\mathbf{p}_i, \ i = 0, \ldots, N-1$, of the unknown vector $\mathbf{p}$. Namely, first we exclude vector $\mathbf{p}_0$ from the first equation of system (3.79) then vector $\mathbf{p}_1$ from the second equation of system (3.79), etc., and finally we exclude vector $\mathbf{p}_{N-1}$ from the Nth equation. The resulting system has $\bar{W}M^N$ unknown entries of vector $\mathbf{p}_N$. The rank of the matrix of the system is $\bar{W}M^N - 1$. Replacing one of the equations of this system by the equation obtained from the normalization condition, we calculate the unique solution $\mathbf{p}_N$ of this system. Then we successively calculate all vectors $\mathbf{p}_i, i = N-1, \ldots, 1, 0$. This procedure can be easily performed by a computer. But it is not sufficiently stable numerically due to the presence of the operation of successive subtraction of matrices or vectors. An alternative algorithm, which is stable in computer realization and is based on the use of the probabilistic meaning of vector $\mathbf{p}$ described in Sect. 2.5, leads to the procedure for calculating the vectors $\mathbf{p_i}, \ i = 0, \ldots, N$ given by the following statement.

Theorem 3.17 *The stationary state probability vectors* $\mathbf{p}_i, \ i = 0, \ldots, N$ *are calculated by the formula*

$$\mathbf{p}_l = \mathbf{p}_0 F_l, l = \overline{1, N}$$

where the matrices F_l *are calculated from recurrent formulas*

$$F_l = (\bar{Q}_{0,l} + \sum_{i=1}^{l-1} F_i \bar{Q}_{i,l})(-\bar{Q}_{l,l})^{-1}, \ l = \overline{1, N-1}, \tag{3.80}$$

$$F_N = (Q_{0,N} + \sum_{i=1}^{N-1} F_i Q_{i,N})(-Q_{N,N})^{-1}, \tag{3.81}$$

where matrices $\overline{Q}_{i,N}$ are given by the backward recursion

$$\overline{Q}_{i,N} = Q_{i,N}, i = \overline{0, N},$$

$$\overline{Q}_{i,l} = Q_{i,l} + \overline{Q}_{i,l+1} G_l, i = \overline{0, l}, l = N-1, N-2, \ldots, 0,$$

the matrices G_i, $i = \overline{0, N-1}$ are given by the backward recursion

$$G_i = (-Q_{i+1,i+1} - \sum_{l=1}^{N-i-1} Q_{i+1,i+1+l} G_{i+l} G_{i+l-1} \ldots G_{i+1})^{-1} Q_{i+1,i},$$

$$i = N-1, N-2, \ldots, 0,$$

and the vector $\mathbf{p}_0$ is the unique solution of the system of linear algebraic equations

$$\mathbf{p}_0 \overline{Q}_{0,0} = \mathbf{0}, \tag{3.82}$$

$$\mathbf{p}_0 \left(\sum_{l=1}^{N} F_l \mathbf{e} + \mathbf{e} \right) = 1. \tag{3.83}$$

Remark 3.5 The procedure for computing vectors $\mathbf{p}_i$, $i = 0, \ldots, N$ can be simplified in order to reduce the computer memory required for its implementation. To do this, we first solve the system of linear algebraic equations (3.82), which has a unique solution up to a constant. Denote this solution by $\tilde{\mathbf{p}}_0$ and denote the constant by γ. Now instead of matrices F_l, $l = \overline{0, N}$, we can work with vectors $\mathbf{f}_l$, given by formulas $\mathbf{f}_l = \gamma \tilde{\mathbf{p}}_0 F_l$. Namely, instead of recursively calculating matrices F_l, $l = \overline{0, N}$ by formulas (3.80) and (3.81), we can calculate the vectors $\mathbf{f}_l$, $l = \overline{0, N}$ recursively by the formulas

$$\mathbf{f}_l = (\tilde{\mathbf{p}}_0 \bar{Q}_{0,l} + \sum_{i=1}^{l-1} \mathbf{f}_i \bar{Q}_{i,l})(-\bar{Q}_{l,l})^{-1}, l = \overline{1, N-1},$$

$$\mathbf{f}_N = (\tilde{\mathbf{p}}_0 Q_{0,N} + \sum_{i=1}^{N-1} \mathbf{f}_i Q_{i,N})(-Q_{N,N})^{-1}.$$

The value of the unknown constant γ can be found now from the normalization condition

$$\sum_{l=0}^{N} \mathbf{f}_l \mathbf{e} = \gamma,$$

and vectors $\mathbf{p}_i,\ i = 0, \ldots, N$, are calculated by formulas $\mathbf{p}_i = \frac{\mathbf{f}_i}{\gamma},\ i = 0, \ldots, N$.

Once we know the vector $\mathbf{p}$, we can calculate the various characteristics of the system performance. For example, one of the basic performance characteristic is the probability P_{loss} of an arbitrary customer loss, and it can be calculated be means of the following statement.

Theorem 3.18 *The probability P_{loss} of an arbitrary customer loss is calculated as follows:*

$$P_{loss} = 1 - \lambda^{-1} \sum_{i=0}^{N-1} \mathbf{p}_i \sum_{k=0}^{N-i} (k + i - N) \tilde{D}_k^{(i)} \mathbf{e}, \tag{3.84}$$

where $\tilde{D}_k^{(i)} = D_k \otimes I_{M^i}, k \geq 0$.

Proof The brief proof of this statement is as follows. According to the formula of total probability, the probability P_{loss} is calculated as

$$P_{loss} = 1 - \sum_{i=0}^{N-1} \sum_{k=1}^{\infty} P_k P_i^{(k)} R^{(i,k)}, \tag{3.85}$$

where P_k is the probability that our arbitrary customer arrives in a size k batch of customers; $P_i^{(k)}$ is the probability that i servers are busy when the size k batch arrives; $R^{(i,k)}$ is the probability that our arbitrary customer shall not be lost given that it arrives in a size k batch of customers and i servers are busy upon its arrival.

Using the $BMAP$ properties described in Sect. 2.1.3, it can be shown that

$$P_i^{(k)} = \frac{\mathbf{p}_i \tilde{D}_k^{(i)} \mathbf{e}}{\boldsymbol{\theta} D_k \mathbf{e}}, i = \overline{0, N-1}, k \geq 1, \tag{3.86}$$

$$P_k = \frac{k \boldsymbol{\theta} D_k \mathbf{e}}{\boldsymbol{\theta} \sum_{l=1}^{\infty} l D_l \mathbf{e}} = k \frac{\boldsymbol{\theta} D_k \mathbf{e}}{\lambda}, k \geq 1, \tag{3.87}$$

$$R^{(i,k)} = \begin{cases} 1 & , k \leq N - i, \\ \frac{N-i}{k}, & k > N - i, i = \overline{0, N-1}. \end{cases} \tag{3.88}$$

Substituting (3.86)–(3.88) into (3.85), after some algebraic transformations we obtain (3.84). □

3.5.2 The Stationary Distribution of the System States for the CR Discipline

According to the definition of the CR discipline, an arbitrary batch of customers is completely lost if the number of idle servers is less than the batch size upon its arrival.

The stationary behavior of the system under consideration with such an admission discipline is described by a continuous-time multidimensional MC $\xi_t, t \geq 0$ with the same state space as the MC studied in the previous subsection. But the generator of this MC is somewhat different.

Lemma 3.5 *The infinitesimal generator Q of the MC $\xi_t, t \geq 0$ has the following block structure:*

$$Q = (Q_{m,m'})_{m,m'=\overline{0,N}} \tag{3.89}$$

$$\begin{pmatrix} D_0 + \sum\limits_{k=N+1}^{\infty} D_k & \mathcal{T}_1^1(0) & \mathcal{T}_2^2(0) & \dots & \mathcal{T}_N^N(0) \\ I_{\overline{W}} \otimes S_0^{\oplus 1} & (D_0 + \sum\limits_{k=N}^{\infty} D_k) \oplus S^{\oplus 1} & \mathcal{T}_1^1(1) \dots & \dots & \mathcal{T}_{N-1}^{N-1}(1) \\ 0 & I_{\overline{W}} \otimes S_0^{\oplus 2} & (D_0 + \sum\limits_{k=N-1}^{\infty} D_k) \oplus S^{\oplus 2} \dots & \dots & \mathcal{T}_{N-2}^{N-2}(2) \\ \dots & \dots & \dots & \dots & \dots \\ 0 & 0 & 0 & \dots & \sum\limits_{k=0}^{\infty} D_k \oplus S^{\oplus N} \end{pmatrix}.$$

Comparing the generators given by formulas (3.78) and (3.89), we see that they differ only in the form of the diagonal blocks and the last block column.

The vector $\mathbf{p}$ of the stationary state probabilities can be calculated using the algorithm given by Theorem 3.17.

Theorem 3.19 *The probability P_{loss} of an arbitrary customer loss for a system with the CR admission discipline is calculated as follows:*

$$P_{loss} = 1 - \lambda^{-1} \sum_{i=0}^{N-1} \mathbf{p}_i \sum_{k=0}^{N-i} k\tilde{D}_k^{(i)} \mathbf{e}.$$

The proof completely coincides with the proof of Theorem 3.18 except that the conditional probability $R^{(i,k)}$ here has the different form

$$R^{(i,k)} = \begin{cases} 1, & k \leq N - i, \\ 0, & k > N - i, \ i = \overline{0, N-1}. \end{cases}$$

3.5.3 The Stationary Distribution of the System States for the CA Discipline

According to this admission discipline, an arbitrary batch of customers is completely admitted to the system if at least one server is idle upon the arrival of the batch. A number of customers corresponding to the number of idle servers immediately start their service. The other customers wait until they can get service.

The stationary behavior of the system under consideration is described by a continuous-time multidimensional MC $\xi_t = \{i_t, \nu_t, \eta_t^{(1)}, \ldots, \eta_t^{(min\{N,i_t\})}\}, \ t \geq 0$. Contrary to the processes describing the behavior of the system in the case of the PA and CR disciplines, the state space of the process $\xi_t, t \geq 0$ is infinite if the batch size is unlimited.

Lemma 3.6 *The infinitesimal generator Q of the MC $\xi_t, t \geq 0$ has the following block structure:*

$$Q_{m,m'} = \begin{cases} D_{m'} \otimes \boldsymbol{\beta}^{\otimes m'}, m = 0, m' = \overline{0, N}, \\ D_{m'} \otimes \boldsymbol{\beta}^{\otimes N}, m = 0, m' > N, \\ I_{\bar{W}} \otimes \mathbf{S}_0^{\oplus(m'+1)}, m = \overline{1, N}, m' = m - 1, \\ I_{\bar{W}} \otimes (\mathbf{S}_0\boldsymbol{\beta})^{\oplus N}, m > N, m' = m - 1, \\ D_0 \oplus S^{\oplus m}, m = \overline{1, N-1}, m' = m, \\ D(1) \oplus S^{\oplus N}, m \geq N, m' = m, \\ D_{m'-m} \otimes I_{M^m} \otimes \boldsymbol{\beta}^{\otimes(m'-m)}, m = \overline{1, N-1}, m < m' < N, \\ D_{m'-m} \otimes I_{M^m} \otimes \boldsymbol{\beta}^{\otimes(N-m)}, m = \overline{1, N-1}, m' \geq N, \\ 0, \textit{otherwise}. \end{cases}$$

Note that matrices $Q_{i,j}$ are of size $\bar{W}M^i \otimes \bar{W}M^j$ for $i \leq N, j \leq N$ and of size $\bar{W}M^N \otimes \bar{W}M^N$ for $i > N, j > N$.

Denote the stationary state probabilities of the MC $\xi_t, t \geq 0$ by

$$p(i, \nu, m^{(1)}, .., m^{min\{(i,N)\}}) = \lim_{t\to\infty} P\{i_t = i, \nu_t = \nu, \eta_t^{(1)} = m^{(1)}, .., \eta_t^{min\{(i,N)\}}\}$$

and consider the row vectors $\mathbf{p}_i, i \geq 0$ of these probabilities renumbered in lexicographic order. The vector $\mathbf{p}_i, i = \overline{0, N-1}$ is of size $\bar{W}M^i$, and vectors $\mathbf{p}_i, i \geq N$ are of size $\bar{W}M^N$.

Since the state space of the process $\xi_t, t \geq 0$, is infinite, the existence of the presented stationary state probabilities is not obvious. But under standard assumptions about the $BMAP$ and the PH-type service process, their existence can be fairly easily proved using the results of Sect. 2.4, since it is easy to verify that the MC $\xi_t, t \geq 0$ belongs to the class of the $M/G/1$-type MCs or the class of quasi-Toeplitz MCs.

Theorem 3.20 *The vectors* $\mathbf{p}_i,\ i \geq 0$ *of the stationary state probabilities are given by the formula*

$$\mathbf{p}_l = \mathbf{p}_0 F_l, l \geq 1,$$

where the matrices F_l *are recursively calculated by the formulas*

$$F_l = (\bar{Q}_{0,l} + \sum_{i=1}^{l-1} F_i \bar{Q}_{i,l})(-\bar{Q}_{l,l})^{-1}, l \geq 1,$$

the matrices $\bar{Q}_{i,l}$ *are given by the formulas*

$$\bar{Q}_{i,l} = Q_{i,l} + \sum_{k=1}^{\infty} Q_{i,l+k} G^{max\{0,l+k-N\}} \cdot G_{min\{N,l+k\}-1} \cdot G_{min\{N,l+k\}-2} \cdot .. \cdot G_l,$$

$$i = 0, \ldots, l, l \geq 1,$$

the matrix G *has the form*

$$G = I_{\bar{W}} \otimes (S_0 \boldsymbol{\beta})^{\oplus N} (-D(1) \oplus S^{\oplus N})^{-1},$$

the matrices $G_i, i = \overline{0, N-1}$ *are calculated by the backward recursion*

$$G_i = -(Q_{i+1,i+1} + \sum_{l=i+2}^{\infty} Q_{i+1,l} G^{max\{0,l-N\}} G_{min\{N,l\}-1}$$

$$\times G_{min\{N,l\}-2} \cdots G_{i+1})^{-1} Q_{i+1,i}, i = N-1, N-2, \ldots, 0,$$

and the vector $\mathbf{p}_0$ *is the unique solution of the system*

$$\mathbf{p}_0 \bar{Q}_{0,0} = \mathbf{0},$$

$$\mathbf{p}_0 (\sum_{l=1}^{\infty} F_l \mathbf{e} + \mathbf{e}) = 1.$$

Theorem 3.21 *The probability* P_{loss} *of an arbitrary customer loss for a system with the* CA *admission discipline is calculated as follows:*

$$P_{loss} = 1 - \lambda^{-1} \sum_{i=0}^{N-1} \mathbf{p}_i \sum_{k=1}^{\infty} k \tilde{D}_k^{(i)} \mathbf{e}.$$

The proof completely coincides with the proof of Theorem 3.18, except that the conditional probability $R^{(i,k)}$ here has the other form

$$R^{(i,k)} = \begin{cases} 1, i \leq N-1, \\ 0, i > N-1. \end{cases}$$

Numerical examples illustrating the calculation of the stationary probabilities of the system states and the loss probability and showing the effect of the $BMAP$ correlation, the variance of the service time distribution, and the type of customer admission discipline are given in [75]. As mentioned above, the invariance property of an arbitrary customer loss with respect to the service time distribution, which is valid for a stationary Poisson input, is not valid in the case of a $BMAP$. This should be taken into account when designing real systems in which the input flow is not a stationary Poisson flow, including modern telecommunications networks.

Chapter 4
Retrial Queuing Systems with Correlated Input Flows and Their Application for Network Structures Performance Evaluation

The theory of retrial queuing systems is an essential part of queuing theory. In such systems, when a customer enters the system and finds the servers busy it does not join a limited or unlimited queue as happens in queuing systems with waiting room and does not leave the system forever as happens in queuing systems with losses. Such a customer leaves the system for a random time interval (it is said to "go to the orbit") and later repeats its attempt to be served. We suggest that the customer repeats its attempts until it receives access to the server.

The importance of this part of the theory is defined by the wide scope of its practical applications. The scope of application lies in the evaluation of performance and the design of phone networks, local computing networks with random multiple access protocols, broadcast networks, and mobile cellular radio networks. The fact of repeated attempts is an integral feature of these and many other real systems, so disregarding the fact (for instance, approximation of queuing systems by relevant queuing systems with losses) may lead to significant inaccuracies in engineering solutions. Notice that random processes describing the behavior of retrial queuing systems are much more complex than similar processes in queuing systems with waiting room and queuing systems with losses. This explains the fact that the theory of retrial queuing systems is less developed.

4.1 $BMAP/SM/1$ Retrial System

The description of a $BMAP/SM/1$ queuing system with waiting room is provided in Sect. 3.2. The input flow is characterized by a matrix generating function $D(z)$, while the service time is characterized by a semi-Markov kernel $B(t)$. In contrast to such a system, we suggest that a customer entering the analyzed system and finding the server busy does not join the queue and does not leave the system forever. That customer goes to the orbit and makes a new effort to get service after a random

A. N. Dudin et al., *The Theory of Queuing Systems with Correlated Flows*,
https://doi.org/10.1007/978-3-030-32072-0_4

amount of time. We assume that the total flow of customers from the orbit in the period of time when there are i customers in the orbit is described by a stationary Poisson process with rate α_i, $i \geq 0$. Below we will analyze several special types of the dependence of α_i on i. But first we analyze the general case.

4.1.1 Stationary Distribution of Embedded Markov Chain

We start the analysis of the system from the Markov chain $\{i_n, \nu_n, m_n\}$, $n \geq 1$ embedded at the service completion epochs t_n, where i_n is the number of customers in the orbit at the moment $t_n + 0$, $i_n \geq 0$, ν_n is the state of the BMAP underlying process ν_n at the moment t_n, and m_n is the state of the service underlying semi-Markov process m_t at the moment $t_n + 0$. Assume that $P\{(i, \nu, m) \to (l, \nu', m')\}$ are the one-step transition probabilities of the chain under consideration. Let us arrange the states of the chain in lexicographic order and compose the matrices of transition probabilities as follows:

$$P_{i,l} = (P\{(i, \nu, m) \to (l, \nu', m')\})_{\nu,\nu'=\overline{0,W},\ m,m'=\overline{1,M}}.$$

Lemma 4.1 *The matrices $P_{i,l}$ are defined as follows:*

$$P_{i,l} = \alpha_i(\alpha_i I - \tilde{D}_0)^{-1} Y_{l-i+1} + (\alpha_i I - \tilde{D}_0)^{-1} \sum_{k=1}^{l-i+1} \tilde{D}_k Y_{l-i-k+1}, \tag{4.1}$$

$$l \geq \max\{0, i-1\}, \quad i \geq 0,$$

$$P_{i,l} = O,\ l < i - 1,$$

where $\tilde{D}_i = D_i \otimes I_M$, $i \geq 0$ and the matrices Y_l are defined as coefficients of the expansion

$$\sum_{l=0}^{\infty} Y_l z^l = \hat{\beta}(-D(z)) = \int_0^{\infty} e^{D(z)t} \otimes dB(t).$$

The proof of the lemma consists of using the formula of total probability taking into account that the matrices

$$\alpha_i(\alpha_i I - \tilde{D}_0)^{-1} = \alpha_i \int_0^{\infty} e^{\tilde{D}_0 t} e^{-\alpha_i t} dt$$

and

$$(\alpha_i I - \tilde{D}_0)^{-1}\tilde{D}_k = \int_0^{\infty} e^{\tilde{D}_0 t} e^{-\alpha_i t} \tilde{D}_k dt$$

define the transition probabilities of the process $\{\nu_n, m_n\}$ during the period of time from the service completion of a customer till the beginning the service of the next customer. Moreover, the first of these matrices corresponds to the case when the next customer is an orbital customer, while the second matrix describes the case when such a customer is a new (primary) customer arriving at the system in a batch of size k, $k \geq 1$.

Analyzing transition probability matrices $P_{i,l}$ of the form (4.1), we notice that when the retrial rate α_i does not depend on i, i.e., $\alpha_i = \gamma$, $i \geq 1$, $\gamma > 0$, such matrices depend on i, l only via the difference $i - l$. In this case, the Markov chain $\{i_n, \nu_n, m_n\}$, $n \geq 1$ belongs to the class of quasi-Toeplitz Markov chains. From the practical point of view, this case has a double interpretation. First variant: if there are i customers in the orbit, each customer can make repeated attempts in time exponentially distributed with parameter $\frac{\gamma}{i}$ regardless of other orbital customers. Second variant: only one of the customers is allowed to make repeated attempts in time exponentially distributed with parameter γ. We will return to this case below. However, researchers in the field of retrial queuing systems mainly consider the case where each of the orbital customers makes repeated attempts regardless of other customers in time exponentially distributed with parameter α. In this case we get the following: $\alpha_i = i\alpha$, $\alpha > 0$. They also consider the following dependence type: $\alpha_0 = 0$, $\alpha_i = i\alpha + \gamma$, $i > 0$, combining both the abovementioned possibilities.

In both cases ($\alpha_i = i\alpha$, $\alpha_i = i\alpha + \gamma$) the transition probability matrices $P_{i,l}$ depend not only on the difference $l - i$ of the values l and i, but also on the value i itself. This means that the analyzed chain does not belong to the class of quasi-Toeplitz Markov chains. Letting i tend to infinity in (4.1) and assuming that $\alpha_i \to \infty$, we see that

$$\lim_{i \to \infty} P_{i,i+l-1} = Y_l, \ l \geq 0$$

and $\sum\limits_{l=0}^{\infty} Y_l$ is a stochastic matrix.

It follows that the chain $\xi_n = \{i_n, \nu_n, m_n\}$, $n \geq 1$ belongs to the class of asymptotically quasi-Toeplitz Markov chains.

Now we will establish the stability condition for the Markov chain ξ_n, $n \geq 1$. We assume the existence of the limit $\lim\limits_{i \to \infty} \alpha_i$ and differentiate two cases:

$$\lim_{i \to \infty} \alpha_i = \infty$$

and

$$\lim_{i\to\infty} \alpha_i = \gamma < \infty.$$

Theorem 4.1 *If* $\lim_{i\to\infty} \alpha_i = \infty$, *the stationary distribution for Markov chains* ξ_n, $n \geq 1$ *exists if the following inequality holds:*

$$\rho < 1, \tag{4.2}$$

where $\rho = \lambda b_1$, λ *is the rate of* $BMAP$ *flow, and* b_1 *is the mean service time.*

If $\lim_{i\to\infty} \alpha_i = \gamma$, *the stationary distribution of the chain* ξ_n, $n \geq 1$ *exists if the following inequality holds:*

$$\mathbf{x}(\hat{\beta}'(1) + (\gamma I - \tilde{D}_0)^{-1}\hat{\beta}(1)\tilde{D}'(1))\mathbf{e} < 1, \tag{4.3}$$

where the vector $\mathbf{x}$ *is the unique solution of the system of linear algebraic equations*

$$\mathbf{x}\left(I - \hat{\beta}(1) - (\gamma I - \tilde{D}_0)^{-1}\hat{\beta}(1)\tilde{D}(1)\right) = \mathbf{0}, \quad \mathbf{x}\mathbf{e} = 1. \tag{4.4}$$

Proof In the case $\lim_{i\to\infty} \alpha_i = \infty$, the matrix generating function $Y(z)$ has the form

$$Y(z) = \hat{\beta}(-D(z)) = \int_0^\infty e^{D(z)t} \otimes dB(t).$$

Thus the proof of formula (4.2) is similar to the proof of Theorem 2.4.

If $\lim_{i\to\infty} \alpha_i = \gamma$, the generating function $Y(z)$ has the following form:

$$Y(z) = \hat{\beta}(-D(z)) + (\gamma I - \tilde{D}_0)^{-1}\tilde{D}(z)\hat{\beta}(-D(z)),$$

and formulas (4.3) and (4.4) follow from Corollary 2.7. □

Below we assume that the parameters of the system satisfy the conditions (4.2) or (4.3) and (4.4). So the stationary probabilities of the Markov chain ξ_n, $n \geq 1$ exist. We denote them by

$$\pi(i, \nu, m) = \lim_{n\to\infty} P\{i_n = i, \nu_n = \nu, m_n = m\}, \ i \geq 0, \nu = \overline{0, W}, m = \overline{1, M}.$$

Let $\boldsymbol{\pi}_i$ be the vector of probabilities of the states with the value i of denumerable components arranged in lexicographic order, $i \geq 0$.

In the case of an arbitrary dependence of the total retrial rate α_i on the number i of orbital customers, the calculation of the vectors $\boldsymbol{\pi}_i,\ i \geq 0$ is possible only based on the algorithm provided in Sect. 2.6.2.

Let us try to analyze in detail the case $\alpha_i = i\alpha + \gamma,\ \alpha \geq 0,\ \gamma \geq 0,\ \alpha + \gamma \neq 0$, using the method of generating functions. Denote $\boldsymbol{\Pi}(z) = \sum\limits_{i=0}^{\infty} \boldsymbol{\pi}_i z^i,\ |z| < 1$.

Theorem 4.2 *If* $\alpha > 0$, *the vector generating function* $\boldsymbol{\Pi}(z)$ *of the stationary distribution of the embedded Markov chain* $\xi_n,\ n \geq 1$ *satisfies the linear matrix differential-functional equation*

$$\boldsymbol{\Pi}'(z) = \boldsymbol{\Pi}(z)\tilde{S}(z) + \boldsymbol{\pi}_0 \gamma \alpha^{-1} z^{-1} \Phi(z). \tag{4.5}$$

If $\alpha = 0$, *the function* $\boldsymbol{\Pi}(z)$ *satisfies the linear matrix functional equation*

$$\boldsymbol{\Pi}(z)\left(zI - \hat{\beta}(-D(z)) - (\gamma I - \tilde{D}_0)^{-1}\tilde{D}(z)\hat{\beta}(-D(z))\right)$$

$$= \boldsymbol{\pi}_0\left((-\tilde{D}_0)^{-1} - (\gamma I - \tilde{D}_0)^{-1}\right)\tilde{D}(z)\hat{\beta}(-D(z)). \tag{4.6}$$

Here

$$\tilde{S}(z) = \Phi^{-1}(z)\Phi'(z) + z^{-1}\Phi^{-1}(z)A\Phi(z) - z^{-1}\alpha^{-1}\tilde{D}_0\Phi(z),$$

$$\Phi(z) = \tilde{D}_0^{-1}\tilde{D}(z)\hat{\beta}(-D(z))(\hat{\beta}(-D(z)) - zI)^{-1},\ A = (\tilde{D}_0 - \gamma I)\alpha^{-1}. \tag{4.7}$$

Notice that the matrix function $\Phi(z)$ defines the following functional equation:

$$\boldsymbol{\Pi}(z) = \boldsymbol{\Pi}(0)\Phi(z)$$

for a $BMAP/SM/1$ queue with an infinite buffer, according to formula (3.34).

Proof The Chapman-Kolmogorov equations for the vectors of stationary probabilities $\boldsymbol{\pi}_i,\ i \geq 0$ have the following form:

$$\boldsymbol{\pi}_j = \sum_{i=0}^{j+1} \boldsymbol{\pi}_i P_{i,j},\ j \geq 0. \tag{4.8}$$

Substituting into (4.8) the transition probability matrices (4.1), multiplying the equations of system (4.8) by corresponding power of z and summing over i, we get the equation

$$\boldsymbol{\Pi}(z)(zI - \hat{\beta}(-D(z))) = \sum_{i=0}^{\infty} \boldsymbol{\pi}_i z^i (\alpha_i I - \tilde{D}_0)^{-1}\tilde{D}(z)\hat{\beta}(-D(z)). \tag{4.9}$$

Now we substitute into (4.9) the explicit expression for the retrial rate $\alpha_i = i\alpha + \gamma$, $i > 0$. In the case $\alpha = 0$ the functional equation (4.6) is obtained in a trivial way. Now let us consider the case $\alpha > 0$. Applying the expansion of a matrix function in spectrum of the matrix $\tilde{D}_0$ (see Appendix A), we can prove the fulfillment of the following statement:

$$\sum_{i=0}^{\infty} \boldsymbol{\pi}_i z^i (\alpha_i I - \tilde{D}_0)^{-1} = \alpha^{-1} \int \boldsymbol{\Pi}(z) z^{-A-I} dz z^A, \tag{4.10}$$

where the matrix A is defined in (4.7).

Substituting (4.10) into (4.9), we get the integral equation for the vector generating function $\boldsymbol{\Pi}(z)$. Differentiating this equation, we get the differential-functional equation (4.5). When $\gamma = 0$, this equation is a linear matrix differential equation. □

If $\alpha = 0$, functional equation (4.6) can be solved using the algorithm presented in Sect. 2.4.2.

If $\alpha > 0$, Eq. (4.5) can be solved easily if the input flow is a stationary Poisson one. In the general case, the solution of equation (4.5) is problematic as the matrices $\tilde{S}(z)$ and $\int \tilde{S}(z)dz$ are not commutative in general, so the solution does not have the form of a matrix exponent. The problem of solving the equation is complicated by two factors: the initial state $\boldsymbol{\pi}_0$ is unknown and the matrix $\tilde{S}(z)$ has singularities in the unit disk of the complex plane. The latter is seen from the form of the matrix $\tilde{S}(z)$ in (4.7) and the known fact that the matrix $\hat{\beta}(-D(z)) - zI$ has W points of singularity in the unit disk. Thus, as in the case of arbitrary dependence of α_i on i, we may recommend to define the vectors of stationary probability using the algorithm presented in Sect. 2.4.2. Note that Eq. (4.5) can be effective in recurrent calculations of factorial moments of the number of customers in the system. In particular, from (4.5) can be derived the following statement:

$$\boldsymbol{\Pi}'(1)\mathbf{e} = \left[\boldsymbol{\pi}_0 \gamma \alpha^{-1} \Phi_0 + \boldsymbol{\Pi}(1)\Phi_0^{-1}(\Phi_1 + (A - \Phi_0 \alpha^{-1} \tilde{D}_0)\Phi_0)\right]\mathbf{e}, \tag{4.11}$$

where the matrices Φ_0 and Φ_1 are the coefficients of the expansion of the function $\Phi(z)$, defined by (4.7):

$$\Phi(z) = \Phi_0 + \Phi_1(z-1) + o(z-1).$$

Relation (4.11) can be used to control the accuracy of calculation of the stationary distribution using the algorithm presented in Sect. 2.6.2.

4.1.2 Stationary Distribution of the System at an Arbitrary Time

In this section, we consider the problem of calculating the stationary distribution of the non-Markovian process $\{i_t, \nu_t, m_t\}$, $t \geq 0$ of the system states at an arbitrary time.

We denote the steady state probabilities by

$$p(i, \nu, m) = \lim_{t\to\infty} P\{i_t = i, \nu_t = \nu, m_t = m\}.$$

Introduce the following notation:

$$\mathbf{p}(i, \nu) = (p(i, \nu, 1), \dots, p(i, \nu, M)), \quad \mathbf{p}_i = (\mathbf{p}(i, 0), \dots, \mathbf{p}(i, W)),$$

$$\mathbf{P}(z) = \sum_{i=0}^{\infty} \mathbf{p}_i z^i, \ |z| < 1.$$

Theorem 4.3 *The generating function $\mathbf{P}(z)$ of the stationary distribution of the system at an arbitrary time is related to the generating function $\mathbf{\Pi}(z)$ of the stationary distribution of the system at the service completion epochs as follows:*

$$\mathbf{P}(z) = \lambda \mathbf{\Pi}(z) \left[z\hat{\beta}^{-1}(-D(z))\nabla^*(z) - I \right] (\tilde{D}(z))^{-1}, \tag{4.12}$$

where $\nabla^*(z) = \int\limits_0^\infty e^{D(z)t} \otimes d\nabla_B(t)$, and the matrix $\nabla_B(t)$ is defined in Theorem 3.6.

Proof Using the approach based on a Markov renewal process, we express the steady state probabilities $\mathbf{p}_i$ via the probabilities $\boldsymbol{\pi}_i$ as follows:

$$\mathbf{p}_0 = \lambda \boldsymbol{\pi}_0 \int\limits_0^\infty \left[e^{D_0 t} \otimes I_M \right] dt = \lambda \boldsymbol{\pi}_0 (-\tilde{D}_0)^{-1},$$

$$\mathbf{p}_i = \lambda \Bigg[\boldsymbol{\pi}_i \int\limits_0^\infty \left[\left(e^{(D_0 - \alpha_i I)t} \right) \otimes I_M \right] dt \tag{4.13}$$

$$+ \sum_{l=1}^{i} \boldsymbol{\pi}_l \int\limits_0^\infty \int\limits_0^t \left[\left(e^{(D_0 - \alpha_l I)v} \alpha_l \right) \otimes I_M \right] dv \, [P(i-l, t-v) \otimes (I - \nabla_B(t-v))] \, dt$$

$$+\sum_{l=0}^{i-1}\boldsymbol{\pi}_l\int_0^{\infty}\int_0^t\sum_{k=1}^{i-l}\Big[\Big(e^{(D_0-\alpha_l I)v}\Big)\otimes I_M\Big][(D_k\otimes I_M)dv]$$

$$\times[P(i-l-k,t-v)\otimes(I-\nabla_B(t-v))]dt\Big]$$

$$=\lambda\Big[\boldsymbol{\pi}_i\Big[(\alpha_i I-D_0)^{-1}\otimes I_M\Big]$$

$$+\sum_{l=1}^{i}\boldsymbol{\pi}_l\Big[\alpha_l(\alpha_l I-D_0)^{-1}\otimes I_M\Big]\int_0^{\infty}[P(i-l,t)\otimes(I-\nabla_B(t))]dt$$

$$+\sum_{l=0}^{i-1}\boldsymbol{\pi}_l\Big[(\alpha_l I-D_0)^{-1}\otimes I_M\Big][\sum_{k=1}^{i-l}D_k\otimes I_M]\int_0^{\infty}[P(i-l-k,t)\otimes(I-\nabla_B(t))]dt\Big],$$

$$i>0.$$

Multiplying the equations obtained by the corresponding powers of z and summing up, we get the following expression for the generating function $\mathbf{P}(z)$:

$$\mathbf{P}(z)=\lambda\Big[\boldsymbol{\pi}_0(-\tilde{D}_0)^{-1}+\sum_{l=1}^{\infty}\boldsymbol{\pi}_l z^l(\alpha_l I-\tilde{D}_0)^{-1} \tag{4.14}$$

$$+[\boldsymbol{\pi}_0(-\tilde{D}_0)^{-1}(\tilde{D}(z)-\tilde{D}_0)+\sum_{l=1}^{\infty}\boldsymbol{\pi}_l z^l+\sum_{l=1}^{\infty}\boldsymbol{\pi}_l z^l(\alpha_l I-\tilde{D}_0)^{-1}\tilde{D}(z)]$$

$$\times\int_0^{\infty}\Big[e^{D(z)t}\otimes(I-\nabla_B(t))\Big]dt\Big].$$

After integration by parts the integral in (4.14) is calculated as follows:

$$\int_0^{\infty}\Big[e^{D(z)t}\otimes(I-\nabla_B(t))\Big]dt=(-\tilde{D}(z))^{-1}(I-\nabla^*(z)).$$

Substituting the last expression into (4.14), after some algebra we get

$$\mathbf{P}(z) = \lambda \left[-\mathbf{\Pi}(z)(\tilde{D}(z))^{-1} + [\mathbf{\Pi}(z)(\tilde{D}(z))^{-1} + \sum_{l=0}^{\infty} \boldsymbol{\pi}_l z^l (\alpha_l I - \tilde{D}_0)^{-1}]\nabla^*(z) \right].$$

Taking into account formula (4.9), this implies formula (4.12). □

Corollary 4.1 *Formula (4.12) for the generating functions* $\mathbf{P}(z)$ *and* $\mathbf{\Pi}(z)$ *for a* $BMAP/SM/1$ *retrial system is of the same form as the analogous formula for the a* $BMAP/SM/1$ *system with waiting room.*

Corollary 4.2 *The scalar generating function* $p(z) = \mathbf{P}(z)\mathbf{e}$ *of the distribution of the number of customers in a* $BMAP/SM/1$ *retrial system is calculated as*

$$p(z) = \lambda(z-1)\mathbf{\Pi}(z)(\tilde{D}(z))^{-1}\mathbf{e}.$$

4.1.3 Performance Measures

Having calculated the stationary distribution of the system states at the embedded epochs and at an arbitrary time, we can define various performance measures of the system. Here we present some of them.

- The vector $\mathbf{q}_0(i)$, the $(\nu(W+1)+m)$th component of which equals the probability that the server is idle at an arbitrary time and there are i customers in the orbit, the underlying process of the $BMAP$ is in the state ν, and the service underlying process is in the state m

$$\mathbf{q}_0(i) = \lambda\boldsymbol{\pi}(i)(\alpha_i I - \tilde{D}_0)^{-1},\ i \geq 0.$$

- The vector $\mathbf{q}_0 = \sum_{i=0}^{\infty} \mathbf{q}_0(i)$ is calculated as follows:

$$\mathbf{q}_0 = \lambda\mathbf{\Pi}(1)\Phi_0^{-1}\tilde{D}_0^{-1}.$$

- Probability $q_0^{(i)}$ that at an arbitrary time the server is idle under the condition that there are i customers in the orbit

$$q_0^{(i)} = \frac{\mathbf{q}_0(i)\mathbf{e}}{\mathbf{p}_i\mathbf{e}},\ i \geq 0.$$

- Probability $p_0^{(a)}$ that an arbitrary customer begins service immediately upon arrival

$$p_0^{(a)} = -\frac{\mathbf{q}_0\tilde{D}_0\mathbf{e}}{\lambda}.$$

- Probability $p_0^{(b)}$ that the first customer in a batch begins service immediately after the arrival of the batch

$$p_0^{(b)} = \frac{-\mathbf{q}_0 D_0 \mathbf{e}}{\lambda_g}, \ \lambda_g = -\boldsymbol{\theta} \tilde{D}_0 \mathbf{e},$$

where $\boldsymbol{\theta}$ is the vector of stationary probabilities of the underlying process of the $BMAP$.
- Probability p_0 that at an arbitrary time the system is idle

$$p_0 = \mathbf{p}_0 \mathbf{e}.$$

- The average number of customers L in the system at an arbitrary time

$$L = \mathbf{P}'(1)\mathbf{e}.$$

4.2 $BMAP/PH/N$ Retrial System

The importance of the investigation of retrial queuing systems has been noted above. In the previous section, we have already considered a $BMAP/SM/1$ retrial system. Single-server queuing systems are often applied to model transmission processes in local communication networks where the transmission of information between the stations is carried out through one common channel, for instance, a bus or a ring. To model other real retrial systems, for instance, mobile communication networks or contact centers, one should be able to calculate performance measures of multi-server retrial queuing systems. However, the investigation of multi-server retrial queuing systems is very complex and does not have an analytical solution even under the simplest assumptions about the input flow (stationary Poisson flow) and the service process (service times are independent exponentially distributed random variables). The results obtained in Sect. 2.7 for continuous-time asymptotical quasi-Toeplitz Markov chains allow us to calculate the performance measures of multi-server retrial queuing systems under general enough assumptions about the input flow and service process.

We will consider a $BMAP$ as an input flow. Note that in the case of single-server retrial queuing systems with a $BMAP$ the relevant analysis is possible using the method of embedded Markov chains or the method of introducing an additional variable even for a general service process such as a semi-Markov one (SM). For multi-server systems both these methods are useless. Thus, the most general process for which the analytical analysis of a multi-server system is still possible is a renewal process in which the service times of customers are independent equally distributed random variables with phase-type distribution, described in Sect. 2.1.

Therefore, a $BMAP/PH/N$ retrial queue can be considered as the most complex multi-server retrial system for analytical investigation, and this section is devoted to the study of such system.

4.2.1 System Description

The system has N identical independent servers. Customers arrive at the system in a $BMAP$. The $BMAP$ has an underlying process $\nu_t,\ t \geq 0$ with state space $\{0, 1, \ldots, W\}$. The behavior of the $BMAP$ is described by matrices $D_k,\ k \geq 0$ of size $W + 1$ or their generating function $D(z) = \sum_{k=0}^{\infty} D_k z^k,\ |z| \leq 1$.

If the arriving batch of primary customers finds several servers idle, the primary customers occupy the corresponding number of servers. If the number of idle servers is insufficient (or all servers are busy) the rest of the batch (or all the batch) goes to the so-called orbit. These customers are called repeated calls. These customers try their luck later, until they can be served. We assume that the total flow of retrials is such that the probability of generating a retrial attempt in the interval $(t, t + \Delta t)$ is equal to $\alpha_i \Delta t + o(\Delta t)$ when the orbit size (the number of customers in the orbit) is equal to $i,\ i > 0,\ \alpha_0 = 0$. The orbit capacity is assumed to be unlimited. We do not fix the explicit dependence of the intensities α_i on i. The following two variants will be dealt with:

- a constant retrial rate : $\alpha_i = \gamma,\ i > 0$;
- an infinitely increasing retrial rate : $\lim_{i\to\infty} \alpha_i = \infty$. This variant includes the classic retrial strategy ($\alpha_i = i\alpha$) and the linear strategy ($\alpha_i = i\alpha + \gamma$). In this section, we will differentiate the case of a constant retrial rate and the general case where the explicit dependency of α_i on i is not defined.

The service time of an arbitrary customer has the PH distribution with underlying process $m_t, t \geq 0$ and irreducible representation $(\boldsymbol{\beta}, S)$ of order M.

4.2.2 Markov Chain Describing the Operation of the System

Let

- i_t be the number of customers in the orbit, $i_t \geq 0$;
- n_t be the number of busy servers, $n_t = \overline{0, N}$;
- $m_t^{(j)}$ be the state of the underlying process of the PH service in the jth busy server, $m_t^{(j)} = \overline{1, M},\ j = \overline{1, n_t}$ (we assume here that the busy servers are numerated in the order in which they are occupied, i.e., the server that begins

the service is appointed the highest number among all busy servers; when some server finishes its service, the servers are correspondingly enumerated);
- ν_t be the state of the underlying process of the $BMAP, \nu_t = \overline{0, W}$,

at the moment $t,\ t \geq 0$.

Consider the multidimensional process

$$\xi_t = (i_t,\, n_t,\, \nu_t,\, m_t^{(1)}, \ldots, m_t^{(n_t)}),\ t \geq 0.$$

We can easily determine that, knowing the state of the process ξ_t at the moment t, we can predict the future behavior of this process, i.e., the process ξ_t is a Markov chain in continuous time. We also can easily determine that the chain is regular and irreducible.

We will denote by level i the set of states of ξ_t where the component i_t takes value i, and other components are arranged in lexicographic order. There are $K = (W+1)\frac{M^{N+1}-1}{M-1}$ states on level i. Notice that the number K may be quite large. For instance, if the state space of underlying Markov chains of the input and service of customers consist only of two elements ($W = 1,\ M = 2$), and the number of servers is five ($N = 5$), then $K = 126$; if $N = 6$, then $K = 254$, etc. The quick growth of the number of states in the level implies definite difficulties in applying the algorithm for the calculation of the stationary distribution for large values of N.

At the same time, the description of the operation of the system in terms of the Markov chain ξ_t is quite demonstrable and useful from the point of view of simplicity of presentation. Below we will outline another method to select a Markov chain describing the operation of the analyzed retrial queuing system. With that choice the number of states on a level can be much smaller.

Lemma 4.2 *The infinitesimal generator of the Markov chain ξ_t has the following form:*

$$Q = \begin{pmatrix} Q_{0,0} & Q_{0,1} & Q_{0,2} & Q_{0,3} & \cdots \\ Q_{1,0} & Q_{1,1} & Q_{1,2} & Q_{1,3} & \cdots \\ O & Q_{2,1} & Q_{2,2} & Q_{2,3} & \cdots \\ O & O & Q_{3,2} & Q_{3,3} & \cdots \\ \vdots & \vdots & \vdots & \vdots & \ddots \end{pmatrix}, \tag{4.15}$$

where the blocks $Q_{i,j}$ define the rates of transition from level i to level j and have the following form:

$$Q_{i,i-1} = \alpha_i \begin{pmatrix} O & I_{\bar{W}} \otimes \boldsymbol{\beta} & O & \cdots & O \\ O & O & I_{\bar{W}M} \otimes \boldsymbol{\beta} & \cdots & O \\ \vdots & \vdots & \vdots & \ddots & \vdots \\ O & O & O & \cdots & I_{\bar{W}M^{N-1}} \otimes \boldsymbol{\beta} \\ O & O & O & \cdots & O \end{pmatrix},\ i \geq 1, \tag{4.16}$$

$$Q_{i,i+k} = \begin{pmatrix} O \dots O & D_{k+N} \otimes \boldsymbol{\beta}^{\otimes N} \\ O \dots O & D_{k+N-1} \otimes I_M \otimes \boldsymbol{\beta}^{\otimes (N-1)} \\ O \dots O & D_{k+N-2} \otimes I_{M^2} \otimes \boldsymbol{\beta}^{\otimes (N-2)} \\ \vdots \ddots \vdots & \\ O \dots O & D_k \otimes I_{M^N} \end{pmatrix}, \quad k \geq 1, \tag{4.17}$$

and the blocks $(Q_{i,i})_{n,n'}$ *of the matrix* $Q_{i,i}$ *are defined as follows:*

$$(Q_{i,i})_{n,n'} = \begin{cases} O, \ n' < n-1, \ n = \overline{2,N}, \\ I_{\bar{W}} \otimes \mathbf{S}_0^{\oplus n}, \ n' = n-1, \ n = \overline{1,N}, \\ D_0 \oplus S^{\oplus n} - \alpha_i (1 - \delta_{n,N}) I_{\bar{W}M^n}, \ n' = n, \ n = \overline{0,N}, \\ D_l \otimes I_{M^n} \otimes \boldsymbol{\beta}^{\otimes l}, \ n' = n + l, \ l = \overline{1, N-n}, \ n = \overline{0,N}, \end{cases} \tag{4.18}$$

$$i \geq 0.$$

Here $\delta_{n,N} = \begin{cases} 1, \ n = N, \\ 0, \ n \neq N \end{cases}$ *is the Kronecker symbol,* $\boldsymbol{\beta}^{\otimes r} \stackrel{def}{=} \underbrace{\boldsymbol{\beta} \otimes \cdots \otimes \boldsymbol{\beta}}_{r}$, $r \geq 1$, $\mathbf{S}_0^{\oplus l} \stackrel{def}{=} \sum_{m=0}^{l-1} I_{M^m} \otimes \mathbf{S}_0 \otimes I_{M^{l-m-1}}, \quad l \geq 1.$

Proof The proof of the lemma is by means of calculation of probabilities of transitions of the components of the Markov chain ξ_t during a time interval of infinitesimal length.

We use the clear probabilistic sense of the matrices presented in the statement of the theorem. The matrix Q, defined by formula (4.15), is a block upper-Hessenberg matrix, i.e., all blocks of the matrix located below the first subdiagonal are equal to zero matrices. This can be explained by the fact that decreasing the number of customers in a small time interval Δt by more than one is possible with probability $o(\Delta t)$ only.

Consider the probability sense of the blocks $Q_{i,j}, \ i \geq 0, \ j \geq \min\{0, i-1\}$. Each of these blocks is a block matrix with blocks $(Q_{i,j})_{n,n'}$, consisting of the transition rates of the Markov chain ξ_t from the states of level i to the states of level j, during which the value of the component n_t is changed from n to $n', \ n, n' = \overline{0,N}$.

The matrices $Q_{i,i-1}$ define the transition rates from the states of level i to the states of level $i-1$. The rates of such transitions are equal to the retrial rate α_i. This explains the presence of the factor α_i in the right-hand part of (4.16). If at the moment of a repeated attempt all the N servers are busy, the repeated attempt will not be successful and the customer will return to the orbit. In this case the transition to level $i-1$ is not possible. So the last block of the matrix $Q_{i,i-1}$ is zero. If the number of busy servers at the moment of the repeated attempt equals $n, n < N$, the transition to level $i-1$ occurs and the number of busy servers increases by one. This provides an explanation for the fact that the blocks $(Q_{i,i-1})_{n,n+1}$ in the first super-diagonal differ from zero, while all the other blocks are zero.

The retrial attempt in a small time interval excludes the probability of transitions of the underlying processes of the $BMAP$ and the PH service in this interval. This explains the presence of the cofactor $I_{\bar{W}M^n}$ in the first super-diagonal block in the nth block row of the matrix $Q_{i,i-1}$. This cofactor is multiplied in the Kronecker way by the vector $\boldsymbol{\beta}$ as at the moment of occupation of the server (assume that this is the j-th server) by a customer arriving from the orbit a selection (according to vector $\boldsymbol{\beta}$) of the initial state of the Markov chain $m_t^{(j)}$ is carried out.

Note, that, in general, the Kronecker multiplication of matrices is useful in the description of the rates of joint transitions of independent components of multidimensional Markov chains.

The matrices $Q_{i,i+k}$ define the transition rates from the states of level i to the states of level $i+k$. Such transitions happen when a batch of size $k+N-n$ arrives at the system, where the number of busy servers at the arrival epoch equals n, $n \le N$; the arrival rate of such batches is defined by the entries of the matrix D_{k+N-n}. k customers of the arriving batch go to the orbit while all the others occupy the $N-n$ idle servers and the number of busy servers becomes equal to N. This explains the fact that all blocks of the matrix $Q_{i,i+k}$, except the blocks $(Q_{i,i+k})_{n,N}$ of the last block column, are zero. At the epoch of the occupation of the $N-n$ idle servers, at each such server a selection (according to the vector $\boldsymbol{\beta}$) of the initial state of the underlying Markov chain of the PH service is carried out. The result of a simultaneous selection at $N-n$ servers is defined by the vector $\boldsymbol{\beta}^{\otimes(N-n)}$. The states of the underlying processes at the n busy servers do not change. Thus, from all that has been said we obtain the "easy-to-read" formula for the block $(Q_{i,i+k})_{n,N}$:

$$(Q_{i,i+k})_{n,N} = D_{k+N-n} \otimes I_{M^n} \otimes \boldsymbol{\beta}^{\otimes(N-n)}.$$

Now we proceed to the explanation of the matrices $Q_{i,i}$.

Non-diagonal entries of the matrix $Q_{i,i}$ define the Markov chain ξ_t transition rates from states of the ith level to states of the same level. Diagonal entries of the matrices $Q_{i,i}$ are the rates of exiting the corresponding states, taken as minus-signed. The blocks $(Q_{i,i})_{n,n'}$ in formula (4.18) are explained as follows:

- In a small time interval only one customer may complete its service. Thus $(Q_{i,i})_{n,n'} = O$, $n' < n-1$.
- The block $(Q_{i,i})_{n,n-1}$ corresponds to the situation when in a small time interval the service was completed at one of the n busy servers. The rates of service completion at one of the n busy servers are defined by the vector $\mathbf{S}_0 \otimes I_{M^{n-1}} + I_M \otimes \mathbf{S}_0 I_{M^{n-2}} + \ldots I_{M^{n-1}} \mathbf{S}_0$, which corresponds to the notation $\mathbf{S}_0^{\oplus n}$.
- The block $(Q_{i,i})_{n,n}$ is diagonal and corresponds to the situation when in a small time interval neither the number of customers in the orbit nor number of busy servers is changed. Non-diagonal sub-blocks of such a block define either the transition rates of the $BMAP$ underlying process (without the transition of

service underlying processes) and are defined by the entries of matrix $D_0 \otimes I_{M^n}$, or the transition rates of the service underlying process the service at one of the n busy servers (without the transition of the $BMAP$ underlying process) and are defined by the entries of the matrix $I_{\bar{W}} \otimes S^{\oplus n}$. The diagonal sub-blocks of the block $(Q_{i,i})_{n,n}$ define the rates of exiting the corresponding states due to the transitions of the $BMAP$ and service underlying processes. Here we assume that there are no retrial attempts from the orbit for the case $n \neq N$. Taking into consideration the fact that $D_0 \otimes I_{M^n} + I_{\bar{W}} \otimes S^{\oplus n} = D_0 \oplus S^{\oplus n}$, we obtain the formula for the block $(Q_{i,i})_{n,n}$ in (4.18).
- The block $(Q_{i,i})_{n,n+l},\ l \geq 1$ corresponds to the situation where in a small time interval a batch of l customers arrives and all such customers start their service immediately. The vector $\boldsymbol{\beta}^{\otimes l}$ defines the initial distribution of the underlying processes of PH service at the l servers that are occupied by the customers of the incoming batch.

□

4.2.3 Ergodicity Condition

As above, we will differentiate the cases of constant retrial rate ($\alpha_i = \gamma,\ i > 0$) and infinitely increasing retrial rate ($\lim\limits_{i\to\infty} \alpha_i = \infty$).

In the first case, we can easily verify that the process ξ_t is a quasi-Toeplitz Markov chain with N block boundary conditions. The ergodicity condition for such a chain can be easily obtained from the results of Sect. 2.4.

Theorem 4.4 *In the case of a constant retrial rate, a necessary and sufficient condition for ergodicity of the Markov chain ξ_t has the following form:*

$$\mathbf{y}((D^{\star}(z))'|_{z=1} - \gamma \hat{I})\,\mathbf{e} < 0, \tag{4.19}$$

where the vector $\mathbf{y}$ is the unique solution of the system of linear algebraic equations

$$\mathbf{y}(I - Y(1)) = \mathbf{0},\ \ \mathbf{ye} = 1. \tag{4.20}$$

Here

$$Y(z) = zI + (C + \gamma\hat{I})^{-1}(\gamma\tilde{I}_{\beta} - \gamma\hat{I}z + zD^{\star}(z)),$$

$$C = \mathrm{diag}\{\,\mathrm{diag}\{\lambda_{\nu},\ \nu = \overline{0, W}\,\} \oplus [\,\mathrm{diag}\{s_m,\ m = \overline{1, M}\,\}\,]^{\oplus r},\ r = \overline{0, N}\,\},$$

$$D^{\star}(z) =$$

$$\begin{bmatrix} D_0 & D_1 \otimes \boldsymbol{\beta}^{\otimes 1} & D_2 \otimes \boldsymbol{\beta}^{\otimes 2} & \dots & D_{N-1} \otimes \boldsymbol{\beta}^{\otimes (N-1)} & \delta_{N-1}(z, \boldsymbol{\beta}) \\ I_{\bar{W}} \otimes \mathbf{S}_0^{\oplus 1} & D_0 \oplus S^{\oplus 1} & D_1 \otimes I_M \otimes \boldsymbol{\beta}^{\otimes 1} & \dots & D_{N-2} \otimes I_M \otimes \boldsymbol{\beta}^{\otimes (N-2)} & \delta_{N-2}(z, \boldsymbol{\beta}) \\ O & I_{\bar{W}} \otimes \mathbf{S}_0^{\oplus 2} & D_0 \oplus S^{\oplus 2} & \dots & D_{N-3} \otimes I_{M^2} \otimes \boldsymbol{\beta}^{\otimes (N-3)} & \delta_{N-3}(z, \boldsymbol{\beta}) \\ \vdots & \vdots & \vdots & \ddots & \vdots & \vdots \\ O & O & O & \dots & I_{\bar{W}} \otimes \mathbf{S}_0^{\oplus N} & D(z) \oplus S^{\oplus N} \end{bmatrix},$$

$$\Delta_m(z, \boldsymbol{\beta}) = z^{-m+1}(D(z) - \sum_{k=0}^{m} D_k z^k) \otimes I_{M^{N-m-1}} \otimes \boldsymbol{\beta}^{\otimes (m+1)}, \; m = \overline{0, N-1},$$

$$\hat{I} = \begin{pmatrix} I_{\bar{W}} & O & \dots & O & O \\ O & I_{\bar{W}M} & \dots & O & O \\ \vdots & \vdots & \ddots & \vdots & \vdots \\ O & O & \dots & I_{\bar{W}M^{N-1}} & O \\ O & O & \dots & O & O_{\bar{W}M^N} \end{pmatrix},$$

$$\tilde{I}_{\boldsymbol{\beta}} = \begin{pmatrix} O & I_{\bar{W}} \otimes \boldsymbol{\beta} & O & \dots & O \\ O & O & I_{\bar{W}M} \otimes \boldsymbol{\beta} & \dots & O \\ \vdots & \vdots & \vdots & \ddots & \vdots \\ O & O & O & \dots & I_{\bar{W}M^{N-1}} \otimes \boldsymbol{\beta} \\ O & O & O & \dots & O \end{pmatrix},$$

$\bar{I} = I - \hat{I}$.

Now let us consider the case of an infinitely increasing retrial rate ($\lim_{i\to\infty} \alpha_i = \infty$). The following statement is valid.

Theorem 4.5 *In case of an infinitely increasing retrial rate the ergodicity condition for the Markov chain ξ_t has the following form:*

$$\rho = \lambda/\bar{\mu} < 1, \tag{4.21}$$

where λ is the fundamental rate of the BMAP, and the value $\bar{\mu}$ is defined by the formula

$$\bar{\mu} = \mathbf{y}\mathbf{S}_0^{\oplus N}\mathbf{e}_{M^{N-1}}, \tag{4.22}$$

where the vector $\mathbf{y}$ *is the unique solution of the system of linear algebraic equations*

$$\mathbf{y}(S^{\oplus N} + \mathbf{S}_0^{\oplus N}(I_{M^{N-1}} \otimes \boldsymbol{\beta})) = \mathbf{0}, \; \mathbf{y}\mathbf{e} = 1. \tag{4.23}$$

Proof In the case under consideration, the matrices $\hat{T}_i$ appearing in the definition of an asymptotically quasi-Toeplitz Markov chain have the following form:

$$\hat{T}_i = C + \alpha_i \hat{I}.$$

We calculate the limiting matrices $Y_k,\ k \geq 0$ and define their generating function $Y(z)$. As a result, we obtain the following formula:

$$Y(z) = \tilde{I}_{\boldsymbol{\beta}} + z\bar{I} + zC^{-1}\bar{I}D^{\star}(z). \tag{4.24}$$

By the substitution of the explicit forms of the matrices $\bar{I}$, $\tilde{I}_{\boldsymbol{\beta}}$, C, and $D^{\star}(z)$ into (4.24), we can easily verify that the matrix $Y(z)$ is reducible, and the normal form of the matrix $Y(1)$ has only one stochastic diagonal block $\tilde{Y}(1)$. The corresponding block $\tilde{Y}(z)$ of the matrix $Y(z)$ has the following form:

$$\tilde{Y}(z) =$$

$$\begin{pmatrix} (\Lambda \oplus S_N)^{-1}(D(z) \oplus S^{\oplus N})z + zI & (\Lambda \oplus S_N)^{-1}(I_{\bar{W}} \otimes \mathbf{S}_0^{\oplus N})z \\ I_{\bar{W}M^{N-1}} \otimes \boldsymbol{\beta} & O_{\bar{W}M^{N-1}} \end{pmatrix}, \tag{4.25}$$

where $\Lambda \oplus S_N \stackrel{def}{=} \text{diag}\{\lambda_\nu,\ \nu = \overline{0, W}\} \oplus [\,\text{diag}\{s_m,\ m = \overline{1, M}\}\,]^{\oplus N}$.

Theorem 3.14 implies that a sufficient ergodicity condition for the Markov chain $\xi_t, t \geq 0$ is the fulfillment of the following inequality:

$$(\det(zI - \tilde{Y}(z)))'\,|_{z=1} > 0. \tag{4.26}$$

Taking into account the block structure of the determinant $\det(zI - \tilde{Y}(z))$, this can be reduced to

$$\det(zI - \tilde{Y}(z)) = \det(\Lambda \oplus S_N)^{-1} z^{\bar{W}M^{N-1}} \det R(z), \tag{4.27}$$

where

$$R(z) = -z(D(z) \oplus S^{\oplus N}) - (I_{\bar{W}} \otimes S_0^{\oplus N})\,(I_{\bar{W}M^{N-1}} \otimes \beta).$$

We can easily verify that the matrix $R(1)$ is an irreducible infinitesimal generator. This property is considered in the following calculations. Differentiating (4.27) at the point $z = 1$, taking into account that $\det R(1) = 0$, and using this in (4.26), we obtain the following inequality:

$$(\det R(z))'\,|_{z=1} > 0. \tag{4.28}$$

As was stated above, (4.28) is equivalent to the following inequality:

$$\mathbf{x}\,R'(1)\,\mathbf{e} > 0, \tag{4.29}$$

where $\mathbf{x}$ is the unique solution of the system of linear algebraic equations

$$\mathbf{x}\,R(1) = \mathbf{0},\ \mathbf{xe} = 1. \tag{4.30}$$

Representing the vector $\mathbf{x}$ in the form $\mathbf{x} = \boldsymbol{\theta} \otimes \mathbf{y}$, we can verify that the vector $\mathbf{x}$ is a solution of system (4.30) provided that the vector $\mathbf{y}$ is a solution of system (4.23). Such a solution is the unique one as the matrix $S^{\oplus N} + \mathbf{S}_0^{\oplus N}(I_{\bar{W}M^{N-1}} \otimes \boldsymbol{\beta})$ is an irreducible infinitesimal generator. □

Assume that the ergodicity conditions (4.19)–(4.20) or (4.21)–(4.23) hold. Then there are steady state probabilities of the Markov chain $\xi_t,\ t \geq 0$:

$$p(i, n, \nu, m^{(1)}, \dots, m^{(n)})$$

$$= \lim_{t \to \infty} P\{i_t = i,\ n_t = n,\ \nu_t = \nu,\ m_t^{(1)} = m^{(1)}, \dots, m_t^{(n)} = m^{(n)}\},$$

$$i \geq 0,\ \nu = \overline{0, W},\ m^{(j)} = \overline{1, M},\ j = \overline{1, n},\ n = \overline{0, N}.$$

Let us enumerate the steady state probabilities in lexicographic order and form the row vectors $\mathbf{p}_i$ of steady state probabilities corresponding to the value i of the first component of the chain, $i \geq 0$.

It is well known that the vectors $\mathbf{p}_i,\ i \geq 0$ satisfy the Chapman-Kolmogorov equations

$$(\mathbf{p}_0, \mathbf{p}_1, \mathbf{p}_2, \dots)Q = \mathbf{0},\ (\mathbf{p}_0, \mathbf{p}_1, \mathbf{p}_2, \dots)\mathbf{e} = 1, \tag{4.31}$$

where the generator Q of the Markov chain $\xi_t,\ t \geq 0$ is defined by formulas (4.15)–(4.18).

To solve the infinite system of equations (4.31), we can use the numerically stable algorithm for calculation of the stationary distribution of asymptotically quasi-Toeplitz Markov chains based on the calculation of an alternative system of equations and described in Sect. 2.7. Such an algorithm is successfully implemented in the framework of the application program package "Sirius++", developed in the research laboratory of applied probability analysis of the Belarusian State University, [29].

4.2.4 Performance Measures

Having calculated the vectors $\mathbf{p}_i, i \geq 0$, we can calculate the various performance measures of the $BMAP/PH/N$ retrial queuing system. Some of them are listed below:

- The joint probability that at an arbitrary time there are i customers in the orbit and n servers busy

$$q_n(i) = [\mathbf{p}_i]_n\mathbf{e},\ n = \overline{0, N},\ i \geq 0.$$

  Here the notation $[\boldsymbol{x}]_n$ has the following sense. Assume that the vector $\boldsymbol{x}$ of length K is defined as $\boldsymbol{x} = (\boldsymbol{x}_0, \dots, \boldsymbol{x}_N)$, where the vector $\boldsymbol{x}_n$ has dimension $(W+1)M^n,\ n = \overline{0, N}$. Then $[\boldsymbol{x}]_n \stackrel{def}{=} \boldsymbol{x}_n$, i.e., $[\mathbf{p}_i]_n$ is the part of the vector $\mathbf{p}_i$ corresponding to the states of the system with i customers in the orbit and n servers busy, $i \geq 0,\ n = \overline{0, N}$.
- The probability that at an arbitrary time n servers are busy under the condition that i customers are in the orbit

$$q_n^{(i)} = \frac{q_n(i)}{\mathbf{p}_i\mathbf{e}},\ n = \overline{0, N}.$$

- The probability that at an arbitrary time n servers are busy

$$q_n = [\mathbf{p}(1)]_n\mathbf{e},\ n = \overline{0, N},\ \mathbf{P}(1) = \sum_{i=0}^{\infty} \boldsymbol{p}_i.$$

- The mean number of busy servers at an arbitrary time

$$\hat{n} = \sum_{n=1}^{N} nq_n.$$

- The mean number of customers in the orbit at an arbitrary time

$$L_{orb} = \sum_{i=1}^{\infty} i\mathbf{p}_i\mathbf{e}.$$

- The stationary distribution of the number of busy servers at the epoch of arrival of a k size batch

$$P_n^{(k)} = \frac{[\mathbf{p}(1)]_n(D_k \otimes I_{M^n})\mathbf{e}}{\boldsymbol{\theta}} D_k\mathbf{e},\ n = \overline{0, N},\ k \geq 1.$$

- The probability that an arbitrary customer will be served immediately after arriving at the system

$$P_{imm} = \frac{1}{\lambda}\sum_{n=1}^{N}[\mathbf{P}(1)]_{N-n}(\sum_{k=0}^{n}(k-n)D_k \otimes I_{M^{N-n}})\mathbf{e}. \tag{4.32}$$

Proof As was noted in Sect. 2.1, under a fixed distribution of the underlying process of the $BMAP$, the probability for an arbitrary customer to arrive in a k size batch is calculated as follows:

$$\frac{kD_k\mathbf{e}}{\boldsymbol{\theta}\sum_{l=1}^{\infty} lD_l\mathbf{e}} = \frac{kD_k\mathbf{e}}{\lambda}.$$

The arbitrary customer will be served immediately after entering the system if it arrives in a k size batch when m servers are busy and $m = \overline{0, N-k}$, or in a k size batch where $k \geq N - m + 1$ but it is one of the first $N - m$ customers within the batch. We assume that an arbitrary customer arriving in a k size batch holds the lth position in the batch, $l = \overline{1, k}$, with probability $\frac{1}{k}$. Applying the formula of total probability, we obtain the following statement:

$$P_{imm} = \frac{1}{\lambda}\sum_{m=0}^{N-1}[\mathbf{P}(1)]_m(\sum_{k=1}^{N-m} kD_k \otimes I_{M^m} + \sum_{k=N-m+1}^{\infty} k\frac{N-m}{k}D_k \otimes I_{M^m})\mathbf{e}.$$

Taking into account the formula

$$(\sum_{k=N-m+1}^{\infty} D_k \otimes I_{M^{N-n}})\mathbf{e} = -(\sum_{k=0}^{N-m} D_k \otimes I_{M^{N-n}})\mathbf{e}$$

and changing the summation index from m to $n = N - m$, we obtain formula (4.32). □

- The probability that an arbitrary batch of customers will be served immediately after entering the system

$$P_{imm}^{b} = \frac{1}{\lambda_b}\sum_{m=1}^{N}[\mathbf{P}(1)]_{N-m}\sum_{k=1}^{m}(D_k \otimes I_{M^{N-m}})\mathbf{e},$$

where λ_b is the arrival rate of batches, $\lambda_b = -\boldsymbol{\theta} D_0\mathbf{e}$.

4.2.5 Case of Impatient Customers

We have assumed above that a customer in the orbit is absolutely persistent, i.e., it repeats its attempts to get service until it finds an idle server. Now let us consider the general case when a customer that has made a retrial attempt and found all servers busy returns to the orbit with probability p and leaves the system forever with probability $1 - p$ $(0 \le p \le 1)$.

In this case, the components of the Markov chain $\xi_t,\ t \ge 0$ describing the operation of the system have the same sense as the corresponding components of a Markov chain describing a retrial queuing system with absolutely persistent customers. The infinitesimal generator has the form (4.15), where the blocks $Q_{i,i+k},\ k \ge 1$ have the form (4.17), the blocks $Q_{i,i-1}$ are obtained via the replacement of the zero matrix in the last row and column in (4.16) with the matrix $(1-p)I_{\bar{W}M^N}$, and the blocks $Q_{i,i}$ are obtained via the replacement of the matrix $\alpha_i(1-\delta_{n,N})I_{\bar{W}M^n}$ with the matrix $\alpha_i(1-p\delta_{n,N})I_{\bar{W}M^n}$ in (4.18).

The considerable difference between systems with absolutely persistent and impatient customers is as follows. When the probability $p = 1$, i.e., the customers are absolutely persistent, the system is not always ergodic. The required ergodicity conditions are given in Theorems 4.4 and 4.5. When $p < 1$, the system is ergodic under any parameters. This can be easily checked by applying the results for an asymptotically quasi-Toeplitz Markov chain presented in Sect. 2.7. In the case under consideration, the limiting matrix Y_0 in the definition of an asymptotically quasi-Toeplitz Markov chain is stochastic and all matrices $Y_k,\ k \ge 1$ are zero. Consequently, $Y'(1) = O$ and inequalities (2.120) hold automatically for any parameters of the system. Another difference of the system with impatient customers is the possibility of customer losses. The loss probability is one of the most important performance measures of the system. It is calculated as follows:

$$P_{loss} = 1 - \frac{1}{\lambda}\sum_{n=1}^{N}[\mathbf{P}(1)]_n \mathbf{S}_0^{\oplus n}\mathbf{e}. \tag{4.33}$$

Let us provide a short proof of formula (4.33). In the stationary regime, the rate of completing the service of customers is defined as

$$\sum_{n=1}^{N}[\mathbf{P}(1)]_n \mathbf{S}_0^{\oplus n}\mathbf{e}.$$

The value λ is the fundamental rate of input flow. The ratio of these two rates defines the probability of successful service of an arbitrary customer, while the complementary probability gives the probability of an arbitrary customer loss.

4.2.6 Numerical Results

To illustrate numerically the results obtained, this section provides graphs demonstrating the dependency of the number of customers in the orbit, L_{orb}, the loss probability P_{loss}, and the probability that an arbitrary customer goes to the orbit, $P_{orb} = 1 - P_{imm}$, on the probability p of returning to the orbit after an unsuccessful retrial attempt, the system load, and the coefficients of correlation and variation of inter-arrival times.

We will consider four different $BMAP$s with the fundamental rate $\lambda = 12$. The first $BMAP$ will be denoted by Exp. This is a stationary Poisson flow with parameter $\lambda = 12$. The coefficients of variation and correlation equal 1 and 0 respectively. The other three $BMAP$s ($BMAP_1$, $BMAP_2$, $BMAP_3$) have the coefficient of variation of inter-arrival time equal to 2. These flows are defined by the matrices $D_0 = \bar{c}\hat{D}_0$ and D_k, $k = \overline{1,4}$, which are determined as $D_k = \bar{c}Dq^{k-1}(1-q)/(1-q^4)$, $k = \overline{1,4}$, where $q = 0.8$ and D is the relevant matrix and the scalar coefficient $\bar{c}$ is assumed equal to 2.4, to ensure the input rate $\lambda = 12$.

$BMAP_1$ is defined by the following matrices:

$$\hat{D}_0 = \begin{pmatrix} -13.334625 & 0.588578 & 0.617293 \\ 0.692663 & -2.446573 & 0.422942 \\ 0.682252 & 0.414363 & -1.635426 \end{pmatrix},$$

$$D = \begin{pmatrix} 11.546944 & 0.363141 & 0.218669 \\ 0.384249 & 0.865869 & 0.080851 \\ 0.285172 & 0.04255 & 0.211089 \end{pmatrix}.$$

For this $BMAP$ the coefficient of correlation is equal to 0.1.

$BMAP_2$ has the coefficient of correlation equal to 0.2, and is defined by the following matrices:

$$\hat{D}_0 = \begin{pmatrix} -15.732675 & 0.606178 & 0.592394 \\ 0.517817 & -2.289674 & 0.467885 \\ 0.597058 & 0.565264 & -1.959665 \end{pmatrix},$$

$$D = \begin{pmatrix} 14.1502 & 0.302098 & 0.081805 \\ 0.107066 & 1.03228 & 0.164627 \\ 0.08583 & 0.197946 & 0.513566 \end{pmatrix}.$$

$BMAP_3$ has the coefficient of correlation equal to 0.3, and is defined by the following matrices:

$$\hat{D}_0 = \begin{pmatrix} -25.539839 & 0.393329 & 0.361199 \\ 0.145150 & -2.232200 & 0.200007 \\ 0.295960 & 0.387445 & -1.752617 \end{pmatrix},$$

$$D = \begin{pmatrix} 24.242120 & 0.466868 & 0.076323 \\ 0.034097 & 1.666864 & 0.186082 \\ 0.009046 & 0.255481 & 0.804685 \end{pmatrix}.$$

We will also consider three $BMAP$s ($BMAP^2$, $BMAP^5$, $BMAP^7$) with the fundamental rate $\lambda = 12$, the coefficient of correlation 0.3, and different coefficients of variation. These $BMAP$s are defined by the matrices D_k, $k = \overline{0,4}$, calculated based on the matrices $\hat{D}_0$ and D using the method described above.

$BMAP^2$ has the coefficient of variation equal to 2. It is identical to $BMAP_3$ introduced above.

$BMAP^5$ has the coefficient of variation equal to 4.68. It is defined by the following matrices:

$$\hat{D}_0 = \begin{pmatrix} -16.196755 & 0.090698 & 0.090698 \\ 0.090698 & -0.545154 & 0.090698 \\ 0.090698 & 0.090699 & -0.313674 \end{pmatrix},$$

$$D = \begin{pmatrix} 15.949221 & 0.066138 & 0 \\ 0.033069 & 0.297622 & 0.033069 \\ 0 & 0.013228 & 0.119049 \end{pmatrix}.$$

$BMAP^7$ has the coefficient of variation equal to 7.34404. It is defined by the following matrices:

$$\hat{D}_0 = \begin{pmatrix} -16.268123 & 0.040591 & 0.040591 \\ 0.040591 & -0.223595 & 0.040591 \\ 0.040591 & 0.040591 & -0.132969 \end{pmatrix},$$

$$D = \begin{pmatrix} 16,161048 & 0.025893 & 0 \\ 0.012947 & 0.116519 & 0.012947 \\ 0 & 0.005179 & 0.046608 \end{pmatrix}.$$

Assume that the number of servers $N = 3$.

We will consider three different PH-distributions (PH_2, PH_5, PH_8) of service time with the same coefficient of variation equal to 1.204, while their rates vary.

These PH-distributions are defined by the vector $\boldsymbol{\beta} = (0.5; 0.5)$ and the matrices S defined through the matrix $S^{(0)} = \begin{pmatrix} -4 & 2 \\ 0 & -1 \end{pmatrix}$ as follows.

PH_2 is defined by the matrix $S = 17.5S^{(0)}$. Then, the service rate μ equals 20, and the system load $\rho = \frac{\lambda}{N\mu}$ equals 0.2.

PH_5 is defined by the matrix $S = 7S^{(0)}$. Here we have $\mu = 8$, $\rho = 0.5$.

PH_8 is defined by the matrix $S = 4.375S^{(0)}$. Here we have $\mu = 5$, $\rho = 0.8$.

Figures 4.1, 4.2, 4.3 illustrate the dependence of probabilities P_{loss}, P_{orb}, and the mean number of customers in the orbit, L_{orb}, on the probability p of returning to the orbit after an unsuccessful attempt and the system load for the $BMAP_2$.

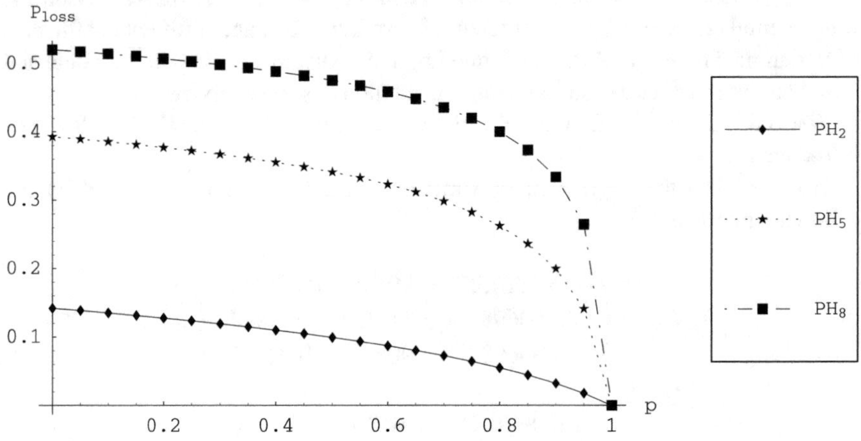

Fig. 4.1 Loss probability P_{loss} as a function of probability p under different system loads

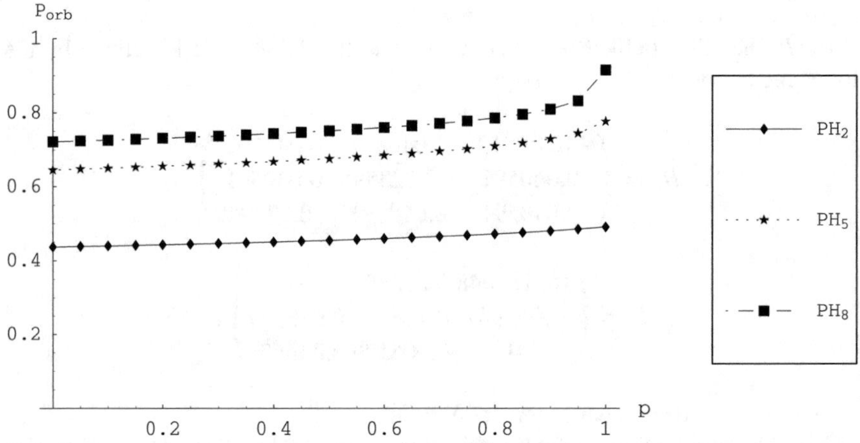

Fig. 4.2 Probability P_{orb} as a function of probability p under different system loads

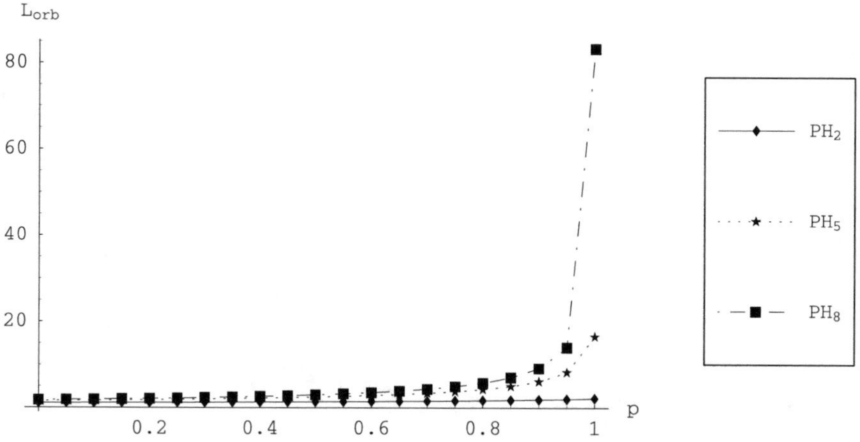

Fig. 4.3 Mean number of customers in the orbit, L_{orb}, as a function of probability p under different system loads

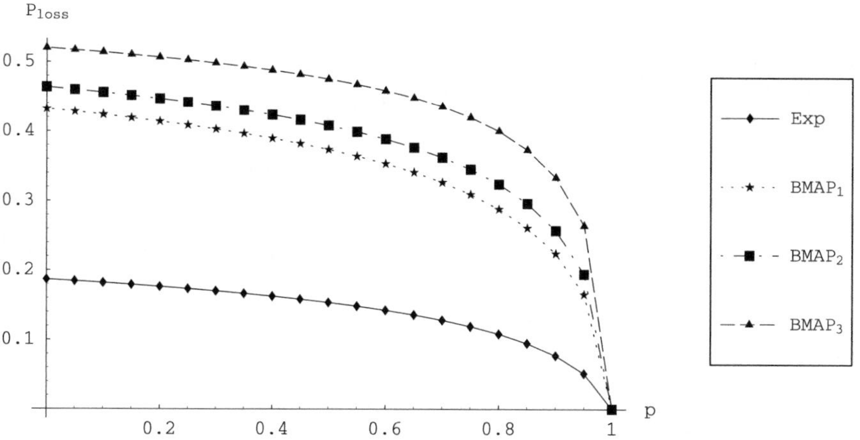

Fig. 4.4 Loss probability P_{loss} as a function of probability p for $BMAP$s with different coefficients of correlation

Figures 4.4 and 4.5 illustrate the dependence of the probability P_{loss} and the mean number of customers in the orbit, L_{orb}, on the probability p of returning to the orbit after an unsuccessful attempt and the coefficient of correlation of the $BMAP$ under the service time distribution PH_8.

From Figs. 4.4 and 4.5 it may be concluded that the correlation in the input flow significantly influences the system performance measures and such measures get worse with greater correlation.

Figures 4.6 and 4.7 illustrate the dependence of the probability P_{loss} and the mean L_{orb} on the probability p and the variation of the $BMAP$ under the PH_8 service time distribution.

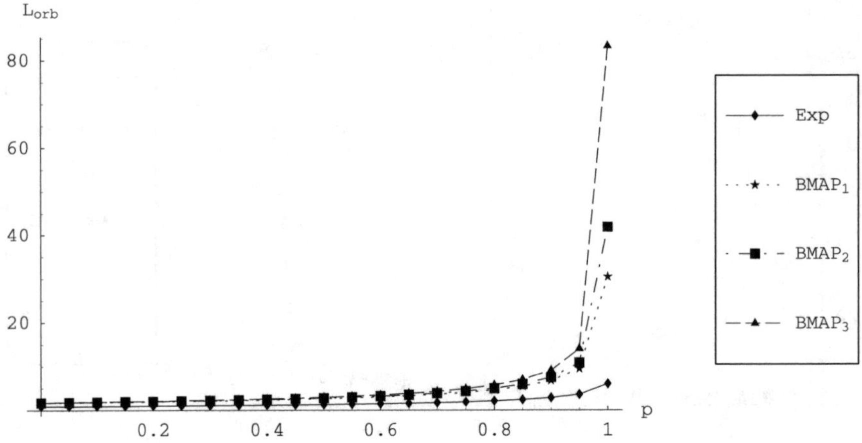

Fig. 4.5 Mean number of customers in the orbit, L_{orb}, as a function of probability p for $BMAP$s with different coefficients of correlation

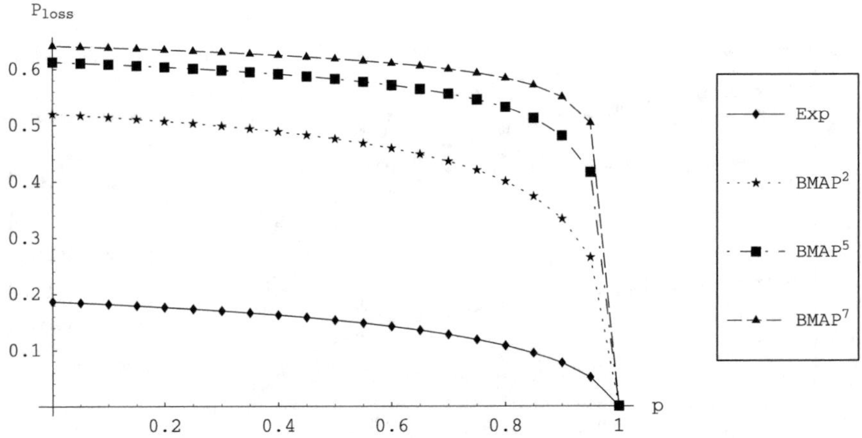

Fig. 4.6 Loss probability P_{loss} as a function of probability p for $BMAP$s with different coefficients of variation

Figures 4.6 and 4.7 demonstrate that the variation in the input flow also significantly influences the system performance measures and such measures get worse with greater growth of variation.

Thus the numerical experiments allow us to conclude that the approximation of a real flow with a stationary Poisson flow may lead to a too-optimistic prediction of system performance measures, if the coefficient of correlation is near zero and the coefficient of variation is not near to 1.

Let us illustrate the possible application of the results obtained in solving a simple optimization problem. Assume that the system incurs a loss as a result of unserved customers leaving the system and due to the long presence of customers in the orbit.

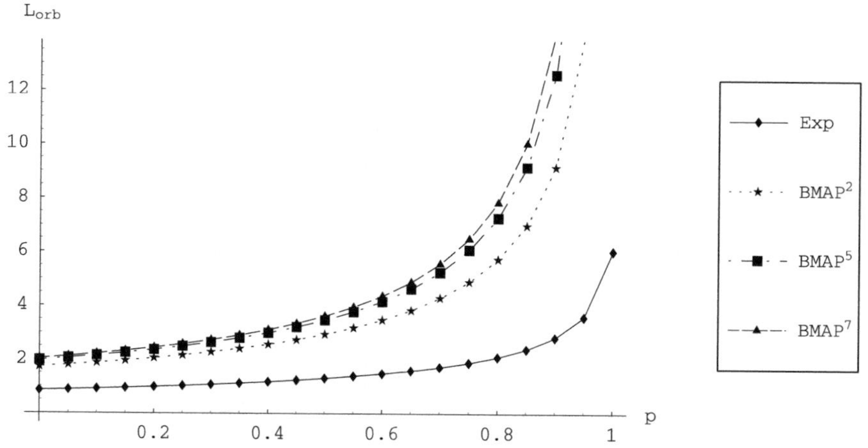

Fig. 4.7 Mean number of customers in the orbit, L_{orb}, as a function of probability p for $BMAP$s with different coefficients of variation

The weighted criterion of the system operation quality has following form:

$$I = c_1 \lambda P_{loss} + c_2 L_{orb}. \tag{4.34}$$

Here c_1 and c_2 are cost coefficients. The coefficient c_1 is the penalty for the loss of one customer, while the coefficient c_2 is the penalty for a presence of the customer in the orbit per unit time.

Assume that it is possible to select the probability p of customer's return to the orbit after an unsuccessful attempt to be served.

Figure 4.8 illustrates the dependence of criterion (4.34) on the probability p under different relations between cost coefficients c_1 and c_2. The input flow is described by $BMAP_3$, while the service process is described by the PH_5 distribution.

According to Fig. 4.8 we may see that if we have a big penalty for loss of customers the management of parameter p may have a significant effect on in the value of quality criterion (4.34).

Consider another numerical example. Assume that the model $BMAP/PH/N$ is applied to make a decision regarding the number of channels providing mobile communication, for instance in an airport terminal. One physical channel can be used to organize eight logical channels. One or two logical channels of each physical channel are reserved for the management of system operations. Thus with one, two, or three physical channels the number of logical channels available to mobile network users is $N = 7, 14$, and 22 respectively. We calculate the main characteristics of the system for one, two, or three physical channels and various assumptions regarding the nature of the input flow. To decrease the dimensionality of the blocks of the generator assume that the service time is exponentially distributed with the parameter (rate) $\mu = 10$. Assume that the strategy for retrials from the

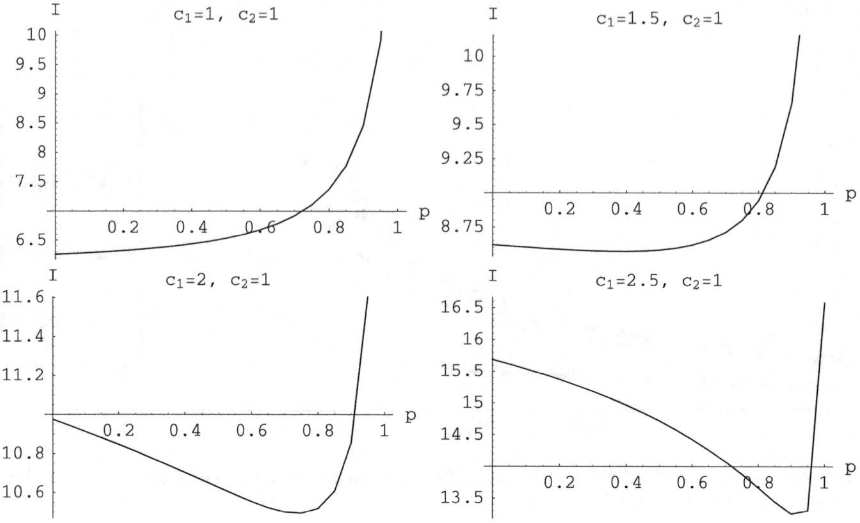

Fig. 4.8 Quality criterion (4.34) as a function of probability p under different cost coefficients c_1 and c_2

orbit is classical and each customer in the orbit generates repeated attempts in time intervals with the exponential distribution with the parameter (rate) $\alpha = 20$.

We will calculate the following system performance characteristics: the mean number of customers in the orbit L_{orb}, the probability of immediate access for service P_{imm}, and the probability P_0 that the orbit is empty. Above we have provided an illustration of the fact that under the same fundamental rate λ the system performance characteristics may significantly depend on the correlation of adjacent inter-arrival intervals. Thus, we will consider five different $BMAP$s ($BMAP_1$, $BMAP_2$, $BMAP_3$, $BMAP_4$, $BIPP$). Each of them has the same fundamental rate $\lambda = 27.5$. The first four flows have the variation coefficient equal to 12.2733, and their coefficients of correlation are $c_{cor} = 0.1; 0.2; 0.3; 0.4$.

$BMAP_1$ has the coefficient of correlation equal to 0.1. It is defined by the following matrices:

$$D_0 = \begin{pmatrix} -5.874 & 0 \\ 0 & -0.129 \end{pmatrix} \quad \text{and} \quad D_k = \begin{pmatrix} 0.586 & 0.002 \\ 0.010 & 0.003 \end{pmatrix}, \ k = 1, \ldots, 10.$$

$BMAP_2$ has the coefficient of correlation equal to 0.2. It is defined by the following matrices:

$$D_0 = \begin{pmatrix} -6.745 & 0 \\ 0 & -0.219 \end{pmatrix} \quad \text{and} \quad D_k = \begin{pmatrix} 0.670 & 0.004 \\ 0.012 & 0.010 \end{pmatrix}, \ k = 1, \ldots, 10.$$

$BMAP_3$ has the coefficient of correlation equal to 0.3. It is defined by the following matrices:

$$D_0 = \begin{pmatrix} -8.771 & 0 \\ 0 & -0.354 \end{pmatrix} \quad \text{and} \quad D_k = \begin{pmatrix} 0.867 & 0.010 \\ 0.012 & 0.024 \end{pmatrix}, \; k = 1, \ldots, 10.$$

$BMAP_4$ has the coefficient of correlation equal to 0.4. It is defined by the following matrices:

$$D_0 = \begin{pmatrix} -17.231 & 0 \\ 0.002 & -0.559 \end{pmatrix} \quad \text{and} \quad D_k = \begin{pmatrix} 1.705 & 0.018 \\ 0.006 & 0.049 \end{pmatrix}, \; k = 1, \ldots, 10.$$

The fifth $BMAP$ is a $BIPP$ (Batch Interrupted Poisson Process) flow. It is defined by the following matrices:

$$D_0 = \begin{pmatrix} -1 & 1 \\ 0 & -10 \end{pmatrix} \quad \text{and} \quad D_k = \begin{pmatrix} 0 & 0 \\ 0.1 & 0.9 \end{pmatrix}, \; k = 1, \ldots, 10.$$

This flow has zero correlation but is characterized by irregularity of arrivals when intervals with intensive arrival of customers alternate with intervals when there are no customers at all

We provide three tables characterizing the values L_{orb}, P_{imm}, and P_0 under various input flows and different numbers of logical channels equal to $N = 7, 14$, and 22 respectively.

Based on the results of the calculations provided in Tables 4.1 and 4.2, we may draw the following conclusions:

- The correlation significantly affects the system operation characteristics. A greater correlation leads to worse characteristics. Thus, in the investigation of real systems it is necessary to create a relevant model of input flow taking into

Table 4.1 Mean number of customers in the orbit, L_{orb}, as a function of the number of servers N for different $BMAP$s

L_{orb}	$BIPP$	$BMAP_1$	$BMAP_2$	$BMAP_3$	$BMAP_4$
$N = 7$	3.1	1.7	2.1	3.7	47.9
$N = 14$	0.3	0.1	0.2	0.3	1.3
$N = 22$	0.02	0	0	0	0.2

Table 4.2 Probability P_{imm} of immediate service as a function of the number of servers N for different $BMAP$s

P_{imm}	$BIPP$	$BMAP_1$	$BMAP_2$	$BMAP_3$	$BMAP_4$
$N = 7$	0.39	0.53	0.48	0.35	0.09
$N = 14$	0.87	0.94	0.92	0.87	0.59
$N = 22$	0.99	1	0.99	0.99	0.92

account the presence or absence of correlation in the flow. This will help avoid design mistakes. For instance, if in the design we want to increase the probability of immediate access of an arbitrary user to the network resources to more than 0.9 under a correlation of 0.1–0.2 it is sufficient to have two physical channels (14 logical channels). A correlation of 0.3 or above will require three physical channels. If the required probability of the immediate access of an arbitrary user to the resources of the mobile network is 0.9, under a coefficient of correlation 0.4 or higher even three physical channels will not be able to provide the service quality required.

- Besides the correlation, the irregularity of customer arrivals (i.e., burstiness) is typical, for instance, for nodes providing mobile communication in the terminals of transport systems, and negatively influences the user service quality as well.

The effect of group arrivals is interesting to examine as well. Consider the following system performance characteristic: $P_{imm}^{(k)}$ the probability of immediate service of an arbitrary customer if it arrived in a group consisting of $k,\ k \geq 1$ customers. It is easy to verify that this probability is calculated as follows:

$$P_{imm}^{(k)} = \sum_{m=0}^{N-k} P_m^{(k)} + \sum_{m=\max\{0,N-k+1\}}^{N-1} \frac{N-m}{k} P_m^{(k)}, \quad k \geq 1,$$

where $P_m^{(k)}$ is the probability that an arbitrary customer arrives at the system in a k size batch and finds m busy servers:

$$P_m^{(k)} = \frac{[\mathbf{P}(1)]_m (D_k \otimes I_{M^m})\mathbf{e}}{\boldsymbol{\theta} D_k \mathbf{e}}, \quad m = \overline{0, N}, k \geq 1.$$

The results of calculation of the probability $P_{imm}^{(k)}$ as a function of the number of servers and the size of the groups is presented in Fig. 4.9. The input data are described in the previous experiment ($BMAP_1$ taken as arrival flow) (Table 4.3).

Figure 4.9 demonstrates that with a small number of servers the value of the probability $P_{imm}^{(k)}$ strongly depends on k. Thus, if $N = 7$ and $k = 1$ this probability is close to 0.8. Averaging the probability by the size of groups we obtain that this probability equals approximately 0.55; the relevant curve in the figure is bold and is defined with the letter a). And if $k = 10$ this probability is less than 0.4. With the growth of the number of servers N this difference begins to smooth out. Nevertheless, the effect of group input should also be taken into account in designing solutions.

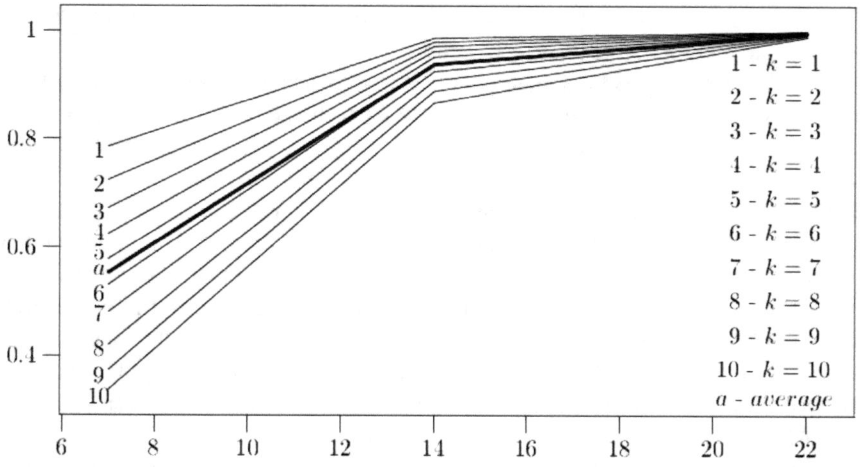

Fig. 4.9 Probability $P_{imm}^{(k)}$ as a function of the number of servers N and the batch size k, $k \geq 1$

Table 4.3 Probability P_0 that the orbit is empty as a function of the number of servers N for different $BMAP$s

P_0	$BIPP$	$BMAP_1$	$BMAP_2$	$BMAP_3$	$BMAP_4$
$N = 7$	0.64	0.67	0.66	0.63	0.64
$N = 14$	0.92	0.95	0.95	0.92	0.85
$N = 22$	0.99	1	1	0.99	0.96

4.3 $BMAP/PH/N$ Retrial System in the Case of a Phase Distribution of Service Time and a Large Number of Servers

4.3.1 *Choice of Markov Chain for Analysis of System*

When constructing the Markov chain ξ_t, $t \geq 0$ describing the operation of the $BMAP/PH/N$ retrial system, the method of taking into account the states of the PH service underlying process is of great importance. To provide better visualization and greater simplicity in introducing Markov chain ξ_t, $t \geq 0$, we use the method providing for dynamic numeration of servers.

Numbers are given only to busy servers. Such servers get numbers in consequential order, i.e., a server starting its service at a moment when n servers are busy gets the number $n + 1$; after the completion of service by any of the servers and assuming the lack of a queue, this server loses its number and the other servers (with larger numbers) are renumbered accordingly. The components of the Markov chain ξ_t, $t \geq 0$ are the states of underlying Markov chains of the service processes in the busy servers at time t. This method of taking into account the states of the service underlying processes is quite natural and comfortable. Under this method the size of the generator blocks of the Markov chain $\xi_t,\ t \geq 0$ is equal to $K = \bar{W}\frac{M^N-1}{M-1}$.

Another possible simple method to take into account the states of service underlying processes assumes that the state of each server of the system is taken into account whether the server is idle or busy. This state is assumed to be equal to m if the state of the service underlying process at the server equals $m,\ m = \overline{1, M}$. If the server is idle, we may artificially take 0 or $M + 1$ as the state of the service underlying process. In this method, the renumbering of servers is not required, but we should perform a randomized selection of the server by which the next customer is to be served if at an arrival epoch there are several idle servers. In this method, the size of the blocks of the generator equals $K_0 = \bar{W}(M + 1)^N$. It is obvious that $K_0 > K$. For instance, if N equals 5, and the state space of underlying input and service processes consist only of two elements ($W = 1,\ M = 2$), then $K = 126$ and $K_0 = 486$.

There is one more method to take into account the states of the service underlying processes on the servers. This method is more efficient in comparison to the two methods above in the situation when the number of servers N is high and the number of states of the service underlying processes, M, is low. The meaning of such a method is simple: instead of taking into account the current service phase we would better take into account the number of servers in each service phase, i.e., for each possible state m of the service underlying process we should take into account the number of servers in the phase m, $m = \overline{1, M}$. In this method the size of the blocks of the generator of the Markov chain $\xi_t,\ t \geq 0$ describing the operation of the $BMAP/PH/N$ retrial system equals $K_1 = \bar{W}\binom{N+M}{M}$. We can easily notice that $K_1 < K$. For instance, in the above example where $K = 126$, we get $K_1 = 42$.

Let us briefly summarize the results for the Markov chain describing the operation of a $BMAP/PH/N$ retrial queuing system and constructed using the just-outlined method of taking into account the states of the service underlying processes.

Instead of the Markov chain

$$\xi_t = (i_t,\ n_t,\ \nu_t,\ m_t^{(1)}, \ldots, m_t^{(n_t)}),\ t \geq 0$$

investigated in Sect. 4.2, we will investigate the multidimensional Markov chain

$$\zeta_t = \{i_t, n_t, \nu_t, h_t^{(1)}, \ldots, h_t^{(M)}\},\ t \geqslant 0,$$

where the components i_t, n_t, ν_t have the same meaning as the corresponding components of the Markov chain $\xi_t,\ t \geq 0$, and $h_t^{(m)}$ is the number of servers in phase m at time t. It is evident that $h_t^{(m)} \in \{0, \ldots, n_t\}, m = \overline{1, M},\ \sum_{m=1}^{M} h_t^{(m)} = n_t$, $n_t = \overline{0, N}$.

Following [99, 100], while forming the levels we will renumber the states of level i in lexicographic order of the components n_t, ν_t and reverse lexicographic order of the components $h_t^{(m)}$, $m = \overline{1, M}$.

To derive the infinitesimal generator of the Markov chain ζ_t, we will use matrices introduced in [99, 100] to describe the transition rates of N parallel Markov chains. We give a list of such matrices with a description of their probability sense.

The matrices $L_{N-n}(N, \tilde{S})$ describe the transition rates of components $\{h_t^{(1)}, \ldots, h_t^{(M)}\}$ of Markov chain ζ_t that occur after the end of service at one of the n busy servers (the reduction by one of the component corresponding to the state from which the service underlying Markov chain at one of the servers transits to the absorbing state).

The matrices $A_n(N, S)$ describe the transition rates of the components $\{h_t^{(1)}, \ldots, h_t^{(M)}\}$ of the Markov chain ζ_t that occur when one of these components transits from one state to another non-absorbing state. Such a transition does not cause any change in the number of busy servers in the system.

The matrices $P_n(\boldsymbol{\beta})$ describe the transition probabilities of the components $\{h_t^{(1)}, \ldots, h_t^{(M)}\}$ of the Markov chain ζ_t that occur after the initialization of the service process at an idle server when the number of busy servers is equal to n.

A detailed description of the matrices $P_n(\boldsymbol{\beta})$, $A_n(N, S)$, $L_{N-n}(N, \tilde{S})$ and the recursive algorithms to calculate the matrices is provided in [99, 100], where such algorithms were introduced, as well as in [63], where they were described in a clearer way, and, for completeness of presentation, at the end of this subsection.

4.3.2 *Infinitesimal Generator and Ergodicity Condition*

Lemma 4.3 *The infinitesimal generator of the Markov chain $\zeta_t, t \geq 0$ has block structure of type (4.15) where the blocks $Q_{i,j}$ are of the following form:*

$$(Q_{i,i})_{n,n'}$$

$$= \begin{cases} I_{\bar{W}} \otimes L_{N-n}(N, \tilde{S}), & n' = n - 1, \\ D_0 \oplus A_n(N, S) + I_{\bar{W}} \otimes \Delta^{(n)} - i\alpha(1 - \delta_{n,N}) I_{\bar{W}K(n)}, & n' = n, \\ D_{n'-n} \otimes P_{n,n'}(\boldsymbol{\beta}), & n' = \overline{n+1, N}, \end{cases}$$

$$(Q_{i,i-1})_{n,n'} = i\alpha I_{\bar{W}} \otimes P_{n,n'}(\boldsymbol{\beta}), n' = n + 1, n = \overline{0, N-1},$$

$$(Q_{i,i+k})_{n,n'} = \begin{cases} D_{N+k-n} \otimes P_{n,N}(\boldsymbol{\beta}), \ n' = N, & n = \overline{0, N-1}, \\ D_k \otimes I_{K(N)}, & n' = n = N, \end{cases} \quad k \geqslant 1,$$

where

- $\Delta^{(n)}$, $n = \overline{0, N}$,—*diagonal matrices providing for the execution of equations* $A\mathbf{e} = \mathbf{0}$,
- $\tilde{S} = \begin{pmatrix} \mathbf{0} & O \\ \mathbf{S}_0 & S \end{pmatrix}$,
- $K(n) = \binom{n+M-1}{M-1}$, $n = \overline{0, N}$,
- $P_{n,n'}(\boldsymbol{\beta}) = P_n(\boldsymbol{\beta})P_{n+1}(\boldsymbol{\beta}) \dots P_{n'-1}(\boldsymbol{\beta})$, $0 \leqslant n < n' \leqslant N$.

The following statement is valid.

Theorem 4.6 *In the case of an infinitely increasing retrial rate the ergodicity condition for the Markov chain ζ_t, $t \geq 0$ has form (4.21), where λ is the fundamental rate of the $BMAP$, and the value $\bar{\mu}$ is defined by the formula*

$$\bar{\mu} = \mathbf{y}L_0(N, \tilde{S})\mathbf{e}, \tag{4.35}$$

where the vector $\mathbf{y}$ *is the unique solution of the system of linear algebraic equations*

$$\mathbf{y}\Big(A_N(N, S) + \Delta^{(N)} + L_0(N, \tilde{S})P_{N-1,N}(\boldsymbol{\beta})\Big) = \mathbf{0}, \ \mathbf{y}\mathbf{e} = 1. \tag{4.36}$$

The proof of the theorem is similar to the proof of Theorem 4.2.

Remark 4.1 Since inequality (4.21) with the value $\bar{\mu}$ given by formulas (4.22)–(4.23) defines the same ergodicity condition as the inequality with the value $\bar{\mu}$ given by formulas (4.35)–(4.36) it is tempting to suggest that the values $\bar{\mu}$ given by formulas (4.22)–(4.23) and (4.35)–(4.36) coincide. It is not possible to demonstrate this fact analytically. Nevertheless, the computer realization of such formulas under various parameters of the input flow and the service process and with a varying number of servers demonstrates that these values coincide. Furthermore, they coincide with the intuitively expected value $\bar{\mu} = N\mu$.

4.3.3 Numerical Examples

As the Markov chain ζ_t, $t \geq 0$ is an asymptotically quasi-Toeplitz Markov chain, its stationary distribution is calculated by applying the algorithm for asymptotically quasi-Toeplitz Markov chains presented in Sect. 2.7. As for a large number of servers N the size K_1 of the blocks $Q_{i,j}$ of the generator of the Markov chain ζ_t, $t \geq 0$ is significantly smaller than the size K of the relevant blocks $Q_{i,j}$ of the generator of the Markov chain ξ_t, $t \geq 0$, we can calculate the stationary distribution of the Markov chain ζ_t, $t \geq 0$ for a significantly larger number of servers N.

Table 4.4 Dependence of block size and calculation time on the number of servers N for the Markov chains ξ_t and ζ_t

Number of servers N	K_1	Calculation time for the chain ξ_t	K	Calculation time for the chain ζ_t
1	6	0 h 0 min 0 s	6	0 h 0 min 0 s
2	12	0 h 0 min 1 s	14	0 h 0 min 2 s
3	20	0 h 0 min 5 s	30	0 h 0 min 14 s
4	30	0 h 0 min 13 s	62	0 h 2 min 9 s
5	42	0 h 0 min 37 s	126	0 h 19 min 13 s
6	56	0 h 1 min 22 s	254	2 h 46 min 52 s
7	72	0 h 2 min 52 s	510	32 h 16 min 26 s

Table 4.5 Dependence of block size and calculation time on the number of servers N for the Markov chain ζ_t

Number of servers N	K_1	Calculation time
7	72	0 h 2 min 52 s
8	90	0 h 5 min 35 s
9	110	0 h 10 min 17 s
10	132	0 h 17 min 48 s
11	156	0 h 29 min 13 s
12	182	0 h 48 min 15 s
13	210	1 h 17 min 51 s
14	240	1 h 45 min 47 s
15	272	2 h 43 min 33 s
16	306	3 h 52 min 45 s
17	342	5 h 21 min 16 s
18	380	8 h 45 min 59 s
19	420	12 h 26 min 18 s
20	462	16 h 35 min 1 s
21	506	21 h 58 min 24 s

Table 4.4 illustrates the dimensions of the blocks of the generator for both Markov chains and the time for their calculation for a testing system with the dimensionality of the $BMAP$ underlying process and service underlying process equal to 2. The calculations were performed on a personal computer AMD Athlon 64 3700+, 2.21 GHz, RAM 2 Gb, with the operating system Microsoft Windows XP installed. The notation $XXhYYmZZs$ means XX hours, YY minutes, and ZZ seconds. The table shows that if $N = 7$ the calculations for the Markov chain $\zeta_t,\ t \geq 0$ take less than 3 min, while the calculations for the Markov chain $\xi_t,\ t \geq 0$ take more than 32 h. If $N > 7$ it is not possible to perform the calculations for the Markov chain $\xi_t,\ t \geq 0$, due to the impossibility of obtaining the operating memory capacity required. For Markov chains $\zeta_t,\ t \geq 0$, we may calculate up to $N = 22$. The information regarding the size K_1 of the corresponding blocks and the calculation time is provided in Table 4.5. It is worth noting the following fact: the size of blocks K equals 510 if $N = 7$. This is almost equal to the size of blocks K_1 ($K_1 = 506$) if $N = 21$. At that size, the calculation time (32 h 16 min 26 s) is

significantly higher (when $N = 21$ it is equal to 21 h 58 min 24 s). This is due to the fact that in the case of the Markov chain $\zeta_t,\ t \geq 0$ the sequence of matrices G_i converges to the matrix G more quickly.

4.3.4 Algorithm for Calculation of the Matrices $P_n(\boldsymbol{\beta})$, $A_n(N, S)$, and $L_{N-n}(N, \tilde{S})$

4.3.4.1 Calculation of the Matrices $L_{N-n}(N, \tilde{S})$

Assume that the number of servers N, the vector $\boldsymbol{\beta}$, and the matrix S, defining the phase-type service process, are fixed.

First we calculate the matrices $L_z(N, \tilde{S}),\ z = \overline{0, N-1}$.

Assume $\tau^{(k)}(S)$ is a matrix obtained from the matrix S by the elimination of its first k rows and k columns. According to the definition $\tau^{(0)}(S) = S$.

Start the cycle (loop) on $j,\ \ j = \overline{1, M-2}$.

Step 1. Calculate the additional matrices $T(j)$.

Calculate the matrices $T(j) = \tau^{M-2-j}(S)$. Later the argument j will be omitted. Introducing $t_{i,l}$ we define the elements of the matrix T. By r we denote the number of rows of the matrix T.

Step 2. Calculate the additional matrices $U_z(N, T),\ z = \overline{1, N}$.

Assume $U_z^{(0)} = t_{1,r},\ z = \overline{1, N}$.

Calculate applying the loop on $w = \overline{1, r-2}$:

$$U_z^{(w)} = \begin{pmatrix} t_{1,r-w}I & U_N^{(w-1)} & 0 & \cdots & 0 & 0 \\ 0 & t_{1,r-w}I & U_{N-1}^{(w-1)} & \cdots & 0 & 0 \\ \vdots & \vdots & \vdots & \ddots & \vdots & \vdots \\ 0 & 0 & 0 & \cdots & t_{1,r-w}I & U_z^{(w-1)} \end{pmatrix},\ z = \overline{1, N}.$$

$U_z(N, T) = zU_z^{(r-2)},\ z = \overline{1, N}$.

Step 3. Calculate the matrices $L_z(N, T),\ z = \overline{0, N-1}$.

Assume $L_z^{(0)} = (N-z)t_{r,1},\ z = \overline{0, N-1}$.

Calculate applying the loop on $w = \overline{1, r-2}$:

$$L_z^{(w)} = \begin{pmatrix} (N-z)t_{r-w,1}I & 0 & \cdots & 0 \\ L_{N-1}^{(w-1)} & (N-z-1)t_{r-w,1}I & \cdots & 0 \\ 0 & L_{N-2}^{(w-1)} & \cdots & 0 \\ \vdots & \vdots & \ddots & \vdots \\ 0 & 0 & \cdots & t_{r-w,1}I \\ 0 & 0 & \cdots & L_z^{(w-1)} \end{pmatrix},\ z = \overline{0, N-1}.$$

Calculate the matrices $L_z(N,T) = L_z^{(r-2)}, \ z = \overline{0, N-1}$.
End of loop on $j, \ \ j = \overline{1, M-2}$.

4.3.4.2 Calculation of the Matrices $A_m(N, S), \ m = \overline{0, N}$

Assume

$$A_m^{(0)} = \begin{pmatrix} 0 & mS_{M-1,M} & 0 & \cdots & 0 & 0 \\ S_{M,M-1} & 0 & (m-1)S_{M-1,M} & \cdots & 0 & 0 \\ 0 & 2S_{M,M-1} & 0 & \cdots & 0 & 0 \\ \vdots & \vdots & \vdots & \ddots & \vdots & \vdots \\ 0 & 0 & 0 & \cdots & 0 & S_{M-1,M} \\ 0 & 0 & 0 & \cdots & mS_{M,M-1} & 0 \end{pmatrix},$$

$$m = \overline{1, N}.$$

For $j = \overline{1, M-2}$, calculate the following matrices recurrently:

$$A_m^{(j)}$$

$$= \begin{pmatrix} 0 & \frac{m}{N}U_N(N,T) & 0 & \cdots & 0 & 0 \\ L_{N-1}(N,T) & A_1^{(j-1)} & \frac{m-1}{N-1}U_{N-1}(N,T) & \cdots & 0 & 0 \\ 0 & L_{N-2}(N,T) & A_2^{(j-1)} & \cdots & 0 & 0 \\ \vdots & \vdots & \vdots & \ddots & \vdots & \vdots \\ 0 & 0 & 0 & \cdots & A_{m-1}^{(j-1)} & \frac{1}{N-m+1}U_{N-m+1}(N,T) \\ 0 & 0 & 0 & \cdots & L_{N-m}(N,T) & A_m^{(j-1)} \end{pmatrix},$$

$$m = \overline{1, N}.$$

Calculate the matrices $A_m(N,S) = A_m^{(M-2)}, \ m = \overline{1, N}, \ A_0(N,S) = 0$.

4.3.4.3 Calculation of the Matrices $P_m(\beta), \ m = \overline{1, N-1}$

Calculate matrices of size $(m+1) \times (m+2)$

$$P_m^{(0)} = \begin{pmatrix} \beta_{M-1} & \beta_M & 0 & \cdots & 0 & 0 \\ 0 & \beta_{M-1} & \beta_M & \cdots & 0 & 0 \\ \vdots & \vdots & \vdots & \ddots & \vdots & \vdots \\ 0 & 0 & 0 & \cdots & \beta_{M-1} & \beta_M \end{pmatrix}, m = \overline{1, N-1}.$$

For $j = \overline{1, M-2}$, assume that $\mathbf{a} = (\beta_{M-j}, \dots, \beta_M)$, and calculate the following matrices:

$$P_m^{(j)} = \begin{pmatrix} \beta_{M-j-1} & \mathbf{a} & 0 & 0 & \cdots & 0 & 0 \\ 0 & \beta_{M-j-1}I & P_1^{(j-1)} & 0 & \cdots & 0 & 0 \\ 0 & 0 & \beta_{M-j-1}I & P_2^{(j-1)} & \cdots & 0 & 0 \\ \vdots & \vdots & \vdots & \vdots & \ddots & \vdots & \vdots \\ 0 & 0 & 0 & 0 & \cdots & \beta_{M-j-1}I & P_m^{(j-1)} \end{pmatrix},$$

$$m = \overline{1, N-1}.$$

Calculate the matrices $P_m(\boldsymbol{\beta}) = P_m^{(M-2)}$, $m = \overline{1, N-1}$, $P_0(\boldsymbol{\beta}) = \boldsymbol{\beta}$.

Chapter 5
Mathematical Models and Methods of Investigation of Hybrid Communication Networks Based on Laser and Radio Technologies

The rapid and continuous increase in the number of users on the Internet and the increase in the volume and quality of information transmitted in broadband wireless networks requires a dramatic increase in the performance of multimedia communication channels. In this regard, in recent years, within the frame of the development of next-generation networks, intensive research is being carried out to improve wireless performance. One of the directions for creating ultra-high-speed (up to 10 Gbit/s) and reliable wireless communications is the development of hybrid systems based on laser and radio technologies.

FSO (Free Space Optics) technology has been widely used in recent times. This technology is based on the transmission of data by modulated radiation in the infrared or visible part of the spectrum through the atmosphere and their subsequent detection by an optical photo-detector device. The main advantages of an atmospheric optical communication link are as follows.

- High bandwidth and quality of digital communication. Modern FSO solutions can provide a transmission speed of digital flows up to 10 Gbit/s with a bit error rate of 10^{-12}, which is currently impossible to achieve with any other wireless technologies;
- High security of the channel from unauthorized access and stealth. No wireless transmission technology can offer such confidentiality of communication as laser. Absence of pronounced external signs (basically, this is electromagnetic radiation) allows not only the transmitted information, but also the very fact of information exchange to be hidden. Therefore, laser systems are often used for a variety of applications where high confidentiality of data transmission is required, including by financial, medical, and military organizations;
- High level of noise immunity and noninterference. FSO equipment is immune to radio interference and does not create interference itself;
- Speed and ease of deployment of an FSO network.

A. N. Dudin et al., *The Theory of Queuing Systems with Correlated Flows*,
https://doi.org/10.1007/978-3-030-32072-0_5

Along with the main advantages of wireless optical systems, their main disadvantages are also known: the dependence of the accessibility of the communication channel on weather conditions and the need to provide direct visibility between the transmitter and the receiver. Unfavorable weather conditions, such as snow or fog, can significantly reduce the effective range of operation of laser atmospheric communication lines. Thus, the attenuation of a signal in an optical channel in a strong fog can reach a critical value of 50–100 dB/km. Therefore, in order to achieve acceptable reliability values of an FSO communication channel, it is necessary to resort to the use of hybrid solutions.

Hybrid radio-optical equipment is based on the use of redundant radio channels (centimeter and/or millimeter range of radio waves) together with an optical channel. Note that the operation of the radio channel of the centimeter range of radio waves is practically independent of the weather. The performance of a millimeter-wave wireless channel is not affected by fog. At the same time, the signal/noise ratio, which determines the quality of a channel's operation, is greatly reduced by heavy rain. This complementary behavior of optical and broadband radio channels has made it possible to put forward the concept of hybrid-carrier systems that function reliably in all weather conditions.

Due to the high demand for high-speed and reliable communication channels, the following architectures of hybrid systems are currently being used to solve the *last mile* problem [3, 24, 39, 88, 91, 120, 122, 123] a high-speed laser channel reserved by a broadband radio channel operating under the IEEE 802.11n protocol in the centimeter band of radio waves (cold or hot reserve); an FSO channel reserved by a radio channel of the millimeter-wave E-band of radio waves (71–76 GHz, 81–86 GHz); an FSO channel and a radio channel of the millimeter band operating in parallel and reserved by an IEEE 802.11n channel, which is in cold reserve.

Practical needs have stimulated theoretical studies on the evaluation of performance and the selection of optimal modes for the operation of hybrid systems using queuing theory models with unreliable service channels. Initially, these papers, see, e.g. [2, 113, 114] used simplified assumptions about the Poisson character of the input flow, the exponential distribution of the packet service time, and the uptime and repair time of the communication channels.

In this chapter we consider a number of queuing systems that can be used to model hybrid communication systems with one or two unreliable channels and various redundancy schemes (cold and hot redundancy). These systems, in contrast to those mentioned above, are considered under much more general assumptions with respect to the nature of the input flow, the flow of breakdowns, and the service and repair time distributions.

Such systems have been considered in the papers [35, 69, 109], which deal with unreliable queues with Markovian arrival processes ($BMAP$ and MAP flows) and PH distributions of packet transmission times via communication channels. Although these assumptions complicate the study of models that adequately describe the operation of hybrid systems, they allow us to take into account the non-stationary, correlated nature of information flows in modern and future 5G networks.

5.1 Analysis of Characteristics of the Data Transmission Process Under Hot-Standby Architecture of a High-Speed FSO Channel Backed Up by a Wireless Broadband Radio Channel

In this section we consider a single-server queuing system consisting of a main unreliable server and a backup (reserve) absolutely reliable server in hot standby. This queue can be used to model a hybrid communication system consisting of an unreliable FSO channel and a reliable radio channel on the millimeter-wave E-band of radio waves 71–76, 81–86 GHz.

5.1.1 Model Description

We consider an unreliable queuing system consisting of an infinite buffer and two heterogeneous servers. By default, we assume that server 1 (main server) is high-speed but unreliable while server 2 (backup or reserve server) is low-speed but absolutely reliable. However, it should be mentioned that the presented analysis is implemented without any use of the fact that the service rate at server 1 is higher than the service rate at server 2.

The backup server is in so-called hot standby. This means the following. If both servers are able to provide the service, they serve a customer independently of each other. The service of a customer is completed when his/her service by either of the two servers is finished. After the service completion, both servers immediately start the service of the next customer, if he/she is present in the system. If the system is idle, the servers wait for the arrival of a new customer. When server 1 fails its repair starts immediately. During the repair period information is transmitted over the backup channel. After the repair period the main server immediately connects to the transmission of information.

Customers arrive at the system in accordance with a $BMAP$ (see Sect. 2.1), which is defined by the underlying process $\nu_t,\ t \geq 0$ with state space $\{0, 1, \ldots, W\}$ and by the matrices $D_k, k \geq 0$ of order $W + 1$. The arrival rate of customers, the arrival rate of groups, and the coefficients of variation and correlation in the $BMAP$ are denoted by $\lambda,\ \lambda_b,\ c_{var}, c_{cor}$, respectively, and are calculated using the formulas given in Chap. 2.

Breakdowns arrive at server 1 in accordance with a MAP with an underlying process $\eta_t, t \geq 0$, which takes values in the set $\{0, 1, \ldots, V\}$ and is defined by the matrices H_0 and H_1. The fundamental rate of this MAP is $h = \boldsymbol{\gamma} H_1 \mathbf{e}$, where the row vector $\boldsymbol{\gamma}$ is the unique solution of the system $\boldsymbol{\gamma}(H_0 + H_1) = \mathbf{0},\ \boldsymbol{\gamma}\mathbf{e} = 1$.

When the server fails, the repair period starts immediately. The duration of this period has a PH-type distribution with an irreducible representation $(\boldsymbol{\tau}, \mathbf{T})$. The repair process is directed by the Markov chain $\vartheta_t,\ \ t \geq 0$ with state space $\{1, \ldots, R, R + 1\}$, where the state $R + 1$ is an absorbing one. The intensities of transitions into the absorbing state are defined by the vector $\mathbf{T}_0 = -T\mathbf{e}$. The repair rate is calculated as $\tau = -(\boldsymbol{\tau} T^{-1}\mathbf{e})^{-1}$. Breakdowns arriving during the repair time are ignored by the system.

The service time of a customer by the kth server, $k = 1, 2$, has PH-type distribution with an irreducible representation $(\boldsymbol{\beta}^{(k)}, \mathbf{S}^{(k)})$. The service process on the kth server is directed by the Markov chain $m_t^{(k)}$, $t \geq 0$ with the state space $\{1, \ldots, M^{(k)}, M^{(k)} + 1\}$, where $M^{(k)} + 1$ is an absorbing state. The transition rates into the absorbing state are defined by the vector $\mathbf{S}_0^{(k)} = -\mathbf{S}^{(k)}\mathbf{e}$. The service rates are calculated as $\mu^{(k)} = -[\boldsymbol{\beta}^{(k)}(\mathbf{S}^{(k)})^{-1}\mathbf{e}]^{-1}$, $k = 1, 2$.

5.1.2 Process of the System States

Let at time t

- i_t be the number of customers in the system, $i_t \geq 0$;
- $r_t = 0$ if server 1 is under repair, $r_t = 1$ if server 1 is not broken, i.e., both the servers are fault-free;
- $m_t^{(k)}$ be the state of the service underlying process at the kth busy server, $m_t^{(k)} = \overline{1, M^{(k)}}$, $k = 1, 2$;
- η_t be the state of the repair underlying process at the server 1, $\eta_t = \overline{1, R}$, if $r_t = 0$;
- ν_t and η_t be the states of the $BMAP$ underlying process of customers and the MAP underlying process of breakdowns, correspondingly, $\nu_t = \overline{0, W}$, $\eta_t = \overline{0, V}$.

The process of the system states is described by the regular irreducible continuous-time Markov chain, $\xi_t, t \geq 0$ with state space

$$\Omega = \{(0, 0, \vartheta, \eta, \nu)\} \bigcup \{(0, 1, \eta, \nu)\} \bigcup \{(i, 0, \vartheta, m^{(2)}, \eta, \nu)\} \bigcup$$

$$\{(i, 1, m^{(1)}, m^{(2)}, \eta, \nu)\}, \ i \geq 1, \ \vartheta = \overline{1, R}, \ m^{(k)} = \overline{1, M^{(k)}}, k = 1, 2,$$

$$\eta = \overline{0, V}, \ \nu = \overline{0, W}.$$

In the following, we will assume that the states of the chain $\xi_t, t \geq 0$ are ordered in lexicographic order. Let Q_{ij}, $i, j \geq 0$ be matrices formed by the rates of the chain transition from the state corresponding to the value i of the component i_n to the state corresponding to the value j of this component. The following statement is true.

Lemma 5.1 *The infinitesimal generator Q of the Markov chain ξ_t, $t \geq 0$ has the block structure*

$$Q = \begin{pmatrix} Q_{0,0} & Q_{0,1} & Q_{0,2} & Q_{0,3} & \cdots \\ Q_{1,0} & Q_1 & Q_2 & Q_3 & \cdots \\ O & Q_0 & Q_1 & Q_2 & \cdots \\ O & O & Q_0 & Q_1 & \cdots \\ \vdots & \vdots & \vdots & \vdots & \ddots \end{pmatrix},$$

where the non-zero blocks have the following form:

$$Q_{0,0} = \begin{pmatrix} T \oplus (H_0 + H_1) \oplus D_0 & \mathbf{T}_0 \otimes I_a \\ \boldsymbol{\tau} \otimes H_1 \otimes I_{\bar{W}} & H_0 \oplus D_0 \end{pmatrix},$$

$$Q_{0,k} = \begin{pmatrix} I_R \otimes \boldsymbol{\beta}^{(2)} \otimes I_{\bar{V}} \otimes D_k & O \\ O & \boldsymbol{\beta}^{(1)} \otimes \boldsymbol{\beta}^{(2)} \otimes I_{\bar{V}} \otimes D_k \end{pmatrix}, \ k \geq 1,$$

$$Q_{1,0} = \begin{pmatrix} I_R \otimes \mathbf{S}_0^{(2)} \otimes I_a & O \\ O & (\mathbf{S}_0^{(1)} \otimes \mathbf{e}_{M^{(2)}} + \mathbf{e}_{M^{(1)}} \otimes \mathbf{S}_0^{(2)}) \otimes I_a \end{pmatrix},$$

$$Q_0 = \begin{pmatrix} I_R \otimes \mathbf{S}_0^{(2)} \boldsymbol{\beta}^{(2)} \otimes I_a & O \\ O & \tilde{S} \otimes I_a \end{pmatrix},$$

$$Q_1 = \begin{pmatrix} T \oplus S^{(2)} \oplus (H_0 + H_1) \oplus D_0 & \mathbf{T}_0 \boldsymbol{\beta}^{(1)} \otimes I_{M^{(2)}a} \\ \mathbf{e}_{M^{(1)}} \boldsymbol{\tau} \otimes I_{M^{(2)}} \otimes H_1 \otimes I_{\bar{W}} & S^{(1)} \oplus S^{(2)} \oplus H_0 \oplus D_0 \end{pmatrix},$$

$$Q_k = \begin{pmatrix} I_{RM^{(2)}\bar{V}} \otimes D_{k-1} & O \\ O & I_{M^{(1)}M^{(2)}\bar{V}} \otimes D_{k-1} \end{pmatrix}, \ k \geq 2.$$

Here $\tilde{S} = \mathbf{S}_0^{(1)} \boldsymbol{\beta}^{(1)} \otimes \mathbf{e}_{M^{(2)}} \boldsymbol{\beta}^{(2)} + \mathbf{e}_{M^{(1)}} \boldsymbol{\beta}^{(1)} \otimes \mathbf{S}_0^{(2)} \boldsymbol{\beta}^{(2)}$, $\bar{W} = W + 1$, $\bar{V} = V + 1$, $a = \bar{W}\bar{V}$.

The proof of the lemma carried out by means of the calculation of probabilities of transitions of the components of the Markov chain ξ_t during a time interval with infinitesimal length.

Analyzing the form of the generator Q, it is easy to see that it has a block upper-Hessenberg structure and blocks $Q_{i,j}$ formed by the transition rates of the chain ξ_t, $t \geq 0$ from states corresponding to the value $i > 1$ of the countable component to states corresponding to the value j of this component depend on i, j only via the difference $j - i$. This means that the Markov chain under consideration belongs to the class of quasi-Toeplitz Markov chains that was investigated in [70].

Corollary 5.1 *The Markov chain $\xi_t, t \geq 0$ belongs to the class of continuous-time quasi-Toeplitz Markov chains.*

In the following it will be useful to have expressions for the matrix generating functions $\tilde{Q}(z) = \sum_{k=1}^{\infty} Q_{0,k} z^k, \; Q(z) = \sum_{k=0}^{\infty} Q_k z^k, \; |z| \leq 1.$

Corollary 5.2 *The matrix generating functions $\tilde{Q}(z), \; Q(z)$ are of the form*

$$\tilde{Q}(z) = z \begin{pmatrix} I_R \otimes \boldsymbol{\beta}^{(2)} \otimes I_{\bar{W}} & O \\ O & \boldsymbol{\beta}^{(1)} \otimes \boldsymbol{\beta}^{(2)} \otimes I_a \end{pmatrix}$$

$$+ \, \mathrm{diag}\{I_{R\bar{V}} \otimes (D(z) - D_0), \; I_{\bar{V}} \otimes (D(z) - D_0)\}, \tag{5.1}$$

$$Q(z) = Q_0 + z \begin{pmatrix} T \oplus S^{(2)} \oplus (H_0 + H_1) \otimes I_{\bar{W}} & T_0 \boldsymbol{\beta}^{(1)} \otimes I_{M^{(2)}} \otimes I_a \\ \mathbf{e}_{M^{(1)}} \boldsymbol{\tau} \otimes I_{M^{(2)}} \otimes H_1 \otimes I_{\bar{W}} & S^{(1)} \oplus S^{(2)} \oplus H_0 \otimes I_{\bar{W}} \end{pmatrix}$$

$$+ \, z\mathrm{diag}\{I_{RM^{(2)}\bar{V}} \otimes D(z), \; I_{M^{(1)}M^{(2)}\bar{V}} \otimes D(z)\}. \tag{5.2}$$

5.1.3 Condition for Stable Operation of the System: Stationary Distribution

Theorem 5.1 *A necessary and sufficient condition for the existence of the stationary distribution of the Markov chain $\xi_t, \; t \geq 0$ is the fulfillment of the inequality*

$$\lambda < \mathbf{x} \, \mathrm{diag}\{\mathbf{e}_R \otimes \mathbf{S}_0^{(2)} \otimes \mathbf{e}_{\bar{V}}, (\mathbf{S}_0^{(1)} \otimes \mathbf{e}_{M^{(2)}} + \mathbf{e}_{M^{(1)}} \otimes \mathbf{S}_0^{(2)}) \otimes \mathbf{e}_{\bar{V}}\}\mathbf{e}, \tag{5.3}$$

where the vector $\mathbf{x}$ *is the unique solution of the system of linear algebraic equations*

$$\mathbf{x}\Gamma = \mathbf{0}, \; \mathbf{x}\mathbf{e} = 1, \tag{5.4}$$

and the matrix Γ has the form

$$\Gamma$$

$$= \begin{pmatrix} T \oplus S^{(2)} \oplus (H_0 + H_1) + I_R \otimes \mathbf{S}_0^{(2)} \boldsymbol{\beta}^{(2)} \otimes I_{\bar{V}} & \mathbf{T}_0 \boldsymbol{\beta}^{(1)} \otimes I_{M^{(2)}} \otimes I_{\bar{V}} \\ \mathbf{e}_{M^{(1)}} \boldsymbol{\tau} \otimes I_{M^{(2)}} \otimes H_1 & \tilde{S} \otimes I_{\bar{V}} + S^{(1)} \oplus S^{(2)} \oplus H_0 \end{pmatrix}.$$

Proof It follows from [70] that a necessary and sufficient condition for the existence of the stationary distribution of the chain $\xi_t, \; t \geq 0$, can be formulated in terms of the matrix generating function $Q(z)$ and has the form of the inequality

$$\mathbf{y}Q^{'}(1)\mathbf{e} < 0, \tag{5.5}$$

where the row vector $\mathbf{y}$ is the unique solution of the system of linear algebraic equations

$$\mathbf{y}Q(1) = \mathbf{0}, \ \mathbf{ye} = 1. \tag{5.6}$$

Let the vector $\mathbf{y}$ be of the form

$$\mathbf{y} = \mathbf{x} \otimes \boldsymbol{\theta}, \tag{5.7}$$

where $\boldsymbol{\theta}$ is the vector of the stationary distribution of the $BMAP$ underlying process $\nu_t,\ t \geq 0$ and $\mathbf{x}$ is a stochastic vector. Substituting expression (5.7) into (5.6), we verify that $\mathbf{y}$ is the unique solution of system (5.6) if the vector $\mathbf{x}$ satisfies system (5.4).

Substituting the vector $\mathbf{y}$ in the form (5.7) into (5.5) and using expression (5.2) to calculate the derivative $Q'(1)$, we reduce inequality (5.5) to the form (5.3). □

Remark 5.1 An intuitive explanation of stability condition (5.3) is as follows. The left-hand side of inequality (5.3) is the rate of customers arriving at the system. The right-hand side of the inequality is the rate of customers leaving the system after service under overload conditions. It is obvious that in a steady state the former rate must be less that the latter one.

Remark 5.2 The system load ρ is found as the ratio of the left-hand side of (5.3) to the right-hand side, i.e.,

$$\rho = \frac{\lambda}{\mathbf{x}\ \mathrm{diag}\{\mathbf{e}_R \otimes \mathbf{S}_0^{(2)} \otimes \mathbf{e}_{\bar{V}},\ (\mathbf{S}_0^{(1)} \otimes \mathbf{e}_{M^{(2)}} + \mathbf{e}_{M^{(1)}} \otimes \mathbf{S}_0^{(2)}) \otimes \mathbf{e}_{\bar{V}}\}\mathbf{e}}.$$

Corollary 5.3 *In the case of a stationary Poisson flow of breakdowns and exponential distributions of service and repair times, the ergodicity condition (5.3) and (5.4) is reduced to the following inequality:*

$$\lambda < \mu_2 + \frac{\tau}{\tau + h}\mu_1. \tag{5.8}$$

Proof In the case under consideration, the vector $\mathbf{x}$ consists of two components, say $\mathbf{x} = (x_1, x_2)$. It is easy to see that inequality (5.3) is reduced to the following one:

$$\lambda < x_1\mu_2 + x_2(\mu_1 + \mu_2). \tag{5.9}$$

System (5.4) is written as

$$\begin{cases} -x_1\tau + x_2 h = 0, \\ x_1\tau - x_2 h = 0, \\ x_1 + x_2 = 1. \end{cases} \tag{5.10}$$

Relations (5.9) and (5.10) imply inequality (5.8). □

In what follows we assume inequality (5.3) is fulfilled.

Denote the stationary state probabilities of the chain $\xi_t,\ t \geq 0$ by

$$p(0, 0, \vartheta, \eta, \nu),\ \ p(0, 1, \eta, \nu),\ \ p(i, 0, \vartheta, m^{(2)}, \eta, \nu),$$

$$p(i, 1, m^{(1)}, m^{(2)}, \eta, \nu),\ i \geq 1,\ \vartheta = \overline{1, R},\ m^{(k)} = \overline{1, M^{(k)}}, k = 1, 2,$$

$$\eta = \overline{0, V},\ \nu = \overline{0, W}.$$

Let us enumerate the steady state probabilities in accordance with the above-introduced order of the states of the chain and form the row vectors $\mathbf{p}_i$ of steady state probabilities corresponding the value i of the first component of the Markov chain, $i \geq 0$.

To calculate the vectors $\mathbf{p}_i,\ i \geq 0$ we use the numerically stable algorithm, see [70], which has been elaborated for calculating the stationary distribution of multidimensional continuous-time quasi-Toeplitz Markov chains. The derivation of this algorithm is based on the censoring technique and the algorithm consists of the following principal steps.

Algorithm 5.1

1. Calculate the matrix G as the minimal non-negative solution of the nonlinear matrix equation

$$\sum_{n=0}^{\infty} Q_n G^n = O.$$

2. Calculate the matrix G_0 from the equation

$$Q_{1,0} + \sum_{n=1}^{\infty} Q_{1,n} G^{n-1} G_0 = O,$$

whence it follows that

$$G_0 = -(\sum_{n=1}^{\infty} Q_{1,n} G^{n-1})^{-1} Q_{1,0}.$$

3. Calculate the matrices $\bar{Q}_{i,l},\ l \geq i, i \geq 0$, using the formulae

$$\bar{Q}_{i,l} = \begin{cases} Q_{0,l} + \sum\limits_{n=l+1}^{\infty} Q_{0,n} G_{n-1} G_{n-2} \dots G_l,\ i = 0,\ l \geq 0, \\ Q_{l-i} + \sum\limits_{n=l+1}^{\infty} Q_{n-i} G_{n-1} G_{n-2} \dots G_l,\ i \geq 1,\ l \geq i, \end{cases}$$

where $G_i = G,\ i \geq 1$.

4. Calculate the matrices Φ_l using the recurrent formulae

$$\Phi_0 = I,\ \Phi_l = \sum_{i=0}^{l-1} \Phi_i \bar{Q}_{i,l}(-\bar{Q}_{l,l})^{-1},\ l \geq 1.$$

5. Calculate the vector $\mathbf{p}_0$ as the unique solution of the system

$$\mathbf{p}_0(-\bar{Q}_{0,0}) = \mathbf{0},\ \ \mathbf{p}_0 \sum_{l=0}^{\infty} \Phi_l \mathbf{e} = 1.$$

6. Calculate the vectors $\mathbf{p}_l,\ l \geq 1$ as follows:

$$\mathbf{p}_l = \mathbf{p}_0 \Phi_l,\ \ l \geq 1.$$

The proposed algorithm is numerically stable because all matrices involved in the algorithm are non-negative.

5.1.4 Vector Generating Function of the Stationary Distribution: Performance Measures

Having calculated the stationary distribution of the system states, $\mathbf{p}_i,\ i \geq 0$ we can find a number of stationary performance measures of the system under consideration. When calculating the performance measures, the following result will be useful, especially in the case when the distribution $\mathbf{p}_i,\ i \geq 0$ is heavy tailed.

Lemma 5.2 *The vector generating function* $\mathbf{P}(z) = \sum\limits_{i=1}^{\infty} \mathbf{p}_i z^i,\ |z| \leq 1$ *satisfies the following equation:*

$$\mathbf{P}(z)Q(z) = z[\mathbf{p}_1 Q_0 - \mathbf{p}_0 \tilde{Q}(z)]. \quad (5.11)$$

In particular, formula (5.11) can be used to calculate the value of the generating function $\mathbf{P}(z)$ and its derivatives at the point $z = 1$ without the calculation of infinite sums. Having calculated these derivatives, we will be able to find moments of the number of customers in the system and some other performance measures of the system. The problem of calculating the value of the vector generating function $\mathbf{P}(z)$ and its derivatives at the point $z = 1$ from Eq. (5.11) is a non-trivial one because the matrix $Q(z)$ is singular at the point $z = 1$.

Let us denote by $f^{(m)}(z)$ the mth derivative of the function $f(z)$, $m \geq 1$, and $f^{(0)}(z) = f(z)$.

Corollary 5.4 *The mth, $m \geq 0$ derivatives of the vector generating function* $\mathbf{P}(z)$ *at the point* $z = 1$ *(so-called vector factorial moments) are recursively calculated as the solution of the system of linear algebraic equations*

$$\mathbf{P}^{(m)}(1) = \left[\left(\mathbf{b}^{(m)}(1) - \sum_{l=0}^{m-1} C_m^l \mathbf{P}^{(l)}(1) Q^{(m-l)}(1)\right)\tilde{I}\right.$$

$$\left. + \frac{1}{m+1}\left[\left(\mathbf{b}^{(m+1)}(1) - \sum_{l=0}^{m-1} C_{m+1}^l \mathbf{P}^{(l)}(1) Q^{(m+1-l)}(1)\right)\mathbf{e}\hat{\mathbf{e}}\right]\right] \mathcal{Q}^{-1}, \qquad (5.12)$$

where

$$\mathcal{Q} = Q(1)\tilde{I} + Q'(1)\mathbf{e}\hat{\mathbf{e}}, \ \hat{\mathbf{e}} = (1, 0, \ldots, 0)^T,$$

$$\mathbf{b}^{(m)}(1) = \begin{cases} \mathbf{p}_1 Q_0 - \mathbf{p}_0 \tilde{Q}(1), \ m = 0, \\ \mathbf{p}_1 Q_0 - \mathbf{p}_0 \tilde{Q}(1) - \mathbf{p}_0 \tilde{Q}'(1), \ m = 1, \\ -\mathbf{p}_0[m\tilde{Q}^{(m-1)}(1) + Q^{(m)}(1)], \ m > 1, \end{cases}$$

$\tilde{I}$ *is a diagonal matrix with diagonal entries* $(0, 1, \ldots, 1)$, *and the derivatives* $Q^{(m)}(1)$, $\tilde{Q}^{(m)}(1)$ *are calculated using formulas (5.1) and (5.2).*

Proof For brevity of presentation, let us denote the vector on the right-hand side of Eq. (5.11) by $\mathbf{b}(z)$. Successively differentiating in (5.11), we obtain the following inhomogeneous system of linear algebraic equations for the components of the vector $\mathbf{P}^{(m)}(1)$:

$$\mathbf{P}^{(m)}(1)Q(1) = \mathbf{b}^{(m)}(1) - \sum_{l=0}^{m-1} C_m^l \mathbf{P}^{(l)}(1) Q^{(m-l)}(1). \qquad (5.13)$$

As was noted above, $Q(1)$ is a singular matrix and has a rank of one less than its order. Thus, it is not possible to develop a recursive scheme for computing the vectors $\mathbf{P}^{(m)}(1)$, $m \geq 0$ directly from (5.13). We will modify system (5.13) to get a system with a non-singular matrix. To this end, we postmultiply the expression for $m + 1$ in (5.13) on both sides by $\mathbf{e}$. Taking into account that $Q(1)\mathbf{e} = \mathbf{0}^T$, we get the following equation for the components of the vector $\mathbf{P}^{(m)}(1)$:

$$\mathbf{P}^{(m)}(1)Q'(1)\mathbf{e} = \frac{1}{m+1}[\mathbf{b}^{(m+1)}(1) - \sum_{l=0}^{m-1} C_{m+1}^l \mathbf{P}^{(l)}(1) Q^{(m+1-l)}(1)]\mathbf{e}.$$

It can be shown that the right-hand side of this equation is not equal to zero. Thus, replacing one of the equations in the system (5.13) (without loss of generality we replace the first equation) with this equation we get the inhomogeneous system of linear algebraic equations for the entries of the vector $\mathbf{P}^{(m)}(1)$:

$$\mathbf{P}^{(m)}(1)\mathcal{Q} = \left[\left(\mathbf{b}^{(m)}(1) - \sum_{l=0}^{m-1} C_m^l \mathbf{P}^{(l)}(1)\mathcal{Q}^{(m-l)}(1)\right)\tilde{I}\right.$$

$$\left. + \frac{1}{m+1}\left[\left(\mathbf{b}^{(m+1)}(1) - \sum_{l=0}^{m-1} C_{m+1}^l \mathbf{P}^{(l)}(1)\mathcal{Q}^{(m+1-l)}(1)\right)\mathbf{e}\hat{\mathbf{e}}\right], \qquad (5.14)$$

which has the unique solution (5.12), if the matrix $\mathcal{Q}$ is non-singular.

Calculate the determinant of this matrix. It can be shown that $\det \mathcal{Q} = [\det Q(z)]'_{z=1}$. In turn, it can be shown (see [70]) that condition (5.3) for the existence of a stationary regime in the system is equivalent to the condition $[\det Q(z)]'_{z=1} < 0$. Since it is assumed that condition (5.3) is satisfied, then $\det \mathcal{Q} = [\det Q(z)]'_{z=1} < 0$, i.e., the matrix $\mathcal{Q}$ is non-singular. From that it follows that system (5.14) has the unique solution (5.12). □

Below, we list some performance measures of the system under consideration:

- Throughput of the system (maximal number of customers that can be processed during a unit of time) is defined by the right-hand side of inequality (5.3).
- Mean number of customers in the system

$$L = \mathbf{P}^{(1)}(1)\mathbf{e}.$$

- Variance of the number of customers in the system

$$V = \mathbf{P}^{(2)}(1)\mathbf{e} + L - L^2.$$

- The proportion of time when server 1 is fault-free

$$P_1 = \mathbf{P}(1)\text{diag}\{O_{RM^{(2)}a}, I_{M^{(1)}M^{(2)}a}\}\mathbf{e} + \mathbf{p}_0\text{diag}\{O_{Ra}, I_a\}\mathbf{e}.$$

- The proportion of time when server 1 is under repair

$$P_0 = \mathbf{P}(1)\text{diag}\{I_{RM^{(2)}a}, O_{M^{(1)}M^{(2)}a}\}\mathbf{e} + \mathbf{p}_0\text{diag}\{I_{Ra}, O_a\}\mathbf{e}.$$

5.1.5 Sojourn Time Distribution

Let $V(x)$ be the stationary distribution function of the sojourn time of an arbitrary customer in the system. Denote the Laplace–Stieltjes transform of this function by $v(s) = \int_0^\infty e^{-sx} dV(x),\ Re\ s \geq 0$.

Theorem 5.2 *The Laplace–Stieltjes transform of the sojourn time stationary distribution is calculated as*

$$v(s) = \lambda^{-1}\{\mathbf{p}_0 \sum_{k=1}^{\infty} Q_{0,k} \sum_{l=1}^{k} \Psi^l(s) + \sum_{i=1}^{\infty} \mathbf{p}_i \sum_{k=2}^{\infty} Q_k \sum_{l=1}^{k-1} \Psi^{i+l}(s)\}\mathbf{e}, \tag{5.15}$$

where

$$\Psi(s) = (sI - \hat{Q})^{-1} Q_0, \quad \hat{Q} = Q(1) - Q_0.$$

Proof The proof is based on the probabilistic interpretation of the Laplace–Stieltjes transform. We assume that, independently of the system operation, a stationary Poisson flow of so-called catastrophes arrives. Let $s, s > 0$ be the intensity of this flow. Then, the Laplace–Stieltjes transform $v(s)$ can be interpreted as the probability of no catastrophe arrival during the sojourn time of an arbitrary customer. It allows us to derive the expression for $v(s)$ by means of probabilistic reasoning.

Let us assume that at the moment of the beginning of a customer service the initial phases of service time at the servers are already determined. Then the matrix of probabilities of no catastrophe arrival during the service time of this customer (and corresponding transitions of the finite components of the Markov chain ξ_t, $t \geq 0$) is evidently calculated as

$$\tilde{\Psi}(s) = \int_0^{\infty} e^{(-sI+\hat{Q})t} Q_{1,0} dt = (sI - \hat{Q})^{-1} Q_{1,0}$$

if at a departure epoch there are no customers in the queue, and as

$$\Psi(s) = \int_0^{\infty} e^{(-sI+\hat{Q})t} Q_0 dt = (sI - \hat{Q})^{-1} Q_0$$

if at a departure epoch there are customers in the queue.

Note that $\tilde{\Psi}(s)\mathbf{e} = \Psi(s)\mathbf{e}$ since $Q_{1,0}\mathbf{e} = Q_0\mathbf{e}$.

Assume that an arbitrary customer arriving in a group of size k is placed in the jth position, $j = \overline{1, k}$, with probability $1/k$. Note that

$$\tilde{\Psi}(s)\mathbf{e} = \Psi(s)\mathbf{e} \tag{5.16}$$

since $Q_{1,0}\mathbf{e} = Q_0\mathbf{e}$. Then, using (5.16) and the law of total probability, we obtain the following expression for $v(s)$:

$$v(s) = \mathbf{p}_0 \sum_{k=1}^{\infty} \frac{k}{\lambda} Q_{0,k} \sum_{l=1}^{k} \frac{1}{k} \Psi^l(s)\mathbf{e}$$

$$+ \sum_{i=1}^{\infty} \mathbf{p}_i \sum_{k=2}^{\infty} \frac{k-1}{\lambda} Q_k \sum_{l=1}^{k-1} \frac{1}{k-1} \Psi^{i+l}(s)\mathbf{e}. \tag{5.17}$$

Formula (5.15) immediately follows from formula (5.17). □

Using Theorem 5.2, one can find moments of arbitrary order of the sojourn time by differentiating (5.15) at the point $s = 0$. The moment of kth order is calculated as $v^{(k)} = \frac{d^k v(s)}{ds^k}|_{s=0},\ k \geq 1$. In particular, the following statement holds.

Corollary 5.5 *The mean sojourn time of an arbitrary customer in the system is calculated as*

$$\bar{v} = -\lambda^{-1}[\mathbf{p}_0 \sum_{k=1}^{\infty} Q_{0,k} \sum_{l=1}^{k} \sum_{m=0}^{l-1} \Psi^m(0)$$

$$+ \sum_{i=1}^{\infty} \mathbf{p}_i \sum_{k=2}^{\infty} Q_k \sum_{l=1}^{k-1} \sum_{m=0}^{i+l-1} \Psi^m(0)]\Psi'(0)\mathbf{e}, \tag{5.18}$$

where

$$\Psi'(0) = -[\hat{Q}(1)]^{-2} Q_0.$$

Proof To obtain formula (5.18), we differentiate (5.15) at the point $s = 0$ and then transform the resulting expression using the fact that the matrix $\Psi(0)$ is stochastic. □

5.2 Analysis of Characteristics of the Data Transmission Process Under Cold-Standby Architecture of a High-Speed FSO Channel Backed Up by a Wireless Broadband Radio Channel

This section is devoted to the investigation of a single-server queuing system with an unreliable server and “cold” reservation. This system differs from the system with “hot” reservation discussed in the previous section in that the backup sever does not transmit information in parallel with the main server, but it is connected to the transmission only during the repair of the main server. The queue under consideration can be applied for modeling of hybrid communication systems with "cold" redundancy where the radio-wave channel is assumed to be absolutely reliable (its work does not depend on the weather conditions) and backs up the FSO channel only in cases when the latter's functioning is interrupted by unfavorable weather conditions. Upon the occurrence of favorable weather conditions the data packets begin to be transmitted over the FSO channel.

5.2.1 Model Description

We consider a queuing system consisting of two heterogeneous servers and infinite waiting room. One of the servers (main server, server 1) is unreliable and the other one (backup server, reserve server, server 2) is absolutely reliable. The latter is in so-called cold standby and connects to the service of a customer only when the main server is under repair. In addition, we assume that switching from the main server to the backup server is not instantaneous, it takes time.

Let us describe the scenario of interaction of server 1 and server 2.

If the main server is fault-free, customers are served by it in accordance with the FIFO discipline. If a breakdown arrives at server 1, the repair of the server and the switch to server 2 begin immediately.

After the switching time has expired, the customer goes to server 2, where it starts its service anew. However, if during the switching time server 1 becomes fault-free, the customer restarts its service on this server.

If the switching time has not expired but the repair period ends, the main server serves anew the customer whose service was interrupted by the arrival of the breakdown, and then serves customers from the queue.

If server 2 completes the service of a customer and the repair period on server 1 has not yet expired, server 2 serves customers from the queue until the end of the repair period. After that, server 1 resumes serving customers. The current customer, if any, is served anew.

Breakdowns arrive at server 1 according to a MAP defined by the $(V+1) \times (V+1)$ matrices H_0 and H_1. The fundamental rate of breakdowms is calculated as $h = \boldsymbol{\gamma} H_1 \mathbf{e}$, where the row vector $\boldsymbol{\gamma}$ is the unique solution of the system $\boldsymbol{\gamma}(H_0 + H_1) = \mathbf{0}$, $\boldsymbol{\gamma}\mathbf{e} = 1$.

The service time of a customer by the kth server, $k = 1, 2$ has PH-type distribution with irreducible representation $(\boldsymbol{\beta}^{(k)}, \mathbf{S}^{(k)})$. The service process on the kth server is directed by the Markov chain $m_t^{(k)}$, $t \geq 0$ with state space $\{1, \dots, M^{(k)}, M^{(k)}+1\}$, where $M^{(k)}+1$ is an absorbing state. The intensities of transitions into the absorbing state are defined by the vector $\mathbf{S}_0^{(k)} = -S^{(k)}\mathbf{e}$. The service rates are calculated as $\mu^{(k)} = -[\boldsymbol{\beta}^{(k)}(S^{(k)})^{-1}\mathbf{e}]^{-1}$, $k = 1, 2$.

The switching time from server 1 to server 2 has PH-type distribution with irreducible representation $(\boldsymbol{\alpha}, A)$. The switching process is directed by the Markov chain $l_t,\ t \geq 0$ with state space $\{1, \ldots, L, L+1\}$, where $L+1$ is an absorbing state. The intensities of transitions into the absorbing state are defined by the vector $A_0 = -A\mathbf{e}$. The switching rate is defined as $a = -[\boldsymbol{\alpha} A^{-1}\mathbf{e}]^{-1}$.

The repair period has PH-type distribution with irreducible representation $(\boldsymbol{\tau}, T)$. The repair process is directed by the Markov chain $\vartheta_t,\ t \geq 0$ with state space $\{1, \ldots, R, R+1\}$, where $R+1$ is an absorbing state. The intensities of transitions into the absorbing state are defined by the vector $\mathbf{T}_0 = -T\mathbf{e}$. The repair rate is $\tau = -(\boldsymbol{\tau} T^{-1}\mathbf{e})^{-1}$.

5.2.2 *Process of the System States*

Let at the moment t

- i_t be the number of customers in the system, $i_t \geq 0$,
- $n_t = 0$ if server 1 is fault-free and $n_t = 1$ if server 1 is under repair;
- $r_t = 0$ if one of the servers is serving a customer and $r_t = 1$ if switching is taking place;
- $m_t^{(k)}$ be the state of the service underlying process at the kth busy server, $m_t^{(k)} = \overline{1, M^{(k)}},\ k = 1, 2$;
- l_t be the state of the switching time process from server 1 to server 2, $l_t = \overline{1, L}$;
- ϑ_t be the state of the repair time underlying process at server 1, $\vartheta_t = \overline{1, R}$;
- ν_t and η_t be the states of the $BMAP$ underlying process and the MAP underlying process respectively, $\nu_t = \overline{0, W},\ \eta_t = \overline{0, V}$.

The process of the system states is described by the regular irreducible continuous-time Markov chain, $\xi_t, t \geq 0$ with state space

$$\Omega = \{(i, n, \nu, \eta),\ i = 0, n = 0, \nu = \overline{0, W}, \eta = \overline{0, V}\}$$

$$\bigcup \{(i, n, \nu, \eta, \vartheta),\ i = 0, n = 1, \nu = \overline{0, W}, \eta = \overline{0, V}, \vartheta = \overline{0, R}\}$$

$$\bigcup \{(i, n, r, \nu, \eta, m^{(1)}),\ i > 0, n = 0,\ r = 0, \nu = \overline{0, W}, \eta = \overline{0, V}, m^{(1)} = \overline{1, M^{(1)}}\}$$

$$\{(i, n, r, \nu, \eta, m^{(2)}, \vartheta),\ i > 0, n = 1,\ r = 0, \nu = \overline{0, W}, \eta = \overline{0, V}, \vartheta = \overline{1, R},$$

$$m^{(2)} = \overline{1, M^{(2)}}\}$$

$$\bigcup \{(i, n, r, \nu, \eta, \vartheta, l),\ i > 0, n = 1,\ r = 1, \nu = \overline{0, W}, \eta = \overline{0, V},$$

$$\vartheta = \overline{1, R}, l = \overline{1, L}\}.$$

In the following, we will assume that the states of the chain $\xi_t, t \geq 0$, are ordered in lexicographic order. Let $Q_{ij},\ i, j \geq 0$ be the matrices formed by the intensities of the chain transitions from the state corresponding to value i of the component i_n to the state corresponding to value j of this component and $Q = (Q_{ij})_{i,j\geq 0}$, be the generator of the chain.

Lemma 5.3 *The infinitesimal generator Q of the Markov chain $\xi_t,\ t \geq 0$ has the block structure*

$$Q = \begin{pmatrix} Q_{0,0} & Q_{0,1} & Q_{0,2} & Q_{0,3} & \cdots \\ Q_{1,0} & Q_1 & Q_2 & Q_3 & \cdots \\ O & Q_0 & Q_1 & Q_2 & \cdots \\ O & O & Q_0 & Q_1 & \cdots \\ \vdots & \vdots & \vdots & \vdots & \ddots \end{pmatrix},$$

where non-zero blocks have the following form:

$$Q_{0,0} = \begin{pmatrix} D_0 \oplus H_0 & I_{\bar{W}} \otimes H_1 \otimes \boldsymbol{\tau} \\ I_a \otimes T_0 & D_0 \oplus H \oplus T \end{pmatrix},$$

$$Q_{0,k} = \begin{pmatrix} D_k \otimes I_{\bar{V}} \otimes \boldsymbol{\beta}^{(1)} & O & O \\ O & D_k \otimes I_{\bar{V}} \otimes I_R \otimes \boldsymbol{\beta}^{(2)} & O \end{pmatrix},\ k \geq 1,$$

$$Q_{1,0} = \begin{pmatrix} I_a \otimes \mathbf{S}_0^{(1)} & O \\ O & I_{aR} \otimes \mathbf{S}_0^{(2)} \\ O & O \end{pmatrix},$$

$$Q_0 = \begin{pmatrix} I_a \otimes \mathbf{S}_0^{(1)} \boldsymbol{\beta}^{(1)} & O & O \\ O & I_{aR} \otimes \mathbf{S}_0^{(2)} \boldsymbol{\beta}^{(2)} & O \\ O & O & O \end{pmatrix},$$

$$Q_1 = \begin{pmatrix} D_0 \oplus H_0 \oplus S^{(1)} & O & I_{\bar{W}} \otimes H_1 \otimes \mathbf{e}_{M^{(1)}} \otimes \boldsymbol{\tau} \otimes \boldsymbol{\alpha} \\ I_a \otimes \mathbf{T}_0 \otimes \mathbf{e}_{M^{(2)}} \otimes \boldsymbol{\beta}^{(1)} & D_0 \oplus H \oplus T \oplus S^{(2)} & O \\ I_a \otimes \mathbf{T}_0 \otimes \mathbf{e}_L \otimes \boldsymbol{\beta}^{(1)} & I_a \otimes I_R \otimes \mathbf{A}_0 \otimes \boldsymbol{\beta}^{(2)} & D_0 \oplus H \oplus T \oplus A \end{pmatrix},$$

$$Q_k = diag\{D_{k-1} \otimes I_{\bar{V}} \otimes I_{M^{(1)}},\ D_{k-1} \otimes I_{\bar{V}} \otimes I_R \otimes I_{M^{(2)}},\ D_{k-1} \otimes I_{\bar{V}} \otimes I_R \otimes I_L\},\ k \geq 2,$$

where $H = H_0 + H_1$.

The proof of the lemma is by means of calculation of the probabilities of transitions of the components of the Markov chain during a time interval with infinitesimal length.

Corollary 5.6 *The Markov chain $\xi_t, t \geq 0$ belongs to the class of continuous-time quasi-Toeplitz Markov chains (QTMC).*

The proof follows from the form of the generator given by Lemma 5.3 and the definition of $QTMC$ given in [70].

In the following it will be useful to have expressions for the generating functions $\tilde{Q}(z) = \sum\limits_{k=1}^{\infty} \tilde{Q}_k z^k,\ Q(z) = \sum\limits_{k=0}^{\infty} Q_k z^k,\ |z| \leq 1.$

Corollary 5.7 *The matrix generating functions $\tilde{Q}(z),\ Q(z)$ are of the form*

$$\tilde{Q}(z) = \begin{pmatrix} (D(z) - D_0) \otimes I_{\bar{V}} \otimes \boldsymbol{\beta}^{(1)} & O & O \\ O & (D(z) - D_0) \otimes I_{\bar{V}_R} \otimes \boldsymbol{\beta}^{(2)} & O \end{pmatrix}, \tag{5.19}$$

$$Q(z) = Q_0 + \mathcal{Q}z + z diag\{D(z) \otimes I_{\bar{V}M^{(1)}},\ D(z) \otimes I_{\bar{V}R},\ D(z) \otimes I_{\bar{V}RL}\}, \tag{5.20}$$

where

$$\mathcal{Q} = \begin{pmatrix} I_{\bar{W}} \otimes H_0 \oplus S^{(1)} & O & I_{\bar{W}} \otimes H_1 \otimes \mathbf{e}_{M^{(1)}} \otimes \boldsymbol{\tau} \otimes \boldsymbol{\alpha} \\ I_a \otimes \boldsymbol{T}_0 \otimes \mathbf{e}_{M^{(2)}} \otimes \boldsymbol{\beta}^{(1)} & I_{\bar{W}} \otimes H \oplus T \oplus S^{(2)} & O \\ I_a \otimes \boldsymbol{T}_0 \otimes \mathbf{e}_L \otimes \boldsymbol{\beta}^{(1)} & I_{aR} \otimes \boldsymbol{A}_0 \otimes \boldsymbol{\beta}^{(2)} & I_{\bar{W}} \otimes H \oplus T \oplus A \end{pmatrix}.$$

5.2.3 Stationary Distribution: Performance Measures

Theorem 5.3 *A necessary and sufficient condition for the existence of the stationary distribution of the Markov chain $\xi_t,\ t \geq 0$ is the fulfillment of the inequality*

$$\lambda < \boldsymbol{\pi}_1 \mathbf{S}_0^{(1)} + \boldsymbol{\pi}_2 \mathbf{S}_0^{(2)}, \tag{5.21}$$

where

$$\boldsymbol{\pi}_1 = \mathbf{x}_1(\mathbf{e}_{\bar{V}M^{(1)}}),\ \boldsymbol{\pi}_2 = \mathbf{x}_2(\mathbf{e}_{\bar{V}RM^{(2)}}), \tag{5.22}$$

and the vector $\mathbf{x} = (\mathbf{x}_1,\ \mathbf{x}_2,\ \mathbf{x}_3)$ is the unique solution of the system

$$\mathbf{x}\Gamma = 0,\ \mathbf{x}\mathbf{e} = 1. \tag{5.23}$$

Here

$$\Gamma = \begin{pmatrix} I_{\bar V} \otimes \mathbf{S}_0^{(1)} \boldsymbol{\beta}^{(1)} & O & H_1 \otimes \mathbf{e}_{M^{(1)}} \otimes \boldsymbol{\tau} \otimes \boldsymbol{\alpha} \\ I_{\bar V} \otimes \boldsymbol{T}_0 \otimes \mathbf{e}_{M^{(2)}} \otimes \boldsymbol{\beta}^{(1)} & I_{\bar V R} \otimes \mathbf{S}_0^{(2)} \boldsymbol{\beta}^{(2)} & O \\ I_{\bar V} \otimes \boldsymbol{T}_0 \otimes \mathbf{e}_L \otimes \boldsymbol{\beta}^{(1)} & I_{\bar V R} \otimes \boldsymbol{A}_0 \otimes \boldsymbol{\beta}^{(2)} & O \end{pmatrix}$$

$$+diag\{H_0 \oplus S^{(1)}, H \oplus T \oplus S^{(2)}, H \oplus T \oplus A\}.$$

Proof The proof is carried out by arguments analogous to those given above in the proof of Theorem 5.1. In this case we represent the vector $\mathbf{y}$ present in the relations (5.5) and (5.6) in the form

$$\mathbf{y} = (\boldsymbol{\theta} \otimes \mathbf{x}_1, \boldsymbol{\theta} \otimes \mathbf{x}_2, \boldsymbol{\theta} \otimes \mathbf{x}_3), \tag{5.24}$$

where $\mathbf{x}_1,\ \mathbf{x}_2,\ \mathbf{x}_3$ are the parts of a stochastic vector $\mathbf{x}$, i.e., $\mathbf{x} = (\mathbf{x}_1,\ \mathbf{x}_2,\ \mathbf{x}_3)$. The vectors $\mathbf{x}_1,\ \mathbf{x}_2,\ \mathbf{x}_3$ are of sizes $\bar V M^{(1)}$, $\bar V R M^{(2)}$, $\bar V R L$, respectively.

Then, taking into account that $\boldsymbol{\theta} \sum\limits_{k=0}^{\infty} D_k = 0$ and $\mathbf{x}$ is a stochastic vector, system (5.6) is reduced to the form (5.23).

Now consider inequality (5.21). Substituting in this inequality the vector $\mathbf{y}$ in the form (5.24), an expression for $Q'(1)$, calculated using formula (5.20), and taking into account that $\boldsymbol{\theta} D'(1)\mathbf{e} = \lambda$, we reduce inequality (5.21) to the following form:

$$\lambda + \mathbf{x} Q^{-} \mathbf{e} < 0, \tag{5.25}$$

where

$$Q^{-} = \begin{pmatrix} H_0 \oplus S^{(1)} & O & H_1 \otimes \mathbf{e}_{M^{(1)}} \otimes \boldsymbol{\tau} \otimes \boldsymbol{\alpha} \\ I_{\bar V} \otimes \boldsymbol{T}_0 \otimes \boldsymbol{\beta}^{(1)} & H \oplus T \oplus S^{(2)} & O \\ I_{\bar V} \otimes \boldsymbol{T}_0 \otimes \mathbf{e}_L \otimes \boldsymbol{\beta}^{(1)} & I_{\bar V} \otimes I_R \otimes \boldsymbol{A}_0 \otimes \boldsymbol{\beta}^{(2)} & H \oplus T \oplus A \end{pmatrix}.$$

Taking into account the relations $H\mathbf{e} = (H_0 + H_1)\mathbf{e} = \mathbf{0}$, $T\mathbf{e} + \boldsymbol{T}_0 = \mathbf{0}$, $A\mathbf{e} + \boldsymbol{A}_0 = \mathbf{0}$, we reduce inequality (5.25) to the form

$$\lambda < \mathbf{x}_1(\mathbf{e}_{\bar V} \otimes I_{M^{(1)}})\mathbf{S}_0^{(1)} + \mathbf{x}_2(\mathbf{e}_{\bar V} \otimes \mathbf{e}_R \otimes I_{M^{(2)}})\mathbf{S}_0^{(2)}.$$

Using the notation (5.22), we reduce this inequality to the form (5.21). □

Remark 5.3 We can give an intuitive explanation of stability condition (5.21). The vectors $\boldsymbol{\pi}_n,\ n = 1, 2$ have the following sense: the entry $\boldsymbol{\pi}_1(m^{(1)})$ of the vector $\boldsymbol{\pi}_1$ is the probability that, under overload conditions, server 1 is fault-free and serves a customer in phase $m^{(1)}$, $m^{(1)} = \overline{1, M^{(1)}}$, and the entry $\boldsymbol{\pi}_2(m^{(2)})$ of the vector $\boldsymbol{\pi}_2$ is the probability that, under overload conditions, server 1 is under repair and server 2 serves a customer in phase $m^{(2)}$, $m^{(2)} = \overline{1, M^{(2)}}$. Then the right-hand side of inequality (5.21) is the rate of customers leaving the system after service under overload conditions while the left-hand side of this inequality is the rate λ of customers arriving at the system. It is obvious that in a steady state the latter rate must be less that the former one.

Corollary 5.8 *In the case of a stationary Poisson flow of breakdowns and an exponential distribution of service and repair times, ergodicity condition (5.21)–(5.23) is reduced to the following inequality:*

$$\lambda < \pi_1\mu_1 + \pi_2\mu_2, \tag{5.26}$$

where

$$\pi_1 = \frac{\alpha+\tau}{h}[1+\frac{\alpha}{\tau}+\frac{\alpha+\tau}{h}]^{-1}, \quad \pi_2 = \frac{\alpha}{\tau}[1+\frac{\alpha}{\tau}+\frac{\alpha+\tau}{h}]^{-1}. \tag{5.27}$$

In what follows we assume inequality (5.21) be fulfilled. Introduce the steady state probabilities of the chain ξ_t :

$$p_0^{(0)}(\nu,\eta) = \lim_{t\to\infty} P\{i_t = 0, n_t = 0, \nu_t = \nu, \eta_t = \eta\}, \ \nu = \overline{0,W}, \eta = \overline{0,V},$$

$$p_0^{(1)}(\nu,\eta) = \lim_{t\to\infty} P\{i_t = 0, n_t = 1, \nu_t = \nu, \eta_t = \eta, \vartheta_t = \vartheta\},$$

$$\nu = \overline{0,W}, \eta = \overline{0,V}, \vartheta = \overline{1,R},$$

$$p_i^{(0,0)}\{(\nu,\eta,m^{(1)})\} = \lim_{t\to\infty} P\{i_t = i, n_t = 0, r_t = 0, \nu_t = \nu, \eta_t = \eta, m_t^{(1)} = m^{(1)}\},$$

$$i > 0, \nu = \overline{0,W}, \eta = \overline{0,V}, m^{(1)} = \overline{1,M^{(1)}},$$

$$p_i^{(1,0)}\{(\nu,\eta,m^{(2)},\vartheta)\} = \lim_{t\to\infty} P\{i_t = i, n_t = 1, r_t = 0, \nu_t = \nu, \eta_t = \eta, \vartheta_t = \vartheta,$$

$$m_t^{(2)} = m^{(2)}\}, i > 0, \nu = \overline{0,W}, \eta = \overline{0,V}, \vartheta = \overline{1,R}, m^{(2)} = \overline{1,M^{(2)}},$$

$$p_i^{(1,1)}\{(\nu,\eta,\vartheta,l)\} = \lim_{t\to\infty} P\{i_t = i, n_t = 1, r_t = 1, \nu_t = \nu, \eta_t = \eta, \vartheta_t = \vartheta, l_t = l\},$$

$$i > 0, \nu = \overline{0,W}, \eta = \overline{0,V}, \vartheta = \overline{1,R}, l = \overline{1,L}.$$

We order the probabilities in lexicographic order of the component i_t and form the vectors of these probabilities $\mathbf{p}_i$, $i \geq 0$.

The vectors $\mathbf{p}_i, i \geq 0$ are calculated using the algorithm given above in this section.

Having calculated the stationary distribution $\mathbf{p}_i$, $i \geq 0$, we can find a number of stationary performance measures of the system. To calculate them, we use the vector generating function $\mathbf{P}(z) = \sum_{i=1}^{\infty} \mathbf{p}_i z^i$, $|z| \leq 1$. This function, like the analogous function in Sect. 5.1.4, satisfies an equation of the form (5.11), where the generating functions $Q(z)$, $\tilde{Q}(z)$ are calculated by formulas (5.19) and (5.20) respectively. Using Eq. (5.11) and formula (5.12), we are able to calculate the vector factorial moments $\mathbf{P}^{(m)}(1)$, $m \geq 0$ of the number of customers in the system.

Now we are able to calculate a number of performance measures of the system under consideration:

- Throughput of the system

$$\varrho = \boldsymbol{\pi}_1 \mathbf{S}_0^{(1)} + \boldsymbol{\pi}_2 \mathbf{S}_0^{(2)}.$$

- Mean number of customers in the system

$$L = \mathbf{P}^{(1)}(1)\mathbf{e}.$$

- Variance of the number of customers in the system

$$V = \mathbf{P}^{(2)}(1)\mathbf{e} + L - L^2.$$

- Probability that the system is empty and server 1 is fault-free

$$P_0^{(0)} = \mathbf{p}_0^{(0)}\mathbf{e}.$$

- Probability that the system is empty and server 1 is under repair

$$P_0^{(1)} = \mathbf{p}_0^{(1)}\mathbf{e}.$$

- Probability that server 1 is serving a customer

$$P_0^{(0,0)} = \mathbf{P}(1)\text{diag}\{I_{aM^{(1)}}, 0_{aR(M^{(2)}+L)}\}\mathbf{e}.$$

- Probability that server 1 is faulty and server 2 is serving a customer

$$P_0^{(1,0)} = \mathbf{P}(1)\text{diag}\{0_{aM^{(1)}}, I_{aRM^{(2)}}, 0_{aRL}\}\mathbf{e}.$$

- Probability that the switching period from server 1 to server 2 is taking place

$$P_0^{(1,1)} = \mathbf{P}(1)\text{diag}\{0_{a(M^{(1)}+RM^{(2)})}, I_{aRL}\}\mathbf{e}.$$

- Probability that server 1 is fault-free

$$P_0 = \mathbf{p}_0^{(0)}\mathbf{e} + \mathbf{P}(1)\text{diag}\{I_{aM^{(1)}}, 0_{aR(M^{(2)}+L)}\}\mathbf{e}.$$

- Probability that server 1 is under repair

$$P_1 = \mathbf{p}_0^{(1)}\mathbf{e} + \mathbf{P}(1)\mathrm{diag}\{0_{aM^{(1)}},\ I_{aR(M^{(2)}+L)}\}\mathbf{e}.$$

Let $V(x)$ be the stationary distribution function of the sojourn time of an arbitrary customer in the system, and $v(s) = \int_0^\infty e^{-sx}dV(x),\ Re(s) \geq 0$ be the Laplace–Stieltjes transform of this function. The Laplace–Stieltjes transform $v(s)$ and its derivatives at the point $s = 0$, determining the moments of sojourn time, can be found by the formulas given in Sect. 5.1.5. In particular, the average mean sojourn time in the system is calculated by the formula (5.18).

5.2.4 Numerical Experiments

Our numerical examples are illustrative and consist of graphs of dependencies of the system performance characteristics on its parameters. Below we present the results of five experiments.

Experiment 5.1 In this experiment, we investigate the behavior of the mean sojourn time $\bar{v}$ as a function of the arrival rate λ under different coefficients of correlation c_{cor} in the $BMAP$.

We consider three different $BMAP$s with the same fundamental rate $\lambda = 5$, and a geometric distribution (with parameter $q = 0.8$) of the number of customers in a batch. The maximal batch size is assumed to be three, so the matrices D_k are defined as

$$D_k = Dq^k(1-q)/(1-q^3), k = \overline{1,3},$$

and then normalized to get the fundamental rate $\lambda = 5$.

The $BMAP$s have different correlation coefficients. The first $BMAP$ is a stationary Poisson process. It is coded as $BMAP_1$ and has $c_{cor} = 0,\ c_{var} = 1$.

The second $BMAP$, which is coded as $BMAP_2$, is characterized by the matrices

$$D_0 = \begin{pmatrix} -3.17040 & 0.939885 \times 10^{-6} \\ 0.939885 \times 10^{-6} & -0.06944 \end{pmatrix}, \quad D = \begin{pmatrix} 3.16070 & 0.009695 \\ 0.05411 & 0.01533 \end{pmatrix}.$$

For this $BMAP$ $c_{cor} = 0.1,\ c_{var} = 3.5$.

The third $BMAP$, which is coded as $BMAP_3$, is characterized by the matrices

$$D_0 = \begin{pmatrix} -86 & 0.01 \\ 0.02 & -2.76 \end{pmatrix}, \quad D = \begin{pmatrix} 85 & 0.99 \\ 0.2 & 2.54 \end{pmatrix}.$$

For this $BMAP$ $c_{cor} = 0.407152, c_{var} = 3.101842$.

The flow of breakdowns is defined by the matrices

$$H_0 = \begin{pmatrix} -8.110725 & 0 \\ 0 & -0.26325 \end{pmatrix}, \; H_1 = \begin{pmatrix} 8.0568 & 0.053925 \\ 0.146625 & 0.116625 \end{pmatrix}$$

For this MAP $h = 0.1$, $c_{cor} = 0.20005$, $c_{var} = 3.51284$.

The service time at server 1 has the Erlang distribution of order 2 with intensity $\mu_1 = 10$ and variation coefficient $c_{var} = 0.707106$. This distribution is defined by

$$\boldsymbol{\beta} = (1, 0)\,, \; S = \begin{pmatrix} -20 & 20 \\ 0 & -20 \end{pmatrix}.$$

The service time at server 2 has the Erlang distribution of order 2 with intensity $\mu_2 = 1$ and variation coefficient $c_{var} = 0.707106$. This distribution is defined by

$$\boldsymbol{\beta}^{(2)} = (1, 0), \; S^{(2)} = \begin{pmatrix} -2 & 2 \\ 0 & -2 \end{pmatrix}.$$

The repair time distribution is hyperexponential of order 2 and is defined by the matrices

$$\boldsymbol{\tau} = (0.98,\ 0.02), \; T^{(1)} = \begin{pmatrix} -100 & 0 \\ 0 & -0.002 \end{pmatrix}.$$

It has intensity $\tau = 0.0999$ and coefficient of variation $c_{var} = 9.948889$.

The switching time distribution is hyperexponential of order 2 and is defined by the matrices

$$\boldsymbol{\alpha} = ((0.05,\ 0.95), \; A = \begin{pmatrix} -1.86075 & 0 \\ 0 & -145.9994 \end{pmatrix}.$$

It has intensity $a = 30$ and coefficient of variation $c_{var} = 5.007242$.

Figure 5.1 shows the mean sojourn time $\bar{v}$ as a function of the arrival rate λ for $BMAP$s with different coefficients of correlation.

Experiment 5.2 In this experiment, we investigate the behavior of the probability $P_{fault\text{-}free}$ as a function of the breakdown rate h under different values of the coefficient of correlation c_{cor} in the MAP of breakdowns.

Here we consider a $BMAP$ with fundamental rate $\lambda = 5$, the coefficient of variation $c_{var} = 3.498651$ and coefficient of correlation $c_{cor} = 0.200504$. The $BMAP$ is defined by the matrices D_0 and $D_k = Dq^{k-1}(1-q)/(1-q^3), k = \overline{1,3}$, where $q = 0.8$,

$$D_0 = \begin{pmatrix} -8.110725 & 0 \\ 0 & -0.26325 \end{pmatrix}, \quad D = \begin{pmatrix} 8.0568 & 0.053925 \\ 0.146625 & 0.116625 \end{pmatrix}.$$

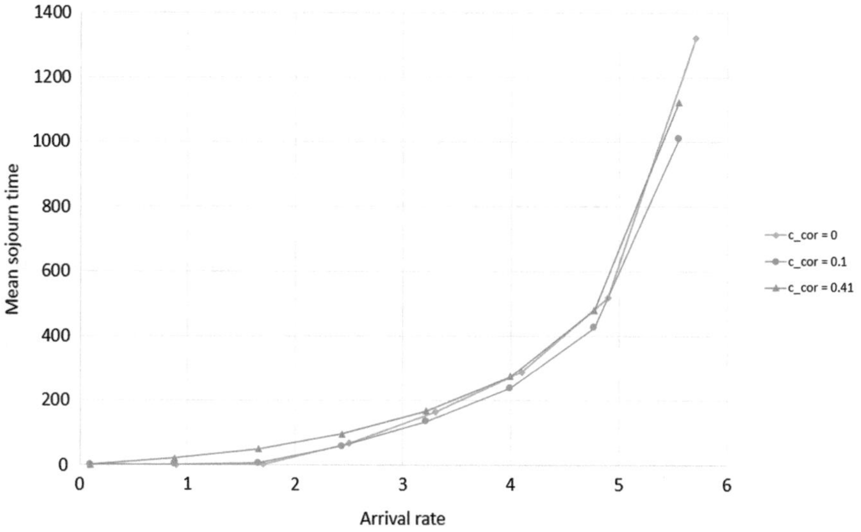

Fig. 5.1 Mean sojourn time $\bar{v}$ as a function of the arrival rate λ for the $BMAP$s with different coefficients of correlation

In this experiment, the service, repair, and switching processes are the same as in Experiment 5.1.

We consider three different MAPs of breakdowns with the same fundamental rate $h = 5$ but different coefficients of variation.

The first MAP is coded as MAP_1. It is the stationary Poisson process. It has $c_{cor} = 0$, $c_{var} = 1$.

The second MAP is coded as MAP_2. It is defined by the matrices

$$H_0 = \begin{pmatrix} -5.34080 & 1.87977 \times 10^{-6} \\ 1.87977 \times 10^{-6} & -0.13888 \end{pmatrix}, \quad H_1 = \begin{pmatrix} 5.32140 & 0.01939 \\ 0.10822 & 0.03066 \end{pmatrix}.$$

In this MAP $c_{cor} = 0.1$, $c_{var} = 3.5$.

The third MAP is coded as MAP_3. It is defined by the matrices

$$H_0 = \begin{pmatrix} -86 & 0.01 \\ 0.02 & -2.76 \end{pmatrix}, \quad H_1 = \begin{pmatrix} 85 & 0.99 \\ 0.2 & 2.54 \end{pmatrix}.$$

In this MAP $c_{corr} = 0.407152$, $c_{var} = 3.1018422$.

Figure 5.2 shows the behavior of the probability $P_{fault\text{-}free}$ as a function of the breakdown rate h under different values of the coefficient of correlation c_{cor} in the MAP of breakdowns.

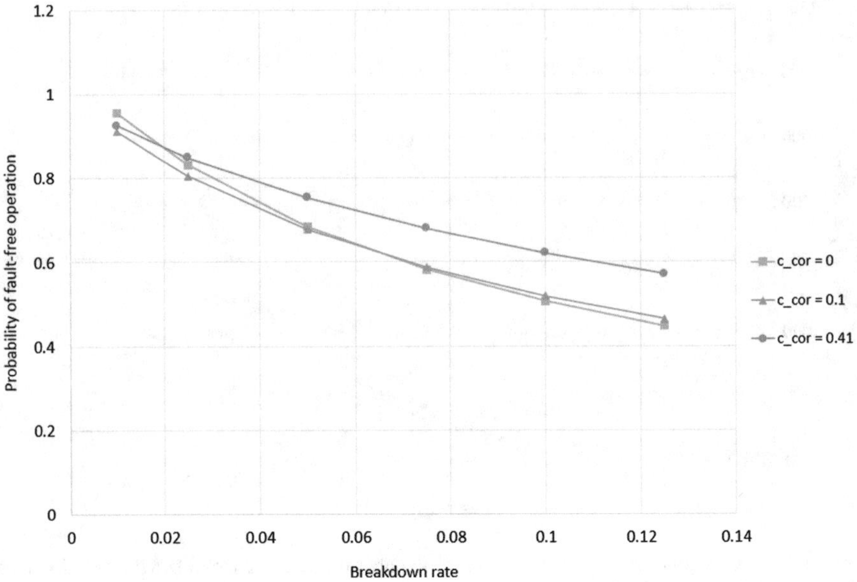

Fig. 5.2 $P_{non\text{-}repair}$ as a function of the breakdown rate h for breakdown MAPs with different coefficients of correlation

Experiment 5.3 In this experiment, we investigate the behavior of the mean sojourn time $\bar{v}$ as a function of the breakdown rate h under different values of the coefficient of variation c_{var} of the repair time.

The input flow is MAP_3 from Experiment 5.1.

To define the flow of breakdowns, we use the matrices D_0, D that were used in the description of the $BMAP$ in Experiment 5.2. In this case the matrices H_0 and H_1 are defined as follows: $H_0 = D_0, \ H_1 = D$. The breakdown rate is $h = 0.1$.

The service times and the switching time are the same as in Experiment 5.1.

In this experiment, we consider three PH distributions of the repair time. For all these distributions the repair rate is $h = 0.1$. The first one is an exponential distribution. It is coded as PH_1 and has $c_{var} = 1$.

The second distribution of the repair time is Erlang of order 2. It is coded as PH_2 and is defined by the matrices

$$\boldsymbol{\tau} = (1, 0), \ T^{(1)} = \begin{pmatrix} -0.05 & 0.05 \\ 0 & -0.05 \end{pmatrix}.$$

For this PH $c_{var} = 0.707106$.

The third distribution is hyperexponential of order 2 with $c_{var} = 9.948889$. It is the same as the distribution of the repair time in Experiment 5.1. We will code it as PH_3.

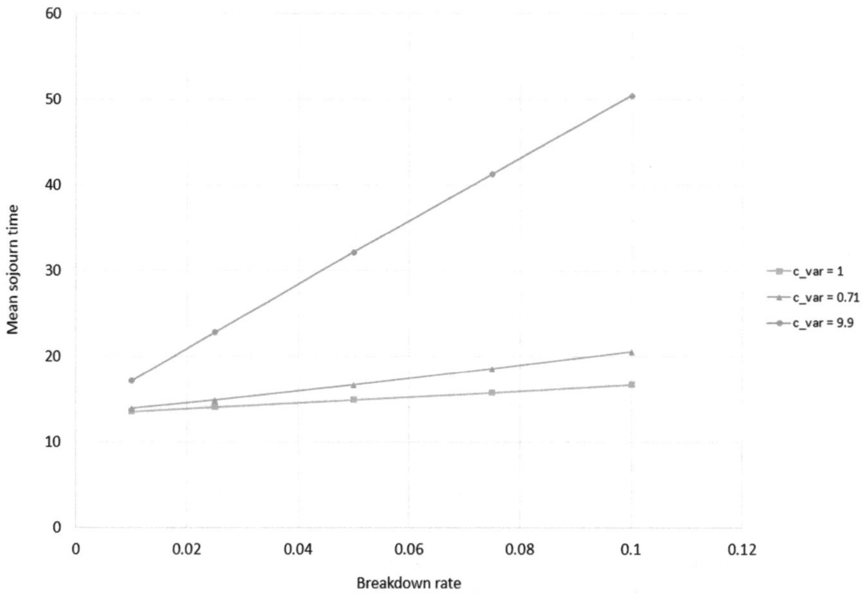

Fig. 5.3 Mean sojourn time $\bar{v}$ as a function of the breakdown rate h under different values of the coefficient of variation c_{var} of the repair time

Figure 5.3 shows the behavior of the mean sojourn time $\bar{v}$ as a function of the breakdown rate h under different values of the coefficient of variation c_{var} of the repair time.

Experiment 5.4 In this experiment, we investigate the behavior of the mean sojourn time $\bar{v}$ as a function of the breakdown rate h under different values of the coefficient of correlation c_{cor} in the MAP of breakdowns.

In this experiment we use the input data of Experiment 5.2.

Figure 5.4 shows the behavior of the mean sojourn time $\bar{v}$ as a function of the breakdown rate h under different values of the coefficient of correlation c_{cor} in the MAP of breakdowns.

Experiment 5.5 In this experiment, we investigate the behavior of the mean sojourn time $\bar{v}$ as a function of the input rate λ under different values of the rate of breakdowns h.

In this experiment, the $BMAP$ of the customers, and distributions of service time, repair time, and switching time are the same as in Experiment 5.2. The MAP flow of breakdowns is defined to be MAP_3 in Experiment 5.2.

Figure 5.5 shows the behavior of the mean sojourn time $\bar{v}$ as a function of the input rate λ under different values of the rate of breakdowns h.

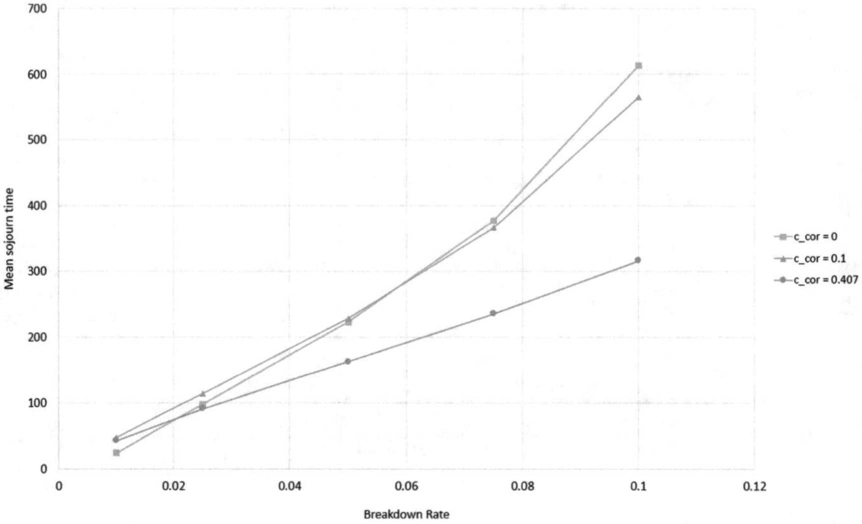

Fig. 5.4 Mean sojourn time $\bar{v}$ as a function of the breakdown rate h under the different values of the coefficient of correlation c_{cor} in the MAP of breakdowns

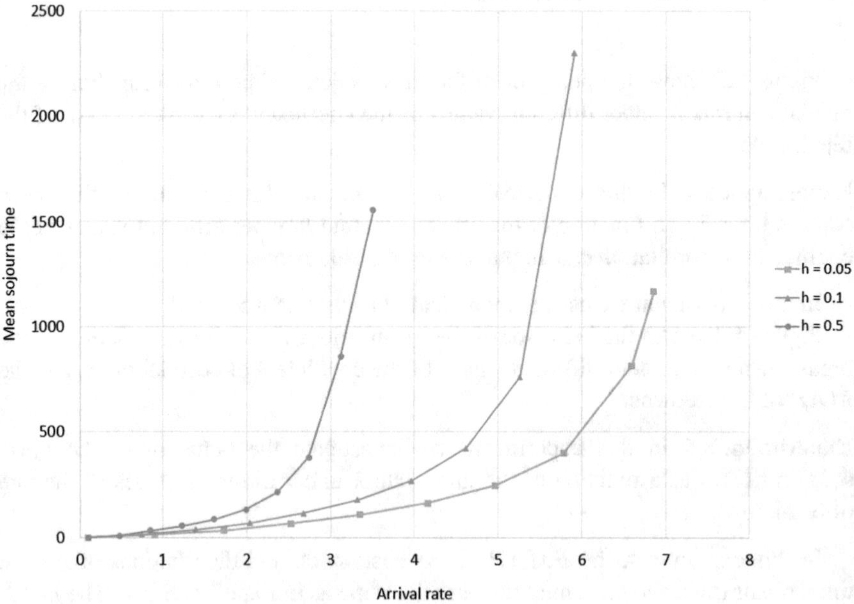

Fig. 5.5 Mean sojourn time $\bar{v}$ as a function of the input rate λ under different values of h, the rate of breakdowns

5.2.5 The Case of an Exponential System

In this section, we investigate a queuing system that is a particular case of the system studied above, in Sects. 5.2.1–5.2.3. Here it is assumed that the input flow and the flow of breakdowns are stationary Poisson, and the service times of the servers and the repair time are exponentially distributed. In addition, we assume that switching from the main server to the backup server (from server 1 to server 2) occurs instantaneously. Note that here, due to the memoryless property of an exponential distribution, we can assume that after switching the current customer does not start the service anew but continues its service at a different speed.

A feature of the study carried out in this subsection is that here we obtain explicit formulas for the generating function of the stationary distribution of the number of customers in the system and the simple Laplace–Stieltjes transform of the sojourn time distribution.

The operation of the system is described by an irreducible Markov chain

$$\xi_t = \{i_t, n_t\},\ t \geq 0,$$

where

- i_t is the number of customers in the system at the moment t, $i_t \geq 0$;
- $n_t = \begin{cases} 1, & \text{if server 1 is fault-free (it is idle or working) at moment} t; \\ 2, & \text{if server 1 is under repair and server 2 serves a customer.} \end{cases}$

Let $Q_{(i,n),(j,n')}$ be the rate of transition of the Markov chain ξ_t from the state (i, n) to the state (j, n'). Define the matrix $Q_{i,j}$ as

$$Q_{i,j} = \begin{pmatrix} Q_{(i,1),(j,1)} & Q_{(i,1),(j,2)} \\ Q_{(i,2),(j,1)} & Q_{(i,2),(j,2)} \end{pmatrix}.$$

Lemma 5.4 *The infinitesimal generator $Q = (Q_{i,j})$ of the Markov chain ξ_t can be represented in the block tridiagonal form*

$$Q = \begin{pmatrix} Q_{0,0} & Q_{0,1} & O & O & \cdots \\ Q_{-1} & Q_0 & Q_1 & O & \cdots \\ O & Q_{-1} & Q_0 & Q_1 & \cdots \\ O & O & Q_{-1} & Q_0 & \cdots \\ \vdots & \vdots & \vdots & \vdots & \ddots \end{pmatrix},$$

where

$$Q_{0,0} = \begin{pmatrix} -(\lambda + h) & h \\ \tau & -(\lambda + \tau) \end{pmatrix}, \; Q_{0,1} = \begin{pmatrix} \lambda & 0 \\ 0 & \lambda \end{pmatrix},$$

$$Q_{-1} = \begin{pmatrix} \mu_1 & 0 \\ 0 & \mu_2 \end{pmatrix}, \; Q_0 = \begin{pmatrix} -(\lambda + \mu_1 + h) & h \\ \tau & -(\lambda + \mu_2 + \tau) \end{pmatrix}, \; Q_1 = \begin{pmatrix} \lambda & 0 \\ 0 & \lambda \end{pmatrix}.$$

It follows from the form of the generator that the chain under consideration belongs to the class of quasi-birth-and-death processes, see, e.g., [93].

Theorem 5.4 *A necessary and sufficient condition for the existence of the stationary distribution of the Markov chain ξ_t is the fulfillment of the inequality*

$$\lambda < \frac{\tau}{h + \tau}\mu_1 + \frac{h}{h + \tau}\mu_2. \tag{5.28}$$

Proof The required condition follows from (5.26) and (5.27) when $\alpha \to \infty$. In this case the system under consideration is equivalent to the system for which Corollary 5.8 is formulated. □

In what follows we assume inequality (5.28) is fulfilled.

Denote the stationary state probabilities of the chain ξ_t, $t \geq 0$ by

$$\alpha_i = \lim_{t \to \infty} P\{i_t = i, n_t = 1\}, \; \beta_i = \lim_{t \to \infty} P\{i_t = i, n_t = 2\}, \; i \geq 0.$$

The system of equilibrium (Chapman–Kolmogorov) equations for these probabilities is evidently written as follows:

$$\alpha_0 (\lambda + h) = \beta_0 \tau + \alpha_0 \mu_1, \tag{5.29}$$

$$\beta_0 (\lambda + \tau) = \beta_0 \mu_2 + \alpha_0 h, \tag{5.30}$$

$$\alpha_i (\lambda + h + \mu_1) = \alpha_{i+1}\mu_1 + \beta_i \tau + \alpha_{i-1}\lambda, \; i > 0, \tag{5.31}$$

$$\beta_i (\lambda + \tau + \mu_2) = \beta_{i+1}\mu_2 + \alpha_{i+1} h + \beta_{i-1}\lambda, \; i > 0. \tag{5.32}$$

Complementing these equations by the normalization equation

$$\sum_{i=0}^{\infty} (\alpha_i + \beta_i) = 1 \tag{5.33}$$

we get a system of linear algebraic equations that has a unique solution.

To calculate this solution, we will use the generating functions method. Introduce the generating functions

$$\alpha(z) = \sum_{i=0}^{\infty} \alpha_i z^i, \ \beta(z) = \sum_{i=0}^{\infty} \beta_i z^i, \ |z| \leq 1.$$

Theorem 5.5 *The generating functions $\alpha(z)$ and $\beta(z)$ have the following form:*

$$\alpha(z) = \frac{\mu_1 \alpha_0 (1 - z) - \tau \beta(z) z}{\lambda z^2 - (\lambda + \mu_1 + h) z + \mu_1}, \tag{5.34}$$

$$\beta(z) = \frac{\alpha_0 h \mu_1 z - \beta_0 \mu_2 \left(\lambda z^2 - z(h + \lambda + \mu_1) + \mu_1\right)}{M_3(z)}, \tag{5.35}$$

where

$$M_3(z) = \lambda^2 z^3 - (h\lambda + \lambda^2 + \lambda\mu_1 + \lambda\mu_2 + \lambda\tau) z^2$$

$$+ (h\mu_2 + \lambda\mu_1 + \lambda\mu_2 + \mu_1\mu_2 + \mu_1\tau) z - \mu_1\mu_2,$$

the probabilities α_0 and β_0 are expressed as follows:

$$\alpha_0 = \frac{\left(\lambda\sigma^2 - (\lambda + \mu_1 + h)\sigma + \mu_1\right)(h\mu_2 - \lambda(h + \tau) + \mu_1\tau)}{\mu_1 (1 - \sigma)(h + \tau)(\mu_1 - \lambda\sigma)}, \tag{5.36}$$

$$\beta_0 = \frac{\alpha_0 h \mu_1 \sigma}{\lambda\sigma^2 - (\lambda + \mu_1 + h)\sigma + \mu_1}, \tag{5.37}$$

and the quantity σ is the unique (real) solution of the cubic equation

$$M_3(z) = 0$$

in the region $|z| < 1$.

Proof It is easy to verify that the system of equations (5.29)–(5.31) is equivalent to Eq. (5.34) for the generating function $\alpha(z)$, and the system (5.30)–(5.32) is equivalent to Eq. (5.35) for the generating function $\beta(z)$.

In order to obtain the generating functions in an explicit form, it is required to find the probabilities α_0, β_0.To find these unknown probabilities, we will use the fact that, if the inequality (5.28) is satisfied, the generating functions $\alpha(z)$, $\beta(z)$ are analytic inside the unit disk $|z| < 1$.

Let us consider the function $\beta(z)$ given by formula (5.35). The denominator of this function is a cubic polynomial. It can be shown that, under fulfillment of ergodicity condition (5.28), this polynomial has a unique and positive real root in the unit disk of the complex plane. Denote this root by σ, $0 < \sigma < 1$. As $\beta(z)$

is an analytical function, the numerator of (5.35) must vanish in the point $z = \sigma$. Equating the numerator to zero, we derive expression (5.37) for β_0.

It follows from (5.34) that

$$h\sum_{i=0}^{\infty}\alpha_i = \tau\sum_{i=0}^{\infty}\beta_i.$$

Using this relation in (5.33), we get

$$\beta(1) = \sum_{i=0}^{\infty}\beta_i = \frac{h}{\tau + h}. \tag{5.38}$$

Substituting the expression (5.38) for $\beta(1)$ and the expression (5.37) for β_0 in (5.35), where we put $z = 1$, we obtain an equation for α_0 that has the solution (5.36). □

The key performance measures of the system are as follows:

- Probability that the system is empty

$$P^{(0)} = \alpha(0) + \beta(0).$$

- Probability $P^{(1)}$ that server 1 is fault-free and probability $P^{(2)}$ that server 1 is under repair

$$P^{(1)} = \frac{\tau}{\tau + h}, \quad P^{(2)} = \frac{h}{\tau + h}.$$

- Probability $P^{(1)}_{serv}$ that server 1 is serving a customer and probability $P^{(2)}_{serv}$ that server 1 is under repair and server 2 is serving a customer

$$P^{(1)}_{serv} = \alpha(1) - \alpha_0, \quad P^{(2)}_{serv} = \beta(1) - \beta_0.$$

- Mean number of customers in the system

$$L = \alpha'(1) + \beta'(1).$$

Remark 5.4 The expressions for probabilities $P^{(1)}$ and $P^{(2)}$ make condition (5.28) intuitively clear. The left-hand side of inequality (5.28) is the mean arrival rate while the right-hand side is the mean departure rate from the system when it is overloaded, because the value of the service rate is equal to μ_1 when server 1 is not broken and is equal to μ_2 when this server is broken.

Now we derive the Laplace–Stieltjes transform of the stationary distribution of the sojourn time of an arbitrary customer in the system.

In the system under consideration, the sojourn time essentially depends on the breakdown arrival process and the duration of the repair period. During the sojourn time time of a customer, it may repeatedly move from server 1 to server 2 and vice versa, which complicates the analysis.

Let $V(t)$ be the stationary distribution function of the sojourn time and $v(s) = \int_0^\infty e^{-st} dV(t)$, $Res \geq 0$ be the Laplace–Stieltjes transform of this distribution.

Theorem 5.6 *The Laplace–Stieltjes transform of the sojourn time of an arbitrary customer in the system is calculated by*

$$v(s) = \sum_{i=0}^{\infty} \mathbf{p}_i \mathbf{v}_i^T(s), \tag{5.39}$$

where

$$\mathbf{p}_i = (\alpha_i, \beta_i),\ i \geq 0,$$

and the row vectors $\mathbf{v}_i(s)$ *are calculated as*

$$\mathbf{v}_i(s) = \mathbf{v}_0(s) \left[A(s) [I - B(s)]^{-1} \right]^i,\ i \geq 1, \tag{5.40}$$

where the row vector $\mathbf{v}_0(s)$ *is of the form*

$$\mathbf{v}_0(s) = \left(\frac{\mu_1(\mu_2 + \tau + s) + h\mu_2}{(\mu_1 + h + s)(\mu_2 + \tau + s) - h\tau}, \right.$$

$$\left. \frac{1}{\mu_2 + \tau + s} [\mu_2 + \tau \frac{\mu_1(\mu_2 + \tau + s) + h\mu_2}{(\mu_1 + h + s)(\mu_2 + \tau + s) - h\tau}] \right), \tag{5.41}$$

and the matrices $A(s)$ *and* $B(s)$ *are given by*

$$A(s) = \begin{pmatrix} \frac{\mu_1}{\mu_1+h+s} & 0 \\ 0 & \frac{\mu_2}{\mu_2+\tau+s} \end{pmatrix},\quad B(s) = \begin{pmatrix} 0 & \frac{\tau}{\mu_2+\tau+s} \\ \frac{h}{\mu_1+h+s} & 0 \end{pmatrix}. \tag{5.42}$$

Proof Let $\mathbf{v}_i(s) = (v_i(s, 1), v_i(s, 2))$, where $v_i(s, n)$, $Re\ s \geq 0$ is the Laplace–Stieltjes transform (LST) of the stationary distribution of the sojourn time of a customer that, upon arrival, finds i customers in the system and server 1 in state n, $i \geq 0$, $n = 1, 2$.

Using the probabilistic interpretation of the LST, we are able to write the following system of linear algebraic equations for $v_0(s, 1)$ and $v_0(s, 2)$:

$$v_0(s,1) = \int_0^\infty e^{-st}e^{-ht}e^{-\mu_1 t}\mu_1\,dt + \int_0^\infty e^{-st}e^{-ht}e^{-\mu_1 t}h\,dt\,v_0(s,2),$$

$$v_0(s,2) = \int_0^\infty e^{-st}e^{-\tau t}e^{-\mu_2 t}\mu_2\,dt + \int_0^\infty e^{-\tau t}e^{-st}e^{-\mu_2 t}\tau\,dt\,v_0(s,1).$$

Integrating in these equations, we obtain the system

$$v_0(s,1) = \frac{\mu_1}{\mu_1+h+s} + \frac{h}{\mu_1+h+s}v_0(s,2),$$

$$v_0(s,2) = \frac{\mu_2}{\mu_2+\tau+s} + \frac{\tau}{\mu_2+\tau+s}v_0(s,1).$$

The solution of this system is as follows:

$$v_0(s,1) = \frac{\mu_1(\mu_2+\tau+s)+h\mu_2}{(\mu_1+h+s)(\mu_2+\tau+s)-h\tau}, \tag{5.43}$$

$$v_0(s,2) = \frac{1}{\mu_2+\tau+s}\left[\mu_2+\tau\frac{\mu_1(\mu_2+\tau+s)+h\mu_2}{(\mu_1+h+s)(\mu_2+\tau+s)-h\tau}\right]. \tag{5.44}$$

Using the probabilistic interpretation of the LST, we write the following system of linear algebraic equation for the functions $v_i(s, 1)$ and $v_i(s, 2)$, $i \geq 1$:

$$v_i(s,1) = \frac{\mu_1}{\mu_1+h+s}v_{i-1}(s,1) + \frac{h}{\mu_1+h+s}v_i(s,2),$$

$$v_i(s,2) = \frac{\mu_2}{\mu_2+\tau+s}v_{i-1}(s,1) + \frac{\tau}{\mu_2+\tau+s}v_i(s,1),\ i \geq 1. \tag{5.45}$$

Introduce the vectors

$$\mathbf{v}_i(s) = (v_i(s,1), v_i(s,2)),\ i \geq 0.$$

Then system (5.45) can be written in the vector form

$$\mathbf{v}_i(s) = \mathbf{v}_{i-1}(s)\,A(s) + \mathbf{v}_i(s)\,B(s),\ i \geq 1, \tag{5.46}$$

where the matrices $A(s)$, $B(s)$ have the form (5.42).

From (5.46), we get the following recursive formula for sequential calculation of the vectors $\mathbf{v}_i(s)$:

$$\mathbf{v}_i(s) = \mathbf{v}_{i-1}(s) A(s) [I - B(s)]^{-1}, i \geq 1.$$

Using this formula, we express all vectors $\mathbf{v}_i(s),\ i \geq 1$ in terms of the vector $\mathbf{v}_0(s)$ defined by formulas (5.43) and (5.44). The resulting expression for the vectors $\mathbf{v}_i(s),\ i \geq 1$ is given by (5.40).

Now, using the formula of total probability, we write the desired Laplace–Stieltjes transform of the sojourn time distribution in the form (5.39). □

Corollary 5.9 *The mean sojourn time of an arbitrary customer is calculated using the formula*

$$Ev = -\mathbf{v}'(0) = -\sum_{i=0}^{\infty} \mathbf{p}_i \frac{d\mathbf{v}_i^T(s)}{ds}|_{s=0}. \tag{5.47}$$

5.3 Analysis of Characteristics of the Data Transmission Process Under Hot-Standby Architecture of a High-Speed FSO Channel Backed Up by a Millimeter Radio Channel

In this section, we consider a two-server queuing system with unreliable heterogeneous servers, which can be used in mathematical modeling of a hybrid communication network consisting of an unreliable FSO (Free Space Optics) channel and an unreliable millimeter radio channel (71–76 GHz, 81–86 GHz). It is assumed that weather conditions unfavorable for one of the channels are not unfavourable for the other channel. The FSO channel cannot transmit data in fog or cloudy weather conditions, and the millimeter radio channel can not transmit during precipitation (rain, snow, etc.). Thus, the hybrid communication system is able to transfer data under almost any weather conditions.

5.3.1 Model Description

We consider a queuing system consisting of two unreliable servers (server 1 and server 2) and an infinite buffer. Customers arrive at the system in accordance with the MAP. The MAP is defined by the underlying process $\nu_t,\ t \geq 0$, with the state space $\{0, \dots, W\}$ and the matrices $D_k,\ k = 0, 1$.

If an arriving customer or the first customer from the queue sees two servers idle and ready for service, he/she starts the service at both servers. If the servers are busy at an arrival epoch or one of the servers is busy while the other server is under repair, the customer is placed at the end of the queue in the buffer and is picked up for a service later on, in accordance with the FIFO discipline. If one of the servers is under repair and the other server is idle, the idle server begins the service of the customer. If the service of a customer at one of the servers is not finished at the end of a repair period on the other server, the latter server immediately connects to the service of the customer. The service of the customer is considered be completed when his/her service by either of the servers is finished.

Breakdowns arrive at the servers according to a MAP defined by the underlying process $\eta_t,\ t \geq 0$, with state space $\{0, \ldots, V\}$ and matrices $H_k,\ k = 0, 1$. An arriving breakdown is directed to server 1 with probability p and to server 2 with complementary probability $1-p$. The fundamental rate of breakdowns is calculated as $h = \boldsymbol{\vartheta} H_1 \mathbf{e}$, where the row vector $\boldsymbol{\vartheta}$ is the unique solution of the system $\boldsymbol{\vartheta}(H_0 + H_1) = \mathbf{0},\ \boldsymbol{\vartheta}\mathbf{e} = 1$.

The service time of a customer by the kth server, $k = 1, 2$ has PH-type distribution with irreducible representation $(\boldsymbol{\beta}^{(k)}, \mathbf{S}^{(k)})$. The service process on the kth server is directed by the Markov chain $m_t^{(k)},\ t \geq 0$, with the state space $\{1, \ldots, M^{(k)}, M^{(k)} + 1\}$, where $M^{(k)} + 1$ is an absorbing state. The intensities of transitions into the absorbing state are defined by the vector $\mathbf{S}_0^{(k)} = -S^{(k)}\mathbf{e}$. The service rates are calculated as $\mu_k = -[\boldsymbol{\beta}^{(k)}(S^{(k)})^{-1}\mathbf{e}]^{-1}$.

The repair period at the kth server, $k = 1, 2$, has PH-type distribution with irreducible representation $(\boldsymbol{\tau}^{(k)}, T^{(k)})$. The repair process at the kth server is directed by the Markov chain $r_t^{(k)}, t \geq 0$, with state space $\{1, \ldots, R_k, R_k + 1\}$, where $R_k + 1$ is an absorbing state. The intensities of transitions into the absorbing state are defined by the vector $\mathbf{T}_0^{(k)} = -T^{(k)}\mathbf{e}$. The repair rate is $\tau_k = -(\boldsymbol{\tau}^{(k)}(T^{(k)})^{-1}\mathbf{e})^{-1}$.

5.3.2 *Process of the System States*

Let at the moment t

- i_t be the number of customers in the system, $i_t \geq 0$,
- $n_t = 0$ if both servers are fault-free, $n_t = k$ if server k is under repair, $k = 1, 2$;
- $m_t^{(k)}$ be the state of the service underlying process at the kth busy server, $m_t^{(k)} = \overline{1, M^{(k)}},\ k = 1, 2$;
- $r_t^{(k)}$ be the state of the repair time underlying process at the kth server, $r_t^{(k)} = \overline{1, R_k}, k = 1, 2$;
- ν_t and η_t be the states of the underlying processes of the MAP of customers and the MAP of breakdowns respectively, $\nu_t = \overline{0, W},\ \eta_t = \overline{0, V}$.

The process of the system states is described by the regular irreducible continuous-time Markov chain, $\xi_t, t \geq 0$, with state space

$$\Omega = \{(i, 0, \nu, \eta),\ i = 0, n = 0,\ \nu = \overline{0, W}, \eta = \overline{0, V}\}$$

$$\bigcup\{(i, 1, \nu, \eta, r^{(1)}),\ i = 0, n = 1,\ \nu = \overline{0, W}, \eta = \overline{0, V}, r^{(1)} = \overline{1, R_1}\}$$

$$\bigcup\{(i, 2, \nu, \eta, r^{(2)}),\ i = 0, n = 2,\ \nu = \overline{0, W}, \eta = \overline{0, V}, r^{(1)} = \overline{1, R_2}\}$$

$$\bigcup\{(i, 0, \nu, \eta, m^{(1)},\ m^{(2)}),\ i > 0, n = 0, \nu = \overline{0, W}, \eta = \overline{0, V},\ m^{(k)} = \overline{1, M_k}, k = 1, 2\}$$

$$\bigcup\{(i, 1, \nu, \eta, m^{(2)}, r^{(1)}),\ i > 0, n = 1,\ \nu = \overline{0, W}, \eta = \overline{0, V}, m^{(2)} = \overline{1, M_2}, r^{(1)} = \overline{1, R_1}\}$$

$$\bigcup\{(i, 2, \nu, \eta, m^{(1)}, r^{(2)}),\ i > 0, n = 2,\ \nu = \overline{0, W}, \eta = \overline{0, V}, m^{(1)} = \overline{1, M_1}, r^{(1)} = \overline{1, R_2}\}.$$

In the following, we assume that the states of the chain $\xi_t, t \geq 0$ are ordered in lexicographic order. Let $Q_{ij},\ i, j \geq 0$ be the matrices formed by the rates of the transition from states corresponding to value i of the component i_n to the states corresponding to the value j of this component. Denote by $Q = (Q_{ij})_{i,j \geq 0}$ the generator of the chain.

Lemma 5.5 *The infinitesimal generator Q of the Markov chain $\xi_t,\ t \geq 0$ has the following block tridiagonal structure:*

$$Q = \begin{pmatrix} Q_{0,0} & Q_{0,1} & O & O & \cdots \\ Q_{1,0} & Q_1 & Q_2 & O & \cdots \\ O & Q_0 & Q_1 & Q_2 & \cdots \\ O & O & Q_0 & Q_1 & \cdots \\ \vdots & \vdots & \vdots & \vdots & \ddots \end{pmatrix},$$

where non-zero blocks have the following form:

$$Q_{0,0} = \begin{pmatrix} D_0 \oplus H_0 & I_{\bar{W}} \otimes pH_1 \otimes \boldsymbol{\tau}^{(1)} & I_{\bar{W}} \otimes (1-p)H_1 \otimes \boldsymbol{\tau}^{(2)} \\ I_a \otimes \boldsymbol{T}_0^{(1)} & D_0 \oplus H \oplus T^{(1)} & O \\ I_a \otimes \boldsymbol{T}_0^{(2)} & O & D_0 \oplus H \oplus T^{(2)} \end{pmatrix},$$

$$Q_{0,1}$$

$$= \begin{pmatrix} D_1 \otimes I_{\bar{V}} \otimes \boldsymbol{\beta}^{(1)} \otimes \boldsymbol{\beta}^{(2)} & O & O \\ O & D_1 \otimes I_{\bar{V}} \otimes \boldsymbol{\beta}^{(2)} \otimes I_{R_1} & O \\ O & O & D_1 \otimes I_{\bar{V}} \otimes \boldsymbol{\beta}^{(1)} \otimes I_{R_2} \end{pmatrix},$$

$$Q_{1,0} = \begin{pmatrix} I_a \otimes \tilde{S}_0 & O & O \\ O & I_a \otimes \mathbf{S}_0^{(2)} \otimes I_{R_1} & O \\ O & O & I_a \otimes \mathbf{S}_0^{(1)} \otimes I_{R_2} \end{pmatrix},$$

$$Q_0 = \begin{pmatrix} I_a \otimes \tilde{S}_0(\boldsymbol{\beta}^{(1)} \otimes \boldsymbol{\beta}^{(2)}) & O & O \\ O & I_a \otimes \mathbf{S}_0^{(2)} \boldsymbol{\beta}^{(2)} \otimes I_{R_1} & O \\ O & O & I_a \otimes \mathbf{S}_0^{(1)} \boldsymbol{\beta}^{(1)} \otimes I_{R_2} \end{pmatrix},$$

$$Q_1$$

$$= \begin{pmatrix} D_0 \oplus H_0 \oplus S^{(1)} \oplus S^{(2)} & I_{\bar{W}} \otimes pH_1 \otimes \mathbf{e}_{M_1} \otimes I_{M_2} \otimes \boldsymbol{\tau}_1 & I_{\bar{W}} \otimes \bar{p}H_1 \otimes I_{M_1} \otimes \mathbf{e}_{M_2} \otimes \boldsymbol{\tau}_2 \\ I_a \otimes \beta^{(1)} \otimes I_{M_2} \otimes \boldsymbol{T}_0^{(1)} & D_0 \oplus H \oplus S^{(2)} \oplus T^{(1)} & O \\ I_a \otimes I_{M_1} \otimes \beta^{(2)} \otimes \boldsymbol{T}_0^{(2)} & O & D_0 \oplus H \oplus S^{(1)} \oplus T^{(2)} \end{pmatrix},$$

$$Q_2 = \begin{pmatrix} D_1 \otimes I_{\bar{V} M_1 M_2} & O & O \\ O & D_1 \otimes I_{\bar{V} M_2 R_1} & O \\ O & O & D_1 \otimes I_{\bar{V} M_1 R_2} \end{pmatrix},$$

where $\bar{p} = 1-p$, $H = H_0+H_1$, $\tilde{S}_0 = -(S_1 \oplus S_2)\mathbf{e}$, $\bar{W} = W+1$, $\bar{V} = V+1$, $a = \bar{W}\bar{V}$.

The proof of the lemma is by means of calculation of probabilities of transitions of the components of the Markov chain ξ_t during a time interval with infinitesimal length.

Corollary 5.10 *The Markov chain* $\xi_t, t \geq 0$ *belongs to the class of quasi-birth-and-death (QBD) processes, see [93].*

5.3.3 Stationary Distribution: Performance Measures

Theorem 5.7 *A necessary and sufficient condition for the existence of the stationary distribution of the Markov chain* ξ_t, $t \geq 0$ *is the fulfillment of the inequality*

$$\lambda < \hat{\delta}_0 \tilde{S}_0 + \hat{\delta}_1 \mathbf{S}_0^{(2)} + \hat{\delta}_2 \mathbf{S}_0^{(1)}, \tag{5.48}$$

where $\hat{\boldsymbol{\delta}}_0 = \boldsymbol{\delta}_0(\mathbf{e}_{\bar{V}} \otimes I_{M_1M_2})$, $\hat{\boldsymbol{\delta}}_1 = \boldsymbol{\delta}_1(\mathbf{e}_{\bar{V}} \otimes I_{M_2} \otimes \mathbf{e}_{R_1})$, $\hat{\boldsymbol{\delta}}_2 = \boldsymbol{\delta}_2(\mathbf{e}_{\bar{V}} \otimes I_{M_1} \otimes \mathbf{e}_{R_2})$, *and the vector* $\boldsymbol{\delta} = (\boldsymbol{\delta}_0,\ \boldsymbol{\delta}_1,\ \boldsymbol{\delta}_2)$ *is the unique solution of the system*

$$\boldsymbol{\delta}\Phi = \mathbf{0},\ \ \boldsymbol{\delta}\mathbf{e} = 1, \tag{5.49}$$

where

$$\Phi =$$

$$\begin{pmatrix} I_{\bar{V}} \otimes \tilde{\mathbf{S}}_0(\boldsymbol{\beta}_1 \otimes \boldsymbol{\beta}_2) + H_0 \otimes I_{M_1M_2} & pH_1 \otimes \mathbf{e}_{M_1} \otimes I_{M_2} \otimes \boldsymbol{\tau}_1 & \bar{p}H_1 \otimes I_{M_1} \otimes \mathbf{e}_{M_2} \otimes \boldsymbol{\tau}_2 \\ I_{\bar{V}} \otimes \boldsymbol{\beta}_1 \otimes I_{M_2} \otimes \boldsymbol{T}_0^{(1)} & I_{\bar{V}} \otimes \mathbf{S}_0^{(2)}\boldsymbol{\beta}_2 \otimes I_{R_1} & O \\ I_{\bar{V}} \otimes I_{M_1} \otimes \boldsymbol{\beta}_2 \otimes \boldsymbol{T}_0^{(2)} & O & I_{\bar{V}} \otimes \mathbf{S}_0^{(1)}\boldsymbol{\beta}_1 \otimes I_{R_2} \end{pmatrix}$$

$$+diag\{I_{\bar{V}} \otimes S_1 \oplus S_2,\ I_{\bar{V}} \otimes S_2 \oplus T^{(1)},\ I_{\bar{V}} \otimes S_1 \oplus T^{(2)}\}.$$

Proof It follows from [93] that a necessary and sufficient condition for the existence of the stationary distribution of the QBD $\xi_t,\ t \geq 0$ is the fulfillment of the following inequality:

$$\mathbf{x}Q_2\mathbf{e} < \mathbf{x}Q_0\mathbf{e}, \tag{5.50}$$

where the vector $\mathbf{x}$ is the unique solution of the system of linear algebraic equations

$$\mathbf{x}(Q_0 + Q_1 + Q_2) = \mathbf{0}, \tag{5.51}$$

$$\mathbf{x}\mathbf{e} = 1. \tag{5.52}$$

Represent the vectors $\mathbf{x}$ in the form

$$\mathbf{x} = (\boldsymbol{\theta} \otimes \boldsymbol{\delta}_0, \boldsymbol{\theta} \otimes \boldsymbol{\delta}_1, \boldsymbol{\theta} \otimes \boldsymbol{\delta}_2), \tag{5.53}$$

where the vectors $\boldsymbol{\delta}_0,\ \boldsymbol{\delta}_1,\ \boldsymbol{\delta}_2$ are of size $\bar{V}M_1M_2,\ \bar{V}M_2R_1$, and $\bar{V}M_1R_2$, respectively.

Substituting the vector $\mathbf{x}$ of form (5.53) into inequality (5.50) and into the system (5.51) and (5.52), after some algebraic transformations we obtain the inequality (5.48) and system (5.49) respectively. □

Corollary 5.11 *In the case of a stationary Poisson flow of breakdowns and an exponential distribution of service and repair times, the stability condition (5.48) and (5.49) is reduced to the following inequality:*

$$\lambda < \frac{\tau_1\tau_2}{\tau_1\tau_1 + ph\tau_2 + (1-p)h\tau_1}(\mu_1 + \mu_2 + \frac{ph}{\tau_1}\mu_2 + \frac{(1-p)h}{\tau_2}\mu_1). \tag{5.54}$$

Proof Taking into account that in the case under consideration $S^{(n)} = -S_0^{(n)} = \mu_n$, $T^{(n)} = -T_0^{(n)} = \tau_n$, $n = 1, 2$, inequality (5.48) reduces to the form

$$\lambda < \delta_0(\mu_1 + \mu_2) + \delta_1\mu_2 + \delta_2\mu_1, \quad (5.55)$$

and system (5.49) reduces to the form

$$\begin{cases} \delta_0 h - \delta_1\tau_1 - \delta_0\tau_2 = 0, \\ \delta_0 ph - \delta_1\tau_1 = 0, \\ \delta_0(1-p)h - \delta_2\tau_2 = 0, \\ \delta_0 + \delta_1 + \delta_2 = 1. \end{cases}$$

The solution of the last system is as follows:

$$\delta_0 = \frac{\tau_1\tau_2}{\tau_1\tau_1 + ph\tau_2 + (1-p)h\tau_1}, \ \delta_1 = \delta_0\frac{ph}{\tau_1}, \ \delta_2 = \delta_0\frac{(1-p)h}{\tau_2}. \quad (5.56)$$

Substituting (5.56) into (5.55), we see that the stability condition has the form (5.54) □

Remark 5.5 Inequality (5.54) is easy to interpret if we note that the values $\delta_0, \delta_1, \delta_2$ are the probabilities that, under overload condition, either the servers are fault-free, or the first server is under repair, or the second server is under repair, respectively. In this case, the right-hand side of (5.54) is the service rate under overload conditions. The condition for the existence of a stationary distribution is that this rate should be greater than the intensity of the input flow λ.

We note that condition (5.48) for the existence of the stationary distribution in the case of a MAP flow of breakdowns and PH distributions of service and repair times is similarly interpreted.

In what follows, we assume inequality (5.48) be fulfilled. Introduce the steady state probabilities of the chain under consideration:

$$p_0^{(0)} = \lim_{t\to\infty} P\{i_t = 0, n_t = 0, \nu_t = \nu, \eta_t = \eta\},$$

$$p_0^{(1)} = \lim_{t\to\infty} P\{i_t = 0, n_t = 1, \nu_t = \nu, \eta_t = \eta, r_t^{(1)} = r^{(1)}\},$$

$$p_0^{(2)} = \lim_{t\to\infty} P\{i_t = 0, n_t = 2, \nu_t = \nu, \eta_t = \eta, r_t^{(2)} = r^{(2)}\},$$

$$p_i^{(0)} = \lim_{t\to\infty} P\{i_t = i, n_t = 0, \nu_t = \nu, \eta_t = \eta, m_t^{(1)} = m^{(1)}, m_t^{(2)} = m^{(2)}\},$$

$$p_i^{(1)} = \lim_{t\to\infty} P\{i_t = i, n_t = 1, \nu_t = \nu, \eta_t = \eta, m_t^{(2)} = m^{(2)}, r_t^{(1)} = r^{(1)}\},$$

$$p_i^{(2)} = \lim_{t \to \infty} P\{i_t = i, n_t = 2, \nu_t = \nu, \eta_t = \eta, m_t^{(1)} = m^{(1)}, r_t^{(2)} = r^{(2)}\},$$

$$i > 0, \nu = \overline{0, W}, \eta = \overline{0, V}, m^{(k)} = \overline{1, M_k}, r^{(k)} = \overline{1, R_k}, \ k = 1, 2.$$

Let us enumerate the steady state probabilities in lexicographic order and form the row vectors $\mathbf{p}_i$ of steady state probabilities corresponding to the value i of the first component of the chain, $i \geq 0$.

To calculate the vectors $\mathbf{p}_i$, $i \geq 0$ we use the algorithm for calculating the stationary distribution of a QBD process, see [93].

Algorithm 5.2

1. Calculate the matrix R as the minimal non-negative solution of the nonlinear matrix equation

$$R^2 Q_0 + R Q_1 + Q_2 = O.$$

2. Calculate the vector $\mathbf{p}_1$ as the unique solution of the system

$$\mathbf{p}_1[Q_1 + Q_{1,0}(-Q_{0,0})^{-1} Q_{0,1} + R Q_0] = \mathbf{0},$$

$$\mathbf{p}_1[\mathbf{e} + Q_{1,0}(-Q_{0,0})^{-1}\mathbf{e} + R(I - R)^{-1}\mathbf{e}] = 1.$$

3. Calculate the vectors $\mathbf{p}_0$, $\mathbf{p}_i$, $i \geq 2$ as follows:

$$\mathbf{p}_0 = \mathbf{p}_1 Q_{1,0}(-Q_{0,0})^{-1}, \quad \mathbf{p}_i = \mathbf{p}_1 R^{i-1}, \ i \geq 2.$$

Having calculated the stationary distribution $\mathbf{p}_i$, $i \geq 0$ we can find a number of stationary performance measures of the system. Some of them are listed below:

- Throughput of the system

$$\varrho = \boldsymbol{\delta}_0 \tilde{\mathbf{S}}_0 + \hat{\boldsymbol{\delta}}_1 \mathbf{S}_0^{(2)} + \hat{\boldsymbol{\delta}}_2 \mathbf{S}_0^{(1)}.$$

In the case of exponential distributions of the service and repair times the throughput of the system is calculated as

$$\varrho = \frac{\tau_1 \tau_2}{\tau_1 \tau_1 + ph\tau_2 + (1-p)h\tau_1}(\mu_1 + \mu_2 + \frac{ph}{\tau_1}\mu_2 + \frac{(1-p)h}{\tau_1}\mu_1).$$

- Mean number of customers in the system $L = \sum\limits_{i=1}^{\infty} i \mathbf{p}_i \mathbf{e}$.
- Variance of the number of customers in the system $V = \sum\limits_{i=1}^{\infty} i^2 \mathbf{p}_i \mathbf{e} - L^2$.

- Probability that the system is idle and both servers are fault-free

$$P_0^{(0)} = \mathbf{p}_0 \begin{pmatrix} \mathbf{e}_a \\ \mathbf{0}_{a(R_1+R_2)} \end{pmatrix}$$

- Probability that the system is idle and server 1 (server 2) is under repair

$$P_0^{(1)} = \mathbf{p}_0 \begin{pmatrix} \mathbf{0}_a \\ \mathbf{e}_{aR_1} \\ \mathbf{0}_{aR_2} \end{pmatrix}, \ P_0^{(2)} = \mathbf{p}_0 \begin{pmatrix} \mathbf{0}_{a(1+R_1)} \\ \mathbf{e}_{aR_2} \end{pmatrix}$$

- Probabilities that there are $i, i > 0$ customers in the system and both server are fault-free (serve a customer) at an arbitrary time and at an arrival epoch

$$P_i^{(0)} = \mathbf{p}_i \begin{pmatrix} \mathbf{e}_{aM_1M_2} \\ \mathbf{0}_{a(M_2R_1+M_1R_2)} \end{pmatrix}, \ P_{arrival}^{(0,i)} = \lambda^{-1}\mathbf{p}_i \begin{pmatrix} I_{\bar{W}} \otimes \mathbf{e}_{\bar{V}M_1M_2} \\ O_{a(M_2R_1+M_1R_2)\times\bar{W}} \end{pmatrix} D_1\mathbf{e}.$$

- Probabilities that there are $i, i > 0$ customers in the system, server 1 is under repair and server 2 is serving a customer at an arbitrary time and at an arrival epoch

$$P_i^{(1)} = \mathbf{p}_i \begin{pmatrix} \mathbf{0}_{aM_1M_2} \\ \mathbf{e}_{aM_2R_1} \\ \mathbf{0}_{aM_1R_2} \end{pmatrix}, \ P_{arrival}^{(1,i)} = \lambda^{-1}\mathbf{p}_i \begin{pmatrix} O_{aM_1M_2\times\bar{W}} \\ I_{\bar{W}} \otimes \mathbf{e}_{\bar{V}M_2R_1} \\ O_{aM_1R_2\times\bar{W}} \end{pmatrix} D_1\mathbf{e}.$$

- Probability that there are $i, i > 0$ customers in the system, server 2 is under repair and server 1 is serving a customer at an arbitrary time and at an arrival epoch

$$P_i^{(2)} = \mathbf{p}_i \begin{pmatrix} \mathbf{0}_{aM_1M_2} \\ \mathbf{0}_{aM_2R_1} \\ \mathbf{e}_{aM_1R_2} \end{pmatrix}, \ P_{arrival}^{(2,i)} = \lambda^{-1}\mathbf{p}_i \begin{pmatrix} O_{a(M_1M_2+M_2R_1)\times\bar{W}} \\ I_{\bar{W}} \otimes \mathbf{e}_{\bar{V}M_1R_2} \end{pmatrix} D_1\mathbf{e}.$$

- Probability that the servers are in state n at an arbitrary time and at an arrival epoch

$$P_n = \sum_{i=0}^{\infty} P_i^{(n)}, \ \hat{P}_n = \sum_{i=0}^{\infty} P_{arrival}^{(n,i)}, \ n = 0, 1, 2.$$

- Probability that an arriving breakdown sees both servers fault-free and will be directed to the kth server

$$P_{break}^{(k)} = h^{-1}[\delta_{1,k}p + \delta_{2,k}(1-p)]\mathbf{p}_i \begin{pmatrix} \mathbf{e}_{\bar{W}} \otimes I_{\bar{V}} \otimes \mathbf{e}_{M_1M_2} \\ O_{a(M_2R_1+M_1R_2)\times\bar{W}} \end{pmatrix} H_1\mathbf{e}, \ k = 1, 2,$$

where $\delta_{i,j}$ is Kronecker's symbol.

5.4 Analysis of Characteristics of the Data Transmission Process Under Cold-Standby Architecture of a High-Speed FSO Channel and Millimeter Channel Backed Up by a Wireless Broadband Radio Channel

In this section, we investigate a two-server queuing system with unreliable heterogeneous servers and a backup reliable server. The system can be used in mathematical modeling of a hybrid communication network consisting of three channels: an FSO channel, a millimeter radio channel, and a broadband radio channel operating under the control of the IEEE 802.11 protocol and used as a backup channel for transmitting information. The main channels—FSO and millimeter radio channel—are subject to the influence of weather conditions and can fail under unfavorable conditions. The FSO channel cannot transmit data in conditions of poor visibility (fog or cloudy weather), and the millimeter radio channel cannot transmit during precipitation (rain, snow, etc.). In the event that both channels fail in poor visibility and precipitation conditions, the information is transmitted over a standby broadband radio channel, which is absolutely reliable but has a much slower transfer rate than the main channels. Thus, the hybrid communication system is able to transmit data practically in all weather conditions using one or another communication channel.

5.4.1 Model Description

We consider a queuing system with waiting room and two unreliable heterogeneous servers, which model FSO and mm-wave channels and a backup reliable server, which models a radio-wave IEEE 802 channel. In the following, the FSO cannel will be named as server 1, the mm-wave channel as server 2, and the radio-wave IEEE 802 channel as server 3.

Customers arrive at the system in accordance with a Batch Markovian Arrival Process ($BMAP$), which is defined by the underlying process $\nu_t,\ t \geq 0$, with finite state space $\{0, \dots, W\}$ and the matrix generating function $D(z) = \sum\limits_{k=0}^{\infty} D_k z^k$, $|z| \leq 1$.

The service time of a customer by the jth server, $j = 1, 2, 3$ has PH-type distribution with irreducible representation $(\boldsymbol{\beta}_j, S_j)$. The service process on the jth server is directed by the Markov chain $m_t^{(j)},\ t \geq 0$, with state space $\{1, \dots, M_j, M_j + 1\}$, where $M_j + 1$ is an absorbing state. The intensities of transitions into the absorbing state are defined by the vector $\mathbf{S}_0^{(j)} = -S_j\mathbf{e}$. The service rate at the jth server is $\mu_j = (\boldsymbol{\beta}_j(-S_j)^{-1}\mathbf{e})^{-1}$, $j = 1, 2, 3$.

Breakdowns arrive at servers 1, 2 according to an $MMAP$ defined by the underlying process η_t, $t \geq 0$, with state space $\{0, \ldots, V\}$ and by the matrices H_0, H_1, H_2. The matrix H_0 defines the intensities of the process η_t, $t \geq 0$ transitions that do not lead to generation of a breakdown. The matrix H_j defines the intensities of the η_t, $t \geq 0$ transitions that are accompanied by the generation of a breakdown directed to the server j, $j = 1, 2$. The rate of breakdowns directed to the jth server is $h_j = \boldsymbol{\vartheta} H_j \mathbf{e}$, $j = 1, 2$, where $\boldsymbol{\vartheta}$ is the unique solution of the system $\boldsymbol{\vartheta}(H_0 + H_1 + H_2) = \mathbf{0}$, $\boldsymbol{\vartheta}\mathbf{e} = 1$.

When a breakdown affects one of the main servers, the repair period at this server starts immediately and the other main server, if it is available, begins the service of the interrupted customer anew. If the latter server is busy or under repair, the customer goes to server 3 and starts its service anew. However, if during the service time of the customer at server 3 one of the main servers becomes fault-free, the customer restarts its service on this server.

The repair period at the jth main server, $j = 1, 2$ has a PH-type distribution with irreducible representation $(\boldsymbol{\tau}_j, T_j)$. The repair process at the jth server is directed by the Markov chain $r_t^{(j)}$, $t \geq 0$, with the state space $\{1, \ldots, R_j, R_j + 1\}$, where $R_j + 1$ is an absorbing state. The repair rate at the jth server is $\tau_j = (\boldsymbol{\tau}_j(-T_j)^{-1}\mathbf{e})^{-1}$, $j = 1, 2$.

5.4.2 Process of the System States

Let at the moment t

- i_t be the number of customers in the system, $i_t \geq 0$,
- $$n_t = \begin{cases} 0, & \text{if both main servers are fault-free (both are busy or idle);} \\ 0_j, & \text{if both main servers are fault-free, the } j\text{th server is busy and the other one is idle, } j = 1, 2; \\ 1, & \text{if server 1 is under repair;} \\ 2, & \text{if server 2 is under repair;} \\ 3, & \text{if both servers are under repair;} \end{cases}$$
- $m_t^{(j)}$ be the state of the service underlying process at the j-th busy server, $j = 1, 2, 3$, $m_t^{(j)} = \overline{1, M_j}$;
- $r_t^{(j)}$ be the state of the repair time underlying process at the j-th busy server, $j = 1, 2$, $r_t^{(j)} = \overline{1, R_j}$;
- ν_t and η_t be the states of the underlying processes of the $BMAP$ and the $MMAP$ respectively, $\nu_t = \overline{0, W}$, $\eta_t = \overline{0, V}$.

The process of the system states is described by the regular irreducible continuous-time Markov chain $\xi_t, t \geq 0$, with the state space

$$\Omega = \{(0, n, \nu, \eta), \ i = 0, n = \overline{0, 3}, \ \nu = \overline{0, W}, \ \eta = \overline{0, V}\}$$

$$\bigcup\{(i, 0_j, \nu, \eta, m^{(j)}), \ i = 1, \ j = 1, 2, n = 0_j, \nu = \overline{0, W}, \eta = \overline{0, V}, \ m^{(j)} = \overline{1, M_j}\}$$

$$\bigcup\{(i, 0, \nu, \eta, m^{(1)}, m^{(2)}), \ i > 1, n = 0, \nu = \overline{0, W}, \eta = \overline{0, V}, \ m^{(1)} = \overline{1, M_1},$$

$$m^{(2)} = \overline{1, M_2}\}$$

$$\bigcup\{(i, 1, \nu, \eta, m^{(2)}, r^{(1)}), \ i \geq 1, n = 1, \ \nu = \overline{0, W}, \eta = \overline{0, V},$$

$$m^{(2)} = \overline{1, M_2}, r^{(1)} = \overline{1, R_1}\}$$

$$\bigcup\{(i, 2, \nu, \eta, m^{(1)}, r^{(2)}), \ i \geq 1, n = 2, \ \nu = \overline{0, W},$$

$$\eta = \overline{0, V}, m^{(1)} = \overline{1, M_1}, r^{(1)} = \overline{1, R_2}\}$$

$$\bigcup\{(i, 3, \nu, \eta, m^{(3)}, r^{(1)}, r^{(2)}), \ i > 0,$$

$$n = 3, \ \nu = \overline{0, W}, \eta = \overline{0, V}, m^{(3)} = \overline{1, M_3}, r^{(j)} = \overline{1, R_j}, \ j = 1, 2\}.$$

In the following, we assume that the states of the chain $\xi_t, t \geq 0$ are ordered in lexicographic order. Let $Q_{ij}, \ i, j \geq 0$ be the matrices formed by the rates of transition from states corresponding to value i of the component i_n to states corresponding to value j of this component. Denote by $Q = (Q_{ij})_{i,j \geq 0}$ the generator of the chain.

Lemma 5.6 *The infinitesimal generator of the Markov chain $\xi_t, \ t \geq 0$ has the following block structure:*

$$Q = \begin{pmatrix} Q_{0,0} & Q_{0,1} & Q_{0,2} & Q_{0,3} & Q_{0,4} \cdots & \\ Q_{1,0} & Q_{1,1} & Q_{1,2} & Q_{1,3} & Q_{1,4} \cdots & \\ O & Q_{2,1} & Q_1 & Q_2 & Q_3 \cdots & \\ O & O & Q_0 & Q_1 & Q_2 \cdots & \\ \vdots & \vdots & \vdots & \vdots & \vdots & \ddots \end{pmatrix},$$

where non-zero blocks $Q_{i,j}$ *are of the following form:*

$$Q_{0,0} =$$

$$\begin{pmatrix} D_0 \oplus H_0 & I_{\bar{W}} \otimes H_1 \otimes \boldsymbol{\tau}_1 & I_{\bar{W}} \otimes H_2 \otimes \boldsymbol{\tau}_2 & O \\ I_a \otimes \boldsymbol{T}_0^{(1)} & D_0 \oplus (H_0 + H_1) \oplus T^{(1)} & O & I_{\bar{W}} \otimes H_2 \otimes I_{R_1} \otimes \boldsymbol{\tau}_2 \\ I_a \otimes \boldsymbol{T}_0^{(2)} & O & D_0 \oplus (H_0 + H_2) \oplus T^{(2)} & I_{\bar{W}} \otimes H_1 \otimes I_{R_2} \otimes \boldsymbol{\tau}_1 \\ O & I_a \otimes I_{R_1} \otimes \boldsymbol{T}_0^{(2)} & I_a \otimes \boldsymbol{T}_0^{(1)} \otimes I_{R_2} & D_0 \oplus H \oplus T^{(1)} \oplus T^{(2)} \end{pmatrix},$$

$$Q_{0,1} =$$

$$diag\{D_1 \otimes I_{\bar{V}} \otimes \boldsymbol{\beta}_1,\ D_1 \otimes I_{\bar{V}} \otimes \boldsymbol{\beta}_2 \otimes I_{R_1},\ D_1 \otimes I_{\bar{V}} \otimes \boldsymbol{\beta}_1 \otimes I_{R_2},\ D_1 \otimes I_{\bar{V}} \otimes \boldsymbol{\beta}_3 \otimes I_{R_1 R_2}\},$$

$$Q_{0,k} =$$

$$diag\{D_k \otimes I_{\bar{V}} \otimes \boldsymbol{\beta}_1 \otimes \boldsymbol{\beta}_2, D_k \otimes I_{\bar{V}} \otimes \boldsymbol{\beta}_2 \otimes I_{R_1}, D_k \otimes I_{\bar{V}} \otimes \boldsymbol{\beta}_1 \otimes I_{R_2}, D_k \otimes I_{\bar{V}} \otimes \boldsymbol{\beta}_3 \otimes I_{R_1 R_2}\},$$

$$k \geq 2,$$

$$Q_{1,0} = \begin{pmatrix} I_a \otimes \boldsymbol{S}_0^{(1)} & O & O & O \\ I_a \otimes \boldsymbol{S}_0^{(2)} & O & O & O \\ O & I_a \otimes \boldsymbol{S}_0^{(2)} \otimes I_{R_1} & O & O \\ O & O & I_a \otimes \boldsymbol{S}_0^{(1)} \otimes I_{R_2} & O \\ O & O & O & I_a \otimes \boldsymbol{S}_0^{(3)} \otimes I_{R_1} \otimes I_{R_2} \end{pmatrix},$$

$$Q_{1,1} = \left(Q_{1,1}^{(1)}\ Q_{1,1}^{(2)} \right),$$

$$Q_{1,1}^{(1)} = \begin{pmatrix} D_0 \oplus H_0 \oplus S_1 & O & I_{\bar{W}} \otimes H_1 \otimes \mathbf{e}_{M_1} \otimes \boldsymbol{\beta}_2 \otimes \boldsymbol{\tau}_1 \\ O & D_0 \oplus H_0 \oplus S_2 & I_{\bar{W}} \otimes H_1 \otimes I_{M_2} \otimes \boldsymbol{\tau}_1 \\ O & I_a \otimes I_{M_2} \otimes \boldsymbol{T}_0^{(1)} & D_0 \oplus (H_0 + H_1) \oplus S_2 \oplus T_1 \\ I_a \otimes I_{M_1} \otimes \boldsymbol{T}_0^{(2)} & O & O \\ O & O & I_a \otimes \boldsymbol{\beta}_2 \otimes \mathbf{e}_{M_3} \otimes I_{R_1} \otimes \boldsymbol{T}_0^{(2)} \end{pmatrix},$$

$$Q_{1,1}^{(2)} = \begin{pmatrix} I_{\bar{W}} \otimes H_2 \otimes I_{M_1} \otimes \boldsymbol{\tau}_2 & O \\ I_{\bar{W}} \otimes H_2 \otimes \mathbf{e}_{M_2} \otimes \boldsymbol{\beta}_1 \otimes \boldsymbol{\tau}_2 & O \\ O & I_{\bar{W}} \otimes H_2 \otimes \mathbf{e}_{M_2} \otimes \boldsymbol{\beta}_3 \otimes I_{R_1} \otimes \boldsymbol{\tau}_2 \\ D_0 \oplus (H_0 + H_2) \oplus S_1 \oplus T_2 & I_{\bar{W}} \otimes H_1 \otimes \mathbf{e}_{M_1} \otimes \boldsymbol{\beta}_3 \otimes \boldsymbol{\tau}_1 \otimes I_{R_2} \\ I_a \otimes \boldsymbol{\beta}_1 \otimes \mathbf{e}_{M_3} \otimes \boldsymbol{T}_0^{(1)} \otimes I_{R_2} & D_0 \oplus H \oplus S_3 \oplus T_1 \oplus T_2 \end{pmatrix},$$

$$Q_{1,k} =$$

$$\begin{pmatrix} D_{k-1} \otimes I_{\bar{V}} \otimes I_{M_1} \otimes \boldsymbol{\beta}_2 & O & O & O \\ D_{k-1} \otimes I_{\bar{V}} \otimes \boldsymbol{\beta}_1 \otimes I_{M_2} & O & O & O \\ O & D_{k-1} \otimes I_{\bar{V}} \otimes I_{M_2 R_1} & O & O \\ O & O & D_{k-1} \otimes I_{\bar{V} M_1 R_2} & O \\ O & O & O & D_{k-1} \otimes I_{\bar{V}} \otimes I_{M_3 R_1 R_2} \end{pmatrix},$$

$$k \geq 2,$$

$$Q_{2,1} = \begin{pmatrix} I_a \otimes I_{M_1} \otimes \mathbf{S}_0^{(2)} & O \\ O & O \end{pmatrix} + \Big(O \mid diag\{I_a \otimes \mathbf{S}_0^{(1)} \otimes I_{M_2},$$

$$I_a \otimes \mathbf{S}_0^{(2)} \boldsymbol{\beta}_2 \otimes I_{R_1},\ I_a \otimes \mathbf{S}_0^{(1)} \boldsymbol{\beta}_1 \otimes I_{R_2},\ I_a \otimes \mathbf{S}_0^{(3)} \boldsymbol{\beta}_3 \otimes I_{R_1 R_2}\} \Big),$$

$$Q_0 =$$

$$\begin{pmatrix} I_a \otimes [\mathbf{S}_0^{(1)} \boldsymbol{\beta}_1 \oplus \mathbf{S}_0^{(2)} \boldsymbol{\beta}_2] & O & O & O \\ O & I_a \otimes \mathbf{S}_0^{(2)} \boldsymbol{\beta}_2 \otimes I_{R_1} & O & O \\ O & O & I_a \otimes \mathbf{S}_0^{(1)} \boldsymbol{\beta}_1 \otimes I_{R_2} & O \\ O & O & O & I_a \otimes \mathbf{S}_0^{(3)} \boldsymbol{\beta}_3 \otimes I_{R_1 R_2} \end{pmatrix},$$

$$Q_1 = \begin{pmatrix} Q_1^{(1,1)} & Q_1^{(1,2)} \\ Q_1^{(2,1)} & Q_1^{(2,2)} \end{pmatrix},$$

$$Q_1^{(1,1)} = \begin{pmatrix} D_0 \oplus H_0 \oplus S_1 \oplus S_2 & I_{\bar{W}} \otimes H_1 \otimes \mathbf{e}_{M_1} \otimes I_{M_2} \otimes \boldsymbol{\tau}_1 \\ I_a \otimes \boldsymbol{\beta}_1 \otimes I_{M_2} \otimes \boldsymbol{T}_0^{(1)} & D_0 \oplus (H_0 + H_1) \oplus S_2 \oplus T_1 \end{pmatrix},$$

$$Q_1^{(1,2)} = \begin{pmatrix} I_{\bar{W}} \otimes H_2 \otimes I_{M_1} \otimes \mathbf{e}_{M_2} \otimes \boldsymbol{\tau}_2 & O \\ O & I_{\bar{W}} \otimes H_2 \otimes \mathbf{e}_{M_2} \otimes \boldsymbol{\beta}_3 \otimes I_{R_1} \otimes \boldsymbol{\tau}_2 \end{pmatrix},$$

$$Q_1^{(2,1)} = \begin{pmatrix} I_a \otimes I_{M_1} \otimes \boldsymbol{\beta}_2 \otimes \boldsymbol{T}_0^{(2)} & O \\ O & I_a \otimes \boldsymbol{\beta}_2 \otimes \mathbf{e}_{M_3} \otimes I_{R_1} \otimes \boldsymbol{T}_0^{(2)} \end{pmatrix},$$

$$Q_1^{(2,2)} = \begin{pmatrix} D_0 \oplus (H_0 + H_2) \oplus S_1 \oplus T_2 & I_{\bar{W}} \otimes H_1 \otimes \mathbf{e}_{M_1} \otimes \boldsymbol{\beta}_3 \otimes I_{R_2} \otimes \boldsymbol{\tau}_1 \\ I_a \otimes \boldsymbol{\beta}_1 \otimes \mathbf{e}_{M_3} \otimes \boldsymbol{T}_0^{(1)} \otimes I_{R_2} & D_0 \oplus H \oplus S_3 \oplus T_1 \oplus T_2 \end{pmatrix},$$

$$Q_{k+1} = \begin{pmatrix} D_k \otimes I_{\bar{V}M_1M_2} & O & O & O \\ O & D_k \otimes I_{\bar{V}M_2R_1} & O & O \\ O & O & D_k \otimes I_{\bar{V}M_1R_2} & O \\ O & O & O & D_k \otimes I_{\bar{V}M_3R_1R_2} \end{pmatrix},$$

$$k \geq 1,$$

where $H = H_0 + H_1 + H_2$, $\otimes$, $\oplus$ *are the symbols of the Kronecker product and sum of matrices,* $\bar{W} = W + 1$, $\bar{V} = V + 1$, $a = \bar{W}\bar{V}$.

The proof of the lemma is by means of calculation the of probabilities of the transitions of the components of the Markov chain ξ_t during a time interval with infinitesimal length.

Corollary 5.12 *The Markov chain* $\xi_t, t \geq 0$, *belongs to the class of continuous-time quasi-Toeplitz Markov chains, see [70].*

In the following it will be useful to have expressions for the generating functions $Q^{(n)}(z) = \sum_{k=2}^{\infty} Q_{n,k}z^k$, $n = 0, 1$ *and* $Q(z) = \sum_{k=0}^{\infty} Q_k z^k$, $|z| \leq 1$.

Corollary 5.13 *The matrix generating functions* $Q(z)$, $Q^{(n)}(z)$, $n = 0, 1$ *are of the form*

$$Q^{(0)}(z) = diag\{(\bar{D}(z) - D_1 z) \otimes I_{\bar{V}} \otimes \boldsymbol{\beta}^{(1)} \otimes \boldsymbol{\beta}^{(2)},\ (\bar{D}(z) - D_1 z) \otimes I_{\bar{V}} \otimes \boldsymbol{\beta}^{(2)} \otimes I_{R_1},$$

$$(\bar{D}(z) - D_1 z) \otimes I_{\bar{V}} \otimes \boldsymbol{\beta}^{(1)} \otimes I_{R_2},\ (\bar{D}(z) - D_1 z) \otimes I_{\bar{V}} \otimes \boldsymbol{\beta}_3 \otimes I_{R_1R_2}\}, \tag{5.57}$$

$$Q^{(1)}(z) = \tag{5.58}$$

$$z \begin{pmatrix} \bar{D}(z) \otimes I_{\bar{V}M_1} \otimes \boldsymbol{\beta}^{(2)} & O & O & O \\ \bar{D}(z) \otimes I_{\bar{V}} \otimes \boldsymbol{\beta}^{(1)} \otimes I_{M_2} & O & O & O \\ O & \bar{D}(z) \otimes I_{\bar{V}M_2R_1} & O & O \\ O & O & \bar{D}(z) \otimes I_{\bar{V}M_1R_2} & O \\ O & O & O & \bar{D}(z) \otimes I_{\bar{V}M_3R_1R_2} \end{pmatrix},$$

$$Q(z) = Q_0 + \mathcal{Q}(z) + z diag\{D(z) \otimes I_{\bar{V}M_1M_2},\ D(z) \otimes I_{\bar{V}M_2M_3R_1},$$

$$D(z) \otimes I_{\bar{V}M_1M_3R_2},\ D(z) \otimes I_{\bar{V}M_3R_1R_2}\}, \tag{5.59}$$

where

$$\bar{D}(z) = D(z) - D_0,$$

and the matrix $\mathcal{Q}$ has the form

$$\mathcal{Q} = \begin{pmatrix} \mathcal{Q}^{(1,1)} & \mathcal{Q}^{(1,2)} \\ \mathcal{Q}^{(2,1)} & \mathcal{Q}^{(2,2)} \end{pmatrix},$$

where

$$\mathcal{Q}^{(1,1)} = \begin{pmatrix} I_{\bar{W}} \otimes H_0 \oplus S^{(1)} \oplus S^{(2)} & I_{\bar{W}} \otimes H_1 \otimes \mathbf{e}_{M_1} \otimes I_{M_2} \otimes \boldsymbol{\tau}_1 \\ I_a \otimes \boldsymbol{\beta}^{(1)} \otimes I_{M_2} \otimes \boldsymbol{T}_0^{(1)} & I_{\bar{W}} \otimes (H_0 + H_1) \oplus S^{(2)} \oplus T_1 \end{pmatrix},$$

$$\mathcal{Q}^{(1,2)} = \begin{pmatrix} I_{\bar{W}} \otimes H_2 \otimes I_{M_1} \otimes \mathbf{e}_{M_2} \otimes \boldsymbol{\tau}_2 & O \\ O & I_{\bar{W}} \otimes H_2 \otimes \mathbf{e}_{M_2} \otimes \boldsymbol{\beta}^{(3)} \otimes I_{R_1} \otimes \boldsymbol{\tau}_2 \end{pmatrix},$$

$$\mathcal{Q}^{(2,1)} = \begin{pmatrix} I_a \otimes I_{M_1} \otimes \boldsymbol{\beta}^{(2)} \otimes \boldsymbol{T}_0^{(2)} & O \\ O & I_a \otimes \boldsymbol{\beta}^{(2)} \otimes \mathbf{e}_{M_3} \otimes I_{R_1} \otimes \boldsymbol{T}_0^{(2)} \end{pmatrix},$$

$$\mathcal{Q}^{(2,2)} = \begin{pmatrix} I_{\bar{W}} \otimes (H_0 + H_2) \oplus S^{(1)} \oplus T_2 & I_{\bar{W}} \otimes H_1 \otimes \mathbf{e}_{M_1} \otimes \boldsymbol{\beta}^{(3)} \otimes I_{R_2} \otimes \boldsymbol{\tau}_1 \\ I_a \otimes \boldsymbol{\beta}^{(1)} \otimes \mathbf{e}_{M_3} \otimes \boldsymbol{T}_0^{(1)} \otimes I_{R_2} & I_{\bar{W}} \otimes H \oplus S^{(3)} \oplus T_1 \oplus T_2 \end{pmatrix}.$$

5.4.3 Stationary Distribution

Theorem 5.8 *A necessary and sufficient condition for the existence of the stationary distribution of the Markov chain ξ_t, $t \geq 0$ is the fulfillment of the inequality*

$$\lambda < \boldsymbol{\pi}_0(\mathbf{S}_0^{(1)} \oplus \mathbf{S}_0^{(2)})\mathbf{e} + \boldsymbol{\pi}_1\mathbf{S}_0^{(2)} + \boldsymbol{\pi}_2\mathbf{S}_0^{(1)} + \boldsymbol{\pi}_3\mathbf{S}_0^{(3)}, \tag{5.60}$$

where

$$\boldsymbol{\pi}_0 = \mathbf{x}_0(\mathbf{e}_{\bar{V}} \otimes I_{M_1 M_2}),\ \boldsymbol{\pi}_1 = \mathbf{x}_1(\mathbf{e}_{\bar{V}} \otimes I_{M_2} \otimes \mathbf{e}_{R_1}),$$

$$\boldsymbol{\pi}_2 = \mathbf{x}_2(\mathbf{e}_{\bar{V}} \otimes I_{M_1} \otimes \mathbf{e}_{R_2}),\ \boldsymbol{\pi}_3 = \mathbf{x}_3(\mathbf{e}_{\bar{V}} \otimes I_{M_3} \otimes \mathbf{e}_{R_1 R_2}),$$

and the vector $\mathbf{x} = (\mathbf{x}_0,\ \mathbf{x}_1,\ \mathbf{x}_2,\ \mathbf{x}_3)$ is the unique solution of the system

$$\mathbf{x}\Gamma = 0,\ \ \mathbf{x}\mathbf{e} = 1. \tag{5.61}$$

Here

$$\Gamma = \begin{pmatrix} \Gamma^{(1,1)} & \Gamma^{(1,2)} \\ \Gamma^{(2,1)} & \Gamma^{(2,2)} \end{pmatrix},$$

where

$$\Gamma^{(1,1)} = \begin{pmatrix} I_{\bar{V}} \otimes (S_0^{(1)}\boldsymbol{\beta}^{(1)} \oplus S_0^{(2)}\boldsymbol{\beta}^{(2)}) & H_1 \otimes \mathbf{e}_{M_1} \otimes I_{M_2} \otimes \boldsymbol{\tau}_1 \\ I_{\bar{V}} \otimes \boldsymbol{\beta}^{(1)} \otimes I_{M_2} \otimes \boldsymbol{T}_0^{(1)} & +I_{\bar{V}} \otimes \mathbf{S}_0^{(2)}\boldsymbol{\beta}^{(2)} \otimes I_{R_1} \end{pmatrix}$$

$$+diag\{H_0 \oplus S^{(1)} \oplus S^{(2)},\ (H_0 + H_1) \oplus S^{(2)} \oplus T_1\},$$

$$\Gamma^{(1,2)} = \begin{pmatrix} H_2 \otimes I_{M_1} \otimes \mathbf{e}_{M_2} \otimes \boldsymbol{\tau}_2 & O \\ O & H_2 \otimes \mathbf{e}_{M_2} \otimes \boldsymbol{\beta}^{(3)} \otimes I_{R_1} \otimes \boldsymbol{\tau}_2 \end{pmatrix},$$

$$\Gamma^{(2,1)} = \begin{pmatrix} I_{\bar{V}} \otimes I_{M_1} \otimes \boldsymbol{\beta}^{(2)} \otimes \boldsymbol{T}_0^{(2)} & O \\ O & I_{\bar{V}} \otimes \boldsymbol{\beta}^{(2)} \otimes \mathbf{e}_{M_3} \otimes I_{R_1} \otimes \boldsymbol{T}_0^{(2)} \end{pmatrix},$$

$$\Gamma^{(2,2)} = \begin{pmatrix} I_{\bar{V}} \otimes \mathbf{S}_0^{(1)}\boldsymbol{\beta}^{(1)} \otimes I_{R_2} & H_1 \otimes \mathbf{e}_{M_1} \otimes \boldsymbol{\beta}^{(3)} \otimes I_{R_2} \otimes \boldsymbol{\tau}_1 \\ I_{\bar{V}} \otimes \boldsymbol{\beta}^{(1)} \otimes \mathbf{e}_{M_3} \otimes \boldsymbol{T}_0^{(1)} \otimes I_{R_2} & I_{\bar{V}} \otimes \mathbf{S}_0^{(3)}\boldsymbol{\beta}_3 \otimes I_{R_1R_2} \end{pmatrix}$$

$$+diag\{(H_0 + H_2) \oplus S^{(1)} \oplus T_2,\ H \oplus S^{(3)} \oplus T_1 \oplus T_2\}.$$

Proof The proof is analogous to the proof of Theorem 5.1. In this case we represent the vector $\mathbf{y}$, present in the relations (5.5) and (5.6), in the form

$$\mathbf{y} = (\boldsymbol{\theta} \otimes \mathbf{x}_0, \boldsymbol{\theta} \otimes \mathbf{x}_1, \boldsymbol{\theta} \otimes \mathbf{x}_2, \boldsymbol{\theta} \otimes \mathbf{x}_3), \tag{5.62}$$

where $\mathbf{x} = (\mathbf{x}_0,\ \mathbf{x}_1,\ \mathbf{x}_2,\ \mathbf{x}_3)$ is a stochastic vector.

Substituting $\mathbf{y}$ in the form (5.62) into system (5.6) and taking into account that $\boldsymbol{\theta} \sum\limits_{k=0}^{\infty} D_k = \mathbf{0}$, we easily reduce system (5.6) to the form (5.61).

Further, substituting into inequality (5.5) the vector $\mathbf{y}$ in the form (5.62) and the expression for $Q'(1)$, calculated using formula (5.59), and taking into account that $\boldsymbol{\theta} D'(1)\mathbf{e} = \lambda$, we reduce this inequality to the form

$$\lambda + \mathbf{x}\mathcal{Q}^{-}\mathbf{e} < 0, \tag{5.63}$$

where the matrix $\mathcal{Q}^{-}$ is obtained from the matrix $\mathcal{Q}$ by removing the expression $I_{\bar{W}}\otimes$ and replacing the multiplier I_a by the multiplier $I_{\bar{V}}$.

Using the relations $H\mathbf{e} = (H_0+H_1+H_2)\mathbf{e} = \mathbf{0},\ S_n\mathbf{e}+\mathbf{S}_0^{(n)} = \mathbf{0},\ T_n\mathbf{e}+\boldsymbol{T}_0^{(n)} = \mathbf{0}$, $n = 1, 2$, we reduce inequality (5.63) to the form

$$\lambda < \mathbf{x}_0(\mathbf{e}_{\bar{V}} \otimes I_{M_1M_2})(\mathbf{S}_0^{(1)} \oplus \mathbf{S}_0^{(2)})\mathbf{e} + \mathbf{x}_1(\mathbf{e}_{\bar{V}} \otimes I_{M_2} \otimes \mathbf{e}_{R_1})\mathbf{S}_0^{(2)}$$

$$+\mathbf{x}_2(\mathbf{e}_{\bar{V}} \otimes I_{M_1} \otimes \mathbf{e}_{R_2})\mathbf{S}_0^{(1)} + \mathbf{x}_3(\mathbf{e}_{\bar{V}} \otimes I_{M_3} \otimes \mathbf{e}_{R_1R_2})\mathbf{S}_0^{(3)}.$$

Using the notation introduced in the statement of the theorem, this inequality takes the form (5.60). □

Remark 5.6 The intuitive explanation of stability condition (5.60) is as follows. Let the system be overloaded. Then the component $\boldsymbol{\pi}_0(m^{(1)}, m^{(2)})$ of the vector $\boldsymbol{\pi}_0$ is the probability that servers 1 and 2 are fault-free and serve customers in phases $m^{(1)}$ and $m^{(2)}$, respectively. The corresponding entry of the column vector $(\mathbf{S}_0^{(1)} \oplus \mathbf{S}_0^{(2)})\mathbf{e}$ is the total service rate of customers at servers 1 and 2 when these servers are in phases $m^{(1)}$ and $m^{(2)}$, respectively. Then the product $\boldsymbol{\pi}_0(\mathbf{S}_0^{(1)} \oplus \mathbf{S}_0^{(2)})\mathbf{e}$ represents the rate of the output flow in periods when customers are served by server 1 and server 2. The other summands of the sum on the right-hand side of inequality (5.60) are interpreted similarly: the second summand is the rate of the output flow from server 2 when server 1 is under repair, the third summand is the rate of the output flow from server 1 when server 2 is under repair, and the fourth summand is the rate of the output flow from server 3 when servers 1 and 2 are under repair. Then the right-hand side of inequality (5.60) is the total rate of the output flow under overload conditions. It is obvious that in steady state the rate of the input flow λ must be less than the rate of the output flow.

Corollary 5.14 *In the case of a stationary Poisson flow of breakdowns, and exponential distributions of the service and repair times, the stationary distribution of the system exists if and only if the inequality*

$$\lambda < \pi_0(\mu_1 + \mu_2) + \pi_1\mu_2 + \pi_2\mu_1 + \pi_3\mu_3 \tag{5.64}$$

holds.

Here the vector $\boldsymbol{\pi} = (\pi_0, \pi_1, \pi_2, \pi_3)$ is the unique solution of the system of linear algebraic equations

$$\boldsymbol{\pi} \begin{pmatrix} -(h_1+h_2) & h_1 & h_2 & 0 \\ \tau_1 & -(h_2+\tau_1) & 0 & h_2 \\ \tau_2 & 0 & -(h_1+\tau_2) & h_1 \\ 0 & \tau_2 & \tau_1 & -(\tau_1+\tau_2) \end{pmatrix} = \mathbf{0},$$

$$\boldsymbol{\pi}\mathbf{e} = 1.$$

Remark 5.7 The physical meaning of stability condition (5.64) becomes immediately understandable if we observe that, under overload conditions, π_0 is the probability that server 1 and server 2 are serving customers and π_j is the probability that only the jth server serves a customer, $j = 1, 2, 3$. Then the right-hand side of inequality (5.64) is the rate of output flow from the system under overload conditions. This rate must be larger than the input rate λ for the stable operation of the system.

In the following we assume that stability condition (5.60) is satisfied. Then there exist stationary probabilities of the system states, which we denote by

$$p_0^{(n)}(\nu,\eta)=\lim_{t\to\infty}P\{i_t=0,n_t=n,\nu_t=\nu,\eta_t=\eta\},\ n=\overline{0,3},\ \nu=\overline{0,W},\eta=\overline{0,V};$$

$$p_1^{(0_n)}\{(\nu,\eta,m^{(k)})\}=\lim_{t\to\infty}P\{i_t=i,n_t=0_n,\nu_t=\nu,\eta_t=\eta,m_t^{(n)}=m^{(n)}\},$$

$$n=1,2,\ \nu=\overline{0,W},\ \eta=\overline{0,V},\ m^{(n)}=\overline{1,M^{(n)}};$$

$$p_1^{(1)}(\nu,\eta,m^{(2)})=\lim_{t\to\infty}P\{i_t=1,n_t=1,\ \nu_t=\nu,\eta_t=\eta,\ m_t^{(2)}=m^{(2)},\ r_t^{(1)}=r^{(1)}\},$$

$$\nu=\overline{0,W},\ \eta=\overline{0,V},\ m^{(2)}=\overline{1,M^{(2)}},r^{(1)}=\overline{1,R^{(1)}};$$

$$p_1^{(2)}(\nu,\eta,m^{(1)})=\lim_{t\to\infty}P\{i_t=1,n_t=2,\ \nu_t=\nu,\eta_t=\eta,\ m_t^{(1)}=m^{(1)},\ r_t^{(2)}=r^{(2)}\},$$

$$\nu=\overline{0,W},\ \eta=\overline{0,V},\ m^{(1)}=\overline{1,M^{(1)}},\ r^{(2)}=\overline{1,R^{(2)}};$$

$$p_1^{(3)}(\nu,\eta,m^{(3)})=\lim_{t\to\infty}P\{i_t=1,n_t=3,\ \nu_t=\nu,\eta_t=\eta,\ m_t^{(3)}=m^{(3)},\ r_t^{(1)}$$

$$=\overline{1,R^{(1)}},r_t^{(2)}=\overline{1,R^{(2)}}\},\ \nu=\overline{0,W},\ \eta=\overline{0,V},\ m^{(3)}=\overline{1,M^{(3)}},\ r^{(1)}=\overline{1,R^{(1)}},$$

$$r^{(2)}=\overline{1,R^{(2)}};$$

$$p_i^{(1)}(\nu,\eta,m^{(2)})=\lim_{t\to\infty}P\{i_t=i,n_t=1,\ \nu_t=\nu,\eta_t=\eta,\ m_t^{(2)}=m^{(2)},\ m_t^{(3)}=m^{(3)},$$

$$r_t^{(1)}=\overline{1,R^{(1)}}\},\ i\geq 2,\nu=\overline{0,W},\ \eta=\overline{0,V},\ m^{(2)}=\overline{1,M^{(2)}},\ m^{(3)}=\overline{1,M^{(3)}},$$

$$r^{(1)}=\overline{1,R^{(1)}};$$

$$p_i^{(2)}(\nu,\eta,m^{(1)})=\lim_{t\to\infty}P\{i_t=i,n_t=2,\ \nu_t=\nu,\eta_t=\eta,\ m_t^{(1)}=m^{(1)},\ m_t^{(3)}=m^{(3)},$$

$$r_t^{(2)}=\overline{1,R^{(2)}}\},\ i\geq 2,\ \nu=\overline{0,W},\ \eta=\overline{0,V},\ m^{(1)}=\overline{1,M^{(1)}},\ m^{(3)}=\overline{1,M^{(3)}},$$

$$r^{(2)}=\overline{1,R^{(2)}};$$

$$p_i^{(3)}(\nu,\eta,m^{(3)})=\lim_{t\to\infty}P\{i_t=i,n_t=3,\ \nu_t=\nu,\eta_t=\eta,\ m_t^{(3)}=m^{(3)},\ r_t^{(1)}$$

$$=\overline{1,R^{(1)}},r_t^{(2)}=\overline{1,R^{(2)}}\},\ i\geq 2,\ \nu=\overline{0,W},\ \eta=\overline{0,V},\ m^{(3)}=\overline{1,M^{(3)}},$$

$$r^{(1)}=\overline{1,R^{(1)}},r^{(2)}=\overline{1,R^{(2)}};$$

$$p_i^{(0)}(\nu, \eta, m^{(1)}, m^{(2)}) = \lim_{t\to\infty} P\{i_t = i, n_t = 0,\ \nu_t = \nu, \eta_t = \eta,$$

$$m_t^{(1)} = m^{(1)},\ m_t^{(2)} = m^{(2)}\}, i > 1,\ \nu = \overline{0, W},\ \eta = \overline{0, V},\ m^{(n)} = \overline{1, M^{(n)}}.$$

For fixed values i, n we arrange the probabilities $p_i^{(n)}$ in lexicographic order and form the vectors of these probabilities

$$\mathbf{p}_0^{(n)}, n = \overline{0, 3};\ \mathbf{p}_1^{(0_1)},\ \mathbf{p}_1^{(0_2)},\ \mathbf{p}_1^{(n)}, n = \overline{1, 3};\ \mathbf{p}_i^{(n)}, n = \overline{0, 3},\ i \geq 2.$$

Next, we form the vectors of stationary probabilities corresponding to the values of the counting component i as

$$\mathbf{p}_0 = (\mathbf{p}_0^{(0)}, \mathbf{p}_0^{(1)}, \mathbf{p}_0^{(2)}, \mathbf{p}_0^{(3)}),$$

$$\mathbf{p}_1 = (\mathbf{p}_1^{(0_1)}, \mathbf{p}_1^{(0_2)}, \mathbf{p}_1^{(1)}, \mathbf{p}_1^{(2)}, \mathbf{p}_1^{(3)}),$$

$$\mathbf{p}_i = (\mathbf{p}_i^{(0)}, \mathbf{p}_i^{(1)}, \mathbf{p}_i^{(2)}, \mathbf{p}_i^{(3)}),\ i \geq 2.$$

The vectors of stationary probabilities $\mathbf{p}_0, \mathbf{p}_1, \mathbf{p}_2, \ldots$ are calculated by an algorithm that is developed, similarly to Algorithm 5.1 described in Sect. 5.1.3, based on the censored Markov chain technique (see [70]). The difference between these algorithms is due to the fact that the spatial homogeneity of the chain considered in Sect. 5.1 is violated for $i = 0$, while the spatial homogeneity of the chain considered in this section is violated for $i = 0$ and for $i = 1$. This difference affects the structure of the generator and, accordingly, the algorithm for calculating the stationary distribution. In the case under consideration, the algorithm has the following form.

Algorithm 5.3

1. Calculate the matrix G as the minimal non-negative solution of the nonlinear matrix equation

$$\sum_{n=0}^{\infty} Q_n G^n = O.$$

2. Calculate the matrix G_1 from the equation

$$Q_{2,1} + \sum_{n=1}^{\infty} Q_n G^{n-1} G_1 = O,$$

whence it follows that

$$G_1 = -(\sum_{n=1}^{\infty} Q_n G^{n-1})^{-1} Q_{2,1}.$$

3. Calculate the matrix G_0 from the equation

$$Q_{1,0} + (Q_{1,1} + \sum_{n=2}^{\infty} Q_{1,n} G^{n-2} G_1) G_0 = O,$$

whence it follows that

$$G_0 = -(Q_{1,1} + \sum_{n=2}^{\infty} Q_{1,n} G^{n-2} G_1)^{-1} Q_{1,0}.$$

4. Calculate the matrices

$$\bar{Q}_{i,l} = \begin{cases} Q_{i,l} + \sum\limits_{n=l+1}^{\infty} Q_{i,n} G_{n-1} G_{n-2} \ldots G_l,\ i = 0, 1,\ l \geq i; \\ Q_{l-i+1} + \sum\limits_{n=l+1}^{\infty} Q_{n-i+1} G_{n-1} G_{n-2} \ldots G_l,\ i > 1,\ l \geq i, \end{cases}$$

where $G_i = G,\ i \geq 2$.

5. Calculate the matrices Φ_i using the recurrent formulae

$$\Phi_0 = I,\ \Phi_i = (\bar{Q}_{0,i} + \sum_{l=1}^{i-1} \Phi_l \bar{Q}_{l,i})(-\bar{Q}_{i,i})^{-1},\ i \geq 1.$$

5. Calculate the vector $\mathbf{p}_0$ as the unique solution of the system

$$\begin{cases} \mathbf{p}_0 \bar{Q}_{0,0} = \mathbf{0}, \\ \mathbf{p}_0 \mathbf{e} + \mathbf{p}_0 \Phi_1 \mathbf{e} + \mathbf{p}_0 \sum\limits_{i=2}^{\infty} \Phi_i \mathbf{e} = 1. \end{cases}$$

7. Calculate the vectors $\mathbf{p}_i,\ i \geq 1$, as follows:

$$\mathbf{p}_i = \mathbf{p}_0 \Phi_i,\ i \geq 1.$$

The proposed algorithm is numerically stable because all matrices involved in the algorithm are non-negative.

5.4.4 Vector Generating Function of the Stationary Distribution: System Performance Measures

Having calculated the vectors $\mathbf{p}_i$, $i \geq 0$, we can calculate various performance measures of the system. In this connection, the following result will be useful.

Lemma 5.7 *The vector generating function* $\boldsymbol{P}(z) = \sum_{i=2}^{\infty} \mathbf{p}_i z^i$, $|z| \leq 1$ *satisfies the following equation:*

$$\boldsymbol{P}(z)Q(z) = \mathbf{b}(z), \tag{5.65}$$

where

$$\mathbf{b}(z)$$

$$= z\{\mathbf{p}_2[Q_0 - Q_{2,1}] + \mathbf{p}_1[Q_{1,0} + Q_{1,1} - Q^{(1)}(z)] + \mathbf{p}_0[Q_{0,0} + Q_{0,1} - Q^{(0)}(z)]\}. \tag{5.66}$$

In particular, formula (5.65) can be used to calculate the values of the function $\boldsymbol{P}(z)$ and its derivatives at the point $z = 1$ without computing infinite sums. This allows us to find the moments of the number of customers in the system and a number of other performance measures of the system. Note that it is impossible to directly calculate the value of $\boldsymbol{P}(z)$ and its derivatives at the point $z = 1$ from (5.65) since the matrix $Q(1)$ is singular. This difficulty can be overcome by using the recurrence formulas given below, in Corollary (5.14).

Denote by $f^{(m)}(z)$ the mth derivative of the function $f(z)$, $m \geq 1$. Let $f^{(0)}(z) = f(z)$.

Corollary 5.15 *The mth, $m \geq 0$ derivatives of the vector generating function* $\mathbf{P}(z)$ *at the point $z = 1$ (so-called vector factorial moments) are recursively calculated as a solution of the system of linear algebraic equations*

$$\begin{cases} \boldsymbol{P}^{(m)}(1)Q(1) = \mathbf{b}^{(m)}(1) - \sum\limits_{l=0}^{m-1} C_m^l \boldsymbol{P}^{(l)}(1)Q^{(m-l)}(1), \\ \boldsymbol{P}^{(m)}(1)Q'(1)\mathbf{e} = \frac{1}{m+1}[\mathbf{b}^{(m+1)}(1) - \sum\limits_{l=0}^{m-1} C_{m+1}^l \boldsymbol{P}^{(l)}(1)Q^{(m+1-l)}(1)]\mathbf{e}, \end{cases}$$

where the derivatives $\mathbf{b}^{(m)}(1)$ *are calculated using formulas (5.65) and (5.66), (5.57)–(5.59).*

The proof of the corollary is similar to the proof of Corollary 5.4.

Now we are able to calculate a number of stationary performance measures of the system. Some of them are listed below:

1. Throughput of the system

$$\varrho = \boldsymbol{\pi}_0(\mathbf{S}_0^{(1)} \oplus \mathbf{S}_0^{(2)})\mathbf{e} + \boldsymbol{\pi}_1\mathbf{S}_0^{(2)} + \boldsymbol{\pi}_2\mathbf{S}_0^{(1)} + \boldsymbol{\pi}_3\mathbf{S}_0^{(3)}.$$

Throughput of the system in the case of exponentially distributed service and repair times

$$\varrho = \pi_0(\mu_1 + \mu_2) + \pi_1\mu_2 + \pi_2\mu_1 + \pi_0\mu_3.$$

2. Mean number of customers in the system $L = [\mathbf{p}_1 + \boldsymbol{P}'(1)]\mathbf{e}$.
3. Variance of the number of customers in the system $V = \boldsymbol{P}^{(2)}(1)\mathbf{e} + L - L^2$.
4. Probability that there are i customers in the system $p_i = \mathbf{p}_i\mathbf{e}$.
5. Probability that there are i customer in the system and both main servers are fault-free

$$P_0^{(0)} = \mathbf{p}_0\begin{pmatrix} \mathbf{e}_a \\ \mathbf{0}^T_{a(R_1+R_2+R_1R_2)} \end{pmatrix}, \quad P_1^{(0)} = \mathbf{p}_1\begin{pmatrix} \mathbf{e}_{a(M_1+M_2)} \\ \mathbf{0}^T_{a(M_2R_1+M_1R_2+M_3R_1R_2)} \end{pmatrix},$$

$$P_i^{(0)} = \mathbf{p}_i\begin{pmatrix} \mathbf{e}_{aM_1M_2} \\ \mathbf{0}^T_{a(M_2R_1+M_1R_2+M_3R_1R_2)} \end{pmatrix}, \ i \geq 2.$$

6. Probability that there are i customers in the system and server 1 (server 2) is under repair

$$P_0^{(1)} = \mathbf{p}_0\begin{pmatrix} \mathbf{0}^T_a \\ \mathbf{e}_{aR_1} \\ \mathbf{0}^T_{aR_2(1+R_1)} \end{pmatrix}, \quad P_0^{(2)} = \mathbf{p}_0\begin{pmatrix} \mathbf{0}^T_{a(1+R_1)} \\ \mathbf{e}_{aR_2} \\ \mathbf{0}^T_{aR_1R_2} \end{pmatrix},$$

$$P_1^{(1)} = \mathbf{p}_1\begin{pmatrix} \mathbf{0}^T_{a(M_1+M_2)} \\ \mathbf{e}_{aM_2R_1} \\ \mathbf{0}^T_{aR_2(M_1+M_3R_1)} \end{pmatrix}, \quad P_1^{(2)} = \mathbf{p}_1\begin{pmatrix} \mathbf{0}^T_{a(M_1+M_1+M_2R_1)} \\ \mathbf{e}_{aM_1R_2} \\ \mathbf{0}^T_{aM_3R_1R_2} \end{pmatrix},$$

$$P_i^{(1)} = \mathbf{p}_i\begin{pmatrix} \mathbf{0}^T_{aM_1M_2} \\ \mathbf{e}_{aM_2R_1} \\ \mathbf{0}^T_{aR_2(M_1+M_3R_1)} \end{pmatrix}, \quad P_i^{(2)} = \mathbf{p}_0\begin{pmatrix} \mathbf{0}^T_{a(M_1M_2+M_2R_1)} \\ \mathbf{e}_{aM_1R_2} \\ \mathbf{0}^T_{aM_3R_1R_2} \end{pmatrix}, \ i \geq 2.$$

7. Probability that there are i customers in the system and both main servers are under repair

$$P_0^{(3)} = \mathbf{p}_0 \begin{pmatrix} \mathbf{0}^T_{a(1+R_1+R_2)} \\ \mathbf{e}_{aR_1R_2} \end{pmatrix}, \quad P_1^{(3)} = \mathbf{p}_1 \begin{pmatrix} \mathbf{0}^T_{a(M_1+M_2+M_2R_1+M_1R_2)} \\ \mathbf{e}_{aM_3R_1R_2} \end{pmatrix},$$

$$P_i^{(3)} = \mathbf{p}_i \begin{pmatrix} \mathbf{0}^T_{a(M_1M_2+M_2R_1+M_1R_2)} \\ \mathbf{e}_{aM_3R_1R_2} \end{pmatrix}, \ i \geq 2.$$

8. Probability that the servers are in the state n

$$P^{(n)} = \sum_{i=0}^{\infty} P_i^{(n)}, \ n = \overline{0, 3}.$$

9. Probability that an arriving size k batch finds i customers in the system and both servers fault-free

$$P_{0,k}^{(0)} = \lambda^{-1} \mathbf{p}_0 \begin{pmatrix} I_{\bar{W}} \otimes \mathbf{e}_{\bar{V}} \\ O_{a(R_1+R_2+R_1R_2)\times \bar{W}} \end{pmatrix} D_k,$$

$$P_{1,k}^{(0)} = \lambda^{-1} \mathbf{p}_1 \begin{pmatrix} I_{\bar{W}} \otimes \mathbf{e}_{\bar{V}(M_1+M_2)} \\ O_{a(M_2R_1+M_1R_2+M_3R_1R_2)\times \bar{W}} \end{pmatrix} D_k,$$

$$P_{i,k}^{(0)} = \lambda^{-1} \mathbf{p}_i \begin{pmatrix} I_{\bar{W}} \otimes \mathbf{e}_{\bar{V}M_1M_2} \\ O_{a(M_2R_1+M_1R_2+M_3R_1R_2)\times \bar{W}} \end{pmatrix} D_k, \ i \geq 2.$$

10. Probability that an arriving size k batch finds i customers in the system and server 1 (server 2) is under repair

$$P_{0,k}^{(1)} = \lambda^{-1} \mathbf{p}_0 \begin{pmatrix} O_{a\times \bar{W}} \\ I_{\bar{W}} \otimes \mathbf{e}_{\bar{V}R_1} \\ O_{a(R_2+R_1R_2)\times \bar{W}} \end{pmatrix} D_k,$$

$$P_{1,k}^{(1)} = \lambda^{-1} \mathbf{p}_1 \begin{pmatrix} O_{a(M_1+M_2)\times \bar{W}} \\ I_{\bar{W}} \otimes \mathbf{e}_{\bar{V}M_2R_1} \\ O_{a(M_1R_2+M_3R_1R_2)\times \bar{W}} \end{pmatrix} D_k,$$

$$P_{i,k}^{(1)} = \lambda^{-1} \mathbf{p}_i \begin{pmatrix} O_{aM_1M_2\times \bar{W}} \\ I_{\bar{W}} \otimes \mathbf{e}_{\bar{V}M_2R_1} \\ O_{a(M_1R_2+M_3R_1R_2)\times \bar{W}} \end{pmatrix} D_k, \ i \geq 2,$$

$$P_{0,k}^{(2)} = \lambda^{-1}\mathbf{p}_0 \begin{pmatrix} O_{a(1+R_1)\times\bar{W}} \\ I_{\bar{W}} \otimes \mathbf{e}_{\bar{V}R_2} \\ O_{aR_1R_2\times\bar{W}} \end{pmatrix} D_k,$$

$$P_{1,k}^{(2)} = \lambda^{-1}\mathbf{p}_1 \begin{pmatrix} O_{a(M_1+M_1+M_2R_1)\times\bar{W}} \\ I_{\bar{W}} \otimes \mathbf{e}_{\bar{V}M_1R_2} \\ O_{aM_3R_1R_2\times\bar{W}} \end{pmatrix} D_k,$$

$$P_{i,k}^{(2)} = \lambda^{-1}\mathbf{p}_0 \begin{pmatrix} O_{a(M_1M_2+M_2R_1)\times\bar{W}} \\ \mathbf{e}_{\bar{V}M_1R_2} \\ O_{aM_3R_1R_2\times\bar{W}} \end{pmatrix} D_k, \ i \geq 2.$$

11. Probability that an arriving size k batch finds i customers in the system and both main servers under repair

$$P_{0,k}^{(3)} = \lambda^{-1}\mathbf{p}_0 \begin{pmatrix} O_{a(1+R_1+R_2)\times\bar{W}} \\ I_{\bar{W}} \otimes \mathbf{e}_{\bar{V}R_1R_2} \end{pmatrix} D_k,$$

$$P_{1,k}^{(3)} = \lambda^{-1}\mathbf{p}_1 \begin{pmatrix} O_{a(M_1+M_2+M_2R_1+M_1R_2)\times\bar{W}} \\ I_{\bar{W}} \otimes \mathbf{e}_{\bar{V}M_3R_1R_2} \end{pmatrix} D_k,$$

$$P_{i,k}^{(3)} = \lambda^{-1}\mathbf{p}_i \begin{pmatrix} O_{a(M_1M_2+M_2R_1+M_1R_2)\times\bar{W}} \\ I_{\bar{W}} \otimes \mathbf{e}_{\bar{V}M_3R_1R_2} \end{pmatrix} D_k, \ i \geq 2.$$

5.5 Numerical Experiments

In this section, we present the results of five numerical experiments. The numerical examples are illustrative and consist of graphs of dependencies of the system performance measures on its parameters.

Experiment 5.6 In this experiment, we investigate the mean number of customers in the system, L, as a function of the arrival rate λ under different values of the rate of breakdowns, h.

Define the input data for this experiment.

$BMAP$ is defined by the matrices D_0, $D_k, k = \overline{1,3}$. The matrix D_0 has the form

$$D_0 = \begin{pmatrix} -1.349076 & 1.09082 \times 10^{-6} \\ 1.09082 \times 10^{-6} & -0.043891 \end{pmatrix}.$$

The matrices D_k are calculated as $D_k = Dq^{k-1}(1-q)/(1-q^3), k = \overline{1,3}, q = 0.8$, where the matrix D is defined as

$$D = \begin{pmatrix} 1.340137 & 0.008939 \\ 0.0244854 & 0.0194046 \end{pmatrix}.$$

This $BMAP$ has coefficient of correlation $c_{cor} = 0.41$ and squared coefficient of variation $c_{var}^2 = 9.5$.

In the course of the experiment, we will change the value of the arrival rate λ by normalizing the matrices $D_k, k = \overline{0,3}$. In this way, any desired value of λ can be obtained while $c_{cor} = 0.41$ and $c_{var}^2 = 9.5$ do not change.

The $MMAP$ of breakdowns is defined by the matrices

$$H_0 = \begin{pmatrix} -8.110725 & 0 \\ 0 & -0.26325 \end{pmatrix},$$

$$H_1 = \frac{1}{3}\begin{pmatrix} 8.0568 & 0.053925 \\ 0.146625 & 0.116625 \end{pmatrix}, \; H_2 = \frac{2}{3}\begin{pmatrix} 8.0568 & 0.053925 \\ 0.146625 & 0.116625 \end{pmatrix},$$

For this $MMAP$ $c_{cor} = 0.2$, $c_{var}^2 = 12.3$.

The PH-type service time distributions on the three servers will be denoted by $PH_1^{(serv)}, PH_2^{(serv)}, PH_3^{(serv)}$. They are Erlang of order 2 with rates $\mu_1 = 10$, $\mu_2 = 7.5$, $\mu_3 = 2$ and squared coefficient of variation $c_{var}^2 = 0.5$.

The PH-type repair time distribution on the two main servers will be denoted by $PH_1^{(repair)}$ and $PH_2^{(repair)}$. We assume that these distributions coincide. They are hyperexponential of order 2 and characterized by the following vectors and matrices

$$\boldsymbol{\tau}^{(1)} = \boldsymbol{\tau}^{(2)} = (0.05,\ 0.95),\ T^{(1)} = T^{(2)} = \begin{pmatrix} -0.003101 & 0 \\ 0 & -0.245 \end{pmatrix}.$$

The repair rate is equal to $\tau = 0.05$, and the squared coefficient of variation is $c_{var}^2 = 25$.

Figure 5.6 depicts the dependence of the mean number of customers in the system, L, on the input rate λ for different rates of breakdowns.

As we should expect, with an increase in the arrival and breakdown rates, the value of L increases. We also see that the rate of increase grows with increasing system load. The numbers corresponding to the points on curves in Fig. 5.6 and the corresponding system load are given in Table 5.1.

Experiment 5.7 In this experiment, we investigate the mean number of customers in the system, L, and variance, V, as functions of the arrival rate λ under $BMAP$s with different coefficients of correlation.

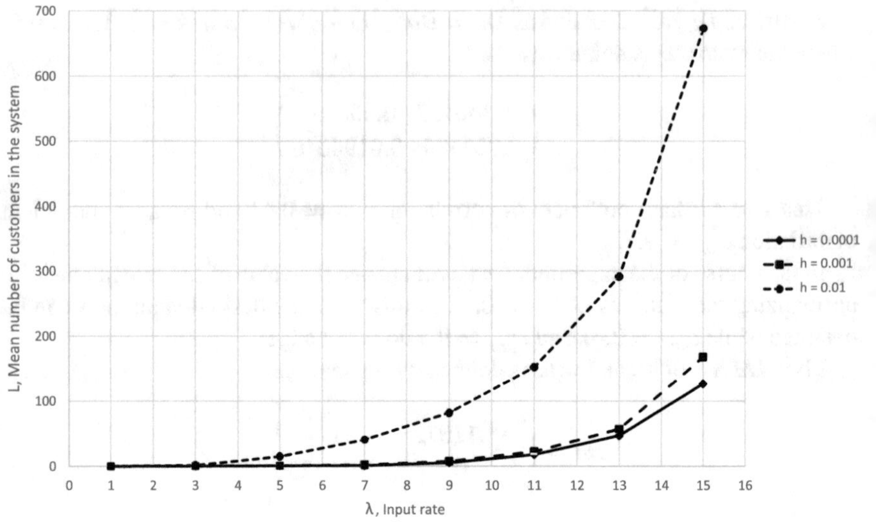

Fig. 5.6 Mean number of customers in the system, L, as a function of the arrival rate λ for different rates of breakdowns ($h = 0.0001$, $h = 0.001$, $h = 0.01$)

Table 5.1 Mean number of customers in the system, L, and the system load ρ as a function of the arrival rate λ for different rates of breakdowns h

λ	1.0	3.0	5.0	7.0	9.0	11.0	13.0	15.0
$h = 0.0001$								
ρ	0.057	0.172	0.286	0.400	0.515	0.629	0.744	0.858
L	0.148	1.191	13.645	37.297	72.677	130.285	239.641	525.405
$h = 0.001$								
ρ	0.058	0.173	0.288	0.404	0.519	0.635	0.750	0.865
L	0.154	1.540	14.879	40.671	81.659	152.103	291.168	673.258
$h = 0.01$								
ρ	0.063	0.189	0.314	0.439	0.566	0.691	0.817	0.943
L	0.418	7.015	34.682	97.493	231.646	517.529	1267.291	5959.399

We consider three different $BMAP$s with the same fundamental rate and different coefficients of correlation.

These $BMAP$s will be named $BMAP1$, $BMAP2$, $BMAP3$.

$BMAP1$ is just a stationary Poisson process. In this case we put $D_0 = -\lambda$, $D_1 = \lambda$. It is well known that a stationary Poisson process has zero correlation, $c_{cor} = 0$.

$BMAP2$ is defined by the matrices D_0 and $D_k = Dq^{k-1}(1-q)/(1-q^3)$, $k = \overline{1,3}$, where $q = 0.8$,

$$D_0 = \begin{pmatrix} -5.34080 & 1.87977 \times 10^{-6} \\ 1.87977 \times 10^{-6} & -0.13888 \end{pmatrix},$$

$$D = \begin{pmatrix} 5.32140 \; 0.01939 \\ 0.10822 \; 0.03066 \end{pmatrix}.$$

This $BMAP2$ has coefficient of correlation $c_{cor} = 0.1$.

$BMAP3$ coincides with $BMAP$ from Experiment 5.6. For this $BMAP$ $c_{cor} = 0.41$.

We vary the fundamental rate λ of these $BMAP$s by multiplying the matrices D_k, $k = \overline{0,3}$, by some positive constants. Any desired value of λ can be obtained in this way while the coefficient of correlation is not changed.

The $MMAP$ of breakdowns is defined by the matrices H_0, H_1, H_2 from Experiment 5.6 normalized to get the breakdown rate $h = 0.001$.

The PH-type distributions of the service and repair times are defined as in Experiment 5.6.

Figures 5.7 and 5.8 show the behavior of the mean number of customers in the system and the variance of the number of customers as functions of the arrival rate λ for the $BMAP$s with different coefficients of correlation.

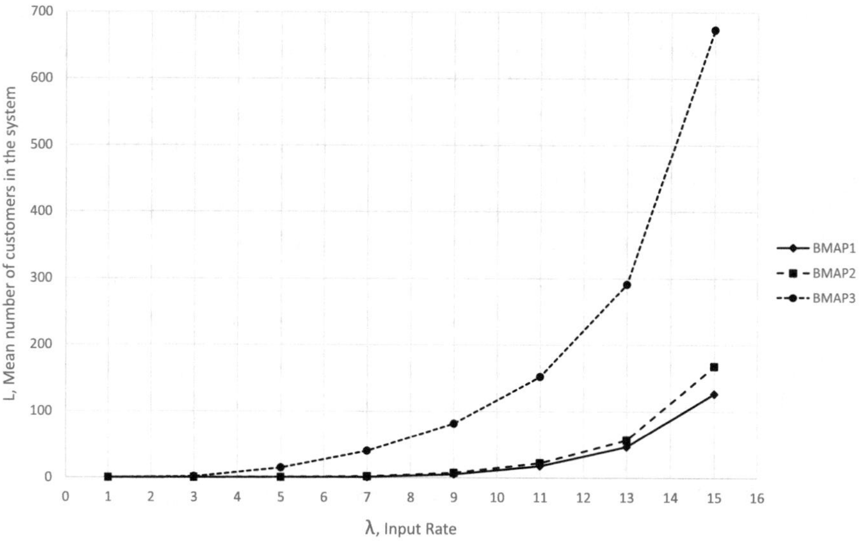

Fig. 5.7 Mean number of customers in the system, L, as a function of the arrival rate λ for $BMAP$s with different coefficients of correlation ($c_{cor} = 0,\ 0.1,\ 0.41$)

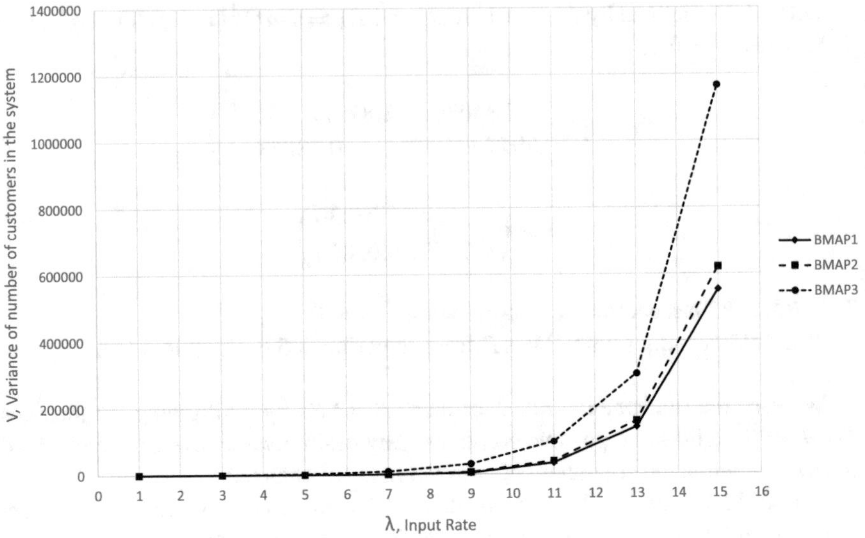

Fig. 5.8 Variance of the number of customers in the system, V, as a function of arrival rate λ for $BMAP$s with different coefficients of correlation ($c_{cor} = 0,\ 0.1,\ 0.41$)

As seen from Figs. 5.7 and 5.8, the values of L and V as expected increase as λ increases. The rate of increase grows with increasing λ. A more interesting observation is that the values of L and V significantly depend on the coefficient of correlation in the input flow. Under the same value of the input rate λ, the mean number of customers and the variance of the number of customers in the system are greater for a greater coefficient of correlation. Besides, the difference in the values of L and V for different coefficients of correlation increases as λ increases. This means that the correlation in the input flow can significantly affect the performance measures of the system and should be taken into account when evaluating the performance. The numbers corresponding to the points on the curves in Figs. 5.7 and 5.8 and the corresponding system load are given in Table 5.2.

Experiment 5.8 In this experiment, we investigate the probabilities $P^{(n)}$, $n = \overline{0, 3}$ that the servers are in the state n as functions of the breakdown rate h.

Recall that the state $n = 0$ means that both servers are fault-free, $n = 1(2)$ means that only server 1 (server 2) is under repair and $n = 3$ means that both servers 1 and 2 are under repair.

In this experiment we will use the following input data.

The $BMAP$ is defined as follows. The matrix D_0 is of the form

$$D_0 = \begin{pmatrix} -8.110725 & 0 \\ 0 & -0.26325 \end{pmatrix}$$

Table 5.2 Mean number of customers in the system L, variance V, and system load ρ as a function of the arrival rate λ for $BMAP$s with different coefficients of correlation

λ	1.0	3.0	5.0	7.0	9.0	11.0	13.0	15.0
$BMAP1 : c_{cor} = 0$								
ρ	0.058	0.173	0.288	0.404	0.519	0.635	0.750	0.865
L	0.104	0.334	0.608	1.030	4.666	12.044	25.595	118.532
V	0.105	0.451	2.086	18.578	3466	22,734	80,912	400,576
$BMAP2 : c_{cor} = 0.1$								
ρ	0.058	0.173	0.288	0.404	0.519	0.635	0.750	0.865
L	0.129	0.436	0.851	1.884	5.292	15.343	34.806	175.56
V	0.249	1.138	4.579	115.53	4569	25,884	90,216	600,057
$BMAP3 : c_{cor} = 0.41$								
ρ	0.058	0.173	0.288	0.404	0.519	0.635	0.750	0.865
L	0.154	1.540	14.879	40.671	81.659	152.103	291.168	673.258
V	0.508	60.73	1583	7988	25,843	73,998	211,357	1,118,030

The maximum number of customers in a group is three, and the matrices D_k are computed as $D_k = Dq^{k-1}(1-q)/(1-q^3), k = \overline{1,3}$, where $q = 0.8$,

$$D = \begin{pmatrix} 8.0568 & 0.053925 \\ 0.146625 & 0.116625 \end{pmatrix}.$$

Then we normalize the matrix $D_k, k = \overline{0,3}$ to get an arrival rate $\lambda = 10$. For this $BMAP$ $c_{cor} = 0.2$, $c_{var}^2 = 12$.

The $MMAP$ of breakdowns and the PH-type distributions of the service and repair times are the same as in Experiment 5.6.

To illustrate the results of the experiment, we will use a decimal logarithmic scale on the X-axis. Figures 5.9, 5.10, 5.11, and 5.12 show graphs for $P^{(0)}$, $P^{(1)}$, $P^{(2)}$, $P^{(3)}$, respectively.

Figure 5.9 shows that for values of h close to zero the probability $P^{(0)}$ tends to unity. Starting at $h = 0.0005$, the probability $P^{(0)}$ begins to decrease, while the probabilities $P^{(n)}, n = \overline{1,3}$ begin to increase.

Experiment 5.9 In this experiment, we investigate the mean number of customers in the system, L, as a function of the breakdown rate h for different coefficients of variation of the repair time.

In the experiment, the $BMAP$ of arrivals, the $MMAP$ of breakdowns, and the PH-type distributions of service times are the same as in Experiment 5.6.

For the sake of simplicity, we will assume that the repair times on server 1 and server 2 have the same PH-type distribution. Consider three repair time distributions named $PH1$, $PH2$, $PH3$ with different coefficients of variation.

$PH1$ is an exponential distribution with parameter 0.05. For this distribution $\tau = 1, T = -0.05$, and $c_{var} = 1$.

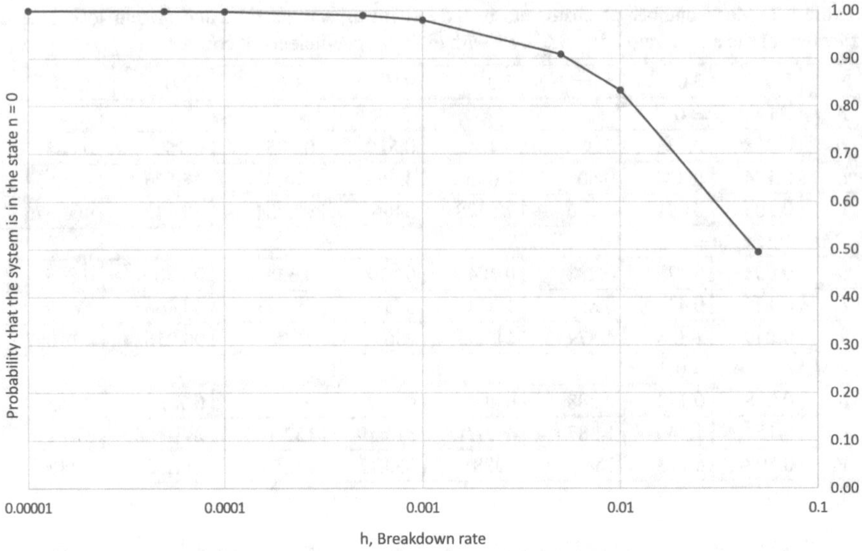

Fig. 5.9 Probability $P^{(0)}$ that both main servers are fault-free as a function of the breakdown rate h

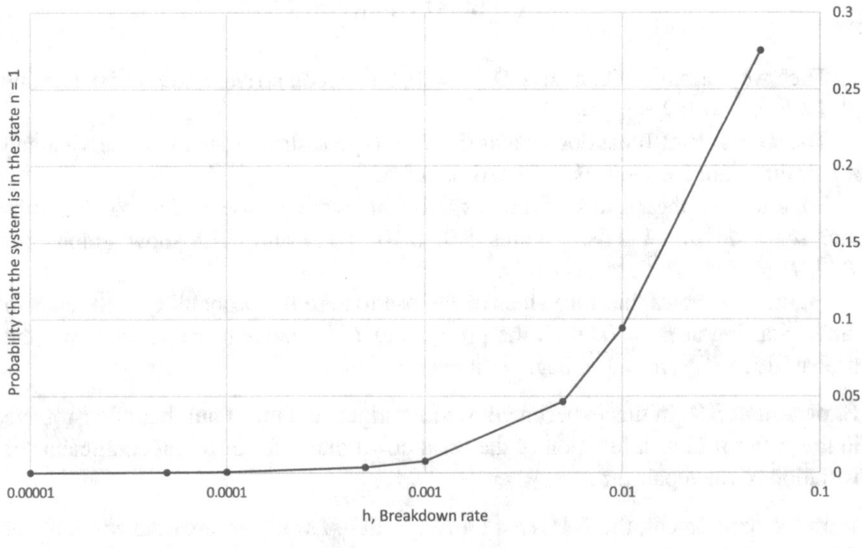

Fig. 5.10 Probability $P^{(1)}$ that only server 1 is under repair as a function of the breakdown rate h

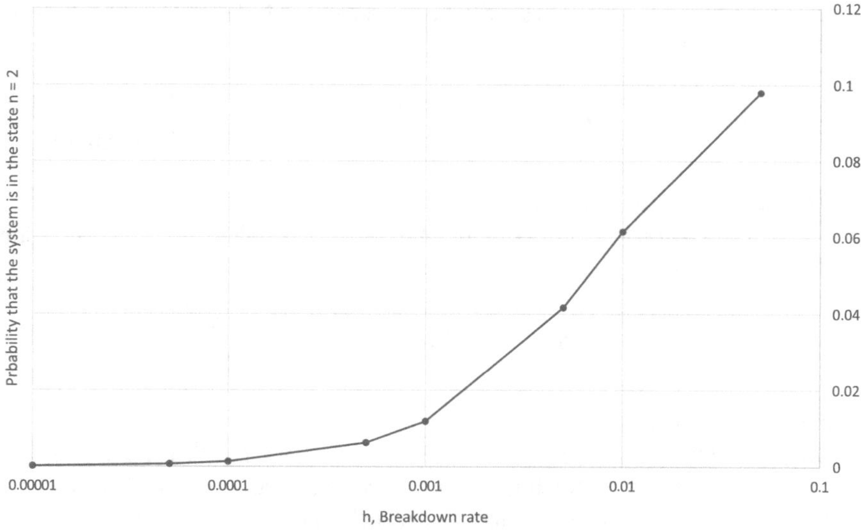

Fig. 5.11 Probability $P^{(2)}$ that only server 2 is under repair as a function of the breakdown rate h

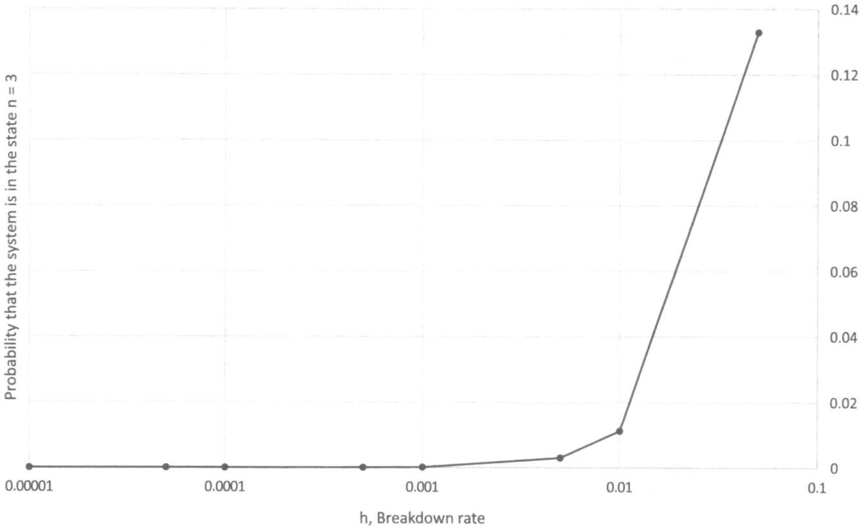

Fig. 5.12 Probability $P^{(3)}$ that both servers are under repair as a function of the breakdown rate h

$PH2$ is a hyper-exponential distribution of order 2. It is defined by the following vector and matrix:

$$\boldsymbol{\tau} = (0.05,\ 0.95),\ \ T = \begin{pmatrix} -0.003101265209245752 & 0 \\ 0 & -0.2450002015316405 \end{pmatrix}.$$

For this distribution $c_{var} = 5$.

$PH3$ is a hyperexponential distribution of order 2. It is defined by the following vector and matrix:

$$\boldsymbol{\tau} = (0.05,\ 0.95),\ \ T = \begin{pmatrix} -100 & 0 \\ 0 & -0.0002 \end{pmatrix}.$$

For this distribution $c_{var} = 9.9$.

We normalize the matrices T defining the $PH1$, $PH2$, $PH3$ distributions in such a way that the corresponding repair rates are equal to $h = 0.05$. Note that the coefficients of variation do not change under such a normalization. Thus, we obtain PH-type distributions with rate $h = 0.05$ and different coefficients of variation.

To illustrate the results of the experiment, we will use a decimal logarithmic scale on the X-axis.

The dependence of the mean number of customers in the system, L, on the breakdown rate h for various coefficients of variation $c_{var} = 1, 5, 9.9$ of the repair time is shown in Fig. 5.13.

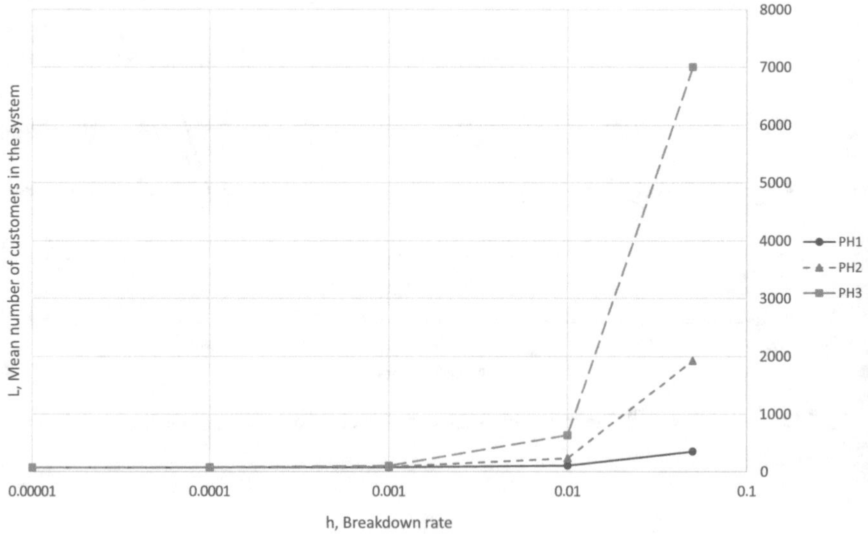

Fig. 5.13 The mean number of customers in the system, L, as a function of repair rate h for repair time with different coefficients of variation ($c_{var} = 1, 5, 9.9$)

Table 5.3 The mean number of customers in the system L, and the system load ρ as a function of repair rate h for repair time with different values of the coefficients of variation ($c_{var} = 1, 5, 9.9$)

h	0.00001	0.0001	0.001	0.01	0.05
$PH1 : c_{var=1}$					
ρ	0.514	0.515	0.519	0.561	0.707
L	71.815	72.085	74.807	104.515	348.21
$PH2 : c_{var=5}$					
ρ	0.514	0.515	0.519	0.566	0.754
L	71.873	72.677	81.659	231.646	1925.314
$PH3 : c_{var=9.9}$					
ρ	0.514	0.515	0.519	0.569	0.7909
L	71.994	73.972	101.003	631.539	7000.653

Figure 5.13 as well as Table 5.3 show that the variation of the repair time significantly affects the mean number of customers in the system, especially for $h \geq 10^{-3}$.

Experiment 5.10 In this experiment, we investigate the throughput of the system ϱ as a function of the breakdown rate h for different repair rate τ.

In the experiment, we will use the same input data as in Experiment 5.6. We will also assume that the repair rates on server 1 and server 2 coincide. We denote their total intensity by τ. The values of h are plotted along the X-axis in accordance with the decimal logarithmic scale.

Figure 5.14 depicts the dependence of the throughput ϱ on the breakdown rate h for different values of the repair rate τ.

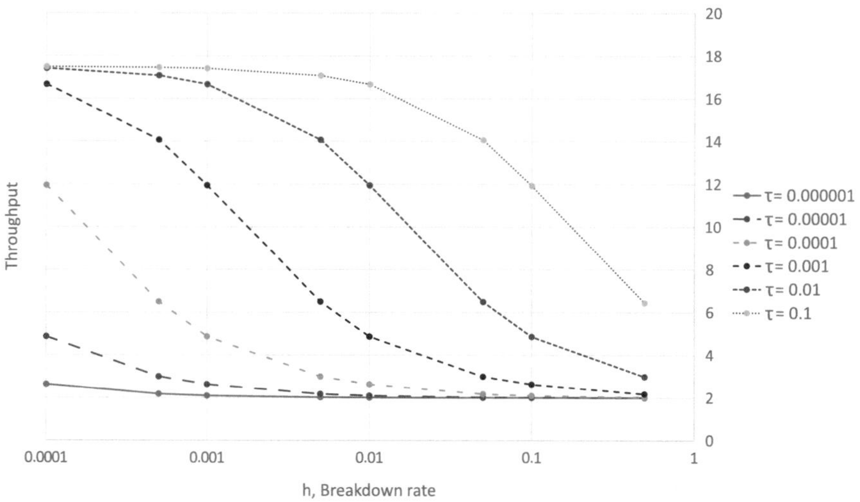

Fig. 5.14 The system throughput ϱ as a function of the breakdown rate h for different values of the repair rate τ ($\tau = 10^n, n = -6, -5, \ldots, -1$)

It can be seen from Fig. 5.14 that when breakdowns occur very rarely (h is close to zero) and the repair is sufficiently fast ($\tau = 10^{-1}$) then $\varrho \approx \mu_1 + \mu_2$, where μ_k is the service rate at the kth server. This result is intuitively understandable when we analyze the operation of the system in the overload regime and confirms the conclusions of Experiment 5.8 that, in this case, the customers are serviced only by the main servers and the participation of the reserve server in the service of customers is almost never required.

It can also be seen from the figure that with increasing breakdown rate and decreasing repair rate, all curves tend to the horizontal asymptote $\varrho = \mu_3$, where μ_3 is the service rate of the reserve server. This behavior of the system throughput is expected since the main servers are almost always under repair and the customers are serviced by the reserve server only. It can be also seen from the figure that the bandwidth does not exceed the horizontal asymptote $\varrho = \mu_1 + \mu_2$.

Chapter 6
Tandem Queues with Correlated Arrivals and Their Application to System Structure Performance Evaluation

Tandem queues are assumed to have several stations (stages, phases) in series. In such systems after receiving service in a certain station, a customer goes to the next station to get service. Tandem queues can be used for modeling real-life networks with a linear topology as well as for the validation of general decomposition algorithms in networks. Thus, tandem queueing systems have found much interest in the literature. Most of early papers on tandem queues are devoted to exponential queuing models. Over the last two decades efforts of many investigators in tandem queues were directed to weakening the distribution assumptions on the arrival pattern and service times. In particular, the arrival process should be able to capture correlation and explosive nature of real traffic in modern communication networks. As it was noted in Sect. 6.2, the correlation and the variability of inter-arrival times in input flows are taken into account via the consideration of a $BMAP$. In this section, we consider a number of two-station and multi-station tandem queues with $BMAP$, general or PH distribution of service times and different configuration of stations.

6.1 Brief Review of Tandem Queues with Correlated Arrivals

Tandem queues adequately simulate the operation of broadband wireless networks with a linear topology (see for example [116, 117, 119]). The most well-studied systems in the literature are dual tandem systems, i.e., systems consisting of two stations. An analytical study of such systems reduces to the study of a multidimensional random process. In the simplest cases, which are quite rare in real systems, the stationary distribution of the state probabilities of this process has a multiplicative form. If the stationary state distribution cannot be presented in this form, and the size of the buffers is unlimited at both tandem stations, the problem of finding the stationary distribution of the tandem state probabilities reduces to

A. N. Dudin et al., *The Theory of Queuing Systems with Correlated Flows*,
https://doi.org/10.1007/978-3-030-32072-0_6

a complex analysis of a random walk in a quarter-plane even under the simplest assumptions about the input flow and the service (a stationary Poisson input of customers, an exponential distribution of the service times on the tandem stations). This analysis is usually carried out by solving the functional equation for the two-dimensional probability generating function. If at least one of the two tandem buffers is finite, the input flow is a MAP, and the service time distributions are of the phase type then the process describing the tandem states belongs to the class of multi-dimensional (vector birth-and-death) processes studied in Sect. 2.2. Therefore, the need to study tandem systems alongside single-server systems with a MAP input and a PH-type distribution of service times was the motivation for M. Neuts to introduce and study the vector birth-and-death processes.

In [84], a tandem of two systems with infinite buffers, a MAP input flow, and an exponential service time distribution is considered. The operation of the system is described by a vector birth-and-death process whose generator is a matrix with blocks of infinitely increasing dimension. The first component of the vector birth-and-death process is the total number of customers at both stations, and the second one specifies all possible variants of the distribution of this number between the two stations. Since the theory of vector birth-and-death processes is strictly applicable only to blocks of fixed dimension, the results of [84] are pretty much heuristic.

Therefore, we only consider dual tandem systems with an arbitrary service time distribution at the first station in which at least one buffer is finite, the input is a MAP or a $BMAP$, and the service time distribution is of PH-type at the second station of the tandem.

It is easier to study systems where the input is a MAP rather than a $BMAP$, an arbitrary distribution of the service time is allowed only at the second station, and the first station buffer is finite. Such an analysis is presented in a series of papers by A. Gomez-Corral (see, for example, [47, 48]). The relative simplicity of the study is explained by the fact that the first station is described by a $MAP/PH/1/N$ queue. For such a queue, the output process (being the incoming flow to the second station) is a MAP and the matrices characterizing this MAP are easily expressed through the matrices defining the input flow and the service process.

If an arbitrary distribution of the service time is allowed at the first station, the system analysis becomes more complicated since the description of the MC embedded at the service completion moments at the first station as an $M/G/1$-type MC requires much a more cumbersome description of the dynamics of the MC's finite components. Therefore, a significant part of the investigation involves the decomposition of the tandem and an approximate analysis of the output flow from the first station, which in this case is not a $BMAP$. Papers offering such an approach are [41, 42, 50, 51, 103].

An exact analysis of tandem systems with a $BMAP/G/1$-type or $BMAP/G/1/N$-type first stations based on the use of the embedded-MC method followed by application of the Markov renewal processes is performed in the following papers. In the paper [74], $BMAP/G/1 \to \bullet/PH/1/M-1$-type and $BMAP/G/1/N \to \bullet/PH/1/M-1$-type tandem queues are investigated where if the buffer of the second station described by a single-server queue with a PH-type service time is full, a customer completing its service at the first station is lost. In the paper [74], such systems are studied under the assumption that a served customer is not lost but blocks the first station server until a free waiting place becomes available at the second station. The paper [7] studies a model with a MAP, finite buffers, an arbitrary service time distribution at both stations, and loss of customers after the first station when the second station buffer is full. In [8], a similar model with blocking of the first server is analyzed. In both papers, the analysis is based on the embedded-MC method and the apparatus of Markov renewal processes. The embedded process is constructed at the moments of service completion at the first station. Due to the assumption that the service time distribution has a general form at the second station, in order to ensure the embedded process is Markovian, it is additionally assumed that at service completion moments at the first station, the service time at the second station starts anew. In [76], the system is similar to the one studied in [74] but it is supposed to have feedback. In the terminology existing in queuing theory, feedback assumes that upon being served at the second station of the tandem, a customer can return to the first station to get service. In [59], a model similar to the one studied in [74] is considered but the first station is supposed to have no buffer. Customers that find a server busy make the repeated attempts to get service (retrial customers). The paper [72] considers systems where the first station is described by a $BMAP/G/1$-type queue. The second station is described by a multi-server queue with exponentially distributed service time and no buffer. A customer completing service at the first station requires a random number of servers to get service at the second station. If the required number of servers is lacking, the customer is either lost or temporarily blocks the operation of the first station server. The paper [61] presents a more detailed study of such a system including an analysis of the time spent by a customer at each tandem station and in the whole tandem. The results are obtained in a simple and compact form. They will be given below in Sect. 6.2. In Sect. 6.3, generalization of these results to the system with semi-Markovian service at the first station is given. In Sect. 6.4, results of Sect. 6.2 are extended to the case when there is a finite buffer between the stations. In Sect. 6.5, the results Sect. 6.2 are extended to the case of the system with retrials at the first station. The results presented in all these sections essentially are partially given in the monograph [73].

The paper [58] analyzes a tandem queue whose second station receives an additional flow of customers (cross-traffic) whose service requires a channel reservation. The paper [60] considers a tandem with the first station described by a $BMAP/G/1$-type queue and the second station described by an arbitrary MC with a finite state space. In the papers [25, 34, 62, 64, 65], dual tandem queues with a $BMAP$ or a MAP input are considered where both stations are multi-server queues. Most of these works contain an analysis of the relevant tandem queues as the models of some real systems (technical support systems, contact centers, etc.)

6.2 $BMAP/G/1 \to \cdot/M/N/0$ Queue with a Group Occupation of Servers of the Second Station

6.2.1 Mathematical Model

We consider a two-station $BMAP/G/1 \to \cdot/M/N/0$ queue. The first station of the queue is a single-server queue with infinite waiting room. The input of customers is a $BMAP$ defined by the underlying process $\nu_t,\ t \geq 0$ with state space $\{0, \dots, W\}$ and the matrix PGF $D(z) = \sum\limits_{k=0}^{\infty} D_k z^k$, $|z| \leq 1$. If a batch of size $k \geq 1$ arrives at the system and finds the first station server idle then one customer of the batch starts its service and the others queue in the waiting space of unlimited size and wait for their service. If the server is busy upon an arrival all customers of the batch take waiting places at the end of the queue.

The service times at the first station are independent and identically distributed random variables with distribution function $B(t)$, LST $\beta(s) = \int\limits_0^{\infty} e^{-st} dB(t)$, and finite first moment $b_1 = \int\limits_0^{\infty} t dB(t)$.

After receiving service at the first station, a customer enters the second station of the system, which is formed by N independent identical servers.

The service at the second station requires a random number ϕ of servers. Here ϕ is an integer-valued random variable with the distribution

$$q_m = P\{\phi = m\},\ q_m \geq 0,\ m = \overline{0, N},\ \sum_{m=0}^{N} q_m = 1.$$

Note that customers arriving within the same batch may require a different number of servers to get service. We assume that $q_0 \neq 1$ since in this case the considered queue can be reduced to a classical one-station $BMAP/G/1$ queue investigated in Sect. 3.1. After a customer starts its service at the second-station servers, each server serves the customer for an exponentially distributed time with parameter μ independently of other servers.

If the required number of idle servers at the second station is not available then a customer cannot be completely served and leaves the system (is lost) with probability p, $0 \leq p \leq 1$, or it waits until the required number of idle servers becomes available with the complementary probability $1 - p$. Its waiting period causes blocking of the first station. As the extreme cases, we have the system with blocking for $p = 0$ and the system with loss for $p = 1$.

6.2.2 Stationary State Distribution of the Embedded Markov Chain

Let t_n be the nth service completion moment at the first station, $n \geq 1$. Consider the process $\xi_n = \{i_n, r_n, \nu_n\}$, $n \geq 1$, where i_n is the number of customers at the first station (excluding a customer causing blocking) at time $t_n + 0$; r_n is the number of busy servers at the second station at time $t_n - 0$; ν_n is the state of the $BMAP$ at time t_n, $i_n \geq 0$, $r_n = \overline{0, N}$, $\nu_n = \overline{0, W}$.

The process $\xi_n, n \geq 1$ is an irreducible MC. Denote by $P\{(i, r, \nu) \to (j, r', \nu')\}$ the transition probabilities of this MC. Renumber all the states in lexicographical order and form the transition probabilities matrix as follows:

$$P\{(i, r) \to (j, r')\} = (P\{(i, r, \nu) \to (j, r', \nu')\})_{\nu, \nu' = \overline{0, W}},$$

$$P_{i,j} = (P\{(i, r) \to (j, r')\})_{r, r' = \overline{0, N}}, \; i, j \geq 0.$$

The matrices $P_{i,j}, \; i, j \geq 0$ define the probabilities of the MC transitions from states with value i of the countable component to states with value j of this component.

Lemma 6.1 *The matrix of transition probabilities of the MC $\xi_n, n \geq 1$ has a block structure of the form (2.59), where*

$$V_i = \sum_{k=1}^{i+1} [-\hat{Q}(\Delta \oplus D_0)^{-1} \tilde{D}_k + (1-p)\mathcal{F}_k \tilde{Q}_3] \Omega_{i-k+1}, \tag{6.1}$$

$$Y_i = \bar{Q}\Omega_i + (1-p) \sum_{k=0}^{i} \mathcal{F}_k \tilde{Q}_3 \Omega_{i-k}, \tag{6.2}$$

$$\Omega_n = \int_0^\infty e^{\Delta t} \otimes P(n, t) dB(t), \qquad \mathcal{F}_n = \int_0^\infty dF(t) \otimes P(n, t), \; n \geq 0,$$

$$\Delta = \begin{pmatrix} 0 & 0 & 0 & \cdots & 0 & 0 \\ \mu & -\mu & 0 & \cdots & 0 & 0 \\ 0 & 2\mu & -2\mu & \cdots & 0 & 0 \\ \vdots & \vdots & \vdots & \ddots & \vdots & \vdots \\ 0 & 0 & 0 & \cdots & N\mu & -N\mu \end{pmatrix}.$$

Here $\tilde{D}_k = I_{N+1} \otimes D_k$, $k \geq 0$, $P(n, t)$, $n \geq 0$ are the coefficients of expansion $e^{D(z)t} = \sum_{n=0}^{\infty} P(n, t) z^n$, $F(t) = (F_{r,r'}(t))_{r,r'=\overline{0,N}}$, where $F_{r,r'}(t) = 0$ if $r \leq r'$, and if $r > r'$, $F_{r,r'}(t)$ is the distribution function given by its LST $f_{r,r'}(s) =$

$\prod_{l=r'+1}^{r} l\mu(l\mu+s)^{-1}$; $\tilde{Q}_m = Q_m \otimes I_{\bar{W}},\ m = \overline{1,3}$;

$Q_m,\ m = \overline{1,3}$ *are the square matrices of order* $N+1$;

$$Q_2 = \operatorname{diag}\{\sum_{m=N-r+1}^{N} q_m,\ r = \overline{0,N}\};$$

$$Q_1 = \begin{pmatrix} q_0 & q_1 & \dots & q_N \\ 0 & q_0 & \dots & q_{N-1} \\ \vdots & \vdots & \ddots & \vdots \\ 0 & 0 & \dots & q_0 \end{pmatrix};\ Q_3 = \begin{pmatrix} 0 & \dots & 0 & q_N \\ 0 & \dots & 0 & q_{N-1} \\ \vdots & \ddots & \vdots & \vdots \\ 0 & \dots & 0 & q_0 \end{pmatrix};$$

$$\bar{Q} = \tilde{Q}_1 + p\tilde{Q}_2,\ \hat{Q} = \tilde{Q}_1 + p\tilde{Q}_2 + (1-p)\int_0^\infty (dF(t) \otimes e^{D_0 t})\tilde{Q}_3.$$

Proof Rewrite the matrices V_i and Y_i of transition probabilities in the block form $V_i = (V_i^{(r,r')})_{r,r'=\overline{0,N}}$, $Y_i = (Y_i^{(r,r')})_{r,r'=\overline{0,N}}$, where blocks $V_i^{(r,r')}$, $Y_i^{(r,r')}$ correspond to the transitions of the process r_n from state r to state r'.

Let $\delta_{r,r'}(t)$ denote the probability that the number of customers at the second station changes from r to r' during time t given that no new customers arrive at this station.

Let us explain the probabilistic sense of the matrices that will be used below.

The (ν, ν')th entry of the matrix $P(n,t)$ is the probability that n customers arrive from the $BMAP$ within the time interval $(0, t]$ and the underlying process $\nu_t = \nu'$ given that $\nu_0 = \nu$, $\nu, \nu' = \overline{0, W}$.

The (ν, ν')th entry of the matrix $\int_0^\infty \delta_{r,r'}(t)P(n,t)dB(t)$ is the probability that during the service time of a customer at the first station n customers arrive, the number of busy servers at the second station changes from r to r' and the $BMAP$ underlying process changes its state from ν to ν'.

Let the process (ν_t, r_t) be in state (ν, r) at an arbitrary time. Then the (ν, ν')th entry of the matrix $\int_0^\infty \delta_{r,r'}(t)e^{D_0 t}D_k dt$ is the probability that the first arrival of the $BMAP$ is a batch of size k, the process ν_t is in state ν' immediately after the batch arrival, and the number of customers at the second station is r' at the moment of arrival.

$F_{r,r'}(t)$ is the distribution function of the time during which the number of busy servers at the second station changes from r to r' given that there is no new arrival at the second station.

The (ν, ν')th entry of the matrix $\int_0^\infty P(n,t)dF_{r,r'}(t)$ is the probability that during the time distributed with distribution function $F_{r,r'}(t)$, n customers arrive at the system and the process ν_t transits from state ν to state ν'.

Taking into account the probabilistic interpretation of the matrices and analyzing the behavior of the system between two consecutive service completion moments at the first station, we obtain the following expressions for the matrices $V_i^{(r,r')}$, $Y_i^{(r,r')}$, $i \geq 0$:

$$V_i^{(r,r')} = \sum_{m=0}^{N-r} q_m \sum_{l=r'}^{r+m} \int_0^\infty \delta_{r+m,l}(t) e^{D_0 t} dt \sum_{k=1}^{i+1} D_k \int_0^\infty \delta_{l,r'}(t) P(i-k+1,t) dB(t)$$

$$+p \sum_{m=N-r+1}^{N} q_m \sum_{l=r'}^{r} \int_0^\infty \delta_{r,l}(t) e^{D_0 t} dt \sum_{k=1}^{i+1} D_k \int_0^\infty \delta_{l,r'}(t) P(i-k+1,t) dB(t)$$

$$+(1-p) \sum_{m=N-r+1}^{N} q_m \Bigg[\int_0^\infty e^{D_0 t} dF_{r,N-m}(t) \sum_{l=r'}^{N} \int_0^\infty \delta_{N,l}(t) e^{D_0 t} dt$$

$$\times \sum_{k=1}^{i+1} D_k \int_0^\infty \delta_{l,r'}(t) P(i-k+1,t) dB(t)$$

$$+ \sum_{k=1}^{i+1} \int_0^\infty P(k,t) dF_{r,N-m}(t) \int_0^\infty \delta_{N,r'}(t) P(i-k+1,t) dB(t) \Bigg], \tag{6.3}$$

$$Y_i^{(r,r')} = \sum_{m=0}^{N-r} q_m \int_0^\infty P(i,t) \delta_{r+m,r'}(t) dB(t)$$

$$+ \sum_{m=N-r+1}^{N} q_m \Bigg[p \int_0^\infty \delta_{r,r'}(t) P(i,t) dB(t)$$

$$+ (1-p) \sum_{k=0}^{i} \int_0^\infty P(k,t) dF_{r,N-m}(t) \int_0^\infty \delta_{N,r'}(t) P(i-k,t) dB(t) \Bigg]. \tag{6.4}$$

In order to derive relations (6.1) and (6.2) from (6.3) and (6.4), we use the definitions for the matrices introduced above and relation $(\delta_{r,r'}(t))_{r,r'=\overline{0,N}} = e^{\Delta t}$. The latter follows from the fact that when no customers arrive at the second station, process r_t describing the evolution of the number of busy servers at the second station is a death process with generator Δ. □

Corollary 6.1 *The process $\xi_n, n \geq 1$ is a QTMC.*

Proof The MC $\xi_n, n \geq 1$ is irreducible and aperiodic, and the matrices of its transition probabilities $P_{i,j},\ i > 0$ are presentable as functions of the difference $j - i$. Thus, this MC belongs to the class of multidimensional QTMCs or the MCs of $M/G/1$-type. □

Corollary 6.2 *The matrix PGFs $V(z)$ and $Y(z)$ have the form*

$$V(z) = \frac{1}{z}[-\hat{Q}(\Delta \oplus D_0)^{-1}(\tilde{D}(z) - \tilde{D}_0) + (1-p)(\mathcal{F}(z) - \mathcal{F}_0)\tilde{Q}_3]\Omega(z), \tag{6.5}$$

$$Y(z) = [\bar{Q} + (1-p)\mathcal{F}(z)\tilde{Q}_3]\Omega(z), \tag{6.6}$$

where

$$\Omega(z)=\int_0^\infty e^{\Delta t}\otimes e^{D(z)t}dB(t),\ \mathcal{F}(z)=\int_0^\infty dF(t)\otimes e^{D(z)t},$$

$$\tilde{D}(z) = \sum_{k=0}^{\infty} \tilde{D}_k z^k,\ |z| \leq 1.$$

Proof Multiplying (6.1) by z^i and summing over $i \geq 0$ after algebraic transformations, we obtain expression (6.5) for $V(z)$. Similarly, by using (6.2), we obtain (6.6). □

Theorem 6.1 *The stationary state distribution of the MC $\xi_n, n \geq 1$ exists if and only if the following inequality holds:*

$$\rho < 1, \tag{6.7}$$

where

$$\rho = \lambda[b_1 + (1-p)\sum_{r=1}^{N} y_r \sum_{m=N-r+1}^{N} q_m \sum_{l=N-m+1}^{r} (l\mu)^{-1}].$$

Here $(y_1, \ldots, y_N)$ is the part of the vector $\mathbf{y} = (y_0, y_1, \ldots, y_N)$ which is the unique solution of the system of linear algebraic equations

$$\mathbf{y}\mathcal{Q}\mathcal{B}^*(0) = \mathbf{y}, \quad \mathbf{ye} = \mathbf{1}, \tag{6.8}$$

where

$$\mathcal{Q} = Q_1 + pQ_2 + (1-p)EQ_3,\ \mathcal{B}^*(s) = \int_0^\infty e^{-st}e^{\Delta t}dB(t).$$

Proof The matrix $Y(1)$ reflects the evolution of the $BMAP$ and the number of busy servers at the second station between neighboring service completion moments at the first station. Taking into account that the process ν_t, $t \geq 0$ is irreducible and it is assumed that $q_0 \neq 1$, it is easy to show that matrix $Y(1)$ is irreducible. Then the QTMC ξ_n, $n \geq 1$ is ergodic if and only if inequality (2.65) holds, where the vector $\mathbf{x}$ is the unique solution of the system of linear algebraic equations (2.66).

We show that (2.65) is equivalent to inequality (6.7).

Rewrite the system (6.8) in the form

$$\mathbf{y}[Q_1 + pQ_2 + (1-p)\int_0^\infty dF(t)Q_3]\int_0^\infty e^{\Delta t}dB(t) = \mathbf{y}, \quad \mathbf{ye} = \mathbf{1},$$

and system (3.67) as

$$\mathbf{x}[(Q_1+pQ_2)\otimes I_{\bar{W}}+(1-p)\int_0^\infty dF(t)Q_3\otimes e^{D(1)t}]\int_0^\infty e^{\Delta t}\otimes e^{D(1)t}dB(t)=\mathbf{x},\ \mathbf{xe}=1. \tag{6.9}$$

By analyzing the two system of equations above, it is easy to show that the vector

$$\mathbf{x} = \mathbf{y} \otimes \boldsymbol{\theta} \tag{6.10}$$

is the unique solution of system (6.9), and hence it is the unique solution of the equivalent system (2.66).

Differentiating (6.6) at the point $z = 1$ and substituting the obtained expression for $Y'(1)$ and the vector $\mathbf{x}$ of the form (6.10) into (2.65), we obtain

$$\lambda[b_1 + (1-p)\mathbf{y}\int_0^\infty t dF(t)Q_3\mathbf{e}] < \mathbf{1}. \tag{6.11}$$

Taking into account that $\int_0^\infty t dF(t)$ is a matrix whose (r, r')th entry is $\sum_{l=r'+1}^{r} (l\mu)^{-1}$ for $r > r'$ and whose other entries are zero, inequality (6.11) can be easily reduced to inequality (6.7). □

Remark 6.1 The vector $\mathbf{y}$ determines the stationary distribution of busy servers at the second station at service completion moments at the first station given that the first station server works without stopping. Then the relation $(1 - p)\sum_{r=1}^{N} y_r \sum_{m=N-r+1}^{N} q_m \sum_{l=N-m+1}^{r} (l\mu)^{-1}$ provides the mean blocking time at the first station when the second station buffer is full, and the value ρ is the system load.

From now on we assume that (6.7) holds. Then there exists a stationary distribution of the MC $\xi_n,\ n \geq 1$, which coincides with its ergodic distribution

$$\pi(i,r,\nu) = \lim_{n\to\infty} P\{i_n = i, r_n = r, \nu_n = \nu\},\ i \geq 0,\ r = \overline{0,N},\ \nu = \overline{0,W}.$$

Introduce the notation for the vectors of stationary probabilities

$$\boldsymbol{\pi}(i,r) = (\pi(i,r,0), \pi(i,r,1), \ldots, \pi(i,r,W)),$$

$$\boldsymbol{\pi}_i = (\boldsymbol{\pi}(i,0), \boldsymbol{\pi}(i,1), \ldots, \boldsymbol{\pi}(i,N)),\ i \geq 0.$$

To calculate the vectors $\boldsymbol{\pi}_i,\ i \geq 0$ we use the algorithms described in Sect. 2.4.

6.2.3 Stationary State Distribution at an Arbitrary Time

Denote the system states at an arbitrary time t by $(i_t,\ k_t,\ r_t,\ \nu_t)$, where i_t is the number of customers at the first station (including a customer causing blocking); k_t is an indicator taking values 0, 1, 2 depending at the first station server state (idle, busy, or blocked); r_t is the number of busy servers at the second station; ν_t is the state of the $BMAP$ at time $t,\ t \geq 0$.

Consider the process $\zeta_t = \{i_t, k_t, r_t, \nu_t\},\ t \geq 0$. This process is not Markovian but its stationary state distribution can be expressed in terms of the stationary state distribution of the embedded MC $\xi_n,\ n \geq 1$ by using results for the Markov renewal processes and semi-regenerative processes, see [22].

Let $p(i,k,r,\nu),\ i \geq 0,\ k = 0,1,2,\ r = \overline{0,N},\ \nu = \overline{0,W}$ be the stationary state probabilities of the process $\zeta_t,\ t \geq 0$. Denote the vectors of these probabilities by

$$\mathbf{p}(i,k,r) = (p(i,k,r,0), \ldots, p(i,k,r,W)),\ \ \mathbf{p}_i(k) = (\mathbf{p}(i,k,0), \ldots, \mathbf{p}(i,k,N)).$$

Theorem 6.2 *Non-zero vectors of the stationary state probabilities* $\mathbf{p}_i(k),\ i \geq 0,\ k = 0,1,2$ *are expressed by the stationary state probability vectors* $\boldsymbol{\pi}_i,\ i \geq 0$ *of the embedded MC* $\xi_n,\ n \geq 1$ *as follows:*

$$\mathbf{p}_0(0) = -\tau^{-1}\boldsymbol{\pi}_0\hat{Q}(\Delta \oplus D_0)^{-1},$$

$$\mathbf{p}_i(1) = \tau^{-1}\{\boldsymbol{\pi}_0\sum_{k=1}^{i}[-\hat{Q}(\Delta \oplus D_0)^{-1}\tilde{D}_k + (1-p)\mathcal{F}_k\tilde{Q}_3]\tilde{\Omega}_{i-k}$$

$$+\sum_{l=1}^{i}\boldsymbol{\pi}_l[\bar{Q}\tilde{\Omega}_{i-l}+(1-p)\sum_{k=0}^{i-l}\mathcal{F}_k\tilde{Q}_3\tilde{\Omega}_{i-k-l}]\},$$

$$\mathbf{p}_i(2)=\tau^{-1}(1-p)\sum_{l=0}^{i-1}\boldsymbol{\pi}_l\sum_{k=0}^{i-l-1}(\mathcal{F}_k+\hat{\delta}_{0,k}I)\tilde{Q}_2\int_0^{\infty}e^{-\mu\mathcal{N}t}\otimes P(i-k-l-1,t)dt,\ i\geq 1,$$

where τ is the mean time interval between two successive service completion moments at the first station:

$$\tau=b_1+\boldsymbol{\pi}_0\hat{Q}(-\tilde{D}_0)^{-1}\mathbf{e}+(1-p)\mathbf{\Pi}(1)(I_{N+1}\otimes\mathbf{e})\int_0^{\infty}tdF(t)Q_3\mathbf{e}, \qquad (6.12)$$

$$\tilde{\Omega}_n=\int_0^{\infty}e^{\Delta t}\otimes P(n,t)(1-B(t))dt,\ n\geq 0.$$

Proof By the definition given in [22], the process $\zeta_t,\ t\geq 0$ is semi-regenerative with the embedded Markov renewal process $\{\xi_n,t_n\},\ n\geq 1$.

The stationary state distribution $p(i,k,r,\nu),\ i\geq 0,\ k=0,1,2,\ r=\overline{0,N},\ \nu=\overline{0,W}$ of the process ζ_t exists if the process $\{\xi_n,t_n\}$ is irreducible and aperiodic, and the inequality $\tau<\infty$ holds. In our case these two conditions hold if the embedded MC $\xi_n,\ n\geq 1$ is ergodic.

First, we find the vectors $\mathbf{p}(i,k,r),\ i\geq 0,\ k=0,1,2,\ r=\overline{0,N}$ of the stationary state probabilities $p(i,k,r,\nu)$. To calculate these vectors, we use the limit theorem for semi-regenerative processes given in [22]. According to this theorem, the stationary state distribution of the process $\zeta_t,\ t\geq 0$ can be expressed in terms of the stationary state distribution of the embedded MC $\xi_n,n\geq 1$.

Introduce into consideration the following matrices of size $\bar{W}\times\bar{W}$:

$$K_{l,r}(i,k,r',t),\ l,i\geq 0,\ k=0,1,2,\ r,r'=\overline{0,N}. \qquad (6.13)$$

The (ν,ν')th entry of these matrices defines the probability that a customer completes its service later than time t and the process $\zeta_t,\ t\geq 0$ is in state (i,r',ν',k) at time t given that the previous customer completed its service at time $t=0$ and the embedded MC ξ_n was in state (l,r,ν) at this time.

According to the limit theorem, the vectors $\mathbf{p}(i,k,r)$ are expressed through the stationary state distribution $\boldsymbol{\pi}(i,r),\ i\geq 0,\ r=\overline{0,N}$ of the MC $\xi_n,n\geq 1$ as follows:

$$\mathbf{p}(i,k,r')=\tau^{-1}\sum_{l=0}^{\infty}\sum_{r=0}^{N}\boldsymbol{\pi}(l,r)\int_0^{\infty}K_{l,r}(i,k,r',t)dt,\ i\geq 0,\ k=0,1,2,\ r'=\overline{0,N}. \qquad (6.14)$$

For the rest of the proof, we need to find expressions for the matrices (6.13) and the value τ.

The mean length τ of the time interval between two neighbouring moments of service completion at the first station is given by the formula

$$\tau = \sum_{i=0}^{\infty}\sum_{r=0}^{N}\sum_{\nu=0}^{W}\pi(i,r,\nu)\tau_{i,r,\nu},$$

where $\tau_{i,r,\nu}$ is the mean time interval between two neighboring service completion moments at the first station given that $\xi_n = (i, r, \nu)$ at the beginning of this interval. By means of algebraic transformations, we can verify that τ satisfies formula (6.12).

Then, analyzing the behavior of the system between two neighbouring service completion moments at the first station and taking into account the probabilistic meaning of the functions and the matrices given in the proof of Lemma 6.1, we obtain the following expressions for non-zero matrices (6.13):

$$K_{0,r}(0,0,r',t) = \sum_{m=\max\{0,r'-r\}}^{N-r} q_m e^{D_0 t}\delta_{r+m,r'}(t)$$

$$+ \sum_{m=N-r+1}^{N} q_m[pe^{D_0 t}\delta_{r,r'}(t) + (1-p)\int_0^t e^{D_0 s}dF_{r,N-m}(s)e^{D_0(t-s)}\delta_{N,r'}(t-s)],$$

$$K_{0,r}(i,1,r',t) = \sum_{m=\max\{0,r'-r\}}^{N-r} q_m \sum_{l=r'}^{r+m}\int_0^t e^{D_0 s}$$

$$\times\delta_{r+m,l}(s)\sum_{k=1}^{i} D_k ds\delta_{l,r'}(t-s)P(i-k,t-s)(1-B(t-s))$$

$$+ \sum_{m=N-r+1}^{N} q_m\{p\sum_{l=r'}^{r}\int_0^t e^{D_0 s}\delta_{r,l}(s)\sum_{k=1}^{i} D_k ds\delta_{l,r'}(t-s)P(i-k,t-s)(1-B(t-s))$$

$$+(1-p)[\int_0^t e^{D_0 s}dF_{r,N-m}(s)\sum_{l=r'}^{N}\int_0^{t-s} e^{D_0 x}\delta_{N,l}(x)\sum_{k=1}^{i} D_k dx$$

$$\times\delta_{l,r'}(t-s-x)P(i-k,t-s-x)(1-B(t-s-x))$$

$$+\sum_{k=1}^{i}\int_0^t P(k,s)dF_{r,N-m}(s)\delta_{N,r'}(t-s)P(i-k,t-s)(1-B(t-s))]\},$$

$$K_{l,r}(i,1,r',t) = \sum_{m=\max\{0,r'-r\}}^{N-r} q_m \int_0^t \delta_{r+m,r'}(s)P(i-l,s)(1-B(s))ds$$

$$+ \sum_{m=N-r+1}^{N} q_m [p \int_0^t \delta_{r,r'}(s)P(i-l,s)(1-B(s))ds$$

$$+(1-p)\sum_{k=0}^{i-l}\int_0^t P(k,s)dF_{r,N-m}(s)\delta_{N,r'}(t-s)$$

$$\times P(i-l-k,t-s)(1-B(t-s))], i \geq l \geq 1,$$

$$K_{l,r}(i,2,r',t) = (1-p) \sum_{m=N+1-\min\{r,r'\}}^{N} q_m \sum_{k=0}^{i-l-1} \int_0^t P(k,s)dF_{r,r'}(s)$$

$$\times P(i-k-l-1,t-s)e^{-r'\mu(t-s)}, \ i \geq 1, \ l \geq 0, \ r,r' = \overline{0,N}.$$

Substituting the obtained expressions for the matrices $K_{l,r}(i,k,r',t)$ in (6.14), after a series of algebraic transformations involving a change of the integration order, we obtain the following formulas:

$$\mathbf{p}(0,0,r') = \tau^{-1}\sum_{r=0}^{N}\boldsymbol{\pi}(0,r)[\sum_{m=\max\{0,r'-r\}}^{N-r} q_m \int_0^\infty \delta_{r+m,r'}(t)e^{D_0 t}dt$$

$$+p \sum_{m=N-r+1}^{N} q_m \int_0^\infty \delta_{r,r'}(t)e^{D_0 t}dt$$

$$+(1-p) \sum_{m=N-r+1}^{N} q_m \int_0^\infty e^{D_0 t}dF_{r,N-m}(t) \int_0^\infty \delta_{N,r'}(t)e^{D_0 t}dt],$$

$$\mathbf{p}(i,1,r') = \tau^{-1}\{\sum_{r=0}^{N}\boldsymbol{\pi}(0,r)\{\sum_{k=1}^{i}[\sum_{m=\max\{0,r'-r\}}^{N-r} q_m \sum_{l=r'}^{r+m}\int_0^\infty \delta_{r+m,l}(t)e^{D_0 t}dt$$

$$+p \sum_{m=N-r+1}^{N} q_m \sum_{l=r'}^{r}\int_0^\infty \delta_{r,l}(t)e^{D_0 t}dt$$

$$+(1-p)\sum_{m=N-r+1}^{N} q_m \int_0^\infty e^{D_0 t} dF_{r,N-m}(t) \sum_{l=r'}^{N} \int_0^\infty \delta_{N,l}(t) e^{D_0 t} dt]$$

$$\times D_k \int_0^\infty \delta_{l,r'}(t) P(i-k,t)(1-B(t))dt$$

$$+(1-p)\sum_{m=N-r+1}^{N} q_m \sum_{k=1}^{i} \int_0^\infty P(k,t) dF_{r,N-m}(t)$$

$$\times \int_0^\infty \delta_{N,r'}(t) P(i-k,t)(1-B(t))dt\}$$

$$+\sum_{l=1}^{i}\sum_{r=0}^{N} \boldsymbol{\pi}(l,r)[\sum_{m=\max\{0,r'-r\}}^{N-r} q_m \int_0^\infty \delta_{r+m,r'}(t) P(i-l,t)(1-B(t))dt$$

$$+p\sum_{m=N-r+1}^{N} q_m \int_0^\infty \delta_{r,r'}(t) P(i-l,t)(1-B(t))dt$$

$$+(1-p)\sum_{m=N-r+1}^{N} q_m \sum_{k=0}^{i-l} \int_0^\infty P(k,t) dF_{r,N-m}(t)$$

$$\times \int_0^\infty \delta_{N,r'}(t) P(i-l-k,t)(1-B(t))dt]\},$$

$$\mathbf{p}(i,2,r') = \tau^{-1}(1-p)\sum_{l=0}^{i-1}\sum_{r=0}^{N} \boldsymbol{\pi}(l,r) \sum_{m=N+1-\min\{r,r'\}}^{N} q_m$$

$$\times \sum_{k=0}^{i-l-1} \int_0^\infty P(k,t) dF_{r,r'}(t) \int_0^\infty P(i-k-l-1,t) e^{-r'\mu t} dt, i \ge 1, \ r' = \overline{0,N}.$$

It is obvious that $\mathbf{p}(0,1,r') = \mathbf{p}(0,2,r') = \mathbf{p}(i,0,r') = \mathbf{0}, \ i \ge 1$.

Combining the vectors $\mathbf{p}(i,k,r'), \ r' = \overline{0,N}$ into the vectors $\mathbf{p}_i(k)$ for all $i \ge 0, \ k = 0,1,2$ and using the matrix denotations introduced earlier, we obtain formulas for vectors $\mathbf{p}_i(k)$. □

Corollary 6.3 *The stationary state probability vectors* $\mathbf{p}_i,\ i \geq 0$ *for process* $\{i_t, r_t, \nu_t\},\ t \geq 0$ *are calculated as follows:*

$$\mathbf{p}_i = \sum_{k=0}^{2} \mathbf{p}_i(k),\ \ i \geq 0.$$

Corollary 6.4 *The PGF* $\mathbf{P}(z) = \sum_{i=0}^{\infty} \mathbf{p}_i z^i,\ |z| \leq 1$ *is expressed through the PGF* $\mathbf{\Pi}(z)$ *as follows:*

$$\mathbf{P}(z)(\Delta \oplus D(z)) = \tau^{-1}\mathbf{\Pi}(z)\{z[I - (1-p)\mathcal{F}(z)\tilde{Q}_2$$

$$\times (\mu\mathcal{N} \oplus (-D(z)))^{-1}(\Delta \oplus D(z))] - [\bar{Q} + (1-p)\mathcal{F}(z)\tilde{Q}_3]\}. \tag{6.15}$$

For brevity, we rewrite (6.15) in the form

$$\mathbf{P}(z)\mathcal{A}(z) = \boldsymbol{b}(z), \tag{6.16}$$

where

$$\mathcal{A}(z) = \Delta \oplus D(z),$$

$$\boldsymbol{b}(z) = \tau^{-1}\mathbf{\Pi}(z)\{z[I - (1-p)\mathcal{F}(z)\tilde{Q}_2(\mu\mathcal{N} \oplus (-D(z)))^{-1}(\Delta \oplus D(z))]$$

$$-[\bar{Q} + (1-p)\mathcal{F}(z)\tilde{Q}_3]\}.$$

By using relation (6.16), we can calculate the vector factorial moments $\mathbf{P}^{(m)} = \frac{d^m \mathbf{P}(z)}{dz^m}|_{z=1},\ m = \overline{0, M}$ via the vector factorial moments $\mathbf{\Pi}^{(m)} = \frac{d^m \mathbf{\Pi}(z)}{dz^m}|_{z=1},\ m = \overline{0, M+1}$. The vectors $\mathbf{\Pi}^{(m)}$ can be calculated directly from the formula $\mathbf{\Pi}^{(m)} = \sum_{i=m}^{\infty} \frac{i!}{(i-m)!}\boldsymbol{\pi}_i,\ m \geq 1$ if we know the stationary distribution $\boldsymbol{\pi}_i,\ i \geq 0$, or we can apply the recurrent procedure to calculate these moments.

Theorem 6.3 *Let* $\mathbf{\Pi}^{(m)} < \infty,\ m = \overline{0, M+1}$. *The vectors* $\mathbf{P}^{(m)},\ m = \overline{0, M}$ *are calculated recurrently:*

$$\mathbf{P}^{(m)}(1) = \Bigg[\Bigg(\boldsymbol{b}^{(m)}(1) - \sum_{l=0}^{m-1} C_m^l \mathbf{P}^{(l)}(1)\mathcal{A}^{(m-l)}(1)\Bigg)\tilde{I} \tag{6.17}$$

$$+\frac{1}{m+1}\Bigg(\boldsymbol{b}^{(m+1)}(1) - \sum_{l=0}^{m-1} C_{m+1}^l \mathbf{P}^{(l)}(1)\mathcal{A}^{(m+1-l)}(1)\Bigg)\mathbf{e}\hat{\mathbf{e}}\Bigg]\tilde{\mathcal{A}}^{-1},$$

where

$$\tilde{\mathcal{A}} = \mathcal{A}(1)\tilde{I} + \mathcal{A}'(1)\mathbf{e}\hat{\mathbf{e}}.$$

Proof Successively differentiating (6.16), we obtain

$$\mathbf{P}^{(m)}(1)\mathcal{A}(1) = \boldsymbol{b}^{(m)}(1) - \sum_{l=0}^{m-1} C_m^l \mathbf{P}^{(l)}(1)\mathcal{A}^{(m-l)}(1). \tag{6.18}$$

The matrix $\mathcal{A}(1)$ is singular, therefore it is not possible to obtain a recursion to calculate the vector factorial moments directly from (6.18).

Let us modify (6.18) as follows. Multiply both sides of expression (6.18) for $m+1$ by $\mathbf{e}$. Next, we replace one of the equations of system (6.18) (without loss of generality, the first one) by the resulting equation, and as a result we have

$$\mathbf{P}^{(m)}(1)\tilde{\mathcal{A}} = \left(\boldsymbol{b}^{(m)}(1) - \sum_{l=0}^{m-1} C_m^l \mathbf{P}^{(l)}(1)\mathcal{A}^{(m-l)}(1)\right)\tilde{I} \tag{6.19}$$

$$+\frac{1}{m+1}\left(\boldsymbol{b}^{(m+1)}(1) - \sum_{l=0}^{m-1} C_{m+1}^l \mathbf{P}^{(l)}(1)\mathcal{A}^{(m+1-l)}(1)\right)\mathbf{e}\hat{\mathbf{e}}.$$

The recurrent procedure (6.17) is obtained directly from (6.19) if the matrix $\tilde{\mathcal{A}}$ is non-singular. We prove that $\det\tilde{\mathcal{A}} \neq 0$.

It can be shown that $\det\tilde{\mathcal{A}} = \det\mathcal{D}\prod_{n=1}^{N}\det(n\mu I - D(1))$, where $\mathcal{D} = D(1)\tilde{I} + D'(1)\mathbf{e}\hat{\mathbf{e}}$. The determinants $\det(n\mu I - D(1)),\ n = \overline{1,N}$ are not zero since the diagonal elements of the matrices $n\mu I - D(1),\ n = \overline{1,N}$ dominate in rows.

Analyzing the structure of the matrix $\mathcal{D}$, we can verify that $\det\mathcal{D} = \boldsymbol{\nabla}D'(1)\mathbf{e}$, where $\boldsymbol{\nabla}$ is the vector of the algebraic adjuncts of the first column of matrix $D(1)$. Since $\boldsymbol{\theta}$ is the invariant vector of the matrix $D(1)$, we have $\boldsymbol{\nabla} = c\boldsymbol{\theta},\ c \neq 0$. Then $\det\mathcal{D} = c\boldsymbol{\theta}D'(1)\mathbf{e} = c\lambda$. Hence, $\det\tilde{\mathcal{A}} \neq 0$. □

6.2.4 *Performance Characteristics of the System*

The mean number of customers at the first station at this station's service completion moments and at an arbitrary time is

$$L = \boldsymbol{\Pi}'(1)\mathbf{e}, \quad \tilde{L} = \mathbf{P}'(1)\mathbf{e}.$$

The stationary state probability vector of the number of busy servers at the second station at service completion moments of the first station and at an arbitrary time is defined as

$$\boldsymbol{r} = \boldsymbol{\Pi}(1)(I_{N+1} \otimes \mathbf{e}_{\bar{W}}), \qquad \tilde{\boldsymbol{r}} = \mathbf{P}(1)(I_{N+1} \otimes \mathbf{e}_{\bar{W}}).$$

The mean number of busy servers at the second station at service completion moments of the first station and at an arbitrary time is defined as

$$N_{busy} = \boldsymbol{r}\mathcal{N}\mathbf{e}, \qquad \tilde{N}_{busy} = \tilde{\boldsymbol{r}}\mathcal{N}\mathbf{e}.$$

The probability that an arbitrary customer leaves the system (causes blocking) after being served at the first station is

$$P_{loss} = p\boldsymbol{\Pi}(1)\tilde{Q}_2\mathbf{e}, \quad P_{block} = (1-p)\boldsymbol{\Pi}(1)\tilde{Q}_2\mathbf{e}.$$

The probabilities P_{idle}, P_{serve}, P_{block} that the first station server is idle, busy, or blocked are

$$P_{idle} = \tau^{-1}\boldsymbol{\pi}_0\hat{Q}(-\tilde{D}_0)^{-1}\mathbf{e}, \ P_{serve} = \tau^{-1}b_1, \ P_{block} = 1 - P_{idle} - P_{serve}.$$

6.2.5 Stationary Sojourn Time Distribution

In this subsection we consider the problems of finding the LST of the stationary state distribution of the virtual and the real sojourn time in the system.

Virtual Sojourn Time The virtual sojourn time in the system consists of the virtual sojourn time at the first station plus the sojourn time at the second station. We suppose that customers are served according to the $FIFO$ discipline (First-In-First-Out).

The virtual sojourn time at the first station consists of:

- *the residual service time* starting from an arbitrary time (the virtual customer arrival moment) and finishing at the next moment of service completion at the first station;
- *the generalized service times* of the customers waiting for service at the first station at the virtual customer arrival time;
- *the generalized service time* of the virtual customer.

Remark 6.2 The generalized service time of a customer consists of this customers service time at the first station and the probable blocking time by the previous customer.

First, we study *a residual service time*. To do this, we consider the process $\chi_t = \{i_t, k_t, r_t, \nu_t, \tilde{\upsilon}_t\}$, $t \geq 0$, where i_t is the number of customers at the first station (including one causing blocking); k_t is an indicator taking values 0, 1, or 2 depending on the state of the first station server at time t (idle, busy, or blocked); ν_t is the state of the $BMAP$ at time t; r_t is the number of the busy servers at the second station before the first service completion moment at the first station after time t; $\tilde{\upsilon}_t$ is the residual service time from time moment t until the first service completion.

According to the definition given in [22], the process χ_t is semi-regenerative with the embedded Markov renewal process $\{\xi_n, t_n\}$, $n \geq 1$.

Denote the stationary state distribution of the process χ_t, $t \geq 0$ by

$$\tilde{V}(i, k, r, \nu, x) = \lim_{t\to\infty} P\{i_t = i, k_t = k, r_t = r, \nu_t = \nu, \tilde{\upsilon}_t < x\}, \tag{6.20}$$

$$i \geq 0,\ k = 0, 1, 2,\ r = \overline{0, N},\ \nu = \overline{0, W},\ x \geq 0.$$

According to [22], the limits (6.20) exist if the process $\{\xi_n, t_n\}$, $n \geq 1$ is irreducible and aperiodic and the inequality $\tau < \infty$ holds. In our case, these conditions hold if inequality (6.7) holds.

Let $\tilde{\boldsymbol{V}}(i, k, x)$ be the row vectors of the stationary state probabilities $\tilde{V}(i, k, r, \nu, x)$ ordered in lexicographical order of the components (r, ν), and $\tilde{\boldsymbol{v}}(i, k, s) = \int\limits_0^\infty e^{-sx} d\tilde{\boldsymbol{V}}(i, k, x)$, $i \geq 0$, $k = 0, 1, 2$ are the corresponding vector LSTs.

Lemma 6.2 *The vector LSTs $\tilde{\boldsymbol{v}}(i, k, s)$ are calculated as follows:*

$$\tilde{\boldsymbol{v}}(0, 1, s) = \mathbf{0},\ \ \tilde{\boldsymbol{v}}(i, 0, s) = \mathbf{0},\ \ i > 0,\ \ \tilde{\boldsymbol{v}}(0, 2, s) = \mathbf{0}, \tag{6.21}$$

$$\tilde{\boldsymbol{v}}(0, 0, s) = -\tau^{-1}\boldsymbol{\pi}_0\hat{Q}(\Delta \oplus D_0)^{-1}[\mathcal{B}^*(s) \otimes I_{\bar{W}}], \tag{6.22}$$

$$\tilde{\boldsymbol{v}}(i, 1, s) = \tau^{-1}\{\boldsymbol{\pi}_0 \sum_{k=1}^{i}[-\hat{Q}(\Delta \oplus D_0)^{-1}\tilde{D}_k + (1-p)\mathcal{F}_k\tilde{Q}_3]$$

$$\times \int\limits_0^\infty (e^{\Delta u} \otimes I_{\bar{W}}) \int\limits_0^u I_{N+1} \otimes P(i-k, y)e^{-s(u-y)} dy dB(u)$$

$$+ \sum_{j=1}^{i} \boldsymbol{\pi}_j[(\tilde{Q}_1 + p\tilde{Q}_2) \int\limits_0^\infty (e^{\Delta u} \otimes I_{\bar{W}}) \int\limits_0^u I_{N+1} \otimes P(i-j, y)e^{-s(u-y)} dy dB(u)$$

$$+(1-p)\sum_{k=0}^{i-j}\mathcal{F}_k\tilde{Q}_3\int_0^{\infty}e^{\Delta u}\otimes P(i-k-j,y)e^{-s(u-y)}dydB(u)]\}, \qquad (6.23)$$

$$\tilde{\boldsymbol{v}}(i,2,s)=\tau^{-1}(1-p)\sum_{j=0}^{i-1}\boldsymbol{\pi}_j(\int_0^{\infty}dF(u)\otimes I_{\bar{W}})\tilde{Q}_3$$

$$\times\int_0^{u}e^{-s(u-y)}[I_{N+1}\otimes P(i-j-1,y)]dy[\mathcal{B}^*(s)\otimes I_{\bar{W}}],\ i>0. \qquad (6.24)$$

Proof Let $\kappa_j^{(i,k,x,t)}(r,\nu;\ r',\nu')$ define the probability that a customer completes its service later than the time moment t, the discrete components of process χ_t take values $(i,\ k,\ r',\ \nu')$ at time t, and the continuous component $\tilde{v}_t < x$ given that the previous customer completed its service at time $t=0$ and the embedded MC ξ_n was in state (j,r,ν) at this moment.

Let us order the probabilities $\kappa_j^{(i,k,x,t)}(r,\nu;\ r',\nu')$ under the fixed values $i,\ j,\ k$ in lexicographic order of components $(r,\nu;\ r',\nu')$ and form the square matrices

$$\tilde{K}_j(i,k,x,t)=(\kappa_j^{(i,k,x,t)}(r,\nu;\ r',\nu'))_{\nu,\nu'=\overline{0,W};r,r'=\overline{0,N}},\ i\ge 0,\ k=0,1,2.$$

Then, using the limit theorem for semi-regenerative processes, see [22], the vectors $\tilde{\boldsymbol{V}}(i,k,x)$ can be expressed through the stationary state distribution $\boldsymbol{\pi}_j,\ j\ge 0$ of the embedded MC $\xi_n,\ n\ge 1$ as follows:

$$\tilde{\boldsymbol{V}}(i,k,x)=\tau^{-1}\sum_{j=0}^{\infty}\boldsymbol{\pi}_j\int_0^{\infty}\tilde{K}_j(i,k,x,t)dt,\ i\ge 0,\ k=0,1,2.$$

The corresponding vector LSTs $\tilde{\boldsymbol{v}}(i,k,s)$ are defined by the formula

$$\tilde{\boldsymbol{v}}(i,k,s)=\tau^{-1}\sum_{j=0}^{\infty}\boldsymbol{\pi}_j\int_0^{\infty}\tilde{K}_j^*(i,k,s,t)dt,\ i\ge 0,\ k=0,1,2, \qquad (6.25)$$

where $\tilde{K}_j^*(i,k,s,t)=\int_0^{\infty}e^{-sx}d\tilde{K}_j(i,k,x,t)$.

Formulas (6.21) follow directly from (6.25) taking into account that $\tilde{K}_j^*(i,k,s,t)=0$ in the region $\{j\ge 0,\ i=0,\ k=1,2\}$ and $\{j\ge 0,\ i>0,\ k=0\}$. The derivation of the expressions for the rest of the matrices $\tilde{K}_j^*(i,k,s,t)$ is based on usage of their probabilistic sense, the probabilistic sense of the LST (see, for example, [80]), and careful analysis of the various scenarios of system behavior during time interval $[0, t]$, which begins at the service completion moment of a customer and finishes before the next customer completes its service.

For example, let $k = 1, i > 0$. Then the matrices $\tilde{K}_j^*(i, 1, s, t)$ have the form

$$\tilde{K}_0^*(i, 1, s, t) = \bar{Q} \int_0^t [I_{N+1} \otimes e^{D_0 x} \sum_{k=1}^{i} D_k dx P(i-k, t-x)]$$

$$\times \int_0^\infty e^{-su} e^{\Delta(t+u)} \otimes I_{\bar{W}} dB(t-x+u)$$

$$+(1-p)\{\int_0^t [dF(x) \otimes \sum_{k=1}^{i} P(k, x) P(i-k, t-x)]\tilde{Q}_3$$

$$\times \int_0^\infty e^{-su} e^{\Delta(t-x+u)} \otimes I_{\bar{W}} dB(t-x+u)$$

$$+ \int_0^t \int_0^y [dF(x) \otimes (e^{D_0 y} \sum_{k=1}^{i} D_k dy P(i-k, t-y))]\tilde{Q}_3$$

$$\times \int_0^\infty e^{-su} e^{\Delta(t-x+u)} \otimes I_{\bar{W}} dB(t-y+u)\}, \; i > 0,$$

$$\tilde{K}_j^*(i, 1, s, t) = \bar{Q}[I_{N+1} \otimes P(i-j, t)] \int_0^\infty e^{-su} e^{\Delta(t+u)} \otimes I_{\bar{W}} dB(t+u)$$

$$+(1-p) \int_0^t [dF(x) \otimes \sum_{k=0}^{i-j} P(k, x) P(i-k-j, t-x)]\tilde{Q}_3$$

$$\times \int_0^\infty e^{-su} e^{\Delta(t-x+u)} \otimes I_{\bar{W}} dB(t-x+u), \; j > 0, \; i > 0.$$

Substituting these expressions into (6.25), after algebraic transformations involving an integration order change, we obtain formula (6.23) for the vectors $\tilde{\mathbf{v}}(i, 1, s), \; i > 0$.

Formulas (6.22) and (6.24) are obtained in a similar way. □

Next, we find the LST of the distribution of the *generalized service time* at the first station. Let $\hat{B}(x) = (\hat{B}(x)_{r,r'})_{r,r'=\overline{0,N}}$, where $\hat{B}(x)_{r,r'} = P\{t_{n+1} - t_n < x, r_{n+1} = r' \mid r_n = r, i_n \neq 0\}$ and $\mathcal{B}(s) = \int\limits_0^\infty e^{-st} d\hat{B}(t)$. For brevity, the matrices $\hat{B}(x)$ and $\mathcal{B}(s)$ are further referred to as the matrix distribution function and the matrix LST of the *generalized service time* distribution. Similar terms will be used for other matrices (vectors) consisting of the distribution functions and LSTs of a distribution.

Lemma 6.3 *The matrix LST of the generalized service time at the first station has the form*

$$\mathcal{B}(s) = [Q_1 + pQ_2 + (1-p)F^*(s)Q_3]\mathcal{B}^*(s), \tag{6.26}$$

where

$$F^*(s) = \int\limits_0^\infty e^{-st} dF(t).$$

Proof Let us analyze the structure of the generalized service time, that is, the time interval between two neighboring service completion moments at the first station that belong to the same busy period.

The generalized service time of a tagged customer is the same as the customer service time at the first station if the previous customer did not block this station. In this case, the matrix LST of the distribution of the generalized sojourn time has the form $(Q_1 + pQ_2)\mathcal{B}^*(s)$. In the case of blocking, the generalized service time consists of the time during which the server is blocked by the previous customer and the tagged customer's service time. Then the corresponding LST has the form $(1-p)\int\limits_0^\infty e^{-st} dF(t) Q_3 \mathcal{B}^*(s)$. □

Now we can derive the relation for the vector LST $\boldsymbol{v}_1(s)$ of the distribution of the virtual sojourn time at the first station. Let $v_1(r, \nu, x)$ be the probability that the $BMAP$ is in state ν at an arbitrary time, the virtual sojourn time at the first station is less than x, and the number of busy servers at the second station is r right before the moment of the virtual sojourn time completion. Then $\boldsymbol{v}_1(s)$ is defined as the vector of LSTs $v_1(r, \nu, s) = \int\limits_0^\infty e^{-sx} dv_1(r, \nu, x)$ written in lexicographic order.

Theorem 6.4 *The vector LST $\boldsymbol{v}_1(s)$ of the stationary distribution of the virtual sojourn time at the first station satisfies the equation*

$$\boldsymbol{v}_1(s)A(s) = \boldsymbol{\pi}_0\Phi(s), \tag{6.27}$$

where

$$A(s) = sI + \sum_{r=0}^{\infty} \mathcal{B}^r(s) \otimes D_r, \tag{6.28}$$

$$\Phi(s) = \tau^{-1}\hat{Q}(\Delta \oplus D_0)^{-1}(\Delta \otimes I_{\bar{W}} - sI)[\mathcal{B}^*(s) \otimes I_{\bar{W}}].$$

Proof As noted above, the virtual sojourn time at the first station at time moment t consists of the residual service time of the current customer, the generalized service time of customers that wait for their service at time t, and the generalized service time of a virtual customer. Note that

- the residual service time and the generalized service times of customers are independent random variables given that the MC $\xi_n,\ n \geq 1$ states are fixed;
- the generalized service times are identically distributed if the number of busy servers at the second station is fixed at the ends of the corresponding time intervals;
- after an arbitrary time moment t the input flow does not affect a virtual sojourn time started at time t.

Taking into account these explanations and using the total probability formula, we obtain the following expression for the vector LST $\boldsymbol{v}_1(s)$:

$$\boldsymbol{v}_1(s)=\tilde{\boldsymbol{v}}(0,0,s)+\sum_{i=1}^{\infty}\tilde{\boldsymbol{v}}(i,1,s)[\mathcal{B}^i(s)\otimes I_{\bar{W}}]+\sum_{i=1}^{\infty}\tilde{\boldsymbol{v}}(i,2,s)[\mathcal{B}^{i-1}(s)\otimes I_{\bar{W}}]. \tag{6.29}$$

Multiplying (6.29) by the matrix $sI + \sum\limits_{r=0}^{\infty} \mathcal{B}^r(s) \otimes D_r$, after a series of cumbersome algebraic transformations we obtain the relation

$$\boldsymbol{v}_1(s)(sI + \sum_{r=0}^{\infty} \mathcal{B}^r(s) \otimes D_r) = \tau^{-1}\{\boldsymbol{\pi}_0 \sum_{i=0}^{\infty}[V_i + \sum_{j=1}^{i+1}\boldsymbol{\pi}_j Y_{i-j+1}][\mathcal{B}^{i+1}(s) \otimes I_{\bar{W}}]$$

$$+\boldsymbol{\pi}_0\hat{Q}[-(\Delta \oplus D_0)^{-1}(sI + \tilde{D}_0) + I][\mathcal{B}^*(s) \otimes I_{\bar{W}}] - \sum_{j=0}^{\infty}\boldsymbol{\pi}_j[\mathcal{B}^{j+1}(s) \otimes I_{\bar{W}}]\}. \tag{6.30}$$

For the further transformations of (6.30), we use the equilibrium equation (2.68). Multiplying the ith equation of (2.68) by $\mathcal{B}^{i+1}(s) \otimes I_{\bar{W}}$ and summing over i, we obtain

$$\sum_{i=0}^{\infty}\boldsymbol{\pi}_i[\mathcal{B}^{i+1}(s)\otimes I_{\bar{W}}] = \boldsymbol{\pi}_0\sum_{i=0}^{\infty}V_i[\mathcal{B}^{i+1}(s)\otimes I_{\bar{W}}] + \sum_{i=0}^{\infty}\sum_{j=1}^{i+1}\boldsymbol{\pi}_j Y_{i-j+1}[\mathcal{B}^{i+1}(s)\otimes I_{\bar{W}}]. \tag{6.31}$$

Using (6.31) in order to simplify (6.30), we obtain (6.27). □

Thus, we obtained Eq. (6.27) for the LST of the sojourn time distribution at the first station. Now we derive the LST of the sojourn time distribution at the second station.

Let $\boldsymbol{v}_2(s)$ be the column vector of the LSTs of the conditional distribution of the customer sojourn time at the second station. The rth entry of this vector is the LST of the sojourn time distribution at the second station given that the number of busy servers at the second station is r before the completion of the customer sojourn time at the first station.

Theorem 6.5 *The vector LST of the stationary sojourn time distribution at the second station has the form*

$$\boldsymbol{v}_2(s) = [Q_4(F^*(s) + I)\hat{I} + pQ_2$$

$$+ (1-p)F^*(s)\text{diag}\{f_{r,0}(s), r = N, N-1, \ldots, 0\}Q_3]\mathbf{e}, \tag{6.32}$$

where

$$Q_4 = \begin{pmatrix} q_0 & q_1 & \cdots & q_{N-1} & q_N \\ q_0 & q_1 & \cdots & q_{N-1} & 0 \\ \vdots & \vdots & \ddots & \vdots & \vdots \\ q_0 & 0 & \ldots & 0 & 0 \end{pmatrix}.$$

Proof A customer sojourn time that requires m idle servers at the second station consists of

(a) a customer service time when at least m servers of the second station are idle immediately after the service completion at the first station;
(b) zero time if the required number of servers are not idle, and the customer leaves the system;
(c) the blocking time and the customer service time if the required number of servers are not idle, and the customer waits for their availability.

It is assumed that the customer is served by m servers independently of each other and its service is completed when all m servers complete the service. This service time distribution is determined by the LST $f_{m,0}(s)$, $m = \overline{1, N}$.

Taking into account the assumption above and items (a)–(c), we obtain relation (6.32) for the vector $\boldsymbol{v}_2(s)$. In this relation, the first term refers to case (a), and the second and the third terms define the LST in cases (b) and (c), respectively. □

Theorem 6.6 *The LST of the stationary distribution of the sojourn time in the system has the form*

$$v(s) = \boldsymbol{v}_1(s)(I_{N+1} \otimes \mathbf{e}_{\bar{W}})\boldsymbol{v}_2(s). \tag{6.33}$$

Proof Formula (6.33) follows from the structure of the virtual sojourn time in the system, which consists of the virtual sojourn time at the first station and the sojourn time at the second one. □

The Real Sojourn Time Let $v_1^{(a)}(s)$ and $v^{(a)}(s)$ be the LSTs of the real sojourn time at the first station and in the whole tandem, respectively.

Theorem 6.7 *The LST of the stationary distribution of the real sojourn time at the first station is calculated as follows:*

$$v_1^{(a)}(s) = \lambda^{-1}\boldsymbol{v}_1(s)\sum_{k=0}^{\infty}[\mathcal{B}^k(s)(\mathcal{B}(s) - I)^{-1} \otimes D_k]\mathbf{e}. \tag{6.34}$$

Proof The real sojourn time at the first station of a tagged customer that arrived in a batch of size k and was the jth customer in the batch consists of

(a) the real sojourn time at the first station of the first customer of the batch which coincides with the virtual sojourn time at the first station;
(b) the generalized service times at the first station of $j - 2$ customers in the batch that arrived together with the tagged customer;
(c) the generalized service time of the tagged customer at the first station.

The vector LST of the first station sojourn time distribution of the first customer in the size k batch containing a tagged customer is determined by the vector $\boldsymbol{v}_1(s)(I_{N+1} \otimes \frac{kD_k\mathbf{e}}{\lambda})$. Assuming that an arbitrary tagged customer arriving in the size k batch is in the jth position in the batch with probability $1/k$, and using the total probability formula, we get

$$v_1^{(a)}(s) = \sum_{k=1}^{\infty}\boldsymbol{v}_1(s)(I_{N+1} \otimes \frac{kD_k\mathbf{e}}{\lambda})\sum_{j=1}^{k}\frac{1}{k}\mathcal{B}^{j-1}(s)\mathbf{e},$$

or

$$v_1^{(a)}(s) = \lambda^{-1}\boldsymbol{v}_1(s)\sum_{k=1}^{\infty}(I_{N+1} \otimes D_k\mathbf{e})\sum_{j=1}^{k}\mathcal{B}^{j-1}(s)\mathbf{e}. \tag{6.35}$$

By means of algebraic transformations, (6.35) reduces to the form (6.34). □

Theorem 6.8 *The LST of the real sojourn time distribution in the whole system is calculated as follows:*

$$v^{(a)}(s) = \lambda^{-1}\boldsymbol{v}_1(s)\sum_{k=0}^{\infty}[\mathcal{B}^k(s)(\mathcal{B}(s) - I)^{-1} \otimes D_k](I_{N+1} \otimes \mathbf{e}_{\bar{W}})\boldsymbol{v}_2(s).$$

6.2.6 Sojourn Time Moments

Denote by $\boldsymbol{v}_1^{(m)}(s)$ the mth derivative of the vector $\boldsymbol{v}_1(s)$, $m \geq 1$. Let $\boldsymbol{v}_1^{(0)}(s) = \boldsymbol{v}_1(s)$.

Theorem 6.9 *Let* $\int\limits_0^{\infty} t^m dB(t) < \infty$, $m = \overline{1, M+1}$. *Then the vectors* $\boldsymbol{v}_1^{(m)}(0)$, $m = \overline{1, M}$ *are calculated recurrently as*

$$\boldsymbol{v}_1^{(m)}(0) = \Bigg[\Bigg(\boldsymbol{\pi}_0\Phi^{(m)}(0) - \sum_{l=0}^{m-1} C_m^l \boldsymbol{v}_1^{(l)}(0)A^{(m-l)}(0)\Bigg)\tilde{I} \tag{6.36}$$

$$+\frac{1}{m+1}\Bigg(\boldsymbol{\pi}_0\Phi^{(m+1)}(0) - \sum_{l=0}^{m-1} C_{m+1}^l \boldsymbol{v}_1^{(l)}(0)A^{(m+1-l)}(0)\Bigg)\mathbf{e}\hat{\mathbf{e}}\Bigg]\tilde{A}^{-1}$$

with initial condition

$$\boldsymbol{v}_1^{(0)}(0) = \boldsymbol{v}_1(0) = [\tau^{-1}\boldsymbol{\pi}_0\hat{Q}(\Delta \oplus D_0)^{-1}(\Delta\mathcal{B}^*(0) \otimes I_{\bar{W}})\tilde{I} + P_{idle}\hat{\mathbf{e}}]\tilde{A}^{-1}, \tag{6.37}$$

where

$$\tilde{A} = A(0)\tilde{I} + A'(0)\mathbf{e}\hat{\mathbf{e}}.$$

Proof Differentiating (6.27), we obtain

$$\boldsymbol{v}_1^{(m)}(0)A(0) = \boldsymbol{\pi}_0\Phi^{(m)}(0) - \sum_{l=0}^{m-1} C_m^l \boldsymbol{v}_1^{(l)}(0)A^{(m-l)}(0),\ m \geq 0. \tag{6.38}$$

It follows from (6.28) that $A(0) = \sum_{r=0}^{\infty} \mathcal{B}^r(0) \otimes D_r$, where $\mathcal{B}(0)$ is an irreducible stochastic matrix. Then, as it is easy to see, $A(0)$ is an irreducible infinitesimal generator, which implies that $A(0)$ is a singular matrix. As a consequence, it is impossible to obtain a recursive procedure to compute the vectors $\boldsymbol{v}_1^{(m)}(0),\ m \geq 0$ directly from (6.38). We transform the system (6.38) to obtain a system with a non-singular matrix. To this aim, we multiply both sides of expression (6.38) for $m+1$ to the right by $\mathbf{e}$. Taking into account the fact that $A(0)\mathbf{e} = \mathbf{0}^T$, we derive

$$\boldsymbol{v}_1^{(m)} A'(0)\mathbf{e} = \frac{1}{m+1}[\boldsymbol{\pi}_0 \Phi^{(m+1)}(0) - \sum_{l=0}^{m-1} C_{m+1}^l \boldsymbol{v}_1^{(l)}(0) A^{(m+1-l)}(0)]\mathbf{e}. \tag{6.39}$$

It can be shown that the right-hand side of expression (6.39) is not zero. It is positive if $m = 2k$, and negative if $m = 2k+1,\ k \geq 0$. Replacing one of the equations of system (6.38) (without loss of generality, the first one) by Eq. (6.39), we obtain the following non-homogeneous system of linear algebraic equations for the vector $\boldsymbol{v}_1^{(m)}(0)$:

$$\boldsymbol{v}_1^{(m)}(0)\tilde{A} = [\boldsymbol{\pi}_0 \Phi^{(m)}(0) - \sum_{l=0}^{m-1} C_m^l \boldsymbol{v}_1^{(l)}(0) A^{(m-l)}(0)]\tilde{I}$$

$$+\frac{1}{m+1}[\boldsymbol{\pi}_0 \Phi^{(m+1)}(0) - \sum_{l=0}^{m-1} C_{m+1}^l \boldsymbol{v}_1^{(l)}(0) A^{(m+1-l)}(0)]\mathbf{e}\hat{\mathbf{e}},\ m \geq 0.$$

This system has a unique solution if the matrix $\tilde{A}$ is non-singular. We prove this by showing that $\det\tilde{A} \neq 0$.

Calculate $\det\tilde{A}$ as follows:

$$\det\tilde{A} = \nabla A'(0)\mathbf{e}, \tag{6.40}$$

where ∇ is the vector of the algebraic adjuncts of the first column of the matrix $A(0)$. Since $A(0)$ is an irreducible infinitesimal generator, the vector ∇ is proportional to the solution of the system

$$\mathbf{x}A(0) = \mathbf{0}, \tag{6.41}$$

i.e.,

$$\nabla = c\mathbf{x},\ c \neq 0. \tag{6.42}$$

Let vector $\mathbf{y}$ be the unique solution of the system of linear algebraic equations

$$\mathbf{y}\mathcal{B}(0) = \mathbf{y},\ \mathbf{y}\mathbf{e} = 1.$$

Then the vector $\mathbf{x} = \mathbf{y} \otimes \boldsymbol{\theta}$ is the solution of system (6.41). This fact can be proved by direct substitution of this vector into (6.41).

It follows from (6.42) that

$$\boldsymbol{\nabla} = c(\mathbf{y} \otimes \boldsymbol{\theta}). \tag{6.43}$$

Substituting expression (6.43) for $\boldsymbol{\nabla}$ in (6.40), we obtain

$$\det\tilde{A} = c(\mathbf{y} \otimes \boldsymbol{\theta})A'(0)\mathbf{e} = c(\mathbf{y} \otimes \boldsymbol{\theta})(sI + \sum_{r=0}^{\infty} \mathcal{B}^r(s) \otimes D_r)'|_{s=0}\mathbf{e}$$

$$= c + c\sum_{r=1}^{\infty} r\mathbf{y}\mathcal{B}'(0)\mathbf{e} \otimes \boldsymbol{\theta} D_r\mathbf{e} = c(1 + \lambda\mathbf{y}\mathcal{B}'(0)\mathbf{e}). \tag{6.44}$$

For a further analysis of $\det\tilde{A}$, we use the ergodicity condition obtained in Theorem 6.1.

Assuming $s = 0$ in (6.26) and taking into account that by definition $Q_1 + pQ_2 + (1-p)F^*(0)Q_3 = \mathcal{Q}$, we obtain $\mathcal{B}(0) = \mathcal{Q}\mathcal{B}^*(0)$, i.e., $\mathcal{B}(0)$ coincides with the matrix of system (6.8). Then the vector $\mathbf{y}$ is the unique solution of system (6.8). It is easy to show that

$$\mathbf{y}\mathcal{B}'(0)\mathbf{e} = -[b_1 + (1-p)\mathbf{y}\int_0^{\infty} t dF(t) Q_3\mathbf{e}]. \tag{6.45}$$

Multiplying (6.45) by λ and comparing the resulting expression to the expression for ρ in the statement of Theorem 6.1, we obtain

$$\lambda\mathbf{y}\mathcal{B}'(0)\mathbf{e} = -\rho. \tag{6.46}$$

It follows from (6.44) and (6.46) that $\det\tilde{A} = c(1-\rho)$. Since the ergodicity condition $\rho < 1$ holds and $c \neq 0$ then $\det\tilde{A} \neq 0$. □

Below, we assume that $\int_0^{\infty} t^k dB(t) < \infty,\ k = \overline{1,2}$.

Corollary 6.5 *The mean sojourn time at the first station is calculated by*

$$\overline{v}_1 = \{[\tau^{-1}\boldsymbol{\pi}_0\hat{Q}(\Delta \oplus D_0)^{-1}((\mathcal{B}^*(0) - \Delta\mathcal{B}^{*\prime}(0)) \otimes I_{\bar{W}}) + \boldsymbol{v}_1(0)A'(0)]\tilde{I}$$

$$+ [P_{idle}b_1 + \frac{1}{2}\boldsymbol{v}_1(0)A''(0)\mathbf{e}]\hat{\mathbf{e}}\}\tilde{A}^{-1}\mathbf{e}, \tag{6.47}$$

where the vector $\boldsymbol{v}_1(0)$ is defined by formula (6.37).

Proof The proof follows from the relation $\overline{v}_1 = -\mathbf{v}_1'(0)\mathbf{e}$ and formula (6.36). □

Theorem 6.10 *The mean sojourn time in the system is calculated as*

$$\overline{v} = \overline{v}_1 + \mathbf{v}_1(0)(I_{N+1} \otimes \mathbf{e}_{\bar{W}})\overline{\mathbf{v}}_2.$$

Here $\mathbf{v}_1(0)$ and $\overline{v}_1$ are defined in (6.37) and (6.47) respectively, and $\overline{\mathbf{v}}_2$ is the vector of the conditional mean sojourn times at the second station, which is calculated as follows:

$$\overline{\mathbf{v}}_2 = -Q_4 \int_0^\infty t dF(t)\hat{\mathbf{e}}^T - (1-p)[\int_0^\infty t dF(t)$$

$$+ E\text{diag}\,\{\sum_{l=1}^{r} (l\mu)^{-1}, r = N, N-1, \ldots, 0\}]Q_3\mathbf{e}. \tag{6.48}$$

Proof Let us differentiate (6.33). Substituting $s = 0$ and changing the sign, we get

$$\overline{v} = -\mathbf{v}_1'(0)(I_{N+1} \otimes \mathbf{e}_{\bar{W}})\mathbf{v}_2(0) - \mathbf{v}_1(0)(I_{N+1} \otimes \mathbf{e}_{\bar{W}})\mathbf{v}_2'(0). \tag{6.49}$$

Substituting $s = 0$ in (6.32), we obtain

$$\mathbf{v}_2(0) = [Q_4(E + I)\hat{I} + pQ_2 + (1-p)EQ_3]\mathbf{e} = \mathbf{e}.$$

This means that the first term in the right-hand side of (6.49) is equal to $-\mathbf{v}_1'(0)\mathbf{e} = \overline{v}_1$. Using the relation $\overline{\mathbf{v}}_2 = -\mathbf{v}_2'(0)\mathbf{e}$ and differentiating (6.32) at $s = 0$, we obtain that $\overline{\mathbf{v}}_2$ has the form (6.48). □

Theorem 6.11 *The mean real sojourn time at the first station is calculated as*

$$\overline{v}_1^{(a)} = -\lambda^{-1}\{\mathbf{v}_1'(0)(\mathbf{e} \otimes \sum_{k=1}^{\infty} kD_k\mathbf{e})$$

$$+\mathbf{v}_1(0)\sum_{k=1}^{\infty}(I_{N+1} \otimes D_k\mathbf{e})[\sum_{n=1}^{k-1}\sum_{l=0}^{n-1} \mathcal{B}^l(0)\mathcal{B}'(0)\mathbf{e}]\}.$$

Proof The proof is based on the relation $\overline{v}_1^{(a)} = -\frac{dv_1^{(a)}(s)}{ds}|_{s=0}$ and formula (6.34). □

Corollary 6.6 *The mean real sojourn time in the system is calculated by the formula*

$$\overline{v}^{(a)} = \overline{v}_1^{(a)} + \lambda^{-1}\boldsymbol{v}_1(0)\sum_{k=1}^{\infty}(I_{N+1} \otimes D_k\mathbf{e})\sum_{n=0}^{k-1}\mathcal{B}^n(0)\overline{\boldsymbol{v}}_2.$$

6.2.7 Numerical Examples

Experiment 6.1 In this experiment, we investigate the effect of the coefficient of correlation of the input flow on the main performance characteristics of the system.

Below we present four MAPs defined by the matrices D_0 and $D_1 = D$. All these MAPs have the same intensity $\lambda = 5$ and different coefficients of correlation.

MAP_1 is a stationary Poisson flow defined by the matrices $D_0 = -5$ and $D = 5$ with the coefficient of correlation $c_{cor} = 0$ and the coefficient of variation $c_{var} = 1$.

MAP_2 has the coefficient of correlation $c_{cor} = 0.1$ and is characterized by the matrices

$$D_0 = \begin{pmatrix} -13.3346 & 0.5886 & 0.6173 \\ 0.6927 & -2.4466 & 0.4229 \\ 0.6823 & 0.4144 & -1.6354 \end{pmatrix}, \; D = \begin{pmatrix} 11.5469 & 0.3631 & 0.2187 \\ 0.3842 & 0.8659 & 0.0809 \\ 0.2852 & 0.0425 & 0.2111 \end{pmatrix}.$$

MAP_3 has the coefficient of correlation $c_{cor} = 0.2$ and is characterized by the matrices

$$D_0 = \begin{pmatrix} -15.7327 & 0.6062 & 0.5924 \\ 0.5178 & -2.2897 & 0.4679 \\ 0.5971 & 0.5653 & -1.9597 \end{pmatrix}, \; D = \begin{pmatrix} 14.1502 & 0.3021 & 0.0818 \\ 0.1071 & 1.032 & 0.1646 \\ 0.0858 & 0.1979 & 0.5136 \end{pmatrix}.$$

MAP_4 has the coefficient of correlation $c_{cor} = 0.3$ and is characterized by the matrices

$$D_0 = \begin{pmatrix} -25.5398 & 0.3933 & 0.3612 \\ 0.1452 & -2.2322 & 0.2000 \\ 0.2960 & 0.3874 & -1.7526 \end{pmatrix}, \; D = \begin{pmatrix} 24.2421 & 0.4669 & 0.0763 \\ 0.0341 & 1.6668 & 0.1861 \\ 0.0090 & 0.2555 & 0.8047 \end{pmatrix}.$$

MAP_2–MAP_4 have the coefficient of variation $c_{var} = 2$.

Based on these MAPs, we construct $BMAP$s with the maximal batch size equal to 5. Each of these $BMAP$s is defined by the matrices $D_k, k = \overline{0,5}$ obtained as follows: the matrix D_0 coincides with the corresponding matrix for the MAP and other matrices are defined as follows: $D_k = D\varkappa^{k-1}(1-\varkappa)/(1-\varkappa^5),\ k = \overline{1,5}$, where $\varkappa = 0.8$. Then we multiply all matrices $D_k,\ k = \overline{0,5}$, by a positive scalar in order to obtain a $BMAP$ with intensity $\lambda = 5$. The $BMAP$ derived from the MAP_n

will be denoted by $BMAP_n,\ n = \overline{1,4}$. Note that the coefficients of correlation of the $BMAP$s constructed in such a way coincide with the corresponding coefficients of the initial MAPs.

The service time at the first station has the order 3 Erlang distribution with parameter 20. The mean service time is $b_1 = 0.15$ and the squared coefficient of variation is $c_{var}^2 = 1/3$.

The number of servers at the second station is $N = 5$, and the probability that a customer served at the first station is lost is $p = 0.5$. The service parameters at the second station are defined by the intensity $\mu = 5$ and the probabilities $q_0 = 0.1,\ q_1 = q_2 = 0.3,\ q_3 = q_4 = q_5 = 0.1$.

Figures 6.1 and 6.2 show the dependence of the mean virtual $\bar{v}$ and the mean real $\bar{v}_a$ sojourn times, the loss probability P_{loss}, and the mean number N_{busy} of busy servers at the second station on the system load ρ. The value ρ varies with the intensity λ, which in turn varies by means of the normalization of the entries of the matrices $D_k,\ k = \overline{0,5}$. Note that the coefficients of variation and correlation of the $BMAP$ flow do not change here.

It can be seen from Figs. 6.1 and 6.2 that for the same value of the system load, the values of these characteristics are significantly different for $BMAP$s with

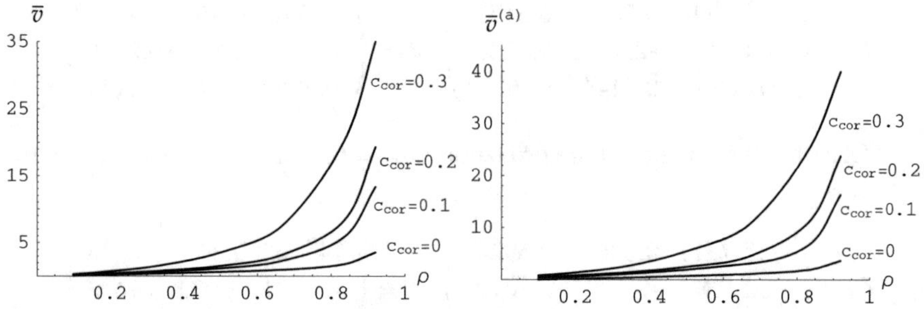

Fig. 6.1 Dependence of the mean virtual and real sojourn time on the system load for $BMAP$s with different coefficients of correlation

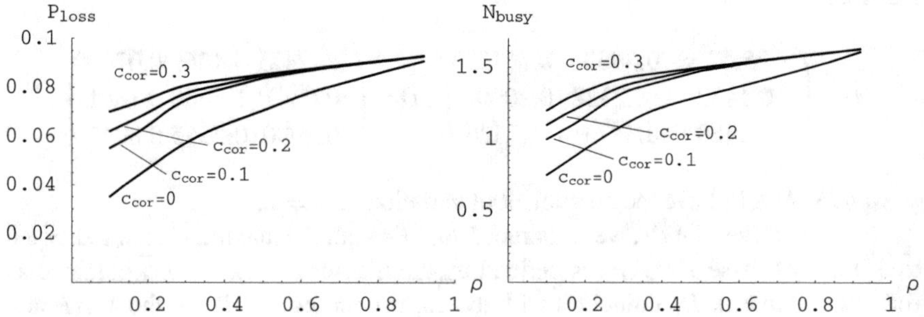

Fig. 6.2 Dependence of the loss probabilities and the mean number of busy servers at the second station on the system load for $BMAP$s with different coefficients of correlation

different correlations. With an increase in the coefficient of correlation all the above characteristics increase, which indicates a deterioration in the quality of service and confirms the necessity of taking into account the effect of correlation of the input flow when assessing the performance of the considered queue.

Experiment 6.2 Let us investigate the effect of the coefficient of the service process variation at the first station on the main performance characteristics.

Consider four service time distributions with the same mean service time $b_1 = 0.1$ but with different coefficients of variation: D is a degenerate distribution; M is an exponential distribution; $HM_2^{(1)}$ and $HM_2^{(2)}$ are order 2 hyperexponential distributions with different parameters. The hyperexponential distributions are given by probabilities (0.05, 0.95) and intensities 0.62025, 48.9998 in case of $HM_2^{(1)}$, and by probabilities (0.98, 0.02) and intensities 10, 000, 0.2 in the case of $HM_2^{(2)}$. The coefficients of variation of the processes D, M, $HM_2^{(1)}$, and $HM_2^{(2)}$ are $0,\ 1,\ 5,\ 9.95$, respectively.

The input flow differs from $BMAP_2$ from the previous experiment the the intensity λ. Here, the matrices $D_k, k = \overline{0, 5}$ are normalized so that $\lambda = 1$. The service intensity μ at the second station is equal to 0.8. The remaining parameters of the system are assumed to be the same as in the previous experiment.

We will change the average service time b_1 for all processes in the interval $[0.1, 0.95]$. Note that the coefficient of variation remains unchanged.

Figures 6.3 and 6.4 show the dependence of the main performance characteristics of the considered system on the mean service time at the first station.

Figures 6.3 and 6.4 show that all characteristics depend significantly on the coefficient of variation of the service process. We also carried out experiments with uniform and Erlang distributions of service time. As the coefficients of variation of these distributions are in the interval (0, 1), the corresponding curves are located between the two lower curves in Figs. 6.3 and 6.4, as expected.

Note that the ergodicity condition $\rho < 1$ does not hold for processes $HM_2^{(1)}$ and $HM_2^{(2)}$ for values b_1 over 0.7 and 0.6, respectively.

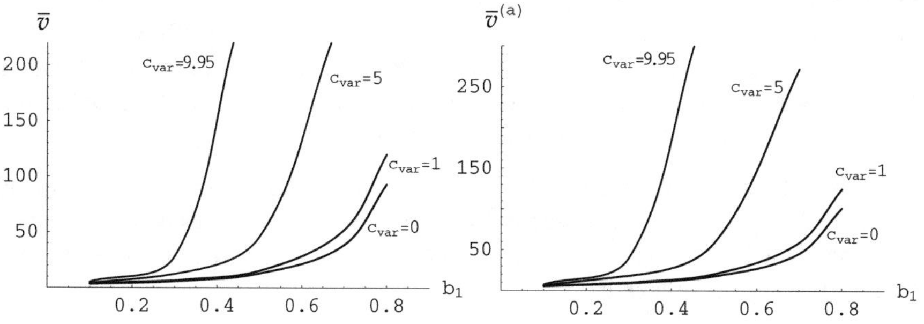

Fig. 6.3 The dependence of the mean virtual and the mean real sojourn times on the mean service time at the first station for the service processes with different coefficients of variation

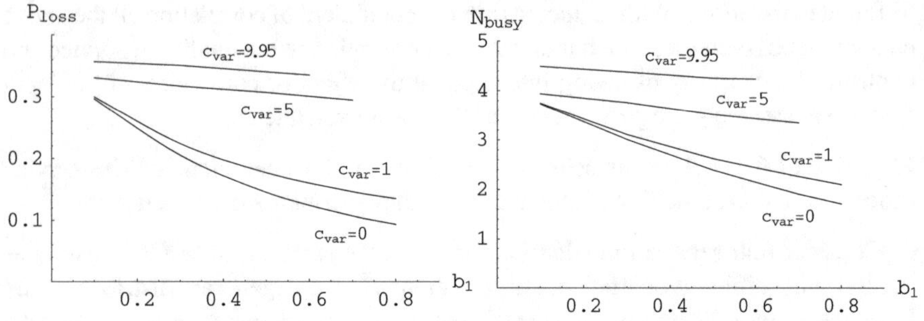

Fig. 6.4 The dependence of the loss probability and the mean number of busy servers at the second station on the mean service time at the first station for service processes with different coefficients of variation

Table 6.1 Values of the system load for different values of the mean service time at the first station and different coefficients of variation

ρ	$c_{var} = 0$	$c_{var} = 1$	$c_{var} = 5$	$c_{var} = 9.95$
$b_1 = 0.1$	0.42754	0.43016	0.45557	0.47940
$b_1 = 0.2$	0.47460	0.48228	0.53677	0.57784
$b_1 = 0.3$	0.53173	0.54515	0.62243	0.67725
$b_1 = 0.4$	0.59698	0.61590	0.71010	0.77695
$b_1 = 0.5$	0.66879	0.69247	0.79902	0.87678
$b_1 = 0.6$	0.74579	0.77339	0.88883	0.97667
$b_1 = 0.7$	0.82687	0.85763	0.97937	**1.07663**
$b_1 = 0.8$	0.91116	0.94440	**1.07051**	**1.17655**

Table 6.1 contains the values of the system load ρ in this experiment under the different values of b_1 and c_{var}. The values of ρ for which the ergodicity condition is not valid are shown in bold. Table 6.2 contains the mean virtual and the mean real sojourn times in this experiment.

Analyzing Figs. 6.2 and 6.4, we can see that the values of P_{loss} and N_{busy} increase as the system load ρ increases but decrease as the mean service time b_1 increases. At first glance, this seems implausible because an increase in b_1 entails an increase in system load, which usually negatively affects the quality of service. Let us explain our results. In Fig. 6.2, the value of ρ increases due to an increase in intensity λ. Obviously, increasing the intensity of the input flow leads to an increase in P_{loss} and N_{busy}. When the mean service time b_1 at the first station increases then the input flow is smoothed and the effect of the correlation in $BMAP$ is reduced for the second station. This positively affects the values P_{loss} and N_{busy}.

Experiment 6.3 In this experiment, we investigate numerically the problem of choosing the optimal number of servers at the second station. Let us find the number of servers N at the second station that minimizes the following quality criterion (the average penalty per time unit):

$$J = J(N) = aN + c_1\lambda P_{loss} + c_2\bar{v}^{(a)}.$$

Table 6.2 Values of the mean virtual and mean real sojourn times for different values of the mean service time at the first station and different coefficients of variation

	$c_{var} = 0$	$c_{var} = 1$	$c_{var} = 5$	$c_{var} = 9.95$
$b_1 = 0.1$				
$\bar{v}$	3.31836	3.35900	4.14502	5.31900
$\bar{v}^{(a)}$	5.34117	5.56531	6.42475	7.61900
$b_1 = 0.3$				
$\bar{v}$	5.78467	6.44377	12.36466	28.03929
$\bar{v}^{(a)}$	9.06977	9.79000	18.36466	40.03929
$b_1 = 0.5$				
$\bar{v}$	12.78808	15.18521	45.44273	307.20933
$\bar{v}^{(a)}$	17.88462	20.60316	58.44273	383.20933
$b_1 = 0.7$				
$\bar{v}$	38.23352	51.20479	252.27566	–
$\bar{v}^{(a)}$	45.47547	58.89069	272.27566	–
$b_1 = 0.8$				
$\bar{v}$	92.87134	120.01298	–	–
$\bar{v}^{(a)}$	101.358218	125.03418	–	–

Here a is the cost of maintaining a server at the second station per time unit, c_1 is the penalty for the loss of one customer after it has been served at the first station per time unit, c_2 is the fee for a customer stay in the system per time unit.

Here and below, when solving integer optimization problems, we use the straightforward enumeration method.

Using a MAP characterized by the matrices

$$D_0 = \begin{pmatrix} -6.74538 & 5.45412 \times 10^{-6} \\ 5.45412 \times 10^{-6} & -0.219455 \end{pmatrix}, \quad D = \begin{pmatrix} 6.700685 & 0.044695 \\ 0.122427 & 0.097023 \end{pmatrix},$$

we construct a $BMAP$ with the matrices $D_k, k = \overline{0, 5}$ in a similar way to the previous experiment, and normalize the matrices so as to obtain intensity $\lambda = 3$. The resulting $BMAP$ has the coefficient of correlation $c_{cor} = 0.2$ and $c_{var}^2 = 12.2732$.

The service time at the first station has the order 3 Erlang distribution with intensity 20. We assume that $p = 0.5, q_0 = 0.1, q_1 = 0.9, q_m = 0, m = \overline{2, N}$.

The cost coefficients are assumed to be $a = 5, c_1 = 50, c_2 = 3$.

The dependence of the cost criterion J on the number N of servers at the second station under the different service intensities μ is shown in Fig. 6.5.

Table 6.3 presents the cost criterion values for this experiment. The optimal value J^* of the cost criterion for each service intensity is shown in bold.

Figure 6.5 and Table 6.3 show that the optimal number of servers N^* increases from three to six as the service intensity decreases from 5 to 1.

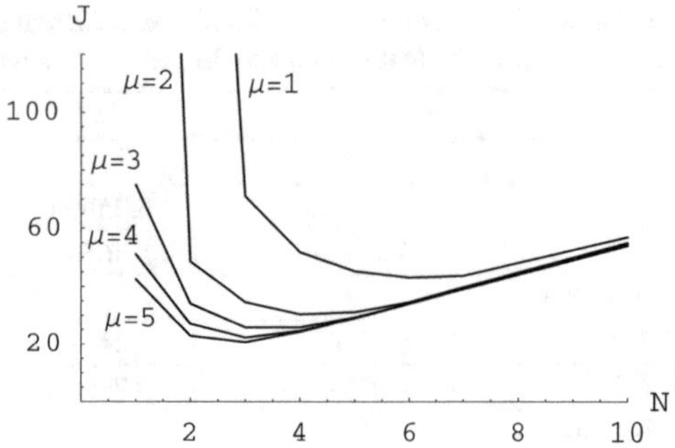

Fig. 6.5 The dependence of the cost criterion on the number of the servers at the second station under different service intensities

Table 6.3 The cost criterion values for different numbers of servers at the second station and the intensity of service at the second station

J	$\mu = 1$	$\mu = 2$	$\mu = 3$	$\mu = 4$	$\mu = 5$
$N = 1$	∞	457.0346	74.9351	51.0311	42.5146
$N = 2$	353.7378	48.6349	34.1130	27.0826	22.9120
$N = 3$	71.0480	34.6565	25.8441	**22.2656**	**20.6657**
$N = 4$	51.7822	**30.4068**	**25.8202**	24.6008	24.2393
$N = 5$	45.2556	31.2080	29.3925	29.1141	29.0630
$N = 6$	**43.0750**	34.6437	34.0980	34.0535	34.0486
$N = 7$	43.6867	39.0715	39.0533	39.0504	39.0476
$N = 8$	47.2564	44.0527	44.0514	44.0495	44.0474
$N = 9$	51.9854	49.0521	49.0511	49.0491	49.0473
$N = 10$	57.0056	55.0458	54.5001	54.0489	54.0472

The relative profit when using the optimal number N^* of servers compared to an arbitrary number N of servers at the second station is determined by the formula $g(N) = \frac{J(N)-J^*}{J^*}100\%$.

Consider the case $\mu = 5$ in detail. As can be seen from Fig. 6.5 and Table 6.3, the optimal value J^* of the cost criterion is 20.6657, and the optimal number of servers is $N^* = 3$. It should be noted that in this case the minimum relative profit is more than 10% if we get the optimal number $N^* = 3$ of servers instead of two servers, and the maximum relative profit is more than 161% if $N^* = 3$ rather than 10.

6.3 $BMAP/SM/1 \to \cdot/M/N/0$ Queue with a Group Occupation of Servers at the Second Station

6.3.1 Mathematical Model

We consider a two-station $BMAP/SM/1 \to \cdot/M/N/0$ queue with a semi-Markovian service process. For a $BMAP/G/1 \to \cdot/M/N/0$ queue described in Sect. 6.2, we assumed that service times at the first station are independent and identically distributed random variables. But in many real systems, service times can be dependent and distributed by different laws. As was noted above, for the modeling of such service processes, it has been proposed in the literature to use the formalism of semi-Markovian (SM) service processes. In our case, we assume that the service times of consecutive customers are given by the sojourn times of a regular ergodic semi-Markovian process $m_t,\ t \geq 0$ in its states. The process has the state space $\{1, 2, \ldots, M\}$ and the semi-Markovian kernel $B(t) = (B_{m,m'}(t))_{m,m'=\overline{1,M}}$. The function $B_{m,m'}(t)$ is interpreted as the probability that the sojourn time of the process in the current state does not exceed t and the next transition is into the state m' given that the current process state is m, $m, m' = \overline{1, M}$. The mean service time b_1 is calculated as

$$b_1 = \boldsymbol{\delta} \int_0^\infty t dB(t)\mathbf{e}, \tag{6.50}$$

where $\boldsymbol{\delta}$ is the invariant vector of the stochastic matrix $B(\infty)$, i.e., this vector satisfies equation

$$\boldsymbol{\delta} B(\infty) = \boldsymbol{\delta},\ \ \boldsymbol{\delta}\mathbf{e} = 1.$$

6.3.2 The Stationary State Distribution of the Embedded Markov Chain

The investigation of the tandem $BMAP/SM/1 \to \cdot/M/N/0$ starts with the introducing of the embedded process $\xi_n = \{i_n, r_n, \nu_n, m_n\}$, $n \geq 1$, where i_n is the number of customers at the first station (excluding one blocking the server) at time moment $t_n + 0$ (t_n is the nth moment of service completion at the first station); r_n is the number of busy servers at the second station at time $t_n - 0$; ν_n is the state of the $BMAP$ at time t_n; m_n is the state of the SM service process at time $t_n + 0$, $i_n \geq 0, r_n = \overline{0, N}, \nu_n = \overline{0, W},\ m_n = \overline{1, M}$.

Lemma 6.4 *The process $\xi_n, n \geq 1$ is a QTMC. The one-step transition probability matrix of this MC has a block structure of the form (2.2), where*

$$V_i = \sum_{k=1}^{i+1} [-\hat{H}(\Delta \oplus \bar{D}_0)^{-1}\hat{D}_k + (1-p)\bar{\mathcal{F}}_k\tilde{H}_3]\bar{\Omega}_{i-k+1},$$

$$Y_i = \bar{H}\bar{\Omega}_i + (1-p)\sum_{k=0}^{i} \bar{\mathcal{F}}_k\tilde{H}_3\bar{\Omega}_{i-k},\ i \geq 0,$$

$$\bar{\Omega}_n = \int_0^\infty e^{\Delta t} \otimes P(n,t) \otimes dB(t), \qquad \bar{\mathcal{F}}_n = \int_0^\infty dF(t) \otimes P(n,t) \otimes I_M,\ n \geq 0.$$

Here

$$\bar{D}_k = D_k \otimes I_M,\ \hat{D}_k = I_{N+1} \otimes \bar{D}_k,\ k \geq 0,$$

$$\tilde{H}_m = Q_m \otimes I_{\bar{W}M},\ m = \overline{1,3},\ \bar{H} = \tilde{H}_1 + p\tilde{H}_2,$$

$$\hat{H} = \tilde{H}_1 + p\tilde{H}_2 + (1-p)\int_0^\infty (dF(t) \otimes e^{D_0 t} \otimes I_M)\tilde{H}_3$$

and the matrices $\Delta,\ Q_m,\ m = \overline{1,3}$ *coincide with those presented above in Sect. 6.2.*

The proof is similar to the proof of Lemma 6.1.

Corollary 6.7 *The matrix PGFs* $V(z)$, $Y(z)$ *have the form*

$$V(z) = \frac{1}{z}[-\hat{H}(\Delta \oplus \bar{D}_0)^{-1}(\hat{D}(z) - \hat{D}_0) + (1-p)(\bar{\mathcal{F}}(z) - \bar{\mathcal{F}}_0)\tilde{H}_3]\bar{\Omega}(z),$$

$$Y(z) = [\bar{H} + (1-p)\bar{\mathcal{F}}(z)\tilde{H}_3]\bar{\Omega}(z), \tag{6.51}$$

where

$$\hat{D}(z) = \sum_{k=0}^{\infty} \hat{D}_k z^k,$$

$$\bar{\Omega}(z) = \int_0^\infty e^{\Delta t} \otimes e^{D(z)t} \otimes dB(t),\ \bar{\mathcal{F}}(z) = \int_0^\infty dF(t) \otimes e^{D(z)t} \otimes I_M,\ |z| \leq 1.$$

Theorem 6.12 *The stationary state distribution of the MC* $\xi_n,\ n \geq 1$ *exists if and only if the inequality*

$$\rho < 1 \tag{6.52}$$

holds. Here

$$\rho = \lambda[b_1 + (1-p)\mathbf{y}\int_0^\infty (t dF(t) Q_3 \otimes I_M)\mathbf{e}],$$

the vector $\mathbf{y}$ *is the unique solution of the following system of linear algebraic equations:*

$$\mathbf{y}(\mathcal{Q} \otimes I_M) \int_0^\infty e^{\Delta t} \otimes dB(t) = \mathbf{y}, \quad \mathbf{ye} = 1, \tag{6.53}$$

matrices Q_3, $\mathcal{Q}$ *are defined in Sect. 6.2, and* b_1 *is the mean service time at the first station and is given by formula (6.50).*

Proof The matrix $Y(1)$ defined by formula (6.51) is irreducible. Hence, in terms of $Y(z)$, a necessary and sufficient condition for the stationary distribution to exist has the form (2.65) and (2.66).

Let the vector $\mathbf{x}$ in (2.66) be of the form

$$\mathbf{x} = \mathbf{y}_1 \otimes \boldsymbol{\theta} \otimes \mathbf{y}_2, \tag{6.54}$$

where $\mathbf{y}_1$ and $\mathbf{y}_2$ are vectors of size $N+1$ and M, respectively, with non-negative entries such that $\mathbf{y} = \mathbf{y}_1 \otimes \mathbf{y}_2$ is the unique solution of system (6.53).

Using direct substitution, we can verify that the vector $\mathbf{x}$ of form (6.54) is the unique solution of system (2.66). We note that the last assumption is correct since the matrix of the first system in (6.53) is irreducible and stochastic.

Then, we differentiate (6.51) at the point $z = 1$ and substitute the expression obtained for $Y'(1)$ and the vector $\mathbf{x}$ of the form (6.54) into inequality (2.65). Then by using the relation $\mathbf{y}(\mathbf{e}_{N+1} \otimes I_M) = \boldsymbol{\delta}$ and by further algebraic transformations, we derive inequality (6.52). □

Let inequality (6.52) hold and let $\pi(i, r, \nu, m)$, $i \geq 0$, $r = \overline{0, N}$, $\nu = \overline{0, W}$, $m = \overline{1, M}$ be the stationary state distribution of the MC ξ_n, $n \geq 1$. Denote by $\boldsymbol{\pi}_i$ the vector of the probabilities $\pi(i, r, \nu, m)$, $i \geq 0$ ordered in lexicographic order. To find the vectors $\boldsymbol{\pi}_i$, $i \geq 0$, we use one of the algorithms presented in Sect. 2.4.

6.3.3 Stationary State Distribution at an Arbitrary Time

Consider the process $\zeta_t = \{i_t, k_t, r_t, \nu_t, m_t\}$, $t \geq 0$ of the system states at an arbitrary time, where i_t is the number of customers at the first station (including a customer which causes blocking); k_t takes the values 0, 1, 2 depending on the state of the server at the first station (idle, busy, or blocked); r_t is the number of busy servers at the second station; ν_t is the state of the $BMAP$; m_t is the state of the SM service process at time t, $t \geq 0$.

Let $p(i, k, r, \nu, m)$, $i \geq 0$, $k = 0, 1, 2$, $r = \overline{0, N}$, $\nu = \overline{0, W}$, $m = \overline{1, M}$ be the stationary state probabilities of the process ζ_t, and $\mathbf{P}_i(k)$ be the vectors of these probabilities ordered in lexicographic order.

Theorem 6.13 *The non-zero vectors* $\mathbf{p}_i(k),\ i \geq 0,\ k = 0, 1, 2$ *of the stationary state probabilities are expressed by the stationary state vectors* $\boldsymbol{\pi}_i,\ i \geq 0$ *of the embedded MC* $\xi_n,\ n \geq 1$, *as follows:*

$$\mathbf{p}_0(0) = -\tau^{-1}\boldsymbol{\pi}_0\hat{H}(\Delta \oplus \bar{D}_0)^{-1},$$

$$\mathbf{p}_i(1) = \tau^{-1}\{\boldsymbol{\pi}_0\sum_{k=1}^{i}[-\hat{H}(\Delta \oplus \bar{D}_0)^{-1}\hat{D}_k + (1-p)\bar{\mathcal{F}}_k\tilde{H}_3]\hat{\Omega}_{i-k}$$

$$+\sum_{l=1}^{i}\boldsymbol{\pi}_l[\bar{H}\hat{\Omega}_{i-l} + (1-p)\sum_{k=0}^{i-l}\bar{\mathcal{F}}_k\tilde{H}_3\hat{\Omega}_{i-k-l}]\},$$

$$\mathbf{p}_i(2) = \tau^{-1}(1-p)\sum_{l=0}^{i-1}\boldsymbol{\pi}_l\sum_{k=0}^{i-l-1}(\bar{\mathcal{F}}_k+\hat{\delta}_{0,k}I)\tilde{H}_2$$

$$\times\int_0^\infty e^{-\mu\mathcal{N}t}\otimes P(i-k-l-1,t)\otimes I_M dt,\ i \geq 1,$$

where

$$\tau = b_1 + \boldsymbol{\pi}_0\hat{H}(-\hat{D}_0)^{-1}\mathbf{e} + (1-p)\mathbf{\Pi}(1)(I_{N+1}\otimes\mathbf{e}_{\bar{W}M})\int_0^\infty t dF(t)Q_3\mathbf{e},$$

$$\hat{\Omega}_n = \int_0^\infty e^{\Delta t}\otimes P(n,t)\otimes(I_M - \nabla_B(t))dt,\ n \geq 0,$$

$$\nabla_B(t) = \text{diag}\{(B(t)\mathbf{e})_j,\ j = \overline{1,M}\}.$$

The proof is similar to the proof of Theorem 6.2.

Corollary 6.8 *The vectors* $\mathbf{p}_i,\ i \geq 0$ *of the stationary state probabilities of the process* $\{i_t, r_t, \nu_t, m_t\},\ t \geq 0$ *are calculated as follows:*

$$\mathbf{p}_i = \sum_{k=0}^{2}\mathbf{p}_i(k),\ i \geq 0.$$

6.3.4 Performance Characteristics of the System

The mean number of customers at the first station at a service completion moment at this station and at an arbitrary time

$$L = \mathbf{\Pi}'(1)\mathbf{e},\ \tilde{L} = \mathbf{P}'(1)\mathbf{e}.$$

The vectors of the stationary distribution of the number of busy servers at the second station at a service completion moment at the first station and at an arbitrary time

$$\boldsymbol{r} = \mathbf{\Pi}(1)(I_{N+1} \otimes \mathbf{e}_{\bar{W}M}), \qquad \tilde{\boldsymbol{r}} = \mathbf{P}(1)(I_{N+1} \otimes \mathbf{e}_{\bar{W}M}).$$

The mean number of busy servers at the second station at a service completion moment at the first station and at an arbitrary time

$$N_{busy} = \boldsymbol{r}\mathcal{N}\mathbf{e}, \qquad \tilde{N}_{busy} = \tilde{\boldsymbol{r}}\mathcal{N}\mathbf{e}.$$

The probability that an arbitrary customer leaves the system (blocks the server) after getting service at the first station

$$P_{loss} = p\mathbf{\Pi}(1)\tilde{H}_2\mathbf{e}, \quad P_{block} = (1-p)\mathbf{\Pi}(1)\tilde{H}_2\mathbf{e}.$$

The probabilities P_{idle}, P_{serve}, P_{block} that the server of the first station is idle, busy, or blocked

$$P_{idle} = \tau^{-1}\boldsymbol{\pi}_0\hat{H}(-\hat{D}_0)^{-1}\mathbf{e},\ P_{serve} = \tau^{-1}b_1,\ P_{block} = 1 - P_{idle} - P_{serve}.$$

6.3.5 Numerical Examples

Let us investigate the effect of the coefficient of correlation of the input flow on the main stationary characteristics of the system performance for different values of the input intensity and the service intensity at the second station.

Consider the four $BMAP$s described in the first experiment of Sect. 6.2.7. These flows have the same intensity $\lambda = 5$ and different coefficient of correlations $c_{cor} = 0, 0.1, 0.2, 0.3$.

The semi-Markovian kernel has the following structure:

$$B(t) = \text{diag}\{B_1(t), B_2(t)\}P,$$

where the transition matrix $P = \begin{pmatrix} 0.6 & 0.4 \\ 0.35 & 0.65 \end{pmatrix}$, $B_1(t)$, and $B_2(t)$ are the order 3 Erlang distribution functions with mean intensities 20 and 50, respectively. The mean service time is $b_1 = 0.102$.

The other system parameters are as follows: $N = 5$, $p = 0.5$, $q_0 = 0.1$, $q_1 = 0.3$, $q_2 = 0.3$, $q_3 = q_4 = q_5 = 0.1$, and $\mu = 5$.

Figure 6.6 shows the dependence of the mean number L of customers at the first station at a service completion moment at this station and the loss probability P_{loss} on the intensity λ.

Figure 6.7 shows the dependence of the mean number L of customers at the first station and the mean number N_{busy} of busy servers at the second station on the service intensity μ.

Based on Figs. 6.6 and 6.7, it can be concluded that the coefficient of correlation is an important characteristic of the input flow and ignoring the correlation can lead to significant errors in assessing the performance of the considered queue.

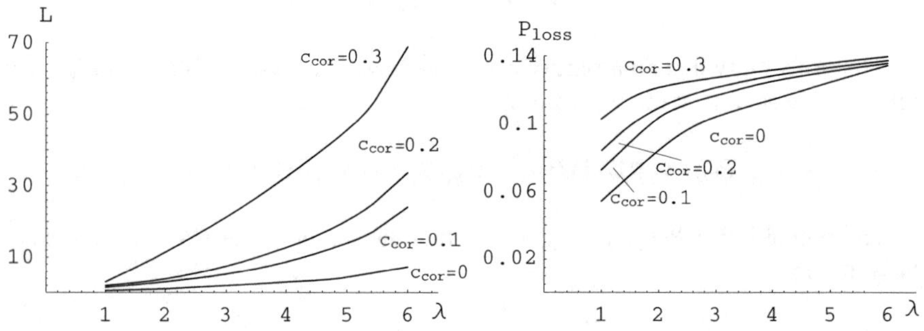

Fig. 6.6 The dependence of L and P_{loss} on the intensity λ for $BMAP$s with different coefficients of correlation

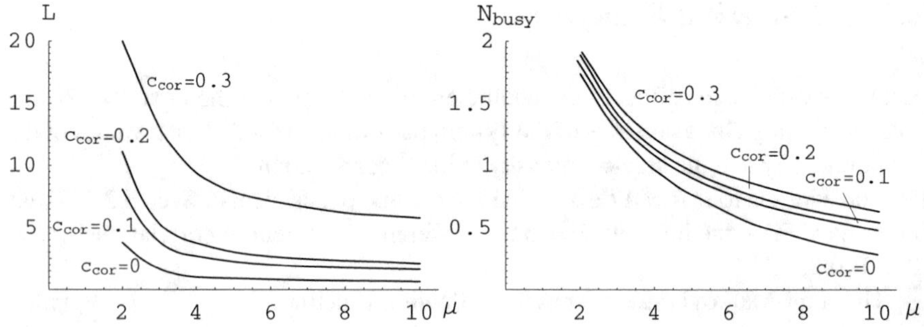

Fig. 6.7 The dependence of L and N_{busy} on the service intensity μ for $BMAP$s with different coefficients of correlation

6.4 $BMAP/G/1 \to \cdot/M/N/R$ Queue

6.4.1 Mathematical Model

Here we consider a $BMAP/G/1 \to \cdot/M/N/R$ queue. This queue differs from the one investigated in Sect. 6.2 by having an intermediate buffer of size $R < \infty$. If all servers at the second station are busy, a customer waits in the buffer for its service. If there are no free waiting places in the buffer, the customer leaves the system partially served (is lost) with probability p, $0 \le p \le 1$, or with probability $1-p$ the server at the first station is blocked and does not serve the next customer until one of servers at the second station becomes free. It is assumed that a customer needs only one server to be served at the second station.

6.4.2 Stationary State Distribution of the Embedded Markov Chain

Let t_n be the nth moment of service completion at the first station, $n \ge 1$. Consider the process $\xi_n = \{i_n, l_n, \nu_n\}$, $n \ge 1$, where i_n is the number of customers at the first station (excluding one blocking the server) at time moment $t_n + 0$; l_n is the number of customers at the second station at time $t_n - 0$; ν_n is the $BMAP$ state at time moment t_n, $i_n \ge 0$, $l_n = \overline{0, N+R}$, $\nu_n = \overline{0, W}$.

Lemma 6.5 *The process $\xi_n, n \ge 1$ is a QTMC. The transition probability matrix has the block form (2.59), where*

$$V_i = \sum_{k=1}^{i+1}[-\hat{Q}(\bar{\Delta} \oplus D_0)^{-1}\tilde{D}_k + (1-p)N\mu Q_2 \otimes \int_0^\infty P(k,t)e^{-N\mu t}dt]\Omega_{i-k+1},$$

$$Y_i = \bar{Q}\Omega_i + (1-p)N\mu(Q_2 \otimes \sum_{k=0}^{i}\int_0^\infty P(k,t)e^{-N\mu t}dt)\Omega_{i-k},\ i \ge 0.$$

Here $\bar{Q} = \tilde{Q}_1 + p\tilde{Q}_2$, $\hat{Q} = \tilde{Q}_1 + p\tilde{Q}_2 + (1-p)N\mu Q_2 \otimes (N\mu I - D_0)^{-1}$, where $\tilde{Q}_m = Q_m \otimes I_{\bar{W}}$, $m = 1, 2$, Q_1 is a square matrix of size $N+R+1$ whose entries $(Q_1)_{j,j+1}$ are 1 and others are zero, and Q_2 is a square matrix of size $N+R+1$ whose entries are all zero except the last diagonal one equal to 1. $\bar{\Delta}$ is a square block-diagonal matrix of size $N+R+1$ whose diagonal blocks are $-\min\{i,N\}\mu, i = \overline{0, N+R}$ and whose sub-diagonal blocks are $\min\{i,N\}\mu, i = \overline{1, N+R}$, $\tilde{D}_k = I_{N+R+1} \otimes D_k, k \ge 0$.

The proof is similar to the proof of Lemma 6.1.

Corollary 6.9 *The matrix PGFs $V(z)$ and $Y(z)$ have the form*

$$V(z) = \frac{1}{z}\{-\hat{Q}(\bar{\Delta} \oplus D_0)^{-1}(\tilde{D}(z) - \tilde{D}_0)$$

$$+(1-p)N\mu Q_2 \otimes [(N\mu I - D(z))^{-1} - (N\mu I - D_0)^{-1}]\}\Omega(z),$$

$$Y(z) = [\bar{Q} + (1-p)Q_2 N\mu \otimes (N\mu I - D(z))^{-1}]\Omega(z). \tag{6.55}$$

Theorem 6.14 *The stationary state distribution of the MC $\xi_n,\ n \geq 1$ exists if and only if inequality*

$$\rho < 1 \tag{6.56}$$

holds. Here

$$\rho = \lambda[b_1 + (1-p)y_{N+R}(N\mu)^{-1}],$$

where y_{N+R} is the last component of the vector $\mathbf{y} = (y_0, y_1, \ldots, y_{N+R})$, which is the unique solution of the system of linear algebraic equations

$$\mathbf{y}(Q_1 + Q_2)\int\limits_0^\infty e^{\bar{\Delta}t}dB(t) = \mathbf{y},\ \ \mathbf{ye} = 1. \tag{6.57}$$

Proof Since the matrix $Y(1)$ defined by formula (6.55) is irreducible, the MC $\xi_n,\ n \geq 1$ is ergodic if and only if inequality (2.65) holds, where the vector $\mathbf{x}$ is the unique solution of system (2.66) of linear algebraic equations.

Let the vector $\mathbf{x}$ be of the form

$$\mathbf{x} = \mathbf{y} \otimes \boldsymbol{\theta}. \tag{6.58}$$

Taking into account the relation $\boldsymbol{\theta} e^{D(1)t} = \boldsymbol{\theta}$, we can verify by direct substitution that the vector $\mathbf{x}$ of form (6.58) is the solution of system (2.66). Substituting the vector $\mathbf{x}$ of form (6.58) into (3.66), by means of algebraic transformations we obtain inequality (6.56). □

Below, we assume that (6.56) holds. Let $\pi(i, l, \nu),\ i \geq 0,\ l = \overline{0, N+R},\ \nu = \overline{0, W}$ be the stationary state distribution of the MC $\xi_n,\ n \geq 1$. The row vectors of the stationary state probabilities $\boldsymbol{\pi}_i,\ i \geq 0$ are calculated by the algorithm described in Sect. 2.4.

6.4.3 Stationary State Distribution at an Arbitrary Time

Consider the process $\zeta_t = \{i_t, k_t, l_t, \nu_t\}, t \geq 0$ of the system states at an arbitrary time, where i_t is the number of customers at the first station, k_t takes values 0, 1, 2 depending on the state of the first station server (idle, busy, or blocked), and ν_t is the state of the $BMAP$ at time moment $t, t \geq 0$.

Let $p(i, k, l, \nu) = \lim_{t\to\infty} P\{i_t = i, k_t = k, l_t = l, \nu_t = \nu\}, i \geq 0, k = 0, 1, 2, l = \overline{0, N+R}, \nu = \overline{0, W}$, be the stationary distribution of the process $\zeta_t, t \geq 0$, and $\mathbf{p}_i(k), i \geq 0, k = 0, 1, 2$, be the vectors of these probabilities.

Theorem 6.15 *The non-zero vectors $\mathbf{p}_i(k), i \geq 0, k = 0, 1, 2$ of the stationary state probabilities are expressed through the stationary distribution vectors $\boldsymbol{\pi}_i, i \geq 0$ of the embedded MC $\xi_n, n \geq 1$ as follows:*

$$\mathbf{p}_0(0) = -\tau^{-1}\boldsymbol{\pi}_0\hat{Q}(\bar{\Delta} \oplus D_0)^{-1},$$

$$\mathbf{p}_i(1) = \tau^{-1}\{\boldsymbol{\pi}_0\sum_{k=1}^{i}[-\hat{Q}(\bar{\Delta}\oplus D_0)^{-1}\tilde{D}_k+(1-p)N\mu Q_2\otimes\int_0^\infty P(k,t)e^{-N\mu t}dt]\tilde{\Omega}_{i-k}$$

$$+\sum_{l=1}^{i}\boldsymbol{\pi}_l[\bar{Q}\tilde{\Omega}_{i-l} + (1-p)N\mu(Q_2 \otimes \sum_{k=0}^{i-l}\int_0^\infty P(k,t)e^{-N\mu t}dt)\tilde{\Omega}_{i-k-l}]\},$$

$$\mathbf{p}_i(2) = \tau^{-1}(1-p)\sum_{l=0}^{i-1}\boldsymbol{\pi}_l(Q_2 \otimes \int_0^\infty e^{-N\mu t}P(i-l-1,t)dt),$$

where

$$\tau = b_1 + \boldsymbol{\pi}_0\hat{Q}(-\tilde{D}_0)^{-1}\mathbf{e} + (1-p)\boldsymbol{\Pi}(1)\tilde{Q}_2\mathbf{e}(N\mu)^{-1}.$$

The proof is similar to the proof of Theorem 6.2.

Corollary 6.10 *The vectors $\mathbf{p}_i, i \geq 0$ of the stationary state probabilities of the process $\{i_t, l_t, \nu_t\}, t \geq 0$ are calculated as follows:*

$$\mathbf{p}_i = \sum_{k=0}^{2}\mathbf{p}_i(k), i \geq 0.$$

6.4.4 Performance Characteristics of the System

The mean number of customers at the first station at a service completion moment of the station and at an arbitrary time

$$L_1 = \mathbf{\Pi}'(1)\mathbf{e}, \quad \tilde{L}_1 = \mathbf{P}'(1)\mathbf{e}.$$

The vectors of the stationary distribution of the number of busy servers at the second station at a service completion moment at the first station and at an arbitrary time

$$L_2 = \mathbf{\Pi}(1)(I_{N+R+1} \otimes \mathbf{e}_{\bar{W}})R_1\mathbf{e}, \qquad \tilde{L}_2 = \mathbf{P}(1)(I_{N+R+1} \otimes \mathbf{e}_{\bar{W}})R_1\mathbf{e},$$

where $R_1 = \text{diag}\{r,\ r = \overline{0, N+R}\}$.

The variance of the number of customers at the first and at the second station at an arbitrary time

$$D_1 = \mathbf{P}''(1)\mathbf{e} - (\tilde{L}_1)^2, \ D_2 = \mathbf{P}''(1)(I_{N+R+1} \otimes \mathbf{e}_{\bar{W}})R_1\mathbf{e} - (\tilde{L}_2)^2.$$

The mean number of busy servers at the second station at a service completion moment at the first station and at an arbitrary time

$$N_{busy} = \mathbf{\Pi}(1)(I_{N+R+1} \otimes \mathbf{e}_{\bar{W}})R_2\mathbf{e}, \qquad \tilde{N}_{busy} = \mathbf{P}(1)(I_{N+R+1} \otimes \mathbf{e}_{\bar{W}})R_2\mathbf{e},$$

where $R_2 = \text{diag}\{r, r = 0, 1, \ldots, N, N, \ldots, N\}$.

The mean number of busy places in the buffer at a service completion moment at the first station and at an arbitrary time

$$N_{buffer} = \mathbf{\Pi}(1)(I_{N+R+1} \otimes \mathbf{e}_{\bar{W}})R_3\mathbf{e}, \tilde{N}_{buffer} = \mathbf{P}(1)(I_{N+R+1} \otimes \mathbf{e}_{\bar{W}})R_3\mathbf{e},$$

where $R_3 = \text{diag}\{0, \ldots, 0, 1, \ldots, R\}$.

The probability that an arbitrary customer leaves the system (blocks the server) after getting service at the first station

$$P_{loss} = p\mathbf{\Pi}(1)\tilde{Q}_2\mathbf{e}, \quad P_{block} = (1-p)\mathbf{\Pi}(1)\tilde{Q}_2\mathbf{e}.$$

The probabilities P_{idle}, P_{serve}, P_{block} that the first station server is idle, busy, or blocked

$$P_{idle} = \tau^{-1}\boldsymbol{\pi}_0\hat{Q}(-\tilde{D}_0)^{-1}\mathbf{e}, \ P_{serve} = \tau^{-1}b_1, \ P_{block}^{(server)} = 1 - P_{idle} - P_{serve}.$$

6.4.5 Numerical Examples

Experiment 6.4 In this experiment, we investigate how the coefficient of correlation of the input flow and the system load affect the stationary characteristics of the system performance.

Consider three of the $BMAP$s described in Experiment 6.1. $BMAP_1$ is a batch Poisson input with a zero coefficient of correlation. $BMAP_2$ and $BMAP_3$ have the coefficients of correlation $c_{cor} = 0.1$ and $c_{cor} = 0.2$, respectively.

The service time at the first station is constant and equals $b_1 = 0.1$. The number of servers at the second station is $N = 6$, the buffer size is $R = 3$, the probability of blocking after the first station service is $p = 0.6$, and the intensity of the second station service $\mu = 2$.

Figure 6.8 shows the dependence of the mean number of customers at the first station, L_1, and at the second station, L_2, on the system load ρ. Figure 6.9 shows the dependence of the probabilities P_{loss} and P_{block}. Figure 6.10 shows the dependence of the variance of the number of customers at the first station, D_1, and at the second station, D_2, on the system load ρ. The value ρ varies by virtue of the changes in the mean intensity λ.

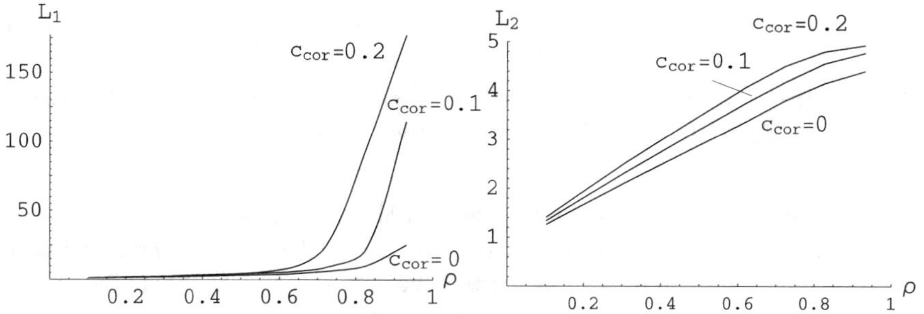

Fig. 6.8 The dependence of the mean number of customers at the first and the second stations on the system load for $BMAP$s with different coefficients of correlation

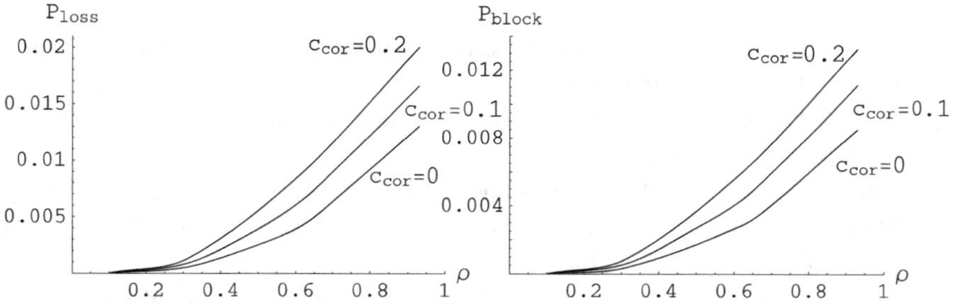

Fig. 6.9 The dependence of the loss and the blocking probabilities on the system load for $BMAP$s with different coefficients of correlation

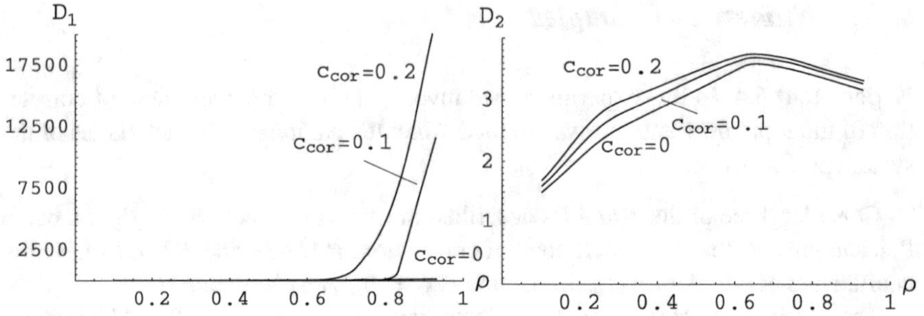

Fig. 6.10 The dependence of the variance of the number of customers at the first and second stations on the system load for $BMAP$s with different coefficients of correlation

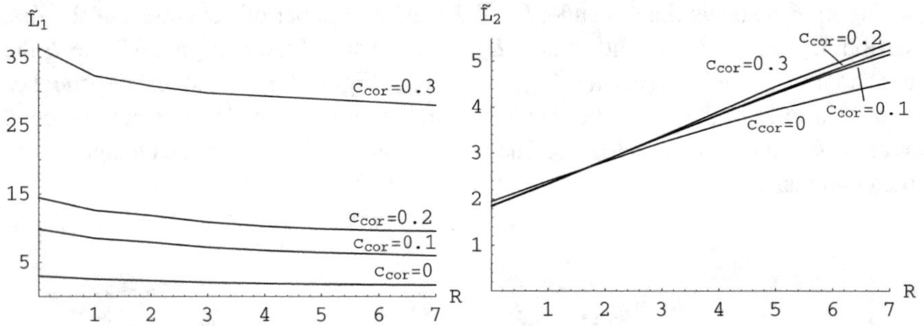

Fig. 6.11 The dependence of the mean number of customers at the first and the second stations on the buffer size for $BMAP$s with different coefficients of correlation

It is seen from the figures that with growth of ρ, the variance D_2 increases to a certain value and then begins to decrease. This is explained by the impossibility of further variance growth with an increase in the number of customers at the second station, due to the limited number of the waiting places at this station.

Experiment 6.5 Let us investigate the effect of the coefficient of correlation of the input flow and the size of the intermediate buffer on the stationary performance characteristics of the system.

Consider the four $BMAP$s described in Experiment 6.1, with maximum batch size $k = 3$, the same intensity $\lambda = 2.5$, and different coefficients of correlation $c_{cor} = 0, 0.1, 0.2, 0.3$. The service time at the first station is constant and equals $b_1 = 0.2$, $N = 3$, $p = 0.5$, $\mu = 1$.

Figures 6.11, 6.12, and 6.13 illustrate the dependence of the main stationary characteristics of the system performance on the buffer size R for $BMAP$s with different coefficients of correlation.

Based on the figures, we can conclude that the values of characteristics $\tilde{L}_1$, $\tilde{N}_{busy}$, $P_{block}^{(server)}$ decrease and the values of $\tilde{L}_2$, $\tilde{N}_{buffer}$, P_{idle} increase as the intermediate buffer size grows for all $BMAP$s. It is also obvious that for a fixed value of the

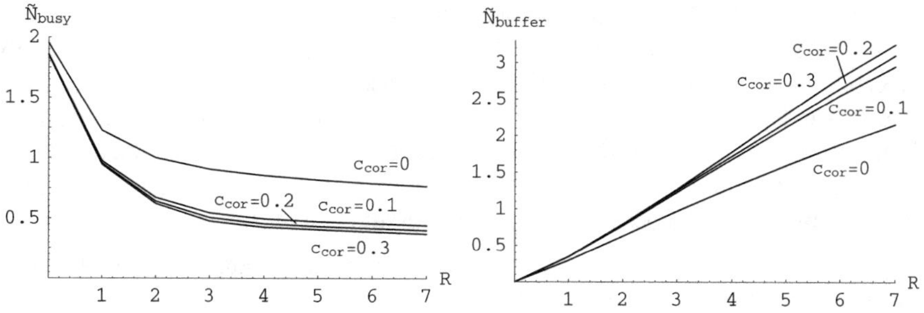

Fig. 6.12 The dependence of the mean number of busy servers at the second station and the busy places in the buffer on the buffer size for $BMAP$s with different coefficients of correlation

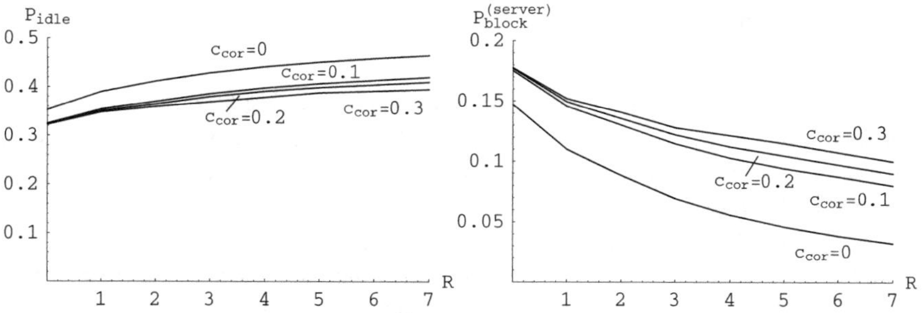

Fig. 6.13 The dependence of probabilities P_{idle} and $P_{block}^{(server)}$ on the buffer size for $BMAP$s with different coefficients of correlation

buffer size R, an increase in the coefficient of correlation of the input flow negatively affects the service process at the first station and at the second one and leads to a deterioration in the quality of the system performance as a whole. Thus, the presented experiments confirm the importance of studying two-station systems with a $BMAP$ for an accurate estimation of the characteristics of real systems.

6.5 $BMAP/G/1 \to \cdot/M/N/0$ Retrial Queue with a Group Occupation of Servers at the Second Station

6.5.1 Mathematical Model

The system considered in this section differs from the $BMAP/G/1 \to \cdot/M/N/0$ queue with waiting investigated in Sect. 6.2 in that a customer arriving at the system when the first station server is busy does not join a queue and leaves the system for an arbitrary time, in other words, goes to an orbit of unlimited size and makes further attempts to get service. In detail, if an arriving batch of primary customers find the first station server idle then one customer starts its service and the others go to the orbit. If the server is busy upon a batch arrival then all customers of the batch go to an orbit. Each customer in the orbit attempts to occupy the server after random time intervals independently of other customers in the orbit. If the number of customers in the orbit is $i,\ i > 0$ at time t then the probability of a repeated attempt from the orbit in the time interval $(t, t + \Delta t)$ is $\alpha_i \Delta t + o(\Delta t)$. For function α_i giving the total intensity of repeated attempts, it is assumed that $\lim\limits_{i\to\infty} \alpha_i = \infty,\ \alpha_0 = 0$. In particular, such a function describes a classical retrial strategy ($\alpha_i = i\alpha$) and a linear strategy ($\alpha_i = i\alpha + \gamma,\ \alpha > 0$).

6.5.2 Stationary State Distribution of the Embedded Markov Chain

Let t_n be the nth moment of service completion at the first station. Consider the process $\xi_n = \{i_n, r_n, \nu_n\},\ n \geq 1$, where i_n is the number of customers in the orbit at time t_n; r_n is the number of busy servers at the second station at time $t_n - 0$; ν_n is the state of the $BMAP$ at time t_n, $i_n \geq 0$, $r_n = \overline{0, N}$, $\nu_n = \overline{0, W}$.

The process $\xi_n,\ n \geq 1$ is an MC. We order its state in lexicographical order and form the matrices $P_{i,j},\ i, j \geq 0$ of the MC transitions from the states corresponding the to i of the denumerable component to states corresponding to value j of this component.

By analyzing the system behavior between two successive service completion moments at the first station, we can show that the following statement holds.

Lemma 6.6 *The matrix of the one-step transition probabilities of the MC $\xi_n,\ n \geq 1$ has the block structure $P = (P_{i,j})_{i,j\geq 0}$ where*

$$P_{i,j} = O,\ j < i - 1,\ P_{i,j} = \bar{Q} A_i[\alpha_i \Omega_{j-i+1} + \sum_{k=1}^{j-i+1} \tilde{D}_k \Omega_{j-i-k+1}]$$

$$+(1-p)[\sum_{n=0}^{j-i+1} \mathcal{F}_n \tilde{Q}_3 A_{i+n} \alpha_{i+n} \Omega_{j-i-n+1}$$

$$+\sum_{n=0}^{j-i} \mathcal{F}_n \tilde{Q}_3 A_{i+n} \sum_{k=1}^{j-i-n+1} \tilde{D}_k \Omega_{j-i-n-k+1}],\ j \geq \max\{0, i-1\},\ i \geq 0, \quad (6.59)$$

$$A_i = \int_0^\infty e^{-\alpha_i t} e^{(\Delta \oplus D_0)t} dt = (\alpha_i I - \Delta \oplus D_0)^{-1},\ i \geq 0.$$

Here the matrices Ω_n, $\mathcal{F}_n$, $n \geq 0$, Δ, $\bar{Q}$, $\tilde{Q}_3$, *and* $\tilde{D}_k$, $k \geq 0$ *coincide with those defined above in Sect. 6.1.*

Proof The lemma is proved by analyzing the possible one-step transitions of the MC ξ_n, $n \geq 1$, taking into account the probabilistic sense of the matrices involved in the right-hand side of (6.59).

The matrix $\bar{Q}$ defines the probabilities that a customer served at the first station either finds the required number of servers at the second station idle and immediately starts its service, or there are not enough idle servers at the second station and the customer leaves the system partially served.

The matrix Δ is the infinitesimal generator of a death process that describes the evolution of the number of busy servers at the second station between two successive service completion moments at the first station.

The matrix $A_i\alpha_i$ defines the probabilities that the service at the first station is initiated by a retrial customer (from the orbit) given that there were i customers in the orbit at the previous service completion moment.

The matrix $A_i\tilde{D}_k$ has a similar probabilistic sense with the only difference that the service at the first station is initiated by a primary customer arriving in a size k batch.

The matrices Ω_n and $\mathcal{F}_n$ give the probabilities that n customers of the $BMAP$ arrive during the service time at the first station and the blocking time of the server of the first station, respectively.

The matrix $(1-p)\mathcal{F}_n\tilde{Q}_3$ defines the probability that a customer served at the first station does not find enough idle servers at the second station, blocks the server of the first station and waits until the required number of idle servers becomes available, and n customers arrive at the system during the blocking time.

Note that all mentioned matrices except the matrix Δ have size $(N+1)\bar{W} \times (N+1)\bar{W}$ and the entry of row $(r\bar{W}+\nu)$ and column $(r'\bar{W}+\nu')$ of each of these matrices gives the probability of an event described when defining the probabilistic sense of the whole matrix and accompanied by the transition of the number of busy servers at the second station from state r to state r' and the transition of the underlying process of the $BMAP$ from state ν to state ν'.

Taking into account the above explanations and using the total probability formula, we obtain the required expression (6.59) for the transition probabilities of the considered MC. □

As can be seen from (6.59), the dependence of the transition probability matrix $P_{i,j}$ on the values i and j of the denumerable component does not reduce to the dependence on $j-i$ as for the case of a QTMC or an $M/G/1$-type MC, see [70, 94] describing many queues with waiting. At the same time, the dependence on i weakens as $i \to \infty$, and the matrices $P_{i,j}$ in the limit depend on i, j only through their difference $j-i$, i.e., the following limits exist:

$$\tilde{Y}_k = \lim_{i\to\infty} P_{i,i+k-1},\ k \geq 0. \tag{6.60}$$

This circumstance supplemented by the fact that the considered MC $\xi_n,\ n \geq 1$ is irreducible and aperiodic allows us to consider this MC as an AQTMC (see [70]).

Let the MC $\eta_n,\ n \geq 1$ be the limiting MC with respect to MC $\xi_n,\ n \geq 1$ and let $\tilde{Y}(z),\ |z| \leq 1$ be the PGF of matrices $\tilde{Y}_k,\ k \geq 0$ of the transition probabilities of this limiting MC. Using (6.60) and Lemma 6.1, we obtain the following statement.

Corollary 6.11 *The MC $\xi_n,\ n \geq 1$ belongs to the class of AQTMCs. The PGF of the transition probability matrices of its limiting MC has the following form:*

$$\tilde{Y}(z) = [\bar{Q} + (1-p)\mathcal{F}(z)\tilde{Q}_3]\Omega(z), \tag{6.61}$$

where

$$\Omega(z) = \sum_{n=0}^{\infty} \Omega_n z^n,\ \ \mathcal{F}(z) = \sum_{n=0}^{\infty} \mathcal{F}_n z^n,\ |z| \leq 1.$$

Theorem 6.16 *A sufficient condition for the MC $\xi_n,\ n \geq 1$, to be ergodic is the fulfillment of the inequality*

$$\rho < 1, \tag{6.62}$$

where

$$\rho = \lambda[b_1 + (1-p)\mathbf{y}\int_0^{\infty} t dF(t) Q_3 \mathbf{e}].$$

Here the vector $\mathbf{y}$ *is the unique solution of the system of linear algebraic equations*

$$\mathbf{y}\mathcal{Q}\int_0^{\infty} e^{\Delta t} dB(t) = \mathbf{y},\ \ \mathbf{y}\mathbf{e} = 1. \tag{6.63}$$

Proof The matrix $\tilde{Y}(1)$ defined by formula (6.61) is irreducible. Thus, we use the form (2.140), (2.141) of the sufficient condition for the existence of the stationary state distribution for a multidimensional AQTMC.

Using relation (6.61) for matrix $\tilde{Y}(z)$, rewrite the system (3.142) in the form

$$\mathbf{x}[(Q_1+pQ_2)\otimes I_{\bar{W}}+(1-p)\int_0^{\infty} dF(t)Q_3\otimes e^{D(1)t}]\int_0^{\infty} e^{\Delta t}\otimes e^{D(1)t} dB(t) = \mathbf{x}, \tag{6.64}$$

$$\mathbf{x}\mathbf{e} = 1.$$

Substituting the vector $\mathbf{x}$ of the form $\mathbf{x} = \mathbf{y}\otimes\boldsymbol{\theta}$ into system (6.64), it is easy to see that this vector is the solution of this system and, correspondingly, of system (2.141).

Note that $\mathbf{y}$ is the unique solution of system (6.63) since the matrix $\mathcal{Q}\int_0^\infty e^{\Delta t}dB(t)$ is stochastic and irreducible.

Now, substituting $\mathbf{x} = \mathbf{y} \otimes \boldsymbol{\theta}$ and the relation for $\tilde{Y}'(1)$ calculated by means of (6.61) into inequality (2.140), we reduce this inequality to the form (6.62) by means of a series of algebraic transformations. □

Remark 6.3 The vector $\mathbf{y}$ gives the stationary distribution of the number of busy servers at the second station at a service completion moment at the first station given that the first station server works without stopping. Then the relation $(1 - p)\mathbf{y}\int_0^\infty tdF(t)Q_3\mathbf{e}$ defines the mean blocking time for the first station server in heavy traffic conditions, and ρ is the system load.

In the following, we assume that inequality (6.62) holds, and we find the stationary distribution of the MC $\xi_n,\ n \geq 1$. Let $\boldsymbol{\pi}_i,\ i \geq 0$ be the vectors giving the required stationary distribution. To find them, we use the algorithm given above for an AQTMC.

6.5.3 Stationary State Distribution at an Arbitrary Time

Consider the process $\zeta_t = \{i_t, r_t, \nu_t, k_t\},\ t \geq 0$ of the system states at an arbitrary time t, where i_t is the number of customers in the orbit, r_t is the number of busy servers at the second station, ν_t is the state of the $BMAP$ underlying process, and k_t is a random variable taking values 0, 1, 2 depending on the state of the first station server (idle, busy, or blocked) at time t.

Let

$$p(i, r, \nu, k) = \lim_{t\to\infty} P\{i_t = i,\ r_t = r,\ \nu_t = \nu,\ k_t = k\}, \tag{6.65}$$

$$i \geq 0,\ r = \overline{0, N},\ \nu = \overline{0, W},\ k = 0, 1, 2$$

be the stationary state distribution of the process $\zeta_t,\ t \geq 0$. Let $\mathbf{p}_i(k)$ be the row vector of the probabilities $p(i, r, \nu, k)$ ordered in lexicographic order of components $(r, \nu),\ i \geq 0,\ k = 0, 1, 2$.

Theorem 6.17 *The non-zero vectors $\mathbf{p}_i(k)$ of the stationary state distribution of the process $\zeta_t,\ t \geq 0$ are expressed through the stationary state distribution $\boldsymbol{\pi}_i,\ i \geq 0$ of the embedded MC $\xi_n,\ n \geq 1$ as follows:*

$$\mathbf{p}_0(0) = \tau^{-1}\boldsymbol{\pi}_0[\bar{Q} + (1-p)\mathcal{F}_0\tilde{Q}_3]A_0, \tag{6.66}$$

$$\mathbf{p}_i(0) = \tau^{-1}[\boldsymbol{\pi}_i\bar{Q}A_i + (1-p)\sum_{l=0}^{i}\boldsymbol{\pi}_l\mathcal{F}_{i-l}\tilde{Q}_3A_i],\ i \geq 1, \tag{6.67}$$

$$\mathbf{p}_i(1) = \tau^{-1}\{\sum_{l=0}^{i} \boldsymbol{\pi}_l[\bar{Q}A_l \sum_{k=1}^{i-l+1} \tilde{D}_k \tilde{\Omega}_{i-l-k+1}$$

$$+(1-p)\sum_{k=0}^{i-l} \mathcal{F}_k \tilde{Q}_3 A_{l+k} \sum_{m=1}^{i-l-k+1} \tilde{D}_m \tilde{\Omega}_{i-l-k-m+1}]$$

$$+\sum_{l=0}^{i+1} \boldsymbol{\pi}_l[\bar{Q}A_l \alpha_l \tilde{\Omega}_{i-l+1} + (1-p)\sum_{k=0}^{i-l+1} \mathcal{F}_k \tilde{Q}_3 A_{l+k} \alpha_{l+k} \tilde{\Omega}_{i-l-k+1}]\},\ i \geq 0, \tag{6.68}$$

$$\mathbf{p}_i(2) = \tau^{-1}(1-p)\sum_{l=0}^{i} \boldsymbol{\pi}_l \sum_{k=0}^{i-l} (\mathcal{F}_k + \hat{\delta}_{0,k} I)\tilde{Q}_2 \int_0^\infty e^{-\mu \mathcal{N} t} \otimes P(i-l-k, t)dt,$$

$$i \geq 0 \tag{6.69}$$

where τ is the mean time interval between two successive service completion moments at the first station:

$$\tau = b_1 + (1-p)\sum_{i=0}^{\infty} \boldsymbol{\pi}_i (I_{N+1} \otimes \mathbf{e}_{\bar{W}}) \int_0^\infty t dF(t) Q_3 \mathbf{e}$$

$$+\sum_{i=0}^{\infty} \boldsymbol{\pi}_i [\bar{Q}(\mathbf{e}_{N+1} \otimes I_{\bar{W}})(\alpha_i I - D_0)^{-1}$$

$$+(1-p)\sum_{n=0}^{\infty} \mathcal{F}_n \tilde{Q}_3 (\mathbf{e}_{N+1} \otimes I_{\bar{W}})(\alpha_{i+n} I - D_0)^{-1}]\mathbf{e}_{\bar{W}}.$$

Proof According to the definition given in [22], the process $\zeta_t,\ t \geq 0$ is a semi-regenerative process with an embedded Markov renewal process $\{\xi_n, t_n\},\ n \geq 1$. Limit (6.65) exists if the process $\{\xi_n, t_n\}$ is irreducible and aperiodic, and if the inequality $\tau < \infty$ holds. In our case, these conditions hold if the embedded MC $\xi_n,\ n \geq 1$ is ergodic. Thus, a sufficient conditions for the limit (6.65) to exist is the validity of (6.62). Expressions (6.66)–(6.69) for vectors that determine the stationary distribution are derived by using the limit theorem for the semi-regenerative processes given in [22], and the probabilistic sense of the matrices involved in these expressions (see the proof of Lemma 6.4). □

Corollary 6.12 *The vectors $\mathbf{p}_i,\ i \geq 0$ of the stationary state probabilities of the process $\{\tilde{i}_t, r_t, \nu_t\}, t \geq 0$, where $\tilde{i}_t$ is the number of customers at the first station* *(both in the orbit and at the server, including a customer blocking the server), are calculated as follows:*

$$\mathbf{p}_0 = \mathbf{p}_0(0), \ \mathbf{p}_i = \mathbf{p}_i(0) + \sum_{k=1}^{2} \mathbf{p}_{i-1}(k), \ i \geq 1.$$

6.5.4 Performance Characteristics of the System

The results obtained make it possible to calculate the stationary distribution of the queuing system under consideration and all its possible performance characteristics. The most important of them are given below.

The mean number of customers at the first station at a service completion moment at this station and at an arbitrary time

$$L = \mathbf{\Pi}'(1)\mathbf{e}, \ \tilde{L} = \mathbf{P}'(1)\mathbf{e}$$

where $\mathbf{\Pi}(z) = \sum_{i=0}^{\infty} \boldsymbol{\pi}_i z^i, \ \mathbf{P}(z) = \sum_{i=0}^{\infty} \mathbf{p}_i z^i, \ |z| \leq 1.$

The vectors of the stationary distribution of the number of busy servers at the second station at a service completion moment at the first station and at an arbitrary time

$$\mathbf{r} = \mathbf{\Pi}(1)(I_{N+1} \otimes \mathbf{e}_{\bar{W}}), \ \tilde{\mathbf{r}} = \mathbf{P}(1)(I_{N+1} \otimes \mathbf{e}_{\bar{W}}).$$

The mean number of busy servers at the second station at a service completion moment at the first station and at an arbitrary time

$$N_{busy} = \mathbf{r}\mathcal{N}\mathbf{e}, \ \tilde{N}_{busy} = \tilde{\mathbf{r}}\mathcal{N}\mathbf{e}.$$

The probability that an arbitrary customer leaves the system (blocks the server) after getting service at the first station

$$P_{loss} = p\mathbf{\Pi}(1)\tilde{Q}_2\mathbf{e}, \quad P_{block} = (1-p)\mathbf{\Pi}(1)\tilde{Q}_2\mathbf{e}.$$

The probabilities P_{idle}, P_{serve}, P_{block} that the first station server is idle, busy, or blocked

$$P_{idle} = \sum_{i=0}^{\infty} \mathbf{p}_i(0)\mathbf{e}, \ P_{serve} = \tau^{-1}b_1, \ P_{block} = 1 - P_{idle} - P_{serve}.$$

The probability that an arbitrary customer arriving at the system immediately receives service (avoiding the orbit)

$$P_{imm} = -\lambda^{-1} \sum_{i=0}^{\infty} \mathbf{p}_i(0)(\mathbf{e}_{N+1} \otimes D_0 \mathbf{e}). \tag{6.70}$$

Let us briefly explain the derivation of formula (6.70). An arbitrary customer receives service immediately if at the moment of its arrival the server of the first station is idle. The probability that an arbitrary customer arrives in a batch of size k and the server is idle at the arrival moment is calculated by the formula $\sum_{i=0}^{\infty} \mathbf{p}_i(0)(I_{N+1} \otimes \sum_{k=1}^{\infty} \frac{kD_k}{\lambda})\mathbf{e}$. We assume that an arbitrary customer arriving in a batch of size k is placed in the first place in the batch with probability $1/k$. Then, by the total probability formula we have

$$P_{imm} = \sum_{i=0}^{\infty} \mathbf{p}_i(0)(I_{N+1} \otimes \sum_{k=1}^{\infty} \frac{kD_k}{\lambda} \frac{1}{k})\mathbf{e}. \tag{6.71}$$

Transforming (6.71) taking into account relation $\sum_{k=1}^{\infty} D_k \mathbf{e} = -D_0 \mathbf{e}$, we get (6.70).

The probability that an arbitrary customer arriving at the system is successfully served at both stations (it avoids the orbit, it is not lost, and it does not block the server at the first station):

$$P_{success} = -\lambda^{-1} \sum_{i=0}^{\infty} \mathbf{p}_i(0)(I_{N+1} \otimes D_0 \mathbf{e}) \int_0^{\infty} e^{\Delta t} dB(t) Q_1 \mathbf{e}. \tag{6.72}$$

Formula (6.72) becomes clear if we take into account that the row vector $-\lambda^{-1} \sum_{i=0}^{\infty} \mathbf{p}_i(0)(I_{N+1} \otimes D_0 \mathbf{e})$ defines the distribution of the number of busy servers at the second station at the moment of arrival of an arbitrary tagged customer that finds the first station server idle, and the column vector $\int_0^{\infty} e^{\Delta t} dB(t) Q_1 \mathbf{e}$ defines the probabilities that a tagged customer after getting service at the first station finds the required number of servers at the second station idle.

6.5.5 Numerical Examples

Experiment 6.6 In this experiment, we investigate how the coefficient of correlation of the input flow affects the basic stationary performance characteristics of the system.

We consider three MAPs defined by the matrices D_0 and $D_1 = D$. These MAPs have the same intensity $\lambda = 7$ but different coefficients of correlation.

MAP_1 is a stationary Poisson flow defined by the matrices $D_0 = -7$ and $D = 7$ and having the coefficient of correlation $c_{cor} = 0$.

MAP_2 has the coefficient of correlation $c_{cor} = 0.1$ and is defined by the matrices

$$D_0 = \begin{pmatrix} -8.22458 & 8.22458 \times 10^{-6} \\ 8.22458 \times 10^{-6} & -0.180152 \end{pmatrix}, \ D = \begin{pmatrix} 8.199422 & 0.02515 \\ 0.140373 & 0.039771 \end{pmatrix}.$$

MAP_3 has the coefficient of correlation $c_{cor} = 0.2$ and is defined by the matrices

$$D_0 = \begin{pmatrix} -9.443532 & 7.63574 \times 10^{-6} \\ 7.63574 \times 10^{-6} & -0.307237 \end{pmatrix}, \ D = \begin{pmatrix} 9.380959 & 0.062565 \\ 0.171397 & 0.135832 \end{pmatrix}.$$

Based on these MAPs, we construct $BMAP$s with the same maximal batch size. These $BMAP$s are defined by the matrices $D_k, k = \overline{0,3}$ obtained as follows: the matrix D_0 coincides with the corresponding matrix of the MAP and other matrices are calculated by the formula $D_k = D\varkappa^{k-1}(1-\varkappa)/(1-\varkappa^3), k = \overline{1,3}$, where $\varkappa = 0.8$. Then all matrices $D_k, k = \overline{0,3}$ are multiplied by some positive constant in order to obtain a $BMAP$ with intensity $\lambda = 7$. The $BMAP$ obtained from MAP_n will be denoted below by $BMAP_n$, $n = 1, 2, 3$.

The service time at the first station has the order 3 Erlang distribution with parameter 60. The mean service time is $b_1 = 0.05$, and the squared coefficient of variation is $c_{var}^2 = 1/3$.

The retrial strategy is assumed to be classic, i.e., function α_i has the form $\alpha_i = i\alpha,\ i \geq 0$.

In the framework of this experiment, $\alpha = 10$. The number of servers at the second station is $N = 7$. The probability that a customer that does not receive immediate access at the second station will be lost equals $p = 0.6$. The service parameters at the second station are determined by intensity $\mu = 3$ and probabilities $q_0 = 0,\ q_1 = 0.9,\ q_2 = q_3 = 0.05$.

Figure 6.14 shows the dependence of the mean number $\tilde{L}$ of customers at the first station and the probability P_{loss} of an arbitrary customer loss after getting service at the first station on the system load ρ.

Figure 6.15 shows the dependence of both the probability P_{imm} of an immediate access to the first station server and the probability $P_{success}$ of an arbitrary customer successfully receiving service at both stations on the system load ρ. The value of ρ varies by virtue of λ, which is changed by normalizing the entries of the matrices $D_k, k = \overline{0,3}$.

The presented figures show that for the same system load ρ, the system characteristics differ significantly for $BMAP$s with different coefficients of correlation. The increase in the coefficient of correlation causes an increase in $\tilde{L}$ and P_{loss} and leads to a decrease in probabilities P_{imm} and $P_{success}$, which negatively affects the quality of service in the considered queue.

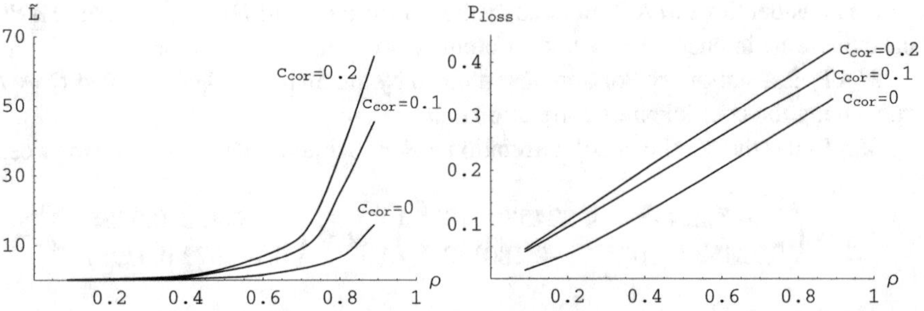

Fig. 6.14 The dependence of the mean number of customers at the first station and the loss probability on the system load for $BMAP$s with different coefficients of correlation

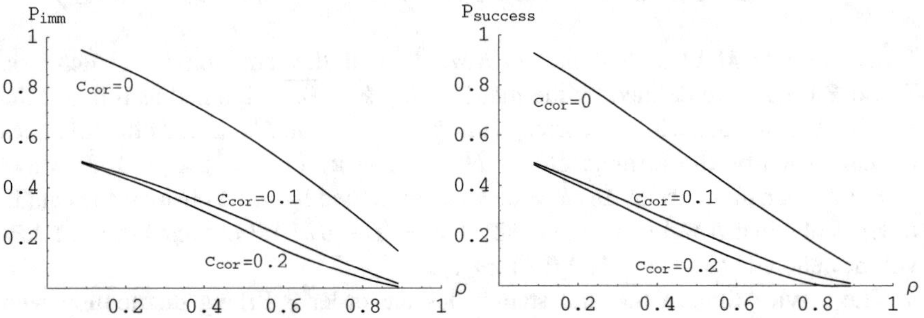

Fig. 6.15 The dependence of probabilities P_{imm} and $P_{success}$ on the system load for $BMAP$s with different coefficients of correlation

The presented experiment confirms the necessity to take into account the influence of correlation in predicting the system behavior since ignoring this important characteristic of the input flow can lead to significant errors in the design and evaluation of the system performance.

Experiment 6.7 In this experiment, we study numerically the problem of the optimal choice of parameter p, i.e., the probability of a customer loss after the first station due to lack of the required number of idle serves at the second station. The expediency of such a problem follows from the following. In order to improve the quality of service in the system under consideration, it is necessary to reduce the probability of a customer loss P_{loss}. In particular, this can be done by reducing the probability p, but this increases the probability of blocking the first station server, which will lead to an increase in the number of customers in the orbit and negatively affect the quality of service.

As a cost criterion of the service quality, consider the average penalty per time unit

$$J = J(p) = c_1\tilde{L} + c_2\lambda P_{loss},$$

where c_1 is the fee for a customer stay in the orbit per time unit, and c_2 is the penalty for one customer loss per time unit.

We consider the problem of finding the optimal probability p which minimizes the cost criterion.

As an input flow, we consider $BMAP_2$ from the previous experiment and take probabilities q_n in the form $q_0 = 0,\ q_1 = 1,\ q_2 = q_3 = 0$, i.e., each customer needs one server to get service at the second station. The other parameters are assumed to be the same as in the previous experiment.

The cost coefficients are as follows: c_1=2, c_2=10.

Figure 6.16 shows the graph of the function $J(p)$ for different values of retrial intensity α: $\alpha = 5, 10, 15, 50$.

As can be seen from the figure, for a fixed value of p, a decrease in the retrial rate leads to an increase in the criterion value J. The greatest difference in the values of the criterion is observed in the region of small values of p. At the same time, the minimum points are very close and are located within the interval (0.315, 0.405). It is interesting to note that the minimum points do not shift when the system load increases (the load decreases as p grows, as shown in Table 6.4).

For a fixed retrial rate α, the relative profit obtained by choosing the optimal parameter p^* compared to an arbitrary value p of this parameter is calculated by the formula $g(p) = \frac{J(p)-J^*}{J^*}100\%$.

We give the values of the relative profit for $\alpha = 10$. In this case, the minimum is reached at $p^* = 0.4$ and is equal to $J^* = 19.3844$. Table 6.2 presents the values of the relative profit $g(p)$ for this case.

As can be seen from Table 6.5, the relative profit is 2–13% even for values close to the optimal p, while the maximum relative profit can reach 54%.

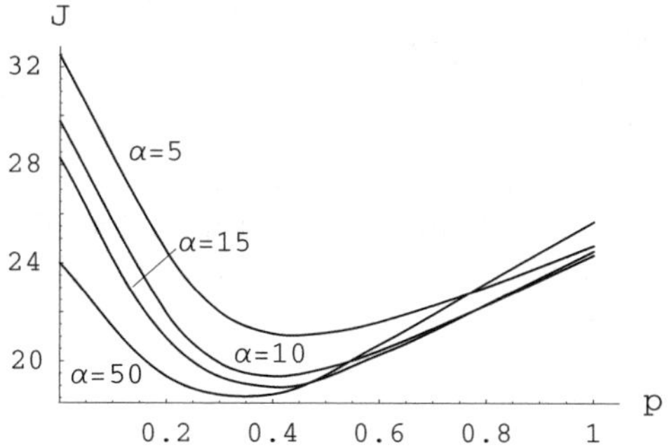

Fig. 6.16 The dependence of the cost criterion J on p for different values of the retrial intensity α

Table 6.4 The system load ρ for different values of p

p	0	0.2	0.3	0.4	0.5	0.6	0.8	1
ρ	0.8195	0.7256	0.6786	0.6317	0.5847	0.5378	0.4439	0.3500

Table 6.5 The relative profit $g(p)$ for different values of p

p	0	0.2	0.3	0.4	0.5	0.6	0.8	1
$g(p)$	54%	13%	3%	0%	2%	5%	15%	26%

6.6 Tandem of Multi-Server Queues Without Buffers

The theory of tandem queuing systems is a link between the theory of queuing systems and the theory of queuing networks. A tandem system can be considered to be the simplest case of a queuing network with a linear topology. Tandem systems are a popular topic for research, see, for example, the papers [4, 44, 50–53, 96].

The majority of works devoted to tandem systems are limited to considering two-station tandems with a stationary Poisson input and an exponentially distributed service times. In this section, we consider a tandem system consisting of an arbitrary finite number of stations represented by multi-server systems, with MAP arrivals and no buffers.

The assumption that the process of customer arrivals is defined as a MAP allows us to take into account the correlated non-stationary nature of information flows in modern telecommunication networks, see, for example, [54, 68].

The service times at the tandem stations have a PH-type distribution, which is known to be significantly more general than an exponential one and allows us to model real service processes well.

A customer arriving at the system under consideration should receive successive service at all tandem stations. However, due to the lack of places to wait at the stations, customers can be lost at each of them. Quality of service in the system is basically determined by the probability of an arbitrary customer successfully receiving service at all stations. However, to assess the effectiveness of a tandem as well as to detect and prevent so-called bottlenecks in the tandem network, loss probabilities in various sections and subsystems of the tandem are used.

As the relevant works, it is worth noting the papers [10, 11]. These papers consider systems with single-server stations, recurrent input of customers, and exponentially distributed service times. Since the inter-arrival times in a recurrent flow are independent random variables distributed according to an arbitrary law, this flow can be considered to be more general than a MAP. At the same time, a MAP can be considered to be a more general flow than a recurrent one because the inter-arrival times can be dependent.

A special case of the system considered in this section was investigated in [77]. It was assumed there that the customer service times at stations have an exponential distribution.

6.6.1 The System Description

We consider a tandem queuing system consisting of $R,\ R > 1$ stations. In terms of Kendall's notation, this system can be described as

$$MAP/PH^{(1)}/N_1/N_1 \to \bullet/PH^{(2)}/N_2/N_2 \to \ldots \to \bullet/PH^{(R)}/N_R/N_R.$$

Station number r, $r = 1, \ldots, R$ is presented by a system with N_r servers with no buffer. The servers of the same station are independent and identical. The input of the first station is a MAP characterized by the state space $\{0, 1, \ldots, W\}$ of the underlying process $\nu_t, t \geq 0$, and by matrices D_0, D_1. The service time of a customer at the rth station has a PH-type distribution with an irreducible representation $(\boldsymbol{\beta}^{(r)}, S^{(r)}), r = \overline{1, R}$.

If a customer arriving at the first station or proceeding to station $r, r = 2, \ldots, R$ after getting service at station $(r - 1)$ finds all servers busy it leaves the system forever.

This section aims to analyze the output process of the tandem stations, the calculation of the stationary state distribution of the whole tandem or its parts (fragments), and the calculation of the loss probabilities associated with a tandem.

6.6.2 Output Flows from the Stations

The process of the system states is described in terms of an irreducible multidimensional continuous-time MC $\xi_t, t \geq 0$:

$$\xi_t = \{n_t^{(R)}, m_{1,t}^{(R)}, \ldots, m_{n^{(R)},t}^{(R)};\ n_t^{(R-1)}, m_{1,t}^{(R-1)}, \ldots,$$

$$m_{n^{(R-1)},t}^{(R-1)}; \ldots, n_t^{(1)}, m_{1,t}^{(1)}, \ldots, m_{n^{(1)},t}^{(1)};\ \nu_t\},\ t \geq 0,$$

where

- $n_t^{(r)}$ is the number of busy servers at station r, $n_t^{(r)} = \overline{0, N^{(r)}}$, $r = \overline{1, R}$;
- $m_{l,t}^{(r)}$ is the state of the PH-type service underlying process at the lth busy server at station r, $m^{(r)} = l, t = \overline{1, M_r}$, $r = \overline{1, R}$, $l = \overline{1, n^{(r)}}$;
- ν_t, $\nu_t \in \{0, \ldots, W\}$ is the state of the MAP underlying process at time t.

The row vector $\mathbf{p}$ of the stationary state probabilities of the MC has size

$$(W + 1) \prod_{r=1}^{R} \frac{M_r^{N_{r+1}} - 1}{M_r - 1},$$

and is calculated as the unique solution of the system of linear algebraic equations

$$\mathbf{p}Q = \mathbf{0},\ \ \mathbf{pe} = 1 \tag{6.73}$$

where the matrix Q is the infinitesimal generator of the MC ξ_t, $t \geq 0$.

This matrix can be constructed by means of the standard technique used in the theory of queuing networks and is not a tough task. But in the case of a large value of R, the task becomes laborious. Therefore, it is interesting to find a way to calculate the stationary distribution of the tandem states as a whole or its parts (fragments) or the marginal stationary state distributions of any station while avoiding writing out an explicit expression for generator Q. It should be noted that the functioning of a tandem fragment consisting of any number of stations and located at the beginning of the tandem does not depend on the states of the other stations. Thus, it is intuitively clear that any decomposition can be used to calculate the stationary state distribution of the tandem and its fragments without a complete construction of a generator.

In this section, we develop a simple, accurate, and convenient method to calculate the marginal stationary state distributions of the tandem fragments as well as the whole tandem, and the corresponding loss probabilities. This method is based on the results of the investigation of the output flows from the tandem stations. These flows belong to the class of MAPs and are described by the following theorem.

Theorem 6.18 *The output flow from station r (which is the input flow to station $(r+1)$), $r \in \{1, 2, \ldots, R-1\}$ belongs to the class of MAPs. This MAP is given by the matrices $D_0^{(r+1)}$ and $D_1^{(r+1)}$ calculated by the following recurrent formulas:*

$$D_0^{(r+1)} = diag\{(S^{(r)})^{\oplus n},\ n = \overline{0, N_r}\} \otimes I_{K_r} \tag{6.74}$$

$$+\begin{pmatrix} D_0^{(r)} & \beta^{(r)} \otimes D_1^{(r)} & O & \ldots & O & O \\ O & I_{M_r} \otimes D_0^{(r)} & I_{M_r} \otimes \beta^{(r)} \otimes D_1^{(r)} & \ldots & O & O \\ O & O & I_{M_r^2} \otimes D_0^{(r)} & \ldots & O & O \\ \vdots & \vdots & \vdots & \ddots & \vdots & \vdots \\ O & O & O & \ldots & I_{M_r^{N_r-1}} \otimes D_0^{(r)} & I_{M_r^{N_r-1}} \otimes \beta^{(r)} \otimes D_1^{(r)} \\ O & O & O & \ldots & 0 & I_{M_r^{N_r}} \otimes (D_0^{(r)} + D_1^{(r)}) \end{pmatrix},$$

$$D_1^{(r+1)} \tag{6.75}$$

$$= \begin{pmatrix} O_{1\times 1} & O & \ldots & O & O & O \\ (S_0^{(r)})^{\oplus 1} & O_{M_r\times M_r} & \ldots & O & O & O \\ O & (S_0^{(r)})^{\oplus 2} & \ldots & O & O & O \\ \vdots & \vdots & \ddots & \vdots & \vdots & \vdots \\ O & O & \ldots & (S_0^{(r)})^{\oplus (N_r-1)} & O_{M_r^{N_r-1}\times M_r^{N_r-1}} & O \\ O & O & \ldots & O & (S_0^{(r)})^{\oplus N_r} & O_{M_r^{N_r}\times M_r^{N_r}} \end{pmatrix} \otimes I_{K_r},$$

$$r = 1, 2, \ldots, R-1$$

with the initial condition

$$D_0^{(1)} = D_0, \ \ D_1^{(1)} = D_1,$$

where the value K_r is calculated as $K_r = (W+1)\prod_{r'=1}^{r-1} \frac{M_r^{N_r+1}-1}{M_{r-1}-1}$.

Proof Let $r = 1$. It is obvious that the process $\{n_t^{(1)}, m_{1,t}^{(1)}, m_{2,t}^{(2)}, \ldots, m_{N_1,t}^{(N_1)}, \nu_t\}$, $t \geq 0$ describing the first station behavior is an MC. We renumber the states of this MC in lexicographical order. Then, as we can easily see, the MC transitions that are not caused by a service completion at the first station (and a customer arrival at the second station) are defined by the matrix $D_0^{(2)}$ calculated by formula (6.74). The MC transitions that are caused by a service completion at the first station (and a new arrival at the second station) are defined by the matrix $D_1^{(2)}$. According to the definition of a MAP, this means that the process of customer arrivals at the second station is a MAP given by the matrices $D_0^{(2)}$ and $D_1^{(2)}$. The further proof of the theorem can be done by an induction. □

Remark 6.4 Below we denote the MAP which is the input flow to station r by $MAP^{(r)}$, $r = \overline{1, R}$. The output flow from station R is denoted by $MAP^{(R+1)}$. The matrices $D_0^{(R+1)}$ and $D_1^{(R+1)}$ defining $MAP^{(R+1)}$ are given by formulas (6.74) and (6.75) for $r = R + 1$.

Using the results of Theorem 6.18, we can calculate the marginal stationary state distribution of the rth tandem station as the stationary state distribution of a $MAP^{(r)}/PH^{(r)}/N_r/N_r$ queue, $r = \overline{1, R}$.

For the convenience of the reader, the algorithm for calculating the stationary distribution of such a system is briefly described in the next subsection. For brevity, we omit the index r in the notation for matrices describing the MAP flow and the PH-type service process.

6.6.3 Stationary State Distribution of a $MAP/PH/N/N$ Queue

The behavior of a $MAP/PH/N/N$ queue is described by an MC $\zeta_t = \{i_t, \nu_t, m_t^{(1)}, .., m_t^{(i_t)}\}$, where i_t is the number of busy servers, $m_t^{(n)}$, $m_t^{(n)} = \overline{0, M}$ is the service phase on the nth busy server, and ν_t, $\nu_t = \overline{0, W}$ is the state of the MAP underlying process at time t.

Renumber the MC states in lexicographical order. Then the infinitesimal generator of this MC is defined as

$$A = (A_{i,j})_{i,j=\overline{0,N}} =$$

$$\begin{pmatrix} D_0 & D_1 \otimes \boldsymbol{\beta} & O & \ldots & O & O \\ I_{\overline{W}} \otimes \mathbf{S}_0^{\oplus 1} & D_0 \oplus S^{\oplus 1} & D_1 \otimes I_M \otimes \boldsymbol{\beta} & \ldots & O & O \\ 0 & I_{\overline{W}} \otimes \mathbf{S}_0^{\oplus 2} & D_0 \oplus S^{\oplus 2} & \ldots & O & O \\ \ldots & \ldots & \ldots & \ldots & \ldots & \ldots \\ 0 & 0 & 0 & \ldots & D_0 \oplus S^{\oplus N-1} & D_1 \otimes I_{M^{N-1}} \otimes \boldsymbol{\beta} \\ 0 & 0 & 0 & \ldots & I_{\overline{W}} \otimes \mathbf{S}_0^{\oplus N} & (D_0 + D_1) \oplus S^{\oplus N} \end{pmatrix}.$$

Let $\mathbf{q}$ be a row vector of the stationary state distribution of the MC ζ_t. This vector is obtained as the unique solution of the system of linear algebraic equations

$$\mathbf{q}A = \mathbf{0}, \ \ \mathbf{q}\mathbf{e} = 1.$$

In the case that the system has large dimension, it is worth using special algorithms to solve it. The most famous numerically stable algorithm is proposed in [75]. It is described below.

Present the vector $\mathbf{q}$ as $\mathbf{q} = (\mathbf{q}_0, \mathbf{q}_1, \ldots, \mathbf{q}_N)$ where the vector $\mathbf{q}_l$ has size $(W + 1)M^l, \ l = 0, \ldots, N$.

Algorithm 6.1 The vectors $\mathbf{q}_l$, $l = 0, \ldots, N$ are calculated as

$$\mathbf{q}_l = \mathbf{q}_0 F_l, \ \ l = \overline{0, N},$$

where the matrices F_l are calculated recurrently as follows:

$$F_l = (\bar{A}_{0,l} + \sum_{i=1}^{l-1} F_i \bar{A}_{i,l})(-\bar{A}_{l,l})^{-1}, l = \overline{1, N-1},$$

$$F_N = (A_{0,N} + \sum_{i=1}^{N-1} F_i A_{i,N})(-A_{N,N})^{-1},$$

the matrices $\overline{A}_{i,N}$ are calculated by the backward recursion

$$\overline{A}_{i,N} = A_{i,N}, i = \overline{0, N},$$

$$\overline{A}_{i,l} = A_{i,l} + \overline{A}_{i,l+1} G_l, i = \overline{0, l}, l = N-1, N-2, \ldots, 0,$$

the matrices $G_i, i = \overline{0, N-1}$ are calculated by the backward recursion

$$G_i = (-A_{i+1,i+1} - \sum_{l=1}^{N-i-1} A_{i+1,i+1+l} G_{i+l} G_{i+l-1} \ldots G_{i+1})^{-1} A_{i+1,i},$$

$$i = N-1, N-2, \ldots, 0,$$

and the vector $\mathbf{q}_0$ is the unique solution of the system

$$\mathbf{q}_0 \overline{A}_{0,0} = 0, \ \ \mathbf{q}_0(\sum_{l=1}^{N} F_l \mathbf{e} + \mathbf{e}) = 1.$$

More detailed information on this algorithm can be found in [75]. Note that subtraction operations are not present in this algorithm, and all inverse matrices exist and are non-negative. Thus, the algorithm is numerically stable.

6.6.4 Stationary State Distribution of the Tandem and Its Fragments

In this subsection, we present methods for calculating the stationary state distribution of a tandem and its fragments based on the results of the output process analysis presented in Theorem 6.18.

Let $\langle r, r+1, \dots, r'\rangle$ denote the tandem fragment consisting of stations r, $(r+1)$, ..., r', $1 \le r \le r' \le R$.

Theorem 6.19 *The stationary state distribution of the tandem fragment $\langle r, r+1, \dots, r'\rangle$ can be calculated as the stationary state distribution of the tandem*

$$MAP^{(r)}/PH^{(r)}/N_r/N_r \to \bullet/PH^{(r+1)}/N_{r+1}/N_{r+1} \to \dots \to \bullet/PH^{(r')}/N_{r'}/N_{r'},$$

where $MAP^{(r)}$ is defined by formulas (6.74) and (6.75).

Corollary 6.13 *The marginal stationary state distribution of the rth tandem station is calculated as the stationary state distribution of a $MAP^{(r)}/PH^{(r)}/N_r/N_r$ queue, $r = \overline{1, R}$.*

Note that each of these marginal distributions can be calculated by Algorithm 6.1 presented in the previous subsection.

Theorem 6.20 *The joint stationary states distribution $\mathbf{p}^{(1,\dots,r)}$ of the first r tandem stations can be calculated as the stationary state distribution of the $MAP^{(r+1)}$ underlying process, i.e., the following formula holds:*

$$\mathbf{p}^{(1,\dots,r)} = \boldsymbol{\theta}^{(r+1)},$$

where the vector $\boldsymbol{\theta}^{(r+1)}$ is the unique solution of the system

$$\boldsymbol{\theta}^{(r+1)}(D_0^{(r+1)} + D_1^{(r+1)}) = \mathbf{0}, \quad \boldsymbol{\theta}^{(r+1)}\mathbf{e} = 1, \; r = \overline{1, R} \tag{6.76}$$

and the matrices $D_0^{(r+1)}$, $D_1^{(r+1)}$ are calculated by recurrent formulas (6.74)–(6.75).

This system can be successfully solved by means of the stable algorithm developed in [75]. □

It is obvious that the vector $\mathbf{p}$ of the stationary state distribution of the whole tandem coincides with the vector $\mathbf{p}^{(1,\dots,R)}$.

Corollary 6.14 *The vector* $\mathbf{p}$ *of the stationary distribution of the tandem states is calculated as the unique solution of the system of linear algebraic equations*

$$\mathbf{p}(D_0^{(R+1)} + D_1^{(R+1)}) = \mathbf{0}, \quad \mathbf{pe} = 1. \tag{6.77}$$

As follows from system (6.77), the matrix $D_0^{(R+1)} + D_1^{(R+1)}$ coincides with the infinitesimal generator Q of the MC $\xi_t, t \geq 0$ describing the tandem behavior, i.e.,

$$Q = D_0^{(R+1)} + D_1^{(R+1)}.$$

Thus, using the results of the analysis of the output flows, we have constructed the matrix Q while avoiding the laborious work required by the direct approach to construct it.

6.6.5 Customer Loss Probability

Having calculated the stationary state distribution of the tandem and its fragments, we can find a number of important stationary performance characteristics of the system. The most important of them are the different probabilities of loss in the tandem. According to the ergodic theorem for Markov chains, see, for example, [104], the probability of losing a customer in a tandem fragment can be calculated as the ratio of the difference between the intensity of the input flow to the fragment and the output flow from this fragment to the intensity of the input flow. Thus, the following theorem holds.

Theorem 6.21 *The arbitrary customer loss probability in the fragment* $\langle r, r+1, \dots, r' \rangle$ *of the tandem is calculated as*

$$P_{loss}^{(r,\dots,r')} = \frac{\lambda_r - \lambda_{r'+1}}{\lambda_r},$$

where λ_l *is the intensity of the customer arrivals at station* l *in* $MAP^{(l)}$,

$$\lambda_l = \boldsymbol{\theta}^{(l)} D_1^{(l)} \mathbf{e}$$

and the vector $\boldsymbol{\theta}^{(l)}$ *is the unique solution of the system*

$$\boldsymbol{\theta}^{(l)}(D_0^{(l)} + D_1^{(l)}) = \mathbf{0}, \ \ \boldsymbol{\theta}^{(l)}\mathbf{e} = 1, \ l = \overline{1, R}.$$

Corollary 6.15 *The probability of an arbitrary customer loss at station* r *is calculated as*

$$P_{loss}^{(r)} = \frac{\lambda_r - \lambda_{r+1}}{\lambda_r}.$$

Corollary 6.16 *The loss probability of an arbitrary customer in the whole tandem is calculated by the formula*

$$P_{loss} = \frac{\lambda_1 - \lambda_{R+1}}{\lambda_1}.$$

6.6.6 Investigation of the System Based on the Construction of a Markov Chain Using the Ramaswami-Lucantoni Approach

The obtained results give a simple and elegant method to calculate the stationary state distribution of a tandem with no intermediate buffers. However, this method does not lead to a decrease in the size of the system of linear algebraic equations that must be solved to calculate the stationary state distribution of some fragments or the whole tandem. In particular, system (6.77) has the same rank K_R as system (6.73).

As can be seen from formulas (6.74)–(6.77), the dimensionality of the systems of linear equations determining the stationary state distributions of a tandem fragment or the whole tandem essentially depends on the number N_r of servers at stations and on the dimension M_r of the state space of the PH-type service process. For large values of N_r and (or) M_r, system (6.73) or system (6.77) identical to it cannot be solved analytically because of the large dimension. This dimension is mainly dependent on the dimensions of the state spaces of the processes $\{m_{1,t}^{(r)}, \ldots, m_{n^{(r)},t}^{(r)}\}$, $r = \overline{1, R}$ describing the states of the PH-type service processes at the tandem stations. The dimension of the state space of the process $\{m_{1,t}^{(r)}, \ldots, m_{n^{(r)},t}^{(r)}\}$ is $M_r^{N_r}$ and can be very large. To reduce this dimension, we will use another approach to describe the PH-type service process at the rth tandem station servers. This approach was proposed in the papers [100], [99] of American authors V. Ramaswami and D. Lucantoni and is based on the features of several Markov processes operating in parallel. With this approach, instead of the process $\{m_{1,t}^{(r)}, \ldots, m_{n^{(r)},t}^{(r)}\}$ we consider the process $\{\vartheta_{1,t}^{(r)}, \ldots, \vartheta_{M_r,t}^{(r)}\}$, where $\vartheta_{l,t}^{(r)}$ is the number of rth station servers that are on the lth service phase, $\vartheta_{l,t}^{(r)} = \overline{1, n^{(r)}}$, $r = \overline{1, R}$, $l = \overline{1, M_r}$. The dimension of the state space of the process $\{\vartheta_{1,t}^{(r)}, \ldots, \vartheta_{M_r,t}^{(r)}\}$ is $\frac{(n^{(r)}+M_r-1)!}{n^{(r)}!(M_r-1)!}$ and can be much smaller than the dimension of the state space of the process $\{m_{1,t}^{(r)}, \ldots, m_{n^{(r)},t}^{(r)}\}$. For example, if $N_r = 10$ and $M_r = 2$ then the corresponding dimensions are 2^{10} and 11, respectively.

Using the Ramaswami-Lucantoni approach, we describe the process of the system states in terms of an irreducible multidimensional continuous-time MC:

$$\tilde{\xi}_t = \{n_t^{(R)}, \vartheta_{1,t}^{(R)}, \ldots, \vartheta_{M_R,t}^{(R)};\ n_t^{(R-1)}, \vartheta_{1,t}^{(R-1)}, \ldots, \vartheta_{M_{R-1},t}^{(R-1)};\ \ldots;$$

$$n_t^{(1)}, \vartheta_{1,t}^{(1)}, \ldots, \vartheta_{M_1,t}^{(1)};\ \nu_t\},\ t \geq 0,$$

where

- $n_t^{(r)}$ is the number of rth station busy servers, $n_t^{(r)} = \overline{0, N^{(r)}}$, $r = \overline{1, R}$;
- $\vartheta_{l,t}^{(r)}$ is the number of rth station servers that are on the lth service phase, $\vartheta_{l,t}^{(r)} = \overline{1, n^{(r)}}$, $r = \overline{1, R}$, $l = \overline{1, M_r}$;
- ν_t, $\nu_t \in \{0, \ldots, W\}$ is the state of the MAP underlying process at time t.

In this case, the row vector of the stationary probabilities of the MC states has dimension

$$(W+1)\prod_{r=1}^{R}(1+\sum_{n=1}^{N_r} C_{n+M_r-1}^{M_r-1}).$$

To formulate a theorem similar to Theorem 6.18, we introduce matrices $P_{n^{(r)}}(\boldsymbol{\beta}^{(r)})$, $A_{n^{(r)}}(N_r, S^{(r)})$, and $L_{N_r-n^{(r)}}(N_r, S^{(r)})$, which describe the intensity of the transitions of the process $\vartheta_t^{(r)} = \{\vartheta_t^{(r)}(1), \ldots, \vartheta_t^{(r)}(M_r)\}$, $r = \overline{1, R}$.

We give a brief explanation of the values of these matrices.

Suppose $L^{(k)}(j) = C_{j+k-1}^{k-1}$. Then

- $L_{N_r-n^{(r)}}(N_r, S^{(r)})$ is the matrix of size $L^{(M_r)}(n^{(r)}) \times L^{(M_r)}(n^{(r)}-1)$ consisting of the intensities of the process $\vartheta_t^{(r)}$ transitions for which one of $n^{(r)}$ busy servers gets free (completes service);
- $P_{n^{(r)}}(\boldsymbol{\beta}^{(r)})$ is the matrix of size $L^{(M_r)}(n^{(r)}) \times L^{(M_r)}(n^{(r)}+1)$ consisting of the intensities of the process $\vartheta_t^{(r)}$ transitions for which the number of busy servers changes from $n^{(r)}$ to $n^{(r)}+1$;
- $A_{n^{(r)}}(N_r, S^{(r)})$ is the matrix of size $L^{(M_r)}(n^{(r)}) \times L^{(M_r)}(n^{(r)})$ consisting of the transition intensities of the process $\vartheta_t^{(r)}$ within its state space without change in the number $n^{(r)}$ of the busy servers.

Below, we assume that $L_{N_r}(N_r, S^{(r)}) = A_0(N_r, S^{(r)}) = 0_{1\times 1}$. The rest of the matrices $P_{n^{(r)}}(\boldsymbol{\beta}^{(r)})$, $A_{n^{(r)}}(N_r, S^{(r)})$, and $L_{N_r-n^{(r)}}(N_r, S^{(r)})$ are calculated by the algorithm given, for example, in [38].

Theorem 6.22 *The output process from station r (the input flow to station $(r+1)$), $r \in \{1, 2, \ldots, R-1\}$ belongs to the class of MAPs. This MAP is defined by the matrices $D_0^{(r+1)}$ and $D_1^{(r+1)}$ calculated by the following recurrent formulas:*

$$D_0^{(r+1)} = diag\{A_n(N_r, S^{(r)}),\ n = \overline{0, N_r}\} \otimes I_{K_r}$$

$$+\begin{pmatrix} D_0^{(r)} & D_1^{(r)} \otimes P_0(\beta^{(r)}) & O & \dots & O & O \\ O & I_{C_{M_r}^{M_r-1}} \otimes D_0^{(r)} & I_{C_{M_r}^{M_r-1}} \otimes D_1^{(r)} \otimes P_1(\beta^{(r)}) & \dots & O & O \\ O & O & I_{C_{M_r+1}^{M_r-1}} \otimes D_0^{(r)} & \dots & O & O \\ \vdots & \vdots & \vdots & \ddots & \vdots & \vdots \\ O & O & O & \dots & I_{C_{N_r+M_r-2}^{M_r-1}} \otimes D_0^{(r)} & I_{C_{N_r+M_r-2}^{M_r-1}} \otimes D_1^{(r)} \otimes P_{N_r-1}(\beta^{(r)}) \\ O & O & O & \dots & 0 & I_{C_{N_r+M_r-1}^{M_r-1}} \otimes (D_0^{(r)} + D_1^{(r)}) \end{pmatrix},$$

$$D_1^{(r+1)} =$$

$$\begin{pmatrix} O_1 & O & \dots & O & O & O \\ L_{N_r-1}(1, S^{(r)}) & O_{C_{M_r}^{M_r-1}} & \dots & O & O & O \\ 0 & L_{N_r-2}(2, S^{(r)}) & \dots & O & O & O \\ \vdots & \vdots & \ddots & \vdots & \vdots & \vdots \\ O & O & \dots & L_1(N_r-1, S^{(r)}) & 0_{C_{N_r+M_r-2}^{M_r-1}} & O \\ O & O & \dots & O & L_0(N_r, S^{(r)}) & 0_{C_{N_r+M_r-1}^{M_r-1}} \end{pmatrix} \otimes I_{K_r},$$

$$r = \overline{1, R-1}$$

with the initial condition

$$D_0^{(1)} = D_0, \ \ D_1^{(1)} = D_1,$$

where K_r is calculated as $K_r = (W+1)\prod_{l=1}^{r-1}(1 + \sum_{n=1}^{N_l} C_{n+M_l-1}^{M_l-1})$.

Theorem 6.22 makes it possible to reduce the order of the matrices $D_0^{(r+1)}$, $D_1^{(r+1)}$, $r \in \{1, 2, \dots, R-1\}$ and use them to calculate the stationary state distributions of tandem fragments and of the entire tandem as well as to calculate all possible loss probabilities using the formulas given above in this section.

Note that we can use some kind of heuristics even if the Ramaswami-Lucantoni method of constructing a generator and the above algorithm for calculating the stationary state distribution, which effectively uses the sparse structure of a generator, do not make it possible to solve the systems of linear algebraic equations for the stationary state distribution of the entire tandem and its fragments. For example, when the matrices $D_0^{(r)}$ and $D_1^{(r)}$ defining the MAP input to the rth station are recursively calculated and the order of these matrices becomes too large for some $r \in \{1, \dots, R-1\}$, then we can try to approximate this MAP by a MAP with the same intensity, a few coinciding moments of the lengths of the inter-arrival times (at least with the same coefficient of variation), and the same coefficient of correlation but with a control process whose state has much smaller dimension, see, for example, [1, 52, 53]. After that, we can continue the recurrence procedure for calculating matrices $D_0^{(r')}$ and $D_1^{(r')}$, $r' > r$ starting from the matrices that define this approximating MAP.

6.7 Tandem of Multi-Server Queues with Cross-Traffic and No Buffers

The tandem queuing system discussed in this section is more general than the one studied in the previous section since it takes into account the presence of correlated cross-traffic incoming to each tandem station in addition to the traffic transmitted from the previous tandem station.

6.7.1 *The System Description*

We consider a tandem queuing system consisting of $R,\ R > 1$ stations. In terms of Kendall's notation, this system can be described as

$$MAP^{(1)}/PH^{(1)}/N_1/N_1 \to \bullet, MAP_2/PH^{(2)}/N_2/N_2 \to \ldots \to \bullet,$$

$$MAP_R/PH^{(R)}/N_R/N_R.$$

Station number r, $r = \overline{1, R}$ has N_r servers and no buffer. Servers of the same station are independent and identical. The input to the first station is a MAP defined as $MAP^{(1)}$. It is characterized by the state space $\{0, 1, \ldots, W_1\}$ of the underlying process $\nu_t^{(1)}, t \geq 0$ and by the matrices $D_0,\ D_1$. The goal of each customer from this flow is to obtain service at all tandem stations.

In addition to the input of customers arriving at station r $(r > 1)$ from station $(r-1)$, it receives an additional MAP flow of customers, denoted by MAP_r. This flow is characterized by the state space $\{0, 1, \ldots, W_r\}$ of the underlying process $\nu_t^{(r)}, t \geq 0$, and by the matrices $H_0^{(r)},\ H_1^{(r)}$. The intensity of customer arrivals from MAP_r is denoted by h_r. The goal of each customer from this flow is to get full service at the rth and all subsequent tandem stations.

The service time of a customer at station r has a PH-type distribution with irreducible representation $(\boldsymbol{\beta}^{(r)}, S^{(r)}), r = \overline{1, R}$.

If a customer arriving at a tandem station from a previous one or from an additional flow finds all the servers busy then it leaves the tandem forever.

6.7.2 Output Flows from the Stations: Stationary State Distribution of the Tandem and Its Fragments

The process of the system states is described in terms of an irreducible multidimensional continuous-time MC

$$\xi_t = \{n_t^{(R)}, m_{1,t}^{(R)}, \ldots, m_{n^{(R)},t}^{(R)}, \nu_t^{(R)};\ n_t^{(R-1)}, m_{1,t}^{(R-1)}, \ldots, m_{n^{(R-1)},t}^{(R-1)}, \nu_t^{(R-1)};$$

$$\ldots, n_t^{(1)}, m_{1,t}^{(1)}, \ldots, m_{n^{(1)},t}^{(1)}, \nu_t^{(1)}\},\ t \ge 0,$$

where

- $n_t^{(r)}$ is the number of busy servers at station r, $n_t^{(r)} \in \{0, \ldots, N^{(r)}\}$, $r \in \{1, \ldots, R\}$;
- $m_{l,t}^{(r)}$ is the phase of the PH-type service process on the lth busy server of station r, $m_{l,t}^{(r)} = \overline{1, M_r}$, $r = \overline{1, R}$, $l = \overline{1, n^{(r)}}$;
- $\nu_t^{(r)}$, $\nu_t^{(r)} \in \{0, \ldots, W_r\}$ is the state of the MAP_r underlying process at time t.

The state space of the considered MC has dimension

$$\prod_{r=1}^{R} \frac{(W_r + 1)(M_r^{N_r+1} - 1)}{M_r - 1}.$$

Theorem 6.23 *The output flow from station r, $r \in \{1, 2, \ldots, R\}$ belongs to the class of MAPs. This MAP is given by the matrices $D_0^{(r)}$ and $D_1^{(r)}$, which are calculated by the following recurrent formulas:*

$$D_0^{(r)} = diag\{(S^{(r)})^{\oplus n},\ n = \overline{0, N_r}\} \otimes I_{K_r} \tag{6.78}$$

$$+ \begin{pmatrix} \tilde{D}_0^{(r)} & \boldsymbol{\beta}^{(r)} \otimes \tilde{D}_1^{(r)} & O & \cdots & O & O \\ O & I_{M_r} \otimes \tilde{D}_0^{(r)} & I_{M_r} \otimes \boldsymbol{\beta}^{(r)} \otimes \tilde{D}_1^{(r)} & \cdots & O & O \\ O & O & I_{M_r^2} \otimes \tilde{D}_0^{(r)} & \cdots & O & O \\ \vdots & \vdots & \vdots & \ddots & \vdots & \vdots \\ O & O & O & \cdots & I_{M_r^{N_r-1}} \otimes \tilde{D}_0^{(r)} & I_{M_r^{N_r-1}} \otimes \boldsymbol{\beta}^{(r)} \otimes \tilde{D}_1^{(r)} \\ O & O & O & \cdots & O & I_{M_r^{N_r}} \otimes (\tilde{D}_0^{(r)} + \tilde{D}_1^{(r)}) \end{pmatrix},$$

$$D_1^{(r)} \tag{6.79}$$

$$
= \begin{pmatrix}
O_{1\times 1} & O & \dots & O & O & O \\
(S_0^{(r)})^{\oplus 1} & 0_{M_r\times M_r} & \dots & O & O & O \\
O & (S_0^{(r)})^{\oplus 2} & \dots & O & O & O \\
\vdots & \vdots & \ddots & \vdots & \vdots & \vdots \\
O & O & \dots & (S_0^{(r)})^{\oplus (N_r-1)} & 0_{M_r^{N_r-1}\times M_r^{N_r-1}} & O \\
O & O & \dots & O & (S_0^{(r)})^{\oplus N_r} & 0_{M_r^{N_r}\times M_r^{N_r}}
\end{pmatrix} \otimes I_{K_r},
$$

$$r = 1, 2, \dots, R,$$

where

$$\tilde{D}_0^{(r)} = D_0^{(r-1)} \oplus H_0^{(r)}, \ \ \tilde{D}_1^{(r)} = D_1^{(r-1)} \oplus H_1^{(r)}, r = \overline{2, R},$$

the initial conditions are given by

$$\tilde{D}_0^{(1)} = D_0, \ \ \tilde{D}_1^{(1)} = D_1,$$

and K_r is calculated as

$$K_r = (W_r + 1) \prod_{l=1}^{r-1} \frac{(W_l + 1)(M_l^{N_l+1} - 1)}{M_l - 1}, \ r = 1, 2, \dots, R.$$

Proof The proof is similar to the proof of Theorem 6.18 for a similar tandem with no cross-traffic. The only difference is that in this case the output flow from station $(r - 1)$ does not coincide with the incoming flow to station r. The latter is a superposition of the output flow from station $(r - 1)$ and an additional MAP to station r. This superposition is again a MAP and is determined by the matrices $\tilde{D}_0^{(r)}, \tilde{D}_1^{(r)}$. □

Remark 6.5 Below the MAP incoming to station r will be denoted by $MAP^{(r)}$, $r = 1, 2, \dots, R$.

Corollary 6.17 *The input flow $MAP^{(r)}$ to station r is defined by the matrices*

$$\tilde{D}_0^{(r)} = D_0^{(r-1)} \oplus H_0^{(r)}, \ \ \tilde{D}_1^{(r)} = D_1^{(r-1)} \oplus H_1^{(r)}, r = \overline{2, R}. \tag{6.80}$$

Using the results of Theorem 6.23 and Corollary 6.17, we can calculate the marginal stationary distribution of the rth station as the stationary state distribution of a $MAP^{(r)}/PH^{(r)}/N_r/N_r$ queue, $r = 1, 2, \dots, R$. The corresponding algorithm is given in Sect. 6.6.3.

Using the results obtained, it is also possible to calculate the stationary state distribution of the whole tandem and its fragments as well as the loss probabilities associated with the tandem. At the same time, statements similar to those in Sect. 6.6

for a tandem with no cross-traffic are valid. For the convenience of a reader, we give these statements below.

Theorem 6.24 *The stationary state distribution of the tandem fragment* $\langle r, r+1, \ldots, r' \rangle$ *can be calculated as the stationary state distribution of the tandem*

$$MAP^{(r)}/PH^{(r)}/N_r/N_r \to \bullet, MAP_{r+1}/PH^{(r+1)}/N_{r+1}/N_{r+1} \to \ldots \to \bullet,$$

$$MAP_{r'}/PH^{(r')}/N_{r'}/N_{r'},$$

where $MAP^{(r)}$ *is given by matrices (6.80).*

Theorem 6.25 *The joint stationary distribution* $\mathbf{p}^{(1,\ldots,r)}$ *of the state probabilities of the first* r *stations of the tandem can be calculated as the stationary state distribution of the underlying process of the* MAP *that is the output flow from the* r*th station, i.e., the following formula holds:*

$$\mathbf{p}^{(1,\ldots,r)} = \boldsymbol{\theta}^{(r)},$$

where the vector $\boldsymbol{\theta}^{(r)}$ *is the unique solution of the system*

$$\boldsymbol{\theta}^{(r)}(D_0^{(r)} + D_1^{(r)}) = \mathbf{0}, \quad \boldsymbol{\theta}^{(r)}\mathbf{e} = 1, \ r = 1, 2, \ldots, R,$$

and the matrices $D_0^{(r)}$, $D_1^{(r)}$ *are calculated by the recurrent formulas (6.78) and (6.79).*

Corollary 6.18 *The vector* $\mathbf{p}$ *of the stationary state distribution of the whole tandem is calculated as the unique solution of the system of linear algebraic equations*

$$\mathbf{p}(D_0^{(R)} + D_1^{(R)}) = \mathbf{0}, \quad \mathbf{pe} = 1.$$

The matrix $D_0^{(R)} + D_1^{(R)}$ *coincides with the infinitesimal generator* Q *of the MC* $\xi_t, t \geq 0$.

6.7.3 Loss Probabilities

Theorem 6.26 *The probability of a customer loss in the fragment* $\langle r, r+1, \ldots, r' \rangle$ *of the tandem is calculated as*

$$P_{loss}^{(r,\ldots,r')} = \frac{\tilde{\lambda}_r - \lambda_{r'}}{\tilde{\lambda}_r},$$

where $\tilde{\lambda}_r$ is the intensity of the input $MAP^{(r)}$ to station r, and λ_r is the intensity of the output flow from station r'.

The intensity $\tilde{\lambda}_r$ is calculated as

$$\tilde{\lambda}_r = \tilde{\boldsymbol{\theta}}^{(r)} \tilde{D}_1^{(r)} \mathbf{e},$$

where the vector $\tilde{\boldsymbol{\theta}}^{(r)}$ is the unique solution of the system

$$\tilde{\boldsymbol{\theta}}^{(r)} (\tilde{D}_0^{(r)} + \tilde{D}_1^{(r)}) = \mathbf{0}, \ \ \tilde{\boldsymbol{\theta}}^{(r)} \mathbf{e} = 1,$$

and the intensity λ_r is calculated by the formula

$$\lambda_r = \boldsymbol{\theta}^{(r)} D_1^{(r)} \mathbf{e},$$

where the vector $\boldsymbol{\theta}^{(r)}$ is the unique solution of the system

$$\boldsymbol{\theta}^{(r)} (D_0^{(r)} + D_1^{(r)}) = \mathbf{0}, \ \ \boldsymbol{\theta}^{(r)} \mathbf{e} = 1, \ \ r = 1, 2, \ldots, R.$$

Corollary 6.19 *The probability $P_{loss}^{(r)}$ of an arbitrary customer loss for station r is calculated by the formula*

$$P_{loss}^{(r)} = \frac{\tilde{\lambda}_r - \lambda_r}{\tilde{\lambda}_r}. \tag{6.81}$$

Corollary 6.20 *The probability P_{loss} of an arbitrary customer loss at the whole tandem is calculated by the formula*

$$P_{loss} = \frac{\tilde{\lambda}_1 - \lambda_R}{\tilde{\lambda}_1}.$$

Theorem 6.27 *The probability of an arbitrary customer loss at the first tandem station is calculated as*

$$P_{loss/1}^{(1)} = \frac{\boldsymbol{\theta} D_1 \mathbf{e}}{\lambda}.$$

The probability that an arbitrary customer arriving from station $(r-1)$ is lost at station r is calculated as

$$P_{loss/1}^{(r)} = \frac{\boldsymbol{\theta}^{(r)} diag\{O_{\frac{M_r^{N_r}-1}{M_r-1}K_r}, \ I_{M_r^{N_r}} \otimes D_1^{(r-1)} \otimes I_{W_r+1}\}\mathbf{e}}{\lambda_{r-1}}, \ r = \overline{2, R}. \tag{6.82}$$

The probability that an arbitrary customer arriving from the additional MAP is lost at station r is calculated as

$$P_{loss/2}^{(r)} = \frac{\boldsymbol{\theta}^{(r)} diag\{O_{\frac{M_r^{N_r}-1}{M_r-1}K_r}, I_{M_r^{N_r}} \otimes I_{\frac{K_r}{W_r+1}} \otimes H_1^{(r)}\}\mathbf{e}}{h_r}, \quad r = \overline{2, R}. \tag{6.83}$$

Proof The numerator of the right-hand side of (6.82) is the intensity of customers arriving at station r from station $(r-1)$ and find all servers of station r busy, and the denominator is the intensity of all customers arriving at station r from station $(r-1)$. By known ergodic considerations, the ratio of these two intensities determines the required probability $P_{loss/1}^{(r)}$. Similar arguments lead to formula (6.83) for probability $P_{loss/2}^{(r)}$. □

Using Theorem 6.27, we can obtain an alternative expression for the total loss probability $P_{loss}^{(r)}$ at station r previously determined by (6.81) in Corollary 6.19. The alternative formula (6.84) and also formulas for some joint probabilities associated with the tandem are given in the following statement.

Corollary 6.21 *The probability that an arbitrary customer from the total input to station r belongs to the output flow from station $(r-1)$ and will be lost since all servers of station r are busy is calculated as*

$$P_{loss,1}^{(r)} = P_{loss/1}^{(r)} \frac{\lambda_{r-1}}{\tilde{\lambda}_r}, \quad r = 2, \ldots, R.$$

The probability that an arbitrary customer from the total input to station r belongs to the additional flow and will be lost since all servers of station r are busy is calculated as follows:

$$P_{loss,2}^{(r)} = P_{loss/2}^{(r)} \frac{h_r}{\tilde{\lambda}_r}, \quad r = 2, \ldots, R.$$

The probability $P_{loss}^{(r)}$ of an arbitrary customer loss at the rth station can be calculated as

$$P_{loss}^{(r)} = P_{loss,1}^{(r)} + P_{loss,2}^{(r)}. \tag{6.84}$$

Corollary 6.22 *In the case $R = 2$, the probability that a customer from the input flow to the first station will not be lost by the system is defined as*

$$P_{succ} = (1 - P_{loss/1}^{(1)})(1 - P_{loss/1}^{(2)}).$$

6.7.4 Investigation of the System Based on the Construction of the Markov Chain Using the Ramaswami-Lucantoni Approach

As can be seen from formulas (6.78) and (6.79), Theorem 6.25, and Corollary 6.18, the dimension of the systems of linear equations determining the stationary state distributions of the tandem fragments and the whole tandem essentially depends on the number N_r of servers at the stations and on the dimension M_r of the state space of the PH-type service underlying process. Using the same arguments as in Sect. 6.6.6, we come to the conclusion that the dimension of the state space of the process describing the service process in the system, and, hence, the dimension of the state space of the MCs describing the tandem fragments and the whole tandem can be significantly reduced if the approach proposed in [99, 100] by American authors V. Ramaswami and D. Lucantoni is used when constructing these MCs.

Using the Ramaswami-Lucantoni approach, we describe the process of the system states in terms of an irreducible multidimensional continuous-time MC:

$$\tilde{\xi}_t = \{n_t^{(R)}, \vartheta_{1,t}^{(R)}, \dots, \vartheta_{M_R,t}^{(R)}, \nu_t^{(R)};\ n_t^{(R-1)}, \vartheta_{1,t}^{(R-1)}, \dots, \vartheta_{M_{R-1},t}^{(R-1)}, \nu_t^{(R-1)};\ \dots;$$

$$n_t^{(1)}, \vartheta_{1,t}^{(1)}, \dots, \vartheta_{M_1,t}^{(1)};\ \nu_t\},\ t \geq 0,$$

where

- $n_t^{(r)}$ is the number of busy servers at station r, $n_t^{(r)} \in \{0, \dots, N^{(r)}\}$, $r \in \{1, \dots, R\}$;
- $\vartheta_{l,t}^{(r)}$ is the number of servers at station r that are on the lth service phase, $\vartheta_{l,t}^{(r)} = \overline{1, n^{(r)}}$, $r = \overline{1, R}$, $l = \overline{1, M_r}$;
- $\nu_t^{(r)}$ is the state of the additional MAP_r underlying process at time t, $\nu_t^{(r)} \in \{0, \dots, W_r\}$.

The state space of this MC has dimension

$$\prod_{r=1}^{R}(W_r + 1)\sum_{n=0}^{N_r} C_{n+M_r-1}^{M_r-1}$$

and, for large values of N_r, M_r, is much smaller than the dimension of the state space of the MC ξ_t constructed above on the basis of the classical approach.

Theorem 6.28 *The output flow from station r, $r \in \{1, 2, \dots, R\}$ belongs to the class of MAPs. This MAP is given by the matrices $D_0^{(r)}$ and $D_1^{(r)}$ calculated by the following recurrent formulas:*

$$D_0^{(r)} = diag\{A_n(N_r, S^{(r)}),\ n = \overline{0, N_r}\} \otimes I_{K_r} \tag{6.85}$$

$$+\begin{pmatrix} \tilde{D}_0^{(r)} & \tilde{D}_1^{(r)} \otimes P_0(\beta^{(r)}) & O & \dots & O & O \\ O & I_{C_{M_r}^{M_r-1}} \otimes \tilde{D}_0^{(r)} & I_{C_{M_r}^{M_r-1}} \otimes \tilde{D}_1^{(r)} \otimes P_1(\beta^{(r)}) & \dots & O & O \\ O & O & I_{C_{M_r+1}^{M_r-1}} \otimes \tilde{D}_0^{(r)} & \dots & O & O \\ \vdots & \vdots & \vdots & \ddots & \vdots & \vdots \\ O & O & O & \dots & I_{C_{N_r+M_r-2}^{M_r-1}} \otimes \tilde{D}_0^{(r)} & I_{C_{N_r+M_r-2}^{M_r-1}} \otimes \tilde{D}_1^{(r)} \otimes P_{N_r-1}(\beta^{(r)}) \\ O & O & O & \dots & 0 & I_{C_{N_r+M_r-1}^{M_r-1}} \otimes (\tilde{D}_0^{(r)} + \tilde{D}_1^{(r)}) \end{pmatrix},$$

$$D_1^{(r)} \tag{6.86}$$

$$= \begin{pmatrix} O_1 & O & \dots & O & O & O \\ L_{N_r-1}(N_r, S^{(r)}) & O_{C_{M_r}^{M_r-1}} & \dots & O & O & O \\ O & L_{N_r-2}(N_r, S^{(r)}) & \dots & O & O & O \\ \vdots & \vdots & \ddots & \vdots & \vdots & \vdots \\ O & O & \dots & L_1(N_r, S^{(r)}) & 0_{C_{N_r+M_r-2}^{M_r-1}} & O \\ O & O & \dots & O & L_0(N_r, S^{(r)}) & 0_{C_{N_r+M_r-1}^{M_r-1}} \end{pmatrix} \otimes I_{K_r},$$

$$r = 1, 2, \dots, R-1,$$

where

$$\tilde{D}_0^{(r)} = D_0^{(r-1)} \oplus H_0^{(r)}, \quad \tilde{D}_1^{(r)} = D_1^{(r-1)} \oplus H_1^{(r)}, r = \overline{2, R},$$

$$\tilde{D}_0^{(1)} = D_0, \quad \tilde{D}_1^{(1)} = D_1,$$

the value K_r is calculated as

$$K_r = \prod_{l=1}^{r-1} (W_l + 1) \sum_{n=0}^{N_l} C_{n+M_l-1}^{M_l-1}, \; r = \overline{1, R},$$

and the matrices $P_{n^{(r)}}(\boldsymbol{\beta}^{(r)})$, $A_{n^{(r)}}(N_r, S^{(r)})$, $L_{N_r-n^{(r)}}(N_r, S^{(r)})$ are defined in Sect. 6.6.6.

Using expressions (6.85) and (6.86) for the matrices $D_0^{(r)}$, $D_1^{(r)}$, we can calculate the stationary state distribution of the tandem and its fragments as well as some loss probabilities for the tandem according to the formulas given above in Theorems 6.24–6.26 and Corollaries 6.17–6.20. The formulas for the conditional loss probabilities given in Theorem 6.27 are modified because the dimensions of the state spaces in the Ramaswami-Lucantoni approach differ from those in the classical approach. Modified formulas are given below in the statement of Theorem 6.29.

Theorem 6.29 *The probability of an arbitrary customer loss at the first station of the tandem is calculated as*

$$P^{(1)}_{loss/1} = \frac{\boldsymbol{\theta} diag\{O_{\sum_{l=0}^{N_1-1} C^{M_l-1}_{l+M_l-1}(W_1+1)}, I_{C^{M_1-1}_{N_1+M_1-1}} \otimes D_1\}\mathbf{e}}{\lambda}.$$

The probability of losing at station r an arbitrary customer arriving from station (r − 1) is calculated as

$$P^{(r)}_{loss/1} = \frac{\boldsymbol{\theta}^{(r)} diag\{O_{\sum_{l=0}^{N_r-1} C^{M_l-1}_{l+M_l-1}K_r}, I_{C^{M_r-1}_{N_r+M_r-1}} \otimes D_1^{(r-1)} \otimes I_{W_r+1}\}\mathbf{e}}{\lambda_{r-1}}, \quad r = \overline{2, R}.$$

The probability of losing at station r an arbitrary customer arriving from the additional flow is calculated as

$$P^{(r)}_{loss/2} = \frac{\boldsymbol{\theta}^{(r)} diag\{O_{\sum_{l=0}^{N_r-1} C^{M_l-1}_{l+M_l-1}K_r}, I_{C^{M_r-1}_{N_r+M_r-1}} \otimes I_{\frac{K_r}{W_r+1}} \otimes H_1^{(r)}\}\mathbf{e}}{h_r}, \quad r = \overline{2, R}.$$

6.8 Tandem of Single-Server Queues with Finite Buffers and Cross-Traffic

In this section, we consider a tandem system consisting of an arbitrary finite number of stations represented by single-server systems with finite buffers and with the first station receiving a MAP input of customers. Each arriving customer has to get a successive service at all tandem stations. In addition, each of the stations receives an additional MAP input of customers that must be served at this station and at all subsequent tandem stations. Due to the limited space for waiting at the stations, the customers of any input flow can be lost. To detect and prevent so-called bottlenecks in the tandem network, it is important to determine the probabilities of the loss of customers arriving at each tandem station as well as the stationary distribution of the probabilities of the number of customers and the sojourn time of a customer at each station and in the whole tandem.

6.8.1 The System Description

We consider a tandem queuing system consisting of R, $R > 1$ stations. In terms of Kendall's notation, this system can be described as

$$MAP_1/PH/1/N_1 \to \cdot, MAP_2/PH/1/N_2 \to \ldots \to \cdot, MAP_R/PH/1/N_R.$$

Station number r, $r = \overline{1, R}$ has a single-server queuing system with a buffer of size $N_r - 1$.

The input flow at the first station is denoted by MAP_1. In addition to the flow of arrivals to station r, $r > 1$ from station $(r - 1)$, station r receives an additional MAP flow of customers denoted by MAP_r.

Denote the matrices describing MAP_1 by D_0, D_1 and the similar matrices describing MAP_r by $H_0^{(r)}$, $H_1^{(r)}$. The intensity of MAP_1 arrivals is denoted by $\tilde{\lambda}_1$ and the intensity of the MAP_r, $r > 1$ arrivals is denoted by h_r.

The service times at station r of the tandem have a PH-type distribution given by irreducible representation $(\boldsymbol{\beta}^{(r)}, S^{(r)})$, where $\boldsymbol{\beta}^{(r)}$ and $S^{(r)}$ are a vector and a matrix of size M_r.

If a customer arriving at station r finds the server busy then it takes a waiting place in the buffer if the latter is not full. Otherwise, the customer leaves the tandem forever (it is lost). The customer service discipline is FIFO (First-In-First-Out).

6.8.2 Output Flows from the Stations

The process of the tandem states is described in terms of an irreducible multidimensional continuous-time MC:

$$\xi_t = \{n_t^{(R)}, \eta_t^{(R)}, \nu_t^{(R)}, n_t^{(R-1)}, \eta_t^{(R-1)}, \nu_t^{(R-1)}, \dots, n_t^{(1)}, \eta_t^{(1)}, \nu_t^{(1)}\},\ t \geq 0,$$

where

- $n_t^{(r)}$ is the number of customers at station r, $n_t^{(r)} \in \{0, \dots, N_r\}$,
- $\eta_t^{(r)}$ is the state of the service process at station r, $\eta_t^{(r)} \in \{1, \dots, M_r\}$,
- $\nu_t^{(r)}$ is the state of the MAP_r underlying process at time t, $\nu_t^{(r)} \in \{0, \dots, W_r\}$, $r \in \{1, \dots, R\}$.

Note that in order to avoid the necessity to separately consider states in which some buffers are empty and a server is idle, we assume that at the end of a service at station r of the tandem, the state of the service underlying process for the next customer is set in accordance with the probabilities specified by the components of the vector $\boldsymbol{\beta}^{(r)}$ regardless of whether there are customers in this station's buffer or not.

The state space of the MC ξ_t, $t \geq 0$ has dimension

$$\prod_{r=1}^{R} (N_r + 1) M_r (W_r + 1).$$

In this section, we develop a simple, accurate, and convenient method proposed in [77] for calculating the marginal stationary probability distributions of the tandem fragments as well as the whole tandem, and the corresponding loss probabilities. We also find the distribution of the customer sojourn time in the tandem.

As for the tandems of multi-server queues described in the previous two sections, we start analysis of the tandem considered in this section with analysis of the output and input flows of the tandem stations, which are proved to belong to the class of MAPs. The corresponding results are formulated in the following theorem and its corollary.

Theorem 6.30 *The output flow from station r, $r \in \{1, 2, \dots, R\}$ belongs to the class of MAPs. This MAP is given by the matrices $D_0^{(r)}$ and $D_1^{(r)}$ calculated by the following recurrent formulas:*

$$D_0^{(r)} = diag\{O, S^{(r)}, \dots, S^{(r)}\} \oplus (D_0^{(r-1)} \oplus H_0^{(r)}) + \tilde{E}_r^+ \otimes (D_1^{(r-1)} \oplus H_1^{(r)}), \tag{6.87}$$

$$D_1^{(r)} = E_r^- \otimes \mathbf{S}_0^{(r)} \boldsymbol{\beta}^{(r)} \otimes I_{K_r}, \; r = 1, 2, \dots, R \tag{6.88}$$

with the initial conditions

$$D_0^{(0)} = D_0, \;\; D_1^{(0)} = D_1, \;\; H_0^{(1)} = H_1^{(1)} = O.$$

Here $\tilde{E}_r^+ = E_r^+ \otimes I_{M_r}$, where E_r^+ and E_r^- are square matrices of size $N_r + 1$ given by the formulas

$$E_r^+ = \begin{pmatrix} 0\,1\,0 \dots 0 \\ 0\,0\,1 \dots 0 \\ \vdots\;\vdots\;\vdots\;\ddots\;\vdots \\ 0\,0\,0 \dots 1 \\ 0\,0\,0 \dots 1 \end{pmatrix}, \quad E_r^- = \begin{pmatrix} 0\,0 \dots 0\,0 \\ 1\,0 \dots 0\,0 \\ 0\,1 \dots 0\,0 \\ \vdots\;\vdots\;\ddots\;\vdots\;\vdots \\ 0\,0 \dots 1\,0 \end{pmatrix},$$

and the values K_r are calculated by the formula

$$K_r = \prod_{r'=1}^{r-1} (N_{r'} + 1) M_{r'} \prod_{r'=1}^{r} (W_{r'} + 1).$$

The proof is similar to the proof of Theorem 6.23.

Corollary 6.23 *The input flow to station r, $r \in \{2, \dots, R\}$ of the tandem belongs to the class of MAPs. This MAP is given by the matrices*

$$\tilde{D}_0^{(r)} = D_0^{(r-1)} \oplus H_0^{(r)}, \quad \tilde{D}_1^{(r)} = D_1^{(r-1)} \oplus H_1^{(r)}, \quad r \in \{2, \dots, R\}, \tag{6.89}$$

where the matrices $D_0^{(r)}$, $D_1^{(r)}$ are calculated by the recurrent formulas (6.87) and (6.88).

Remark 6.6 By assuming $r = 1$ in formula (6.89), we get the natural relations

$$\tilde{D}_k^{(1)} = D_k,\ k = 0, 1.$$

Remark 6.7 Below, we will denote the total incoming MAP to station r by $MAP^{(r)}$, $r = 1, 2, \ldots, R$. Note that the notation of $MAP^{(1)}$ means the same as the previously introduced notation of MAP_1. Both of these symbols are used for the incoming flow to the first station.

Using the results of Theorem 6.30, we can calculate the marginal stationary distribution of station r as the stationary state distribution of a $MAP^{(r)}/PH/1/N_r$ queue, $r = \overline{1, R}$.

For the convenience of the reader, in the next subsection, we present an algorithm for calculating the stationary state distribution of this type of system. For brevity, we omit index r in the notation of the matrices describing the MAP and the service process and also in the notation of the number of waiting places in the system.

6.8.3 Stationary State Distribution and the Customer Sojourn Time Distribution for a $MAP/PH/1/N$ Queue

The behavior of a $MAP/PH/1/N$ queue can be described by a three-dimensional MC $\zeta_t = \{n_t, \eta_t, \nu_t\}$, where n_t is the number of customers in the system, $n_t \in \{0, \ldots, N\}$, η_t is the current state of the service underlying process, $\eta_t \in \{1, \ldots, M\}$, and ν_t, $\nu_t \in \{0, \ldots, W\}$ is the state of the MAP underlying process at time t.

Renumber the MC states in lexicographical order. Then the infinitesimal generator of this MC has the form

$$A = \begin{pmatrix} \hat{D}_0 & \hat{D}_1 & O & \ldots & O & O \\ \hat{\mathbf{S}} & S \oplus D_0 & \hat{D}_1 & \ldots & O & O \\ O & \hat{\mathbf{S}} & S \oplus D_0 & \ldots & O & O \\ \vdots & \vdots & \vdots & \ddots & \vdots & \vdots \\ O & O & O & \ldots & S \oplus D_0 & \hat{D}_1 \\ O & O & O & \ldots & \hat{\mathbf{S}} & S \oplus (D_0 + D_1) \end{pmatrix},$$

where $\hat{D}_k = I_M \otimes D_k,\ k = 0, 1, \hat{\mathbf{S}} = \mathbf{S}_0\boldsymbol{\beta} \otimes I_{W+1}$.

Let $\mathbf{q}$ be the row vector of the stationary state distribution of the MC states. This vector is defined as the unique solution of the system of linear algebraic equations

$$\mathbf{q}A = \mathbf{0},\ \ \mathbf{q}\mathbf{e} = 1.$$

If the system has large dimension, it is worth using a special algorithm to solve it taking into account the probabilistic meaning of the entries of the matrix A

(see [75]). This algorithm essentially takes into account the sparse structure of the matrix A and is described as follows.

Let the vector $\mathbf{q}$ be presented as $\mathbf{q} = (\mathbf{q}_0, \mathbf{q}_1, \ldots, \mathbf{q}_N)$, where the vectors $\mathbf{q}_i$, $i = 0, \ldots, N$ have size $M(W+1)$.

Algorithm 6.2 The stationary state distribution vectors $\mathbf{q}_i$, $i = 0, \ldots, N$ are calculated as

$$\mathbf{q}_l = \mathbf{q}_0 F_l, \; l = 1, \ldots, N,$$

where the matrices F_l are calculated from the following recurrent formulas:

$$F_0 = I, \;\; F_i = -F_{i-1}\hat{D}_1(S \oplus D_0 + (S \oplus D_1)\delta_{i,N} + \hat{D}_1 G_i)^{-1}, \;\; i = 1, \ldots, N,$$

and the matrices $G_i, i = \overline{0, N-1}$ are calculated by the following backward recursion:

$$G_i = -(S \oplus D_0 + \hat{D}_1 G_{i+1})^{-1}\hat{\mathbf{S}}, \;\; i = N-2, N-3, \ldots, 0$$

with the initial condition

$$G_{N-1} = -(S \oplus (D_0 + D_1))^{-1}\hat{\mathbf{S}},$$

and the vector $\mathbf{q}_0$ is the unique solution of the system of linear algebraic equations

$$\mathbf{q}_0(\hat{D}_0 + \hat{D}_1 G_0) = \mathbf{0}, \quad \mathbf{q}_0 \sum_{l=0}^{N} F_l \mathbf{e} = 1.$$

Proposition 6.1 *The LST $v(s)$ of the distribution of the customer sojourn time in a $MAP/PH/1/N$ queue is given as follows:*

$$v(s) = \lambda^{-1}\left[\sum_{i=0}^{N-1} \mathbf{q}_i(I_M \otimes D_1\mathbf{e})(sI - S)^{-1}\mathbf{S}_0(\beta(s))^i + \mathbf{q}_N(\mathbf{e}_M \otimes D_1\mathbf{e})\right].$$

The proof is easily provided by using the law of total probability and the probabilistic meaning of the LST in terms of catastrophes. Note that the entries of the column vector $(sI - S)^{-1}\mathbf{S}_0$ give the probabilities that a catastrophe does not happen during a residual service time.

Proposition 6.2 *The mean customer sojourn time* v_1 *in a* $MAP/PH/1/N$ *queue is calculated as follows:*

$$v_1 = \lambda^{-1} \sum_{i=0}^{N-1} \mathbf{q}_i (I_M \otimes D_1 \mathbf{e})((-S)^{-1} + ib_1 I)\mathbf{e}.$$

6.8.4 The Stationary Distribution of the Tandem and Its Fragments

Theorem 6.31 *The stationary distribution of the fragment* $\langle r, r+1, \ldots, r' \rangle$ *of the tandem under consideration can be calculated as the stationary state distribution of the tandem* $MAP^{(r)}/PH/1/N_r \to \cdot/PH/1/N_{r+1} \to \ldots \to \cdot/PH/1/N_{r'}$*, where* $MAP^{(r)}$ *is defined by formulas (6.87)–(6.89).*

Corollary 6.24 *The vector* $\mathbf{p}^{(r)}$ *of the marginal stationary state distribution of station r is calculated as the stationary state distribution of a* $MAP^{(r)}/PH/1/N_r$ *queue,* $r = 1, 2, \ldots, R$*, and can be calculated by the algorithm described in Sect. 6.8.3.*

Theorem 6.32 *The joint stationary distribution* $\mathbf{p}^{(1,\ldots,r)}$ *of the probabilities of the states of the first r stations of the tandem can be calculated as the stationary state distribution of the output* $MAP^{(r)}$ *flow from station r, i.e., the following formula holds:*

$$\mathbf{p}^{(1,\ldots,r)} = \boldsymbol{\theta}^{(r)},$$

where the vector $\boldsymbol{\theta}^{(r)}$ *is the unique solution of the system of linear algebraic equations*

$$\boldsymbol{\theta}^{(r)}(D_0^{(r)} + D_1^{(r)}) = \mathbf{0}, \quad \boldsymbol{\theta}^{(r)}\mathbf{e} = 1, \; r = 1, 2, \ldots, R,$$

and the matrices $D_0^{(r)}$, $D_1^{(r)}$ *are calculated by recurrent formulas (6.87) and (6.88).*

In the case of high dimensionality, this system can be successfully solved by using the stable Algorithm 6.2 described in Sect. 6.8.3 above. It is obvious that the vector $\mathbf{p}$ of the stationary distribution of the whole tandem coincides with vector $\mathbf{p}^{(1,\ldots,R)}$.

Corollary 6.25 *The vector* $\mathbf{p}$ *of the stationary state distribution of the tandem is calculated as the unique solution of the system of linear algebraic equations*

$$\mathbf{p}(D_0^{(R)} + D_1^{(R)}) = \mathbf{0}, \quad \mathbf{p}\mathbf{e} = 1. \tag{6.90}$$

As follows from system (6.90), the matrix $D_0^{(R)} + D_1^{(R)}$ coincides with the infinitesimal generator Q of the MC $\xi_t, t \geq 0$ describing the tandem operation.

Thus, using the results of the analysis of the output flows from tandem stations, we constructed the matrix Q while avoiding the time-consuming work required by the direct approach to construct this matrix.

6.8.5 Loss Probabilities

Theorem 6.33 *The probability $P_{loss}^{(r)}$ of an arbitrary customer loss at station r is calculated as*

$$P_{loss}^{(r)} = \frac{\tilde{\lambda}_r - \lambda_r}{\tilde{\lambda}_r}, \quad r = 1, 2, \ldots, R. \tag{6.91}$$

Here $\tilde{\lambda}_r$ is the intensity of the total $MAP^{(r)}$ flow incoming to station r,

$$\tilde{\lambda}_r = \tilde{\boldsymbol{\theta}}^{(r)} \tilde{D}_1^{(r)} \mathbf{e},$$

where $\boldsymbol{\theta}^{(r)}$ is the unique solution of the system

$$\tilde{\boldsymbol{\theta}}^{(r)} (\tilde{D}_0^{(r)} + \tilde{D}_1^{(r)}) = \mathbf{0}, \quad \tilde{\boldsymbol{\theta}}^{(r)} \mathbf{e} = 1,$$

and λ_r is the intensity of the output flow from station r,

$$\lambda_r = \boldsymbol{\theta}^{(r)} D_1^{(r)} \mathbf{e},$$

where the vector $\boldsymbol{\theta}^{(r)}$ is the unique solution of the system

$$\boldsymbol{\theta}^{(r)} (D_0^{(r)} + D_1^{(r)}) = \mathbf{0}, \quad \boldsymbol{\theta}^{(r)} \mathbf{e} = 1.$$

Theorem 6.34 *The probability that an arbitrary customer arriving from station $(r - 1)$ will be lost at station r is calculated as*

$$P_{loss/1}^{(r)} = \frac{\tilde{\boldsymbol{\theta}}^{(r)} (D_1^{(r-1)} \otimes I_{W_r+1}) \mathbf{e}}{\lambda_{r-1}}, \quad r = 1, 2, \ldots, R. \tag{6.92}$$

Here $D_1^{(0)} = D_1$ and λ_0 is the intensity of the input flow to the tandem.

The probability that an arbitrary customer from an additional input flow will be lost at station r is calculated as

$$P_{loss/2}^{(r)} = \frac{\tilde{\boldsymbol{\theta}}^{(r)} (I_{K_r(N_r+1)} \otimes H_1^{(r)}) \mathbf{e}}{h_r}, \quad r = 2, \ldots, R. \tag{6.93}$$

Proof The numerator of the right-hand side of (6.92) is the intensity of the input of customers that arrive at station r from station $(r-1)$ and find all servers of station r busy, and the denominator is the intensity of the flow of all customers arriving at station r from station $(r-1)$. By known ergodic considerations, the ratio of these two intensities determines the required probability $P^{(r)}_{loss/1}$. Similar arguments lead to formula (6.93) for the probability $P^{(r)}_{loss/2}$. □

Using Theorem 6.34, we can obtain an alternative expression for the total loss probability $P^{(r)}_{loss}$ at station r previously defined by (6.91) in Theorem 6.33. An alternative formula (6.96) as well as the formulas for some joint probabilities associated with the tandem are given in the following statement.

Corollary 6.26 *The probability that an arbitrary customer from the total flow to station r belongs to the station $(r-1)$ output flow and will be lost due to the lack of idle servers at station r is calculated as*

$$P^{(r)}_{loss,1} = P^{(r)}_{loss/1}\frac{\lambda_{r-1}}{\tilde{\lambda}_r}, \quad r = 1, 2, \ldots, R. \tag{6.94}$$

The probability that an arbitrary customer from the total input flow to station r belongs to the additional input flow and will be lost due to the lack of idle servers at station r is calculated as follows:

$$P^{(r)}_{loss,2} = P^{(r)}_{loss/2}\frac{h_r}{\tilde{\lambda}_r}, \quad r = 2, \ldots, R. \tag{6.95}$$

The probability $P^{(r)}_{loss}$ of an arbitrary customer loss at station r can be calculated as

$$P^{(r)}_{loss} = P^{(r)}_{loss,1} + P^{(r)}_{loss,1}. \tag{6.96}$$

Proof Formulas (6.94)–(6.96) are derived from the obvious probabilistic considerations. □

Corollary 6.27 *In the case $R = 2$, the probability that a customer from the input flow will not be lost in the system is defined as*

$$P_{succ} = (1 - P^{(1)}_{loss/1})(1 - P^{(2)}_{loss/1}).$$

6.8.6 Stationary Distribution of Customer Sojourn Time in the Tandem Stations and in the Whole Tandem

To accurately calculate the stationary distribution of an arbitrary customer sojourn time in the tandem, it is necessary to trace the process of passing this customer through the tandem stations. An arbitrary customer sojourn time in the tandem depends on the tandem state at the moment the customer arrives at the first tandem station and on the underlying processes of cross-traffic at the tandem stations at which the tagged customers has not yet arrived. Note the fee that has to be paid to obtain the generator of the Markov process describing the behavior of the system in a more or less simple recurrent form. This fee is that this process is described as

$$\xi_t = \{n_t^{(R)}, \eta_t^{(R)}, \nu_t^{(R)}, n_t^{(R-1)}, \eta_t^{(R-1)}, \nu_t^{(R-1)}, \ldots, n_t^{(1)}, \eta_t^{(1)}, \nu_t^{(1)}\},\ t \geq 0,$$

i.e., the state of the tandem stations is described not in the order of increasing station number but in the order of their decrease. In principle, this is not a significant drawback since the way to find the marginal probability distribution of the station states is described above, and the expressions are given for the various loss probabilities in the tandem under consideration. After all, if we need to know the stationary probabilities of the process

$$\tilde{\xi}_t = \{n_t^{(1)}, \eta_t^{(1)}, \nu_t^{(1)}, n_t^{(2)}, \eta_t^{(2)}, \nu_t^{(2)}, \ldots, n_t^{(R)}, \eta_t^{(R)}, \nu_t^{(R)}\},\ t \geq 0$$

then vectors of the stationary state distribution of the process $\tilde{\xi}_t$ can be derived by the corresponding permutations of the components of the stationary probability vectors of the process ξ_t. In principle, we could also do the same to find the generator of the process $\tilde{\xi}_t$ by the appropriate permutations of the components of the above process ξ_t generator. However, such permutations inevitably lead to the loss of the specificity of the block structure of a generator. If the generator of the process ξ_t has a block three-diagonal structure then the algorithm given in Sect. 6.8.3 can be effectively used to calculate the stationary distribution. At the same time, the generator of the process $\tilde{\xi}_t$ obtained by permutations does not have a pronounced block structure, which drastically reduces our ability to calculate the characteristics of the tandem on a computer even with a relatively small number of stations and small buffer sizes.

By the above reasoning, it seems logical to propose exact or approximate formulas for calculating the loss probability of an arbitrary customer arriving at the first tandem station, the LST of the sojourn time distribution of such a customer at tandem stations and in the whole tandem and the corresponding mean times (delays at the tandem stations and in the whole tandem) not based on a strict analysis of the process of passing through a tandem by a tagged customer.

Formula (6.94) gives probability $P_{loss,1}^{(r)}$ that an arbitrary customer from the total input flow arriving at station r belongs to the output process from station $(r-1)$ and will be lost due to the lack of idle servers at station r. Although along with the customers arriving at the first tandem station, the output flow from station $(r-1)$

includes other customers (arriving at from cross-traffic the previous stations), it can be assumed that $P^{(r)}_{loss,1}$ is the probability that a customer arriving at the first tandem station will be lost at station r. Under this assumption, the following statement holds.

Theorem 6.35 *The probability P_{loss} that a customer arriving at the first station of the tandem will be lost while passing through the tandem stations is given by*

$$P_{loss} = \sum_{r=0}^{R-1} \prod_{k=1}^{r} (1 - P^{(k)}_{loss,1}) P^{(r+1)}_{loss,1}.$$

Theorem 6.36 *The LST $v^{(r)}(s)$ of the distribution of the sojourn time at station r of a customer arriving from station $(r-1)$, $r \in \{1, \ldots, R\}$ is given as follows:*

$$v^{(r)}(s) = P^{(r)}_{loss,1} + \lambda^{-1}_{r-1} \sum_{i=0}^{N_r-1} \mathbf{p}^{(r)}_i (I_{M_r} \otimes (D^{(r-1)}_1 \otimes I_{W_r+1})\mathbf{e})(sI - S^{(r)})^{-1}\mathbf{S}^{(r)}_0 (\beta^{(r)}(s))^i$$

where

$$\beta^{(r)}(s) = \boldsymbol{\beta}^{(r)}(sI - S^{(r)})^{-1}\mathbf{S}^{(r)}_0, \ \mathrm{Re}\, s > 0,$$

$\lambda_0 = \lambda.$

The proof follows from Proposition 6.1.

Corollary 6.28 *The LST $v^{(r)}_a(s)$ of the distribution of the sojourn time at station r of a customer arriving from station $(r-1)$ and not lost at the rth station, $r \in \{1, \ldots, R\}$ is given as follows:*

$$v^{(r)}_a(s)$$

$$= (1-P^{(r)}_{loss,1})^{-1}\lambda^{-1}_{r-1} \sum_{i=0}^{N_r-1} \mathbf{p}^{(r)}_i (I_{M_r} \otimes (D^{(r-1)}_1 \otimes I_{W_r+1})\mathbf{e})(sI - S^{(r)})^{-1}\mathbf{S}^{(r)}_0 (\beta^{(r)}(s))^i.$$

Corollary 6.29 *The mean sojourn time $v^{(r)}_1$ at station r for a customer arriving from the $(r-1)$th station, $r \in \{1, \ldots, R\}$ is given as follows:*

$$v^{(r)}_1 = \lambda^{-1}_{r-1} \sum_{i=0}^{N_r-1} \mathbf{p}^{(r)}_i (I_{M_r} \otimes (D^{(r-1)}_1 \otimes I_{W_r+1})\mathbf{e})((-S^{(r)})^{-1} + ib^{(r)}_1 I)\mathbf{e}.$$

The proof follows from Proposition 6.2.

Theorem 6.37 *The LST $v(s)$ of the sojourn time distribution of a customer arriving at the first station is given as follows:*

$$v(s) = P_{loss}$$

$$+\prod_{r=1}^{R}\left[\lambda_{r-1}^{-1}\sum_{i=0}^{N_r-1}\mathbf{p}_i^{(r)}(I_{M_r}\otimes(D_1^{(r-1)}\otimes I_{W_r+1})\mathbf{e})(sI-S^{(r)})^{-1}\mathbf{S}_0^{(r)}(\beta^{(r)}(s))^i\right].$$

Corollary 6.30 *The LST $v_a(s)$ of the sojourn time distribution of a customer arriving at the first station and not lost in the system is given as follows:*

$$v_a(s) = (1-P_{loss})^{-1}$$

$$\times\prod_{r=1}^{R}\left[\lambda_{r-1}^{-1}\sum_{i=0}^{N_r-1}\mathbf{p}_i^{(r)}(I_{M_r}\otimes(D_1^{(r-1)}\otimes I_{W_r+1})\mathbf{e})(sI-S^{(r)})^{-1}\mathbf{S}_0^{(r)}(\beta^{(r)}(s))^i\right].$$

Corollary 6.31 *The mean sojourn time v_1 of a customer arriving at the first station is given as follows:*

$$v_1 = \sum_{r=1}^{R}\left[\lambda_{r-1}^{-1}\sum_{i=0}^{N_r-1}\mathbf{p}_i^{(r)}(I_{M_r}\otimes(D_1^{(r-1)}\otimes I_{W_r+1})((-S^{(r)})^{-1}+ib_1^{(r)}I)\mathbf{e}\right].$$

Corollary 6.32 *The mean sojourn time V of a customer arriving at the first station and not lost by the system is given as follows:*

$$V = (1-P_{loss})^{-1}v_1.$$

Appendix A
Some Information from the Theory of Matrices and Functions of Matrices

A.1 Stochastic and Sub-stochastic Matrices: Generators and Subgenerators

Let $A = (a_{ij})_{i,j=\overline{1,n}}$ be the square matrix of size $n \times n$.

The characteristic polynomial of the matrix A is the polynomial of the form $\det(\lambda I - A)$. The characteristic polynomial has degree n. The equation of the form

$$\det(\lambda I - A) = 0 \tag{A.1}$$

is called the characteristic equation.

The roots of the characteristic equation are called the characteristic numbers of the matrix A.

The square matrix $P = (p_{ij})_{i,j=\overline{1,n}}$ is called stochastic if for all $i,\ i = \overline{1,n}$ the inequalities $p_{ij} \geq 0,\ j = \overline{1,n}$ and equations $\sum\limits_{j=1}^{n} p_{ij} = 1$ hold.

The square matrix $\tilde{P} = (\tilde{p}_{ij})_{i,j=\overline{1,n}}$ is called sub-stochastic if for all $i,\ i = \overline{1,n}$, $\tilde{p}_{ij} \geq 0,\ j = \overline{1,n}$ we have $\sum\limits_{j=1}^{n} \tilde{p}_{ij} \leq 1$ and there exists at least one value i for which $\sum\limits_{j=1}^{n} \tilde{p}_{ij} < 1$.

The square matrix $Q = (q_{ij})_{i,j=\overline{1,n}}$ is called a generator if for all $i,\ i = \overline{1,n}$ we have $q_{ii} < 0,\ q_{ij} \geq 0,\ j = \overline{1,n},\ j \neq i$, and the equation $\sum\limits_{j=1}^{n} q_{ij} = 0$ holds.

A. N. Dudin et al., *The Theory of Queuing Systems with Correlated Flows*,
https://doi.org/10.1007/978-3-030-32072-0

The square matrix $Q = (\tilde{q}_{ij})_{i,j=\overline{1,n}}$ is called a subgenerator if for all $i,\ i = \overline{1,n}$, $\tilde{q}_{ii} < 0,\ \tilde{q}_{ij} \geq 0,\ j = \overline{1,n},\ j \neq i$, we have $\sum_{j=1}^{n} \tilde{q}_{ij} \leq 0$, and there exists at least one value i for which $\sum_{j=1}^{n} \tilde{q}_{ij} < 0$.

It is easy to see that if the matrix P is stochastic then the matrix $P - I$ is a generator. If the matrix $\tilde{P}$ is sub-stochastic then the matrix $\tilde{P} - I$ is a subgenerator. Therefore, the properties of the stochastic matrices and the generators, the sub-stochastic matrices and the subgenerators are related.

Lemma A.1 *A non-negative matrix is stochastic if and only if it has an eigenvector* $(1, 1, \ldots, 1)$ *corresponding to the eigenvalue 1. The eigenvalue 1 is maximal for a stochastic matrix.*

It follows from this lemma that if the matrix A is stochastic then the matrix $I - A$ is singular.

Lemma A.2 *A generator has an eigenvector* $(1, 1, \ldots, 1)$ *corresponding to the eigenvalue 0.*

Theorem A.1 (Hadamard Theorem) *If the entries a_{ij} of the matrix A are such that for all $i,\ i = \overline{1,n}$ the inequalities $|a_{ii}| > \sum_{j=1, j\neq i}^{n} |a_{ij}|$ are valid, i.e., the diagonal entries dominate, then the matrix A is non-singular.*

Theorem A.2 (O. Tausski Theorem) *If the matrix A is irreducible, for all $i,\ i = \overline{1,n}$ the inequalities $|a_{ii}| \geq \sum_{j=1, j\neq i}^{n} |a_{ij}|$ hold, and for at least one i the inequality is exact, then the matrix A is non-singular.*

Lemma A.3 *Let $\tilde{P}$ be a sub-stochastic matrix. If there exists a non-zero non-negative matrix $\bar{P}$ such that the matrix $P = \tilde{P} + \bar{P}$ is stochastic and irreducible then the matrix $\tilde{P}$ is non-singular.*

Lemma A.4 *Let $\tilde{Q}$ be a subgenerator. If there exists a non-zero non-negative matrix $\bar{Q}$ such that the matrix $Q = \tilde{Q} + \bar{Q}$ is an irreducible generator then the matrix $\tilde{Q}$ is non-singular. All its eigenvalues have a negative real part.*

Lemmas A.3 and A.4 are proved similarly. Let us prove Lemma A.4.

Proof If the subgenerator $\tilde{Q}$ is such that the inequalities $\sum_{j=1}^{n} \tilde{q}_{ij} \leq 0$ strictly hold for all $i,\ i = \overline{1,n}$, then the nonsingularity of the subgenerator follows from the Hadamard theorem. The fact that the eigenvalues of the matrix $\tilde{Q}$ have a negative real part is proved by contradiction. Let there exist an eigenvalue λ such that $Re\ \lambda \geq 0$. Then $\det(\lambda I - \tilde{Q}) = 0$. But this is impossible by the Hadamard theorem since all the diagonal elements of the matrix $\lambda I - \tilde{Q}$ are known to dominate since this is valid already for matrix $\tilde{Q}$.

Let not all inequalities $\sum_{j=1}^{n} \tilde{q}_{ij} \leq 0$ hold strictly. Having at least one i, $i = \overline{1,n}$ such that inequality $\sum_{j=1}^{n} \tilde{q}_{ij} \leq 0$ strictly holds follows from the lemma assumptions that the matrix $\bar{Q}$ is non-negative and has positive elements, and the matrix $Q = \tilde{Q} + \bar{Q}$ is a generator and therefore has zero sums of elements in all rows.

If the matrix $\tilde{Q}$ is irreducible, the lemma follows immediately from the theorem of O. Tausski.

Now let the matrix $\tilde{Q}$ be reducible. Then by permuting rows and columns, it can be reduced to the canonical normal form

$$\tilde{Q} = \begin{pmatrix} Q^{(1)} & O & \dots & O & O \\ O & Q^{(2)} & \dots & O & O \\ \vdots & \vdots & \ddots & \vdots & \vdots \\ O & O & \dots & Q^{(m)} & O \\ Q^{(m+1,1)} & Q^{(m+1,2)} & \dots & Q^{(m+1,m)} & Q^{(m+1)} \end{pmatrix}, \tag{A.2}$$

where the matrices $Q^{(r)}$, $r = \overline{1, m+1}$ are irreducible, and among the matrices $Q^{(m+1,r)}$, $r = \overline{1,m}$ there is at least one non-zero matrix.

It is known that

$$\det\tilde{Q} = \prod_{r=1}^{m+1} \det Q^{(r)}.$$

The matrix $Q^{(m+1)}$ is non-singular by the theorem of O. Tausski. Since it is irreducible then there is a non-exact domination of diagonal entries in all rows and exact domination in at least one row, which follows from the fact that among the matrices $Q^{(m+1,r)}$, $r = \overline{1,m}$ there is at least one non-zero matrix.

The matrices $Q^{(r)}$, $r = \overline{1,m}$ are also nonsingular by the theorem of O. Tausski. The presence of exact domination in at least one row follows from the fact that otherwise the matrix $Q^{(r)}$ is a generator, which implies that the canonical normal form of the matrix Q has a structure similar to (A.2), which contradicts the assumption of its irreducibility. Thus, $\det\tilde{Q}$ is a product of nonzero determinants and is itself nonzero. That is, the matrix $\tilde{Q}$ is nonsingular.

The statement that the real part of the eigenvalues is negative is completely similar to the case considered above when exact domination holds for all rows of the matrix $\tilde{Q}$. □

Corollary A.1 *The matrix D_0 in the definition of a $BMAP$ is nonsingular. All its eigenvalues have a negative real part.*

A matrix A is called stable if all its eigenvalues have a negative real part. A matrix B is semistable if the real parts of all its eigenvalues are negative or equal to zero.

Lemma A.5 *If a matrix A is stable then it is non-singular.*

Lemma A.6 *If the matrix A is such that $A^n \to O$ as $n \to \infty$ then the matrix $I - A$ has an inverse matrix, and*

$$(I - A)^{-1} = \sum_{k=0}^{\infty} A^k.$$

Lemma A.7 *Let P be the matrix of transition probabilities of an irreducible MC, and the stochastic matrix A its limit matrix, that is, $P^n \to A$ as $n \to \infty$. Then the matrix A has the form*

$$A = \begin{pmatrix} \boldsymbol{\alpha} \\ \boldsymbol{\alpha} \\ \vdots \\ \boldsymbol{\alpha} \end{pmatrix},$$

where $\boldsymbol{\alpha}$ is the left stochastic eigenvector of the matrix P corresponding to the eigenvalue 1, that is, this vector is the unique solution of the system $\boldsymbol{\alpha} P = \boldsymbol{\alpha},\ \boldsymbol{\alpha}\mathbf{e} = 1$. The vector $\boldsymbol{\alpha}$ is called the limiting vector of the irreducible MC.

The following relations hold: $PA = AP = A$.

The matrix Z given by the relation $Z = (I-(P-A))^{-1}$ is called the fundamental matrix of an irreducible MC with matrix P of transition probabilities, limiting matrix A, and limiting vector $\boldsymbol{\alpha}$.

Taking into account the fact that $A^k = A,\ k \geq 1$, the matrix Z is also defined by the relations

$$Z = \sum_{n=0}^{\infty} (P - A)^n = I + \sum_{n=1}^{\infty} (P^n - A).$$

Lemma A.8 *The following formulas are valid:*

$$PZ = ZP,\ Z\mathbf{e} = \mathbf{e},\ \boldsymbol{\alpha} Z = \boldsymbol{\alpha}, I - Z = A - PZ.$$

Lemma A.9 *Let the matrix P be an irreducible stochastic matrix and $\boldsymbol{\alpha}$ be a left stochastic eigenvector of the matrix P.*

Then the matrix $I - P + \mathbf{e}\boldsymbol{\alpha}$ is non-singular.

Proof The statement of the lemma follows immediately from Lemma A.6 if we show that

$$(P - \mathbf{e}\boldsymbol{\alpha})^n \to O$$

as $n \to \infty$.

According to the notation introduced,

$$\mathbf{e}\boldsymbol{\alpha} = A,\ PA = A,\ P^n \to A \text{ as } n \to \infty.$$

Taking these relations into account and using the Newton binomial formula, we have

$$(P - A)^n = \sum_{k=0}^{n} C_n^k P^{n-k}(-1)^k A^k = P^n + \sum_{k=1}^{n} C_n^k P^{n-k}(-1)^k A^k$$

$$= P^n + \sum_{k=1}^{n} C_n^k (-1)^k A = P^n - A \to O.$$

□

Lemma A.10 *Let $A_i,\ i \geq 0$ be sub-stochastic matrices of size $K \times K$ and their matrix PGF $A(z) = \sum_{i=0}^{\infty} A_i z^i$ be such that $A(1)$ is an irreducible stochastic matrix and $A'(z)|_{z=1} < \infty$. Then there exists the vector $\boldsymbol{\Delta} = (\Delta_1, \ldots, \Delta_K)^T$ whose components have signs coinciding with the sign of determinant $(\det(zI - A(z)))'|_{z=1}$ (or zero if $(\det(zI - A(z)))'|_{z=1} = 0$) such that the system of linear algebraic equations for the components of the vector $\mathbf{u} = (u_1, \ldots, u_K)^T$*

$$(I - A(1))\mathbf{u} = (A(z) - zI)'|_{z=1}\mathbf{e} + \boldsymbol{\Delta} \tag{A.3}$$

has an infinite set of solutions.

Proof Since the matrix $A(1)$ is stochastic and irreducible then the rank of the matrix $I - A(1)$ is $K - 1$. Let $D_m(\boldsymbol{\Delta})$ be the determinant obtained by replacing the mth column of $\det(I - A(1))$ with the column of free terms of system (A.3). Then system (A.3) has a solution if and only if

$$D_m(\boldsymbol{\Delta}) = 0,\ m = \overline{1, K}.$$

Expanding the determinant $D_m(\boldsymbol{\Delta})$ along the mth column, we obtain the system of linear algebraic equations for the vector $\boldsymbol{\Delta}$ entries:

$$\boldsymbol{\nabla}_m \boldsymbol{\Delta} = -\boldsymbol{\nabla}_m (A(z) - zI)'|_{z=1}\mathbf{e},\ m = \overline{1, K}. \tag{A.4}$$

Here $\boldsymbol{\nabla}_m$ is a row vector of the algebraic adjuncts of the mth column of the determinant $\det(I - A(1))$. It is easy to see that the matrix of system (A.4) is the matrix $\text{Adj}(I - A(1))$ of the algebraic adjuncts of the determinant $\det(I - A(1))$. It is known that the dimension of the space of the right zero vectors of a stochastic irreducible matrix is equal to one. From the relations

$$(I - A(1))\text{Adj}(I - A(1)) = \det(I - A(1))I = 0$$

it follows that each of the columns of the matrix $\mathrm{Adj}(I - A(1))$ is a right zero vector of the matrix $I - A(1)$. Obviously, the vector $\mathbf{e}$ is also a right zero vector of this matrix. Then all the columns of the matrix $\mathrm{Adj}(I - A(1))$ coincide up to constants c_k with the vector $\mathbf{e}$, that is

$$\mathrm{Adj}(I - A(1))_{(k)} = c_k \mathbf{e},\ k = \overline{1, K},$$

and the matrix $\mathrm{Adj}(I - A(1))$ has the form

$$\mathrm{Adj}(I - A(1)) = \begin{pmatrix} c_1 \ c_2 \cdots c_n \\ c_1 \ c_2 \cdots c_n \\ \cdot \ \ \cdot \ \cdots \ \cdot \\ c_1 \ c_2 \cdots c_n \end{pmatrix}.$$

It is known that in the case of an irreducible stochastic matrix $A(1)$, all algebraic adjuncts of the determinant $\det(I - A(1))$ are positive, that is,

$$c_k > 0,\ k = \overline{1, K}. \tag{A.5}$$

It is easy to show that the right-hand side of (A.4) is equal to $(\det(zI - A(z)))'|_{z=1}$. It is also obvious that

$$\nabla_m = (c_1, c_2, \ldots, c_n),\ m = \overline{1, K}.$$

Then system (A.4) is equivalent to one equation

$$\sum_{k=1}^{n} c_k \Delta_k = (\det(zI - A(z)))'|_{z=1}. \tag{A.6}$$

It follows from (A.5), (A.6) that there exists a vector $\boldsymbol{\Delta}$ whose sign coincides with the sign of the right-hand side of (A.6) (or equals zero if the right-hand side of (A.6) is 0) satisfying Eq. (A.6). This vector is also a solution of system (A.4). Substituting this solution into (A.3), we obtain a system of linear algebraic equations for the components of the vector $\mathbf{u}$ that has an infinite set of solutions. □

A.2 Functions of Matrices

Functions of matrices can be defined in different ways. One of them is the following. Let $f(\lambda)$ be a function. If $f(\lambda)$ is a polynomial, that is,

$$f(\lambda) = a_0 + a_1\lambda + \ldots + a_K\lambda^K,$$

then for any matrix A the function $f(A)$ of this matrix is defined by the relation

$$f(A) = a_0 I + a_1 A + \ldots + a_K A^K.$$

Let $\tilde{\lambda}_k$, $k = \overline{1, n_1}$ be the eigenvalues of an $n \times n$ matrix A, where n_1 is the number of distinct eigenvalues, and $\tilde{r}_k$, $k = \overline{1, n_1}$ are their multiplicities, $\sum\limits_{k=1}^{n_1} \tilde{r}_k = n$, $\tilde{r}_k \geq 1$.

The set of numbers $\tilde{\lambda}_k$ form the spectrum of the matrix A. The maximum of the modulus of the eigenvalues of the matrix A is called the spectral radius of this matrix and is denoted by $\varrho(A)$. If the matrix A is non-negative then the eigenvalue at which the maximum is attained is real and corresponds to a nonnegative eigenvector. The spectral radius of a stochastic matrix is 1 and the corresponding eigenvector is $\mathbf{e}$. If we compute the additional minors of order $(n - 1)$ for all entries of the matrix $\lambda I - A$ and denote their greatest common divisor by $D_{n-1}(\lambda)$ then the polynomial

$$\psi(\lambda) = \frac{\det(\lambda I - A)}{D_{n-1}(\lambda)}$$

is called the minimal characteristic polynomial of the matrix A with roots λ_k, $k = \overline{1, n_1}$ of multiplicities r_k, $r_k \leq \tilde{r}_k$.

Then, for any function $f(\lambda)$ such that for all λ_k, $k = \overline{1, n_1}$ there exist $f(\lambda_k), \ldots, f^{(r_k-1)}(\lambda_k)$, $k = \overline{1, n_1}$, the function $f(A)$ exists and is defined as follows:

$$f(A) = \sum_{k=1}^{n_1} \sum_{j=0}^{r_k-1} f^{(j)}(\lambda_k) Z_{kj}, \tag{A.7}$$

where Z_{kj} are some matrices that do not depend on the choice of the function f. These matrices are called constituent matrices of the given matrix A and are defined as

$$Z_{kj} = (j!)^{-1}(A - \lambda_k I)^j Z_{k0}, \ j \geq 0.$$

For illustration we give an example of calculating the matrices Z_{kj}. Let the matrix A be of the form

$$A = \begin{pmatrix} 2 & -1 & 1 \\ 0 & 1 & 1 \\ -1 & 1 & 1 \end{pmatrix}.$$

The characteristic equation is written as

$$\det(\lambda I - A) = (\lambda - 1)^2(\lambda - 2) = 0.$$

This means that the matrix A has the eigenvalues $\lambda_1 = 1$ of multiplicity $r_1 = 2$ and $\lambda_2 = 2$ of multiplicity $r_2 = 1$.

The greatest common divisor is $D_2(\lambda) = 1$. Therefore, the characteristic and minimal polynomials are equal. Equation (A.7) takes the form

$$f(A) = f(1)Z_{10} + f'(1)Z_{11} + f(2)Z_{20}. \tag{A.8}$$

Let us formulate a system of linear algebraic equations to find the matrices Z_{ij} by choosing different functions $f(\lambda)$ in Eq. (A.8). Taking the function $f(\lambda) = 1$, we obtain the equation

$$I = Z_{10} + Z_{20}. \tag{A.9}$$

By taking $f(\lambda) = \lambda - 1$, we get

$$A - I = Z_{11} + Z_{20}. \tag{A.10}$$

By taking the function $f(\lambda) = (\lambda - 1)^2$, we get

$$(A - I)^2 = Z_{20}. \tag{A.11}$$

Solving the system of Eqs. (A.9)–(A.11) and substituting the obtained expressions for the matrices Z_{ij} into Eq. (A.8), we obtain the formula

$$f(A) = f(1)\begin{pmatrix} 1 & 0 & 0 \\ 1 & 0 & 0 \\ 1 & -1 & 1 \end{pmatrix} + f'(1)\begin{pmatrix} 1 & -1 & 1 \\ 1 & -1 & 1 \\ 0 & 0 & 0 \end{pmatrix} + f(2)\begin{pmatrix} 0 & 0 & 0 \\ -1 & 1 & 0 \\ -1 & 1 & 0 \end{pmatrix}.$$

This formula makes it easy to calculate various functions of matrix A. For example,

$$e^A = e^1\begin{pmatrix} 1 & 0 & 0 \\ 1 & 0 & 0 \\ 1 & -1 & 1 \end{pmatrix} + e^1\begin{pmatrix} 1 & -1 & 1 \\ 1 & -1 & 1 \\ 0 & 0 & 0 \end{pmatrix} + e^2\begin{pmatrix} 0 & 0 & 0 \\ -1 & 1 & 0 \\ -1 & 1 & 0 \end{pmatrix}.$$

Formula (A.7) is sometimes called the formula for the matrix expansion on the spectrum. It is useful in verifying certain non-obvious properties of functions from matrices. We illustrate this with the example of the following statement proof.

Corollary A.2 *For the matrix D_0 from the definition of a $BMAP$, the integral $\int_0^\infty e^{D_0 t}\,dt$ exists and equals $(-D_0)^{-1}$.*

Proof Let $\lambda_k, k = \overline{1, n_1}$ be roots of multiplicities r_k of the minimal characteristic polynomial of the matrix D_0. Then, according to (A.7), for any function $f(\lambda)$ such that this function and the corresponding number of its derivatives exist at the points $\lambda_k, k = \overline{1, n_1}$, the following formula holds:

$$f(D_0) = \sum_{k=1}^{n_1} \sum_{j=0}^{r_k-1} f^{(j)}(\lambda_k) Z_{kj}, \tag{A.12}$$

where Z_{kj} are the constituent matrices for the matrix D_0.

We take the function $f(\lambda) = \int\limits_0^\infty e^{\lambda t} dt$ as a function $f(\lambda)$. It follows from Corollary A.1 that $Re\lambda < 0$. Hence, this function and its derivatives exist at the points $\lambda_k, k = \overline{1, n_1}$. Thus,

$$\int\limits_0^\infty e^{D_0 t} dt = \sum_{k=1}^{n_1} \sum_{j=0}^{r_k-1} \int\limits_0^\infty (e^{\lambda_k t})^{(j)} dt Z_{kj} = \sum_{k=1}^{n_1} \sum_{j=0}^{r_k-1} \int\limits_0^\infty t^j e^{\lambda_k t} dt Z_{kj}$$

$$= \sum_{k=1}^{n_1} \sum_{j=0}^{r_k-1} \frac{j!}{(-\lambda_k)^{j+1}} Z_{kj} = \sum_{k=1}^{n_1} \sum_{j=0}^{r_k-1} \left(\frac{1}{-\lambda_k} \right)^{(j)} Z_{kj} = (-D_0)^{-1}. \tag{A.13}$$

□

The following useful property of the functions of matrices should be noted. If the complex function $h(\lambda) = g(\varphi(\lambda))$ exists on the spectrum of the matrix A (that is, this function and the corresponding number of its derivatives exist at all points that are eigenvalues of the matrix) then the formula $h(A) = g(\varphi(A))$ holds.

A.3 Norms of Matrices

To estimate the proximity between matrices, the concept of a matrix norm is often used. Let A be a square matrix of size $n \times n$ with entries $a_{i,j}$, $i, j = \overline{1, n}$. The function $||A||$ of the matrix A taking values in the set of real numbers is called the norm if the following axioms hold:

(1) $||A|| \geq 0$ (non-negativity);
(1a) $||A|| = 0$ if and only if $A = 0$ (positivity);
(2) $||cA|| = |c|||A||$ for all complex numbers c (absolute homogeneity);
(3) $||A + B|| \leq ||A|| + ||B||$ (the triangle inequality);
(4) $||AB|| \leq ||A||||B||$ (the ring property).

The most popular norms for matrices are the following:

- The norm l_1 defined as $||A||_{l_1} = \sum_{i=1}^{n} \sum_{j=1}^{n} |a_{i,j}|$;
- The norm l_2 (Euclidean norm, Frobenius norm, Schur norm, Hilbert-Schmidt norm) defined as $||A||_{l_2} = (\sum_{i=1}^{n} \sum_{j=1}^{n} |a_{i,j}|^2)^{\frac{1}{2}}$;
- The norm l_∞ defined as $||A||_{l_\infty} = \max_{i,j=\overline{1,n}} |a_{i,j}|$;
- The maximum absolute column sum norm $||A||_1$ defined as

$$||A||_1 = \max_{j=\overline{1,n}} \sum_{i=1}^{n} |a_{i,j}|;$$

- The maximum absolute row sum norm $||A||_\infty$ defined as

$$||A||_\infty = \max_{i=\overline{1,n}} \sum_{j=1}^{n} |a_{i,j}|.$$

The following statement holds. For any matrix norm $||A||$ of any matrix A, $\varrho(A) \leq ||A||$ holds, where $\varrho(A)$ is the spectral radius of the matrix A.

A.4 The Kronecker Product and Sum of the Matrices

Consider the matrices $A = (a_{ij})_{i=\overline{1,n},j=\overline{1,m}}$ of order $n \times m$ and $B = (b_{ij})_{i=\overline{1,r},j=\overline{1,s}}$ of order $r \times s$.

The matrix

$$A \otimes B = \begin{pmatrix} a_{11}B & a_{12}B & \dots & a_{1n}B \\ \vdots & \vdots & \ddots & \vdots \\ a_{m1}B & a_{m2}B & \dots & a_{mn}B \end{pmatrix}$$

of order $nr \times ms$ is called the Kronecker product of these matrices.

The Kronecker Product Properties

1. $(A + B) \otimes C = A \otimes C + B \otimes C$.
2. $A \otimes (B + C) = A \otimes B + A \otimes C$.
3. $(\alpha A) \otimes B = A \otimes \alpha B = \alpha A \otimes B, \alpha = const$.
4. $A \otimes (B \otimes C) = (A \otimes B) \otimes C$.
5. $(A \otimes B)^T = A^T \otimes B^T$.
6. If matrices A and B are square and non-singular then $(A \otimes B)^{-1} = A^{-1} \otimes B^{-1}$.

7. In general, the matrices $(A \otimes B)$ and $(B \otimes A)$ are not equal. But there exist permutation matrices P and Q such that

$$(A \otimes B) = P(B \otimes A)Q.$$

If the matrices A and B are square then $P = Q^T$.

8. If matrices A, B, C, D are such that the products of matrices AB and CD make sense then the formula

$$(AB) \otimes (CD) = (A \otimes C)(B \otimes D),$$

is called the mixed-product rule. This rule is easily extended to the case of the product of many matrices.

9. If matrices A and B are square of sizes n and m, respectively then

$$det A \otimes B = (det A)^m (det B)^n,$$

$$tr A \otimes B = (tr A)(tr B),$$

where $tr A$ is the trace of the matrix A, which equals the sum of the main diagonal entries of the matrix A,

$$rank A \otimes B = (rank A)(rank B).$$

10. If matrices A and B are square, and $\mathbf{x}$ and $\mathbf{y}$ are their eigenvectors corresponding to the eigenvalues λ and μ, respectively, then the vector $\mathbf{x} \otimes \mathbf{y}$ is an eigenvector of the matrix $A \otimes B$ corresponding to the eigenvalue $\lambda\mu$.

Let A and B be square matrices of size $n \times n$ and $m \times m$, respectively. The Kronecker sum of these matrices is the matrix

$$A \oplus B = A \otimes I_m + I_n \otimes B.$$

A useful property of the Kronecker sum of matrices is the following:

$$e^A \otimes e^B = e^{A \oplus B}.$$

If the matrix A is stable and the matrix G is semistable then the integral $\int_0^\infty e^{At} \otimes e^{Gt} dt$ exists and equals $-(A \oplus G)^{-1}$.

References

1. Alfa, A.S., Diamond, J.E.: On approximating higher order MAPs with MAPs of order two. Queueing Syst. **34**, 269–288 (2000)
2. Arnon, S.: Advanced Optical Wireless Communication Systems. Cambridge University Press, Cambridge (2012)
3. Atakora, M., Chenji, H.: Optimal multicasting in hybrid RF/FSO DTNs. In: 2016 IEEE Global Communications Conference, GLOBECOM 2016 Proceedings, pp. 1–6. IEEE, Los Alamitos (2016)
4. Balsamo, S., De Nilto Persone, V., Inverardi, P.: A review on queueing network models with finite capacity queues for software architectures performance prediction. Perform. Eval. **51**, 269–288 (2003)
5. Basharin, G.P., Bocharov, P.P., Kogan, Ya.A.: Queueing Analysis for Computing Networks: Theory and Computational Methods. Nauka, Moscow (1989) (in Russian)
6. Bellman, R.: Introduction to Matrix Analysis. McGraw-Hill, London (1960)
7. Bocharov, P.P., Manzo, R., Pechinkin, A.V.: Analysis of a two-phase queueing system with a Markov arrival process and losses. J. Math. Sci. **1313**, 5606–5613 (2005)
8. Bocharov, P.P., Manzo, R., Pechinkin, A.V.: Analysis of a two-phase queueing system with a Markov arrival process and blocking. J. Math. Sci. **132**(5), 578–589 (2006)
9. Bocharov, P.P., D'Apice, C., Pechinkin, A.V.: Queueing Theory. Walter de Gruyter, Berlin (2011)
10. Bromberg, M.A.: Multi-phase systems with losses with exponential servicing. Autom. Remote. Control. **40**, 27–31 (1979)
11. Bromberg, M.A., Kokotushkin, V.A., Naumov, V.A.: Service by a cascaded network of instruments. Autom. Remote. Control. **38**, 60–64 (1977)
12. Bronstein, O.I., Dukhovny, I.M.: Priority Models in Computer Systems. Nauka, Moscow (1976) (in Russian)
13. Brugno, A., Dudin, A.N., Manzo, R.: Retrial queue with discipline of adaptive permanent pooling. App. Math. Model. **50**, 1–16 (2017)
14. Brugno, A., D'Apice, C., Dudin, A.N., Manzo, R.: Analysis of a $MAP/PH/1$ queue with flexible group service. Appl. Math. Comput. Sci. **27**(1), 119–131 (2017)
15. Brugno, A., Dudin, A.N., Marzo, R.: Analysis of a strategy of adaptive group admission of customers to single server retrial system. J. Amb. Intel. Hum. Comp. **9**, 123–135 (2018)
16. Burke, P.J.: The output of a queueing system. Oper. Res. **4**, 699–704 (1956)
17. Chakravarthy, S.R.: The batch Markovian arrival process: a review and future work. In: Krishnamoorthy, A., Raju, N., Ramaswami, V. (eds.) Advances Probability Theory stochastic. Notable, Branchburg (2011)

A. N. Dudin et al., *The Theory of Queuing Systems with Correlated Flows*,
https://doi.org/10.1007/978-3-030-32072-0

18. Chakravarthy, S., Dudin, A.N.: A multi-server retrial queue with $BMAP$ arrivals and group services. Queueing Syst. **42**, 5–31 (2002)
19. Chakravarthy, S., Dudin, A.N.: Analysis of a retrial queueing model with MAP arrivals and two types of customers. Math. Comput. Modell. **37**, 343–364 (2003)
20. Chakravarthy, S., Dudin, A.N.: Multi-threshold control for the $BMAP/SM/1/K$ queue with group services. J. Appl. Math. Stochast. Anal. **16**(4), 327–348 (2003)
21. Chakravarthy, S.R., Dudin, A.N.: A queueing model for crowdsourcing. J. Oper. Res. Soc. **68**(3), 221–236 (2017)
22. Cinlar, E.: Introduction to Stochastic Processes. Prentice Hall, Upper Saddle River (1975)
23. Daigle, J.N.: Queueing Theory for Telecommunications. Addison-Wesley, Boston (1992)
24. Douik, A., Dahrouj, H., Al-Naffouri, T.Y., Alouini, M.S.: Hybrid radio/free-space optical design for next generation backhaul systems. IEEE Trans. Commun. **64**(6), 2563–2577 (2016)
25. Dudin, S.A., Dudina, O.S.: Help desk center operating model as a two-phase queueing system. Probl. Inf. Transm. **49**(1), 58–72 (2013)
26. Dudin, A.N., Klimenok, V.I.: Queueing Systems with Correlated Arrivals. Belarussian State University, Minsk (2000) (in Russian)
27. Dudin, A., Nazarov, A.: On a tandem queue with retrials and losses and state dependent arrival, service and retrial rates. Int. J. Oper. Res. **29**(2), 170–182 (2017)
28. Dudin, A.N., Shaban, A.A.: Analysis of the $BMAP/SM/1/N$ type system with randomized choice of customers admission discipline. In: Dudin, A., Gortsev, A., Nazarov, A., Yakupov, R. (eds.) Information Technologies and Mathematical Modelling - Queueing Theory and Applications, ITMM 2016. Springer, Berlin (2016)
29. Dudin, A.N., Klimenok, V.I., Klimenok, I.A., et al.: Software "SIRIUS+" for evaluation and optimization of queues with the $BMAP$-input. In: Latouche, G., Taylor, P. (eds.) Advances in Matrix Analytic Methods for Stochastic Models, pp. 115–133. Notable, Branchburg (2000)
30. Dudin, A.N., Medvedev, G.A., Melenets, Yu.V.: Practicum on a Computer Theory Queuing. Universitetskoe, Minsk (2000) (in Russian)
31. Dudin, A.N., Listopad, N.I., Tsarenkov, G.V.: Improved algorithm for optimizing the operation of an Internet site. In: Problems in Designing Information and Telecommunication Systems, pp. 28–43. BSU Publishing, Minsk (2001) (in Russian)
32. Dudin, A.N., Klimenok, V.I., Tsarenkov, G.V.: A single-server queueing system with batch Markov arrivals, semi-Markov service, and finite buffer: its characteristics. Autom. Remote. Control. **63**(8), 1285–1297 (2002)
33. Dudin, A.N., Shaban, A.A., Klimenok, V.I.: Analysis of a queue in the $BMAP/G/1/N$ system. Int. J. Simul. Syst. Sci. Technol. **6**(1–2), 13–23 (2005)
34. Dudin, A., Dudin, S., Dudina, O.: Tandem queueing system $MAP/M/N/K - N \rightarrow \bullet M/R/\infty$ with impatient customers as a model of remote technical support. In: Proceedings of the 2nd Baltic Congress on Future Internet Communications, No. BCFIC 2012. 1569570915, pp. 1–6 (2012)
35. Dudin, A., Klimenok, V., Vishnevsky, V.: Analysis of unreliable single server queuing system with hot back-up server. In: Plakhov, A., Tchemisova, T., Freitas A. (eds.) Optimization in the Natural Sciences, EmC-ONS 2014. Springer, Berlin (2015)
36. Dudin, A., Deepak, T.G., Varghese, C.J., Krishnamoorty, A., Vishnevsky, V.: On a BMAP/G/1 retrial system with two types of search of customers from the orbit. Commun. Comput. Inf. Sci. **800**, 1–12 (2017)
37. Dudin, A., Lee, M., Dudina, O., Lee, S.: Analisys of priority retrial queue with many types of customers and servers reservation as a model of cognitive radio system. IEEE Trans. Commun. **65**(1), 186–199 (2017)
38. Dudina, O., Kim, C.S., Dudin, S.: Retrial queueing system with Markovian arrival flow and phase type service time distribution. Comput. Ind. Eng. **66**, 360–373 (2013)
39. Esmail, M.A., Fathallah, H., Alouini, M.S.: Outdoor FSO communications under fog: attenuation modeling and performance evaluation. IEEE Photonics J. **8**(4), 1–22 (2016)
40. Feller, W.: An Introduction to Probability Theory and Its Applications. Wiley, Hoboken (1957)

41. Ferng, H.W., Chang, J.F.: Departure processes of $BMAP/G/1$ queues. Queueing Systems **39**, 109–135 (2001)
42. Ferng, H.W., Chang, J.F.: Connection-wise end-to-end performance analysis of queueing networks with $MMPP$ inputs. Perform. Eval. **43**, 39–62 (2001)
43. Gantmacher, F.R.: The Theory of Matrices. Chelsea, New York (1960)
44. Gnedenko, B.V., Koenig, D.: Handbuch der Bedienungstheorie. Akademie-Verlag, Berlin (1983)
45. Gnedenko, B.V., Kovalenko, I.N.: Introduction to Queueing Theory, 2nd edn. Birkhauser, Cambridge (1989)
46. Gnedenko, B.V., Danielyan, E.A., Dimitrov, B.N., Klimov, G.P., Matvejev, V.F.: Priority Queueing Systems. Moscow State University, Moscow (1973) (in Russian)
47. Gomez-Corral, A.: A tandem queue with blocking and Markovian arrival process. Queueing Systems **41**, 343–370 (2002)
48. Gomez-Corral, A.: On a tandem G-network with blocking. Adv. Appl. Probab. **34**(3), 626–661 (2002)
49. Graham, A.: Kronecker Products and Matrix Calculus with Applications. Ellis Horwood, Chichester (1981)
50. Heindl, A.: Decomposition of general tandem networks with $MMPP$ input. Perform. Eval. **44**, 5–23 (2001)
51. Heindl, A.: Decomposition of general queue networks with $MMPP$ inputs and customer losses. Perform. Eval. **51**, 117–136 (2003)
52. Heindl, A., Telek, M.: Output models of $MAP/PH/1(/K)$ queues for an efficient network decomposition. Perform. Eval. **49**, 321–339 (2002)
53. Heindl, A., Mitchell, K., van de Liefvoort, A.: Correlation bounds for second order MAPs with application to queueing network decomposition. Perform. Eval. **63**, 553–577 (2006)
54. Heyman, D., Lucantoni, D.: Modelling multiple IP traffic streams with rate limits. IEEE/ACM Trans. Netw. **11**, 948–958 (2003)
55. Jaiswal, N.K.: Priority Queues. V. 50 of Mathematics in Science and Engineering. Academic, Cambridge (1968)
56. Kemeni, J.G., Snell, J.L., Knapp, A.W.: Denumerable Markov Chains. Van Nostrand, New York (1966)
57. Khinchine, A.: Work on the Mathematical Theory of Queueing Systems. Fizmatgiz, Moscow 1963 (in Russian)
58. Kim, C.S., Klimenok, V., Taramin, O.: A tandem retrial queueing system with two Markovian flows and reservation of channels. Comput. Oper. Res. **37**(7), 1238–1246 (2010)
59. Kim, C.S., Park, S.H., Dudin, A., Klimenok, V., Tsarenkov, G.: Investigation of the $BMAP/G/1 \rightarrow \bullet/PH/1/M$ tandem queue with retrials and losses. App. Math. Model. **34**(10), 2926–2940 (2010)
60. Kim, Ch., Dudin, A., Klimenok, V.: Tandem retrial queueing system with correlated arrival flow and operation of the second station described by a Markov chain. In: Kwiecien, A., Gaj, P., Stera, P. (eds.) Computer Networks, CN 2012. Springer. Berlin (2012)
61. Kim, C.S., Klimenok, V.I., Taramin, O.S., Dudin, A.: A tandem $BMAP/G/1 \rightarrow \bullet/M/N/0$ queue with heterogeneous customers. Math. Probl. Eng. **2012**, Article ID 324604, 26 (2012)
62. Kim, C.S., Dudin, A., Dudin, S., Dudina, O.: Tandem queueing system with impatient customers as a model of call center with interactive voice response. Perform. Eval. **70**, 440–453 (2013)
63. Kim, C.S., Dudin, S., Taramin, O., Baek, J.: Queueing system $MMAP/PH/N/N+R$ with impatient heterogeneous customers as a model of call center. App. Math. Model. **37**(3), 958–976 (2013)
64. Kim, C.S., Dudin, A., Dudina, O., Dudin, S.: Tandem queueing system with infinite and finite intermediate buffers and generalized phase-type service time distribution. Eur. J. Oper. Res. **235**, 170–179 (2014)
65. Kim, C.S., Klimenok, V., Dudin, A.: Priority tandem queueing system with retrials and reservation of channels as a model of call center. Comput. Ind. Eng. **96**, 61–71 (2016)

66. Kim, J., Dudin, A., Dudin, S., Kim, C.: Analysis of a semi-open queueing network with Markovian arrival process. Perform. Eval. **120**, 1–19 (2018)
67. Kleinrock, L.: Queueing Systems, Volume II: Computer Applications. Wiley, New York (1976)
68. Klemm, A., Lindermann, C., Lohmann, M.: Modelling IP traffic using the batch Markovian arrival process. Perform. Eval. **54**, 149–173 (2003)
69. Klimenok, V.I.: Two-server queueing system with unreliable servers and markovian arrival process. Commun. Comput. Inform. Sci. **800**, 42–55 (2017)
70. Klimenok, V.I., Dudin, A.N.: Multi-dimensional asymptotically quasi-Toeplitz Markov chains and their application in queueing theory. Queueing Systems **54**, 245–259 (2006)
71. Klimenok, V., Dudina, O.: Retrial tandem queue with controlled strategy of repeated attempts. Qual. Technol. Quant. Manag. **14**(1), 74–93 (2017)
72. Klimenok, V.I., Taramin, O.S.: Tandem service system with batch Markov flow and repeated calls. Autom. Remote. Control. **71**(1), 1–13 (2010)
73. Klimenok, V.I., Taramin, O.S.: Two-Phase Queueing Systems with Correlated Arrivals. RIVSH, Minsk (2011) (in Russian)
74. Klimenok, V.I., Breuer, L., Tsarenkov, G.V., Dudin, A.N.: The $BMAP/G/1/N \to PH/1/M-1$ tandem queue with losses. Perform. Eval. **61**, 17–60 (2005)
75. Klimenok, V.I., Kim, C.S., Orlovsky, D.S., Dudin, A.N.: Lack of invariant property of Erlang $BMAP/PH/N/0$ model. Queueing Systems **49**, 187–213 (2005)
76. Klimenok, V., Kim, C.S., Tsarenkov, G.V., Breuer, L., Dudin, A.N.: The $BMAP/G/1 \to \cdot/PH/1/M$ tandem queue with feedback and losses. Perform. Eval. **64**, 802–818 (2007)
77. Klimenok, V., Dudin, A., Vishnevsky, V.: On the stationary distribution of tandem queue consisting of a finite number of stations. In: Kwiecien, A., Gaj, P., Stera, P. (eds.) Computer Networks, CN 2012. Springer, Berlin (2012)
78. Klimenok, V., Dudina, O., Vishnevsky, V., Samuylov, K.: Retrial tandem queue with BMAP-input and semi-Markovian service process. In: Vishnevsky, V., Samouylov, K., Kozyrev, D. (eds.) Distributed Computer and Communication Networks, DCCN 2017. Springer, Berlin (2017)
79. Klimenok, V., Dudin, A., Samouylov, K.: Analysis of the $BMAP/PH/N$ queueing system with backup servers. App. Math. Model. **57**, 64–84 (2018)
80. Klimov, G.P.: Stochastic Queueing Systems. Moscow, Nauka (1966) (in Russian)
81. Lakatos, L., Szeidl, L., Telek, M.: Introduction to Queueing Systems with Telecommunication Application. Springer, New York (2013)
82. Leland, W.E., Taqqu Murad, S., Willinger, W., Wilson, D.V.: On the self-similar nature of ethernet traffic. IEEE/ACM Trans. Networking **2**(1), 1–15 (1994)
83. Lema, M.A., Pardo, E., Galinina, O., Andreev, S., Dohler, M.: Flexible dual-connectivity spectrum aggregation for decoupled uplink and downlink access in 5G heterogeneous systems. IEEE J. Sel. Areas Commun. **34**(11), 2851–2865 (2016)
84. Lian, Z., Liu, L.: A tandem network with MAP inputs. Oper. Res. Lett. **36**, 189–195 (2008)
85. Lucantoni, D.M.: New results on the single server queue with a batch Markovian arrival process. Commun. Stat. Stoch. Model. **7**(1), 1–46 (1991)
86. Lucantoni, D.M.: The $BMAP/G/1$ queue: A tutorial. In: Donatiello, L., Nelson, R. (eds.) Performance Evaluation of Computer and Communications Systems, pp. 330–358. Springer, Berlin (1993)
87. Lucantoni, D.M., Neuts, M.F.: Some steady-state distributions for the $MAP/SM/1$ queue. Commun. Stat. Stoch. Model. **10**, 575–598 (1994)
88. Makki, B., Svensson, T., Eriksson, T., Alouini, M.S.: On the performance of RF-FSO links with and without hybrid ARQ. IEEE Trans. Wirel. Commun. **I5**(7), 4928–4943 (2016)
89. Mitrani, I.: The spectral expansion solution method for Markov processes on lattice strips. In: Dshalalow, J.H. (ed.) Advances in Queueing: Theory, Methods and Open Problems, pp. 337–352. CRC Press, Boca Raton (1995)
90. Moustafa, M.D.: Input-output Markov processes. Proc. Koninkl. Net. Akad. Wetensch. **A60**, 112–118 (1957)

91. Najafi, M., Jamali, V., Ng, D.W.K., Schober, R.: C-RAN with hybrid RF/FSO fronthaul links: joint optimization of RF time allocation and fronthaul compression. In: GLOBECOM 2017 IEEE Global Communications Conference, pp. 1–7. IEEE, Los Alamitos (2017)
92. Neuts, M.F.: A versatile Markovian point process. J. Appl. Probab. **16**(4), 764–779 (1979)
93. Neuts, M.F.: Matrix-Geometric Solutions in Stochastic Models: An Algorithmic Approach. Johns Hopkins University, Baltimore (1981)
94. Neuts, M.F.: Structured Stochastic Matrices of $M/G/1$ Type and Their Applications. Marcel Dekker, New York (1989)
95. Niknam, S., Nasir, A.A., Mehrpouyan, H., Natarajan, B.: A multiband OFDMA heterogeneous network for millimater wave 5G wireless applications. IEEE Access **4**, 5640–5648 (2016)
96. Perros, H.G.: A bibliography of papers on queueing networks with finite capacity queues. Perform. Eval. **10**, 255–260 (1989)
97. Press, W.H., Flannery, B.P., Teukolsky, S.A., Vetterling, W.T.: Numerical Recipes. Cambridge University Press, Cambridge (1986)
98. Ramaswami, V.: The $N/G/1$ queue and its detailed analysis. Adv. Appl. Probab. **12**(1), 222–261 (1980)
99. Ramaswami, V.: Independent Markov processes in parallel. Commun. Stat. Stoch. Model. **1**, 419–432 (1985)
100. Ramaswami, V., Lucantoni, D.: Algorithm for the multi-server queue with phase-type service. Commun. Stat. Stoch. Model. **1**, 393–417 (1985)
101. Schwartz, M.: Computer-Communication Network Design and Analysis. Prentice Hall, New Jersey (1977)
102. Sennot L.I., Humblet P.A., Tweedie R.L. Mean drifts and non-ergodicity of Markov chains. Oper. Res. **31**, 783–789 (1983)
103. Shioda, S.: Departure process of the $MAP/SM/1$ queue. Queueing Systems **44**, 31–50 (2003)
104. Skorokhod, A.: Foundations of Probability Theory and the Theory of Random Processes. Izd. Kiev Univ, Kiev (1974) (in Russian)
105. Syski, R.: A personal view of queueing theory. In: Dshalalow, J.H. (ed.) Frontiers in Queueing, pp. 3–18. CRC Press, Boca Raton (1998)
106. Tsybakov, B.S.: Model of teletraffic based on a self-similar random process. Radio Eng. **4**(5), 24–31 (1999) (in Russian)
107. Vishnevsky, V.M.: Theoretical Foundations of Computer Network Design. Technosphere, Moscow (2003) (in Russian)
108. Vishnevsky, V.M., Dudin, A.N.: Queueing systems with correlated arrival flows and their applications to modeling telecommunication. Autom. Remote Control **78**(8), 1361–1403 (2017)
109. Vishnevsky, V.M., Klimenok, V.I.: Unreliable queueing system with cold redundancy. In: Gaj, P., Kwiecien, A., Stera, P. (eds.) Computer Networks, CN 2015. Springer, Berlin (2015)
110. Vishnevsky, V., Semenova, O.: Polling Systems: Theory and Applications for Broadband Wireless Networks. LAMBERT, Riga (2012)
111. Vishnevsky, V.M., Zhozhikashvili, V.A.: Queueing Networks: Theory and Application to Communication Networks. Radio i Svyaz', Moscow (1988) (in Russian)
112. Vishnevsky, V.M., Portnoy, S.L., Shakhnovich, I.V.: Encyclopedia of WiMAX. In: The Path to 4G. Technosphere, Moscow (2010) (in Russian)

113. Vishnevsky, V.M., Semenova, O.V., Sharov, S.Yu.: Modeling and analysis of a hybrid communication channel based on free-space optical and radio-frequency technologies. Autom. Remote. Control. **74**, 521–528 (2013)
114. Vishnevsky, V.M., Kozyrev, D.V., Semenova, O.V.: Redundant queuing system with unreliable servers. Proceedings of the 6th International Congress on Ultra Modern Telecommunications and Control Systems, pp. 383–386 . IEEE Xplore Digital Library, New York (2014)
115. Vishnevsky, V., Larionov, A., Frolov, S.: Design and scheduling in 5G stationary and mobile communication systems based on wireless millimeter-wave mesh networks. In: Vishnevsky, V., Kozyrev, D., Larionov, A. (eds.) Distributed Computer and Communication Networks, DCCN 2013. Springer, Berlin (2014)
116. Vishnevsky, V.M., Dudin, A.N., Kozyrev, D.V., Larionov, A.A.: Methods of performance evaluation of broadband wireless networks along the long transport routes. In: Vishnevsky, V., Kozyrev, D. (eds.) Distributed Computer and Communication Networks, DCCN 2015. Springer, Berlin (2016)
117. Vishnevsky, V.M., Krishnamoorthy, A., Kozyrev, D.V., Larionov, A.A.: Review of methodology and design of broadband wireless networks with linear topology. Indian J. Pure Appl. Math. **47**(2), 329–342 (2016)
118. Vishnevsky, V., Larionov, A.A., Ivanov, R.E.: Applying graph-theoretic approach for time-frequency resource allocation in 5G MmWave Backhaul network. In: IEEE Xplore Digital Library, pp. 71–78. IEEE, Piscataway (2016)
119. Vishnevsky, V.M., Larionov, A.A., Ivanov, R.E.: An open queueing network with a correlated input arrival process for broadband wireless network performance evaluation. In: Dudin, A., Gortsev, A., Nazarov, A., Yakupov, R. (eds.) Information Technologies and Mathematical Modelling - Queueing Theory and Applications, ITMM 2016. Springer, Berlin (2016)
120. Wu, Y.,Yang, Q., Park, D., Kwak, K.S.: Dynamic link selection and power allocation with reliability guarantees for hybrid FSO/RF systems. IEEE Access **5**, 13654–13664 (2017)
121. Yashkov, S.F.: Queueing Analysis in a Computer System. Radio i Svyaz', Moscow (1989) (in Russian)
122. Zhang, Y., Chu, Y.-J., Nguyen, T.: Coverage algorithms for WiFO: a hybrid FSO-WiFi femtocell communication system. In: 26th International Conference on Computer Communication and Networks (ICCCN), pp. 1–6. IEEE, Los Alamitos (2017)
123. Zhou, K., Gong, C., Wu, N., Xu, Z.: Distributed channel allocation and rate control for hybrid FSO/RF vehicular ad hoc networks. J. Opt. Commun. Netw. **9**(8), 669–681 (2017)

Printed in the United States
by Baker & Taylor Publisher Services